Advances and Novel Approaches in Discrete Optimization

Advances and Novel Approaches in Discrete Optimization

Editor

Frank Werner

MDPI • Basel • Beijing • Wuhan • Barcelona • Belgrade • Manchester • Tokyo • Cluj • Tianjin

Editor
Frank Werner
Otto-von-Guericke-Universität
Magdeburg, Germany

Editorial Office
MDPI
St. Alban-Anlage 66
4052 Basel, Switzerland

This is a reprint of articles from the Special Issue published online in the open access journal *Mathematics* (ISSN 2227-7390) (available at: https://www.mdpi.com/journal/mathematics/special_issues/Advance_Novel_Approaches_Discrete_Optimization).

For citation purposes, cite each article independently as indicated on the article page online and as indicated below:

LastName, A.A.; LastName, B.B.; LastName, C.C. Article Title. *Journal Name* **Year**, *Article Number*, Page Range.

ISBN 978-3-03943-222-6 (Hbk)
ISBN 978-3-03943-223-3 (PDF)

© 2020 by the authors. Articles in this book are Open Access and distributed under the Creative Commons Attribution (CC BY) license, which allows users to download, copy and build upon published articles, as long as the author and publisher are properly credited, which ensures maximum dissemination and a wider impact of our publications.

The book as a whole is distributed by MDPI under the terms and conditions of the Creative Commons license CC BY-NC-ND.

Contents

About the Editor vii

Preface to "Advances and Novel Approaches in Discrete Optimization" ix

Lili Zuo, Zhenxia Sun, Lingfa Lu and Liqi Zhang
Single-Machine Scheduling with Rejection and an Operator Non-Availability Interval
Reprinted from: *Mathematics* **2019**, *7*, 668, doi:10.3390/math7080668 1

Hongjun Wei, Jinjiang Yuan and Yuan Gao
Transportation and Batching Scheduling for Minimizing Total Weighted Completion Time
Reprinted from: *Mathematics* **2019**, *7*, 819, doi:10.3390/math7090819 9

Wenhua Li, Weina Zhai and Xing Chai
Online Bi-Criteria Scheduling on Batch Machines with Machine Costs
Reprinted from: *Mathematics* **2019**, *7*, 960, doi:10.3390/math7100960 19

Nodari Vakhania
Dynamic Restructuring Framework for Scheduling with Release Times and Due-Dates
Reprinted from: *Mathematics* **2019**, *7*, 1104, doi:10.3390/math7111104 31

Wenhua Li , Libo Wang, Xing Chai and Hang Yuan
Online Batch Scheduling of Simple Linear Deteriorating Jobs with Incompatible Families
Reprinted from: *Mathematics* **2020**, *8*, 170, doi:10.3390/math8020170 73

Alexander A. Lazarev, Nikolay Pravdivets and Frank Werner
On the Dual and Inverse Problems of Scheduling Jobs to Minimize the Maximum Penalty
Reprinted from: *Mathematics* **2020**, *8*, 1131, doi:10.3390/math8071131 85

Yuri N. Sotskov, Natalja M. Matsveichuk and Vadzim D. Hatsura
Schedule Execution for Two-Machine Job-Shop to Minimize Makespan with Uncertain
Processing Times
Reprinted from: *Mathematics* **2020**, *8*, 1314, doi:10.3390/math8081314 101

Haidar Ali, Muhammad Ahsan Binyamin, Muhammad Kashif Shafiq and Wei Gao
On the Degree-Based Topological Indices of Some Derived Networks
Reprinted from: *Mathematics* **2019**, *7*, 612, doi:10.3390/math7070612 153

Bin Yang, Vinayak V. Manjalapur, Sharanu P. Sajjan, Madhura M. Mathai and Jia-Bao Liu
On Extended Adjacency Index with Respect to Acyclic, Unicyclic and Bicyclic Graphs
Reprinted from: *Mathematics* **2019**, *7*, 652, doi:10.3390/math7070652 171

Michal Staš
Join Products $K_{2,3} + C_n$
Reprinted from: *Mathematics* **2020**, *8*, 925, doi:10.3390/math8060925 181

Yuriy Shablya, Dmitry Kruchinin and Vladimir Kruchinin
Method for Developing Combinatorial Generation Algorithms Based on AND/OR Trees
and Its Application
Reprinted from: *Mathematics* **2020**, *8*, 962, doi:10.3390/math8060962 191

Urmila Pyakurel, Hari Nandan Nath, Stephan Dempe and Tanka Nath Dhamala
Efficient Dynamic Flow Algorithms for Evacuation Planning Problems with Partial Lane Reversal
Reprinted from: *Mathematics* **2019**, *7*, 993, doi:10.3390/math7100993 . . . . **213**

Shenshen Gu and Yue Yang
A Deep Learning Algorithm for the Max-Cut Problem Based on Pointer Network Structure with Supervised Learning and Reinforcement Learning Strategies
Reprinted from: *Mathematics* **2020**, *8*, 298, doi:10.3390/math8020298 . . . . **243**

Yajaira Cardona-Valdés, Samuel Nucamendi-Guillén, Rodrigo E. Peimbert-García, Gustavo Macedo-Barragán and Eduardo Díaz-Medina
A New Formulation for the Capacitated Lot Sizing Problem with Batch Ordering Allowing Shortages
Reprinted from: *Mathematics* **2020**, *8*, 878, doi:10.3390/math8060878 . . . . **263**

Alexander Pankratov, Tatiana Romanova and Igor Litvinchev
Packing Oblique 3D Objects
Reprinted from: *Mathematics* **2020**, *8*, 1130, doi:10.3390/math8071130 . . . . **279**

Krishan Arora, Ashok Kumar, Vikram Kumar Kamboj, Deepak Prashar, Sudan Jha, Bhanu Shrestha and Gyanendra Prasad Joshi
Optimization Methodologies and Testing on Standard Benchmark Functions of Load Frequency Control for Interconnected Multi Area Power System in Smart Grids
Reprinted from: *Mathematics* **2020**, *8*, 980, doi:10.3390/math8060980 . . . . **295**

Peter Drahoš, Michal Kocúr, Oto Haffner, Erik Kučera and Alena Kozáková
RISC Conversions for LNS Arithmetic in Embedded Systems
Reprinted from: *Mathematics* **2020**, *8*, 1208, doi:10.3390/math8081208 . . . . **319**

About the Editor

Frank Werner studied Mathematics from 1975–1980 and graduated from the Technical University Magdeburg (Germany) with honors. He defended his Ph.D. thesis on the solution of special scheduling problems in 1984 with 'summa cum laude' and his habilitation thesis in 1989. In 1992, he received a grant from the Alexander von Humboldt Foundation. Currently, he works as an extraordinary professor at the Faculty of Mathematics of the Otto von Guericke University Magdeburg (Germany). He is the author or editor of seven books, among them the textbooks 'Mathematics of Economics and Business' and 'A Refresher Course in Mathematics', and he has published more than 280 papers in international journals. He serves on the Editorial Board of 17 journals, in particular, he is the Editor-in-Chief of Algorithms and an Associate Editor of the *International Journal of Production Research* and *Journal of Scheduling*. He has served as a member of the Program Committee of more than 80 international conferences. His research interests are operations research, combinatorial optimization, and scheduling.

Preface to "Advances and Novel Approaches in Discrete Optimization"

Discrete optimization is an important area of applied mathematics which lies at the intersection of several disciplines and covers both theoretical and practical aspects. This book is the result of a Special Issue entitled 'Advances and Novel Approaches in Discrete Optimization'. In the call for papers for this issue, I asked for submissions presenting new theoretical results, structural investigations, new models, and algorithmic approaches as well as new applications of discrete optimization problems. Among the possible subjects were integer programming, combinatorial optimization, optimization problems of graphs and networks, scheduling, logistics, and transportation, to name but a few.

In response to the call for papers, 43 submissions were received. All submissions have been reviewed, as a rule, by at least three experts in the discrete optimization area. Finally, 17 papers were accepted for this Special Issue, all of which are of high quality and reflect the great interest in the area of discrete optimization. This corresponds to an acceptance rate of 39.5%. The authors of these publications represent 13 different countries: China, Pakistan, India, Nepal, Germany, Mexico, USA, Australia, Slovakia, Russia, Korea, Ukraine, and Belarus.

This book contains both theoretical works and practical applications in the field of discrete optimization. Although many different aspects of discrete optimization have been addressed by the submissions, among the accepted papers, a major part deals with scheduling problems as well as graphs and networks. We hope that researchers and practitioners will find much inspiration for their future work in the exciting area of discrete optimization. Next, all published articles in this book are briefly surveyed in the order of their sequence in the book.

The first seven articles deal with scheduling problems. In the first article, Zuo et al. consider two single-machine scheduling problems with possible job rejection and a non-availability interval of the operator simultaneously. The objective is to minimize the sum of either the makespan or total weighted completion time of the accepted jobs and the total cost for the rejected jobs. The authors suggest a pseudo-polynomial solution algorithm as well as a fully polynomial-time approximation scheme.

In the next article, Wei et al. deal with the problem of transportation and batching scheduling. A single vehicle is considered, and the goal is to minimize total weighted completion time. The main results of this paper are the proof that the problem is NP-hard in the strong sense for any batch capacity of at least 3 as well as a polynomial-time 3-approximation algorithm for the case of a batch capacity of at least 2.

In the third article, Li et al. consider a bi-criteria online scheduling problem on parallel batch machines. The batch capacity is unbounded, the processing times of all jobs and batches are equal to one, and the objective is to minimize the maximum machine cost subject to a minimum makespan. The authors consider two types of cost functions and present two best possible online algorithms for the problem under consideration.

Vakhania investigates a single-machine scheduling problem with given release dates, due dates, and divisible processing times. The objective is to minimize maximum lateness. He suggests a general method which also leads to useful structural properties of this problem and helps to identify polynomially solvable cases. In particular, for the case of mutually divisible job processing times, a polynomial-time algorithm results, and this case turns out to be maximal polynomially solvable one of this problem with nonarbitrary processing times.

Li et al. consider an online scheduling problem with parallel batch machines and linearly deteriorating jobs. The batch capacity is unbounded, and the objective is to minimize the makespan. For the special case of $m = 1$, a best possible online algorithm with a competitive ratio of $(1 + \alpha_{max})^f$ is given, where f denotes the number of job families and α_{max} gives the maximal deterioration rate of a job. Furthermore, for $m = f \geq 1$, a best possible online algorithm with a competitive ratio of $1 + \alpha_{max}$ is also derived.

Then Lazarev et al. consider the single-machine problem with given release dates and the objective of minimizing the maximum job penalty. While this problem is NP-hard in the strong sense, they introduce a dual and an inverse problem, which can both be polynomially solved. The optimal function value of the dual problem is incorporated as a lower bound into a branch and bound algorithm for the original problem. The authors present computational results with this enumerative algorithm for hard benchmark instances with up to 20 jobs. Most of the instances considered can be solved very fast by the proposed algorithm.

In the last scheduling article, Sotskov et al. consider the two-machine job-shop scheduling problem with interval processing times and makespan minimization. The focus of this paper is how to execute a schedule in a best way. The paper uses the concept of a minimal dominant set of schedules. An online algorithm of the complexity $O(n^2)$ has been developed, where n denotes the number of jobs. Detailed numerical results are given for instances with up to 100 jobs and different maximal percentage errors in the processing times.

The next articles deal with graph-theoretic subjects and applications of graphs. Ali et al. consider degree-based topological indices and some derived graphs. The goal of this paper is to investigate the chemical behavior of these graphs by means of the topological indices. In particular, the authors find the exact results for the Forgotten index, the Balaban index, the reclassified Zagreb indices, the ABC_4 index, and the GA_5 index of Hex-derived networks of type 3.

Then, Yang et al. consider the extended adjacency index of a molecular graph. In particular, the authors show some graph transformations which increase or decrease this index. Then, they derive the extremal acyclic, unicyclic, and bicyclic graphs with a minimum and a maximum extended adjacency index, respectively.

Stas investigates a graph-theoretic subject, namely the crossing number of a graph which is the minimum number of edge crossings over all drawings of the graph in the plane. In particular, he presents this number for the join product $K_{2,3} + C_n$, where $K_{2,3}$ is the complete bipartite graph and C_n is a cycle on n vertices. The methods applied by the author use several combinatorial properties on cyclic permutations.

Then, Shablya et al. look for new combinatorial generation algorithms. They give basic general methods and investigate one of them based on AND/OR trees. They apply the method of compositae from the theory of generating functions. To show the effectiveness of the suggested modifications, they also derive new ranking and unranking algorithms for several combinatorial sets.

The remaining articles deal with interesting applications of discrete optimization in several research fields. Pyakurel et al. present efficient algorithms to solve dynamic flow problems with constant attributes as well as generalized problems with partial contraflow reconfiguration in the context of evacuation planning. In particular, a strongly polynomial time algorithm for calculating an approximate solution of the quickest partial contraflow problem on two terminal networks is derived. Numerical results are given for the road network of Kathmandu (Nepal) as the evacuation network.

Then, Gu and Yang consider the max-cut problem. They develop a unique method combining a pointer network and two deep learning strategies, namely supervised learning and reinforcement learning. The pointer network model includes a long short-term memory network and an encoder–decoder. It turns out that their model can be used to solve large-scale max-cut problems heuristically, where for high-dimensional cases, reinforcement learning turned out to be superior to supervised learning.

Cordona-Valdes et al. deal with the multi-product, multi-period capacitated lot sizing problem. Determining the optimal lot size allows shortages resulting in a penalty cost. Two mixed-integer formulations are developed: one model allows shortages, and the other one enforces the fulfillment of the demand. The developed models have been applied to a Mexican fashion retail company within a case study. Both formulations significantly reduced the final inventory costs.

Pankratov et al. deal with packing problems of irregular 3D objects. By using the phi-function technique, the problem is reduced to the solution of a nonlinear programming model and solved by a multi-start strategy with finding local extreme points. The algorithm has been tested on some benchmark instances.

Then, Arora et al. present an optimized analysis and planning for power generation and management. They describe several optimization methodologies. In particular, binary variations of the moth flame optimizer and the Harris hawks optimizer are analyzed and tested on 23 benchmark functions, e.g., unimodal, multi-modal ones and functions with fixed dimension. The comparison and simulation results demonstrate that their implemented algorithm delivered better results towards the load frequency control problem of a smart grid arrangement compared to earlier methods.

In the last article, Drahos et al. present a method for a conversion between the logarithmic number system (LNS) and floating point (FLP) representations using reduced instruction set computing (RISC). After giving an overview on FLP and LNS number representations, two algorithms of the RISC conversion between both systems using the 'looping in sectors' procedure are presented. The proposed algorithms deliver a very small maximum relative conversion error, and the authors mention also some interesting applications such as camera systems or car control units.

Finally, I would like to thank all authors for submitting their work to this Special Issue and also all referees for their support by giving timely and insightful reports. My special thanks go to the staff of the journal Mathematics for their skilled and pleasant cooperation during the preparation of this issue.

Frank Werner
Editor

Article

Single-Machine Scheduling with Rejection and an Operator Non-Availability Interval

Lili Zuo [1], Zhenxia Sun [1], Lingfa Lu [1,*] and Liqi Zhang [2]

[1] School of Mathematics and Statistics, Zhengzhou University, Zhengzhou 450001, China
[2] College of Information and Management Science, Henan Agricultural University, Zhengzhou 450002, China
* Correspondence: lulingfa@zzu.edu.cn

Received: 8 July 2019; Accepted: 24 July 2019; Published: 26 July 2019

Abstract: In this paper, we study two scheduling problems on a single machine with rejection and an operator non-availability interval. In the operator non-availability interval, no job can be started or be completed. However, a *crossover* job is allowed such that it can be started before this interval and completed after this interval. Furthermore, we also assume that job rejection is allowed. That is, each job is either accepted and processed in-house, or is rejected by paying a rejection cost. Our task is to minimize the sum of the makespan (or the total weighted completion time) of accepted jobs and the total rejection cost of rejected jobs. For two scheduling problems with different objective functions, by borrowing the previous algorithms in the literature, we propose a pseudo-polynomial-time algorithm and a fully polynomial-time approximation scheme (FPTAS), respectively.

Keywords: scheduling with rejection; machine non-availability; operator non-availability; dynamic programming; FPTAS

1. Introduction

In this section, we introduce some models on scheduling with (machine or operator) non-availability intervals, scheduling with rejection, and scheduling with rejection and non-availability intervals, respectively.

1.1. Scheduling with Non-Availability Intervals

In most scheduling problems, it is assumed that the machines are available at all times. However, in some industrial settings, the assumption might not be true. Some machines or the operator might be unavailable in some time intervals. Recently, some researchers have studied some scheduling problems with the non-availability intervals. Two models of the non-availability interval were studied mainly: one is the machine non-availability (MNA) intervals due to the machine maintenances and the other is the operator non-availability (ONA) intervals because the operator is resting from work. The difference between MNA intervals and ONA intervals is that a crossover job can exist in the ONA interval. However, no job can be processed in the MNA interval.

To the best of our knowledge, the earliest scheduling problem with MNA intervals was studied by Schmidt [1]. He considered a parallel-machine scheduling problem in which each machine has different MNA intervals. The task is to find a feasible preemptive schedule if it exists. A polynomial-time algorithm is presented for the above problem. We first introduce some single-machine scheduling problems with an MNA interval (a, b). The corresponding problem can be denoted by $1|MNA(a, b)|f$, where "$MNA(a, b)$" means that there is an MNA interval (a, b) and f is the objective function to be minimized. For problem $1|MNA(a, b)|\sum C_j$, Adiri et al. [2] proved that the problem is NP-hard and then presented a $\frac{5}{4}$-approximation algorithm. For problem $1|MNA(a, b)|C_{max}$, Lee [3] showed that this problem is binary NP-hard and then provided a $\frac{4}{3}$-approximation algorithm. For problem

$1|\text{MNA}(a,b)|L_{\max}$, Kacem [4] designed a $\frac{3}{2}$-approximation algorithm and a fully polynomial-time approximation scheme (FPTAS). If there are $k \geq 2$ MNA intervals $(a_1, b_1), (a_2, b_2), \cdots, (a_k, b_k)$ on the machine, the corresponding problem $1|\text{MNA}(a_i, b_i)|C_{\max}$ is strongly NP-hard (see [3]) when k is arbitrary. Breit et al. [5] showed that, for any $\rho \geq 1$ and $k \geq 2$, there is no ρ-approximation algorithm for problem $1|\text{MNA}(a_i, b_i)|C_{\max}$ unless P = NP.

When there are $m \geq 2$ parallel machines $M_1, \cdots, M_m$ and each machine M_i has an MNA interval (a_i, b_i), the corresponding problem is denoted by $Pm|\text{MNA}(a_i, b_i)|f$. For problem $Pm|\text{MNA}(0, b_i)|C_{\max}$, i.e., each machine M_i has a machine release time b_j, Lee [6] provided a modified LPT (MLPT) algorithm with a tight approximation ratio $\frac{4}{3}$. Kellerer [7] improved this bound $\frac{4}{3}$ to $\frac{5}{4}$ by a dual approximation algorithm using a bin packing approach. For problem $Pm|\text{MNA}(0, b_i)|\sum C_j$, Schmidt [8] showed that the SPT rule is optimal. Lee [6] also studied the problem $Pm|\text{MNA}(a_i, b_i)|C_{\max}$ with the assumption that one machine is always available. He showed that the approximation ratios of LS (List Scheduling) and LPT are m and $\frac{m+1}{2}$, respectively. Furthermore, Liao et al. [9] considered the same problem with $m = 2$ and developed exact algorithms based on the TMO algorithm for problem $P2||C_{\max}$. Aggoune [10] studied the flow-shop scheduling problem with several MNA intervals on each machine. A heuristic algorithm is provided to approximately solve this problem. Burdett and Kozan [11] also addressed some MNA intervals in railway scenarios. They introduced new fixed jobs for the MNA intervals. Some constructive heuristics and meta-heuristic algorithms were proposed in this paper. For more new models and results about this topic, the reader is referred to the survey by Ma et al. [12].

Brauner et al. [13] first studied the scheduling problems with an ONA interval. Similarly, this scheduling model can be denoted by $1|\text{ONA}(a, b)|f$. For problem $1|\text{ONA}(a, b)|C_{\max}$, Brauner et al. [13] proved that it is binary NP-hard and provided an FPTAS. For problem $1|\text{ONA}(a, b)|L_{\max}$, Kacem et al. [14] proposed an FPTAS by borrowing the FPTAS for problem $1|\text{MNA}(a, b)|L_{\max}$. Chen et al. [15] considered the problem $1|\text{ONA}(a, b)|\sum C_j$ and presented a $\frac{20}{17}$-approximation algorithm. Wan and Yuan [16] further considered the problem $1|\text{ONA}(a, b)|\sum w_j C_j$. They designed a pseudo-polynomial-time dynamic programming (DP) algorithm and then converted the DP algorithm into an FPTAS. Burdett et al. [17] considered the flexible job shop scheduling with operators (FJSOP) for coal export terminals. A hybrid meta-heuristic and a lot of numerical testings were designed for the above problem.

1.2. Scheduling with Rejection

In many practical cases, processing all jobs may occur high inventory or tardiness costs. However, rejecting some jobs can save time and reduce costs. When a job is rejected, a corresponding rejection cost is required. The decision maker needs to determine which jobs should be accepted (and a feasible schedule for accepted jobs), and which jobs should be rejected, such that the production cost and the total rejection cost are minimized. Thus, both from the practical and theoretical point of view, scheduling models with rejection are very interesting. In addition, an important application also occurs in scheduling with outsourcing. If the outsourcing cost is treated as the rejection cost, scheduling with rejection and scheduling with outsourcing are in fact equivalent.

Scheduling models with rejection were first introduced by Bartal et al. [18]. They considered several off-line and on-line scheduling problems on m parallel machines. The task is to minimize the sum of the makespan of accepted jobs and the total rejection cost of rejected jobs. For the on-line version, they designed an on-line algorithm with the best-possible competitive ratio of 2.618. For the off-line version, they provided an FPTAS when m is fixed, and a PTAS when m is arbitrary.

Next, we only introduce some results on the single-machine scheduling with rejection. The corresponding problem can be denoted by $1|\text{rej}|f + e(R)$, where f is the objective function on the set A of accepted jobs and $e(R)$ is the total rejection cost on the set R of rejected jobs. For problem $1|r_j, \text{rej}|C_{\max} + e(R)$, Cao and Zhang [19] proved that this problem is NP-hard and designed a PTAS. However, they also pointed out that it is open whether this problem is ordinary or strongly NP-hard. Zhang et al. [20] showed that this problem is binary NP-hard by providing two different pseudo-polynomial-time algorithms. Finally, they also provided a 2-approximation algorithm and an FPTAS for the above problem. For problem $1|\text{rej}|L_{\max} + e(R)$, Sengupta [21] proved that this problem is binary NP-hard. He also proposed two dynamic programming algorithms and converted one of the two algorithms into an FPTAS. Engels et al. [22] studied the problem $1|\text{rej}|\sum w_j C_j + e(R)$. They showed that this problem is binary NP-hard and then provided an FPTAS. They also showed that, when $w_j = 1$, the problem $1|\text{rej}|\sum C_j + e(R)$ is polynomial-time solvable. Recently, Shabtay et al. [23] presented a comprehensive survey for the off-line scheduling problems with rejection. For other models and results on scheduling with rejection, the reader is referred to the survey by Shabtay et al. [23].

1.3. Scheduling with Rejection and Non-Availability Intervals

There are only two articles which considered "scheduling with rejection" and "machine non-availability intervals" together. Zhong et al. [24], and Zhao and Tang [25] considered the problems $1|\text{MNA}(a,b), \text{rej}|C_{\max} + e(R)$ and $1|\text{MNA}(a,b), \text{rej}|\sum w_j C_j + e(R)$, respectively. Both of them presented a pseudo-polynomial dynamic programming algorithm and an FPTAS for the corresponding problem. In addition, Li and Chen [26] investigated several scheduling problems with rejection and a deteriorating maintenance activity on a single machine. In their model, the starting time of the maintenance activity (non-availability intervals) is not fixed and the duration is a linear increasing function of its starting time. Some (pseudo-)polynomial-time algorithms are presented for some different objective functions. However, to the best of our knowledge, no article considered "scheduling with rejection" and "operator non-availability intervals" simultaneously. In this paper, we are the first to consider scheduling with rejection and an operator non-availability interval.

2. Problem Formulation

The single-machine scheduling with rejection and an operator non-availability interval can be stated formally as follows. There are n jobs $J_1, J_2, \cdots, J_n$ and a single machine. Each job J_j is available at time 0 and has a processing time p_j, a weight w_j and a rejection cost e_j. Each job J_j is either rejected and the rejection cost e_j has to be paid, or accepted and then processed non-preemptively on the machine. There is an operator non-availability interval (a,b) on the machine, where we assume that $0 < a < b$. This implies that, in any feasible schedule π, no accepted job J_j can be started or be completed in the interval (a,b). However, a crossover job is allowed such that it can start before this interval and complete after this interval. Without loss of generality, we assume that all the parameters a, b, p_j, w_j and e_j are non-negative integers. Let A and R be the sets of accepted jobs and rejected jobs, respectively. Denote by $C_{\max} = \max\{C_j : J_j \in A\}$, $\sum w_j C_j = \sum_{J_j \in A} w_j C_j$ and $e(R) = \sum_{J_j \in R} e_j$ the makespan of accepted jobs, the total weighted completion time of accepted jobs and the total rejection cost of rejected jobs, respectively. Our task is to find a feasible schedule such that $C_{\max} + e(R)$ or $\sum w_j C_j + e(R)$ is minimized. By using the notation for scheduling problems, the corresponding problems are denoted by $1|\text{ONA}(a,b), \text{rej}|C_{\max} + e(R)$ and $1|\text{ONA}(a,b), \text{rej}|\sum w_j C_j + e(R)$, respectively.

Two similar problems related to the above problems are $1|\text{MNA}(a,b), \text{rej}|C_{\max} + e(R)$ and $1|\text{MNA}(a,b), \text{rej}|\sum w_j C_j + e(R)$. Zhong et al. [24] and Zhao and Tang [25] presented a pseudo-polynomial dynamic programming algorithm and an FPTAS for the corresponding problem, respectively. In this paper, by borrowing the algorithms in [24,25], we also presented a pseudo-polynomial dynamic programming algorithm and an FPTAS for the problems $1|\text{ONA}(a,b), \text{rej}|C_{\max} + e(R)$ and $1|\text{ONA}(a,b), \text{rej}|\sum w_j C_j + e(R)$, respectively.

3. Pseudo-Polynomial-Time Algorithms

Brauner et al. [13] and Chen et al. [15] showed that problems $1|\text{ONA}(a,b)|C_{\max}$ and $1|\text{ONA}(a,b)|\sum C_j$ are NP-hard. Thus, two problems $1|\text{ONA}(a,b), \text{rej}|C_{\max} + e(R)$ and $1|\text{ONA}(a,b), \text{rej}|\sum w_j C_j + e(R)$ are also NP-hard. In this section, we show that the above problems can be solved in pseudo-polynomial time. That is, both of the problems, $1|\text{ONA}(a,b),\text{rej}|C_{\max}+e(R)$ and $1|\text{ONA}(a,b),\text{rej}|\sum w_j C_j + e(R)$, are binary NP-hard.

For problems $1|\text{MNA}(a,b),\text{rej}|C_{\max}+e(R)$ and $1|\text{MNA}(a,b),\text{rej}|\sum w_j C_j + e(R)$, Zhong et al. [24] and Zhao and Tang [25] presented a pseudo-polynomial dynamic programming algorithm, respectively. By enumerating the crossover job and its starting time, we show that problem $1|\text{ONA}(a,b),\text{rej}|f+e(R)$ can be decomposed into many subproblems of $1|\text{MNA}(a,b),\text{rej}|f+e(R)$. Consequently, by borrowing the algorithms in [24,25], we also presented a pseudo-polynomial algorithm for problems $1|\text{ONA}(a,b),\text{rej}|C_{\max}+e(R)$ and $1|\text{ONA}(a,b),\text{rej}|\sum w_j C_j + e(R)$, respectively.

Lemma 1. *If algorithm A^{MNA} solves problem $1|\text{MNA}(a,b),\text{rej}|f+e(R)$ in $O(T)$ time, then Algorithm 1 yields an optimal schedule for problem $1|\text{ONA}(a,b),\text{rej}|f+e(R)$ in $O(naT)$ time.*

Algorithm 1 A^{ONA}

Step 1: Let A^{MNA} be a dynamic programming algorithm for problem $1|\text{MNA}(a,b),\text{rej}|f+e(R)$, where $f \in \{C_{\max}, \sum w_j C_j\}$.

Step 2: Applying algorithm A^{MNA} to the problem $1|\text{MNA}(a,b),\text{rej}|f+e(R)$, we can obtain a schedule π_0.

Step 3: For each job J_j with $j=1,\cdots,n$ and each integer t with $0 \leq t \leq a$ and $t+p_j \geq b$, we first schedule J_j in time interval $[t, t+p_j]$. Set $a' = t$, $b' = t + p_j$ and instance $I' = \{J_1,\cdots,J_n\} \setminus J_j$. Applying algorithm A^{MNA} to the instance I' of problem $1|\text{MNA}(a',b'),\text{rej}|f+e(R)$, we can obtain a schedule $\pi(j,t)$.

Step 4: Choose the schedule among all schedules in $\{\pi_0\} \cup \{\pi(j,t) : 1 \leq j \leq n, 0 \leq t \leq a \text{ and } t+p_j \geq b\}$ with the smallest objective value.

Proof. Firstly, we show that Algorithm 1 yields an optimal schedule for problem $1|\text{ONA}(a,b),\text{rej}|f+e(R)$. Let π^* be an optimal schedule for problem $1|\text{ONA}(a,b),\text{rej}|f+e(R)$. We distinguish two cases into our discussion.

Case 1: No crossover job exist in π^*.

In this case, interval (a,b) is completely forbidden. That is, an ONA interval is equivalent to an MNA interval. Thus, in this case, π_0 is also an optimal schedule for problem $1|\text{ONA}(a,b),\text{rej}|f+e(R)$.

Case 2: There is a crossover job in π^*.

Let J_j be the crossover job in π^* and let t be the starting time of J_j in π^*. Clearly, we have $0 \leq t \leq a$ and $t+p_j \geq b$. In this case, no job in $I' = \{J_1,\cdots,J_n\} \setminus J_j$ can be processed in the interval $(t, t+p_j)$. Set $a' = t$ and $b' = t+p_j$. The remaining problem is equivalent to the instance I' of the problem $1|\text{MNA}(a',b'),\text{rej}|f+e(R)$. Thus, in this case, $\pi(j,t)$ is also an optimal schedule for problem $1|\text{ONA}(a,b),\text{rej}|f+e(R)$.

From the above discussions, Algorithm 1 always yields an optimal schedule for problem $1|\text{ONA}(a,b),\text{rej}|f+e(R)$. Note that algorithm A^{MNA} is called at most $O(na)$ times. Thus, the time complexity of Algorithm 1 is exactly $O(naT)$. □

Note that Zhong et al. [24] presented an $O(n\sum_{j=1}^n p_j)$-time dynamic programming algorithm for problem $1|\text{MNA}(a,b),\text{rej}|C_{\max}+e(R)$. Zhao and Tang [25] presented an $O(na\sum_{j=1}^n p_j)$-time algorithm for problem $1|\text{MNA}(a,b),\text{rej}|\sum w_j C_j + e(R)$. Thus, we have the following corollaries.

Corollary 1. *Algorithm 1 solves problem* $1|ONA(a,b), rej|C_{max} + e(R)$ *in* $O(n^2 a \sum_{j=1}^{n} p_j)$ *time and solves problem* $1|ONA(a,b), rej| \sum w_j C_j + e(R)$ *in* $O(n^2 a^2 \sum_{j=1}^{n} p_j)$ *time.*

Example 1. *A simple example for problem* $1|ONA(a,b), rej|C_{max} + e(R)$ *is constructed as follows: Given a positive number* $k > 0$, *we have three jobs* J_1, J_2, J_3 *with* $(p_1, e_1) = (2k, k)$, $(p_2, e_2) = (4k, 2k)$ *and* $(p_3, e_3) = (7k, +\infty)$. *We also assume that there is a single machine with an ONA interval* (a, b), *where* $a = 4k$ *and* $b = 9k$. *Note that* $p_1 < p_2 < b_k - a_k = 5k < p_3$. *Thus,* J_3 *is the unique candidate for the crossover job. Note further that* $e_3 = +\infty$. *Thus,* J_3 *must be accepted in any optimal schedule. We distinguish two cases in our discussion.*

Case 1: J_3 is not a crossover job.

In this case, since $p_3 = 7k > a = 4k$, J_3 must be processed at or after time $b = 9k$. Thus, the optimal schedule is to process J_2 and J_3 in $[0, 4k]$ and $[9k, 16k]$, respectively, and reject J_1. That is, the $C_{max} + e(R)$ value is $16k + k = 17k$.

Case 2: J_3 is the crossover job and J_3 starts its processing at time t.

Since $t \in [0, 4k]$ and $t + p_3 \geq b = 9k$, we have $2k \leq t \leq 4k$. When $t \in [2k, 4k)$, only J_1 can be processed before J_3. Thus, the $C_{max} + e(R)$ value is at least $t + 7k + 2k \geq 11k$. The minimum value $C_{max} + e(R) = 11k$ can be reached by processing J_1 and J_3 in $[0, 9k]$ and reject J_2. When $t = 4k$, only one job between J_1 and J_2 can be processed before J_3. Thus, the $C_{max} + e(R)$ value is at least $4k + 7k + k \geq 12k$. The minimum value $C_{max} + e(R) = 12k$ can be reached by processing J_2 and J_3 in $[0, 11k]$ and reject J_1.

By combining all cases, the optimal $C_{max} + e(R)$ value is $11k$, which can be reached by processing J_1 and J_3 in $[0, 9k]$ and reject J_2.

4. Approximation Schemes

In this section, by borrowing the FPTASs in [24,25], we also presented an FPTAS for problems $1|ONA(a,b), rej|C_{max} + e(R)$ and $1|ONA(a,b), rej| \sum w_j C_j + e(R)$, respectively. Given an $\epsilon = \frac{1}{E}$ for some positive integer E, we set $t_i = i\epsilon a$ for each $i = 1, 2, \cdots, E$. To propose an FPTAS, we delay the starting of the crossover job slightly such that the crossover job starts its processing at some time t_i. Such a delay may increase the objective value by $1 + \epsilon$ times, we say that it produces a $(1 + \epsilon)$-loss.

Lemma 2. *If a crossover job exists in an optimal schedule, with a* $(1 + \epsilon)$-*loss, we can assume that the crossover job starts its processing at some time* t_i.

Proof. Let π^* be an an optimal schedule for problem $1|ONA(a,b), rej|C_{max} + e(R)$ or problem $1|ONA(a,b), rej| \sum w_j C_j + e(R)$. Let A^* and R^* be the set of accepted jobs and the set of rejected jobs in π^*, respectively. In addition, we also assume that $J_j \in A^*$ is the jth processed job in π^*. Without loss of generality, we assume that $J_k \in A^*$ is the crossover job in π^*. Furthermore, we assume that the starting time of J_k is t with $0 \leq t \leq a$ and $t + p_k \geq b$. If $t \neq t_i$ for each $i = 1, 2, \cdots, E$, then there is some i with $1 \leq i \leq E$ such that $t_{i-1} < t < t_i$, where $t_0 = 0$. By delaying the processing of $J_k, J_{k+1}, \cdots, J_{|A^*|}$ by a length of $t_i - t$, we can obtain a schedule π. Note that $t_i \leq a$ and $t_i + p_k \geq t + p_k \geq b$. Thus, π is feasible for problems $1|ONA(a,b), rej|C_{max} + e(R)$ and $1|ONA(a,b), rej| \sum w_j C_j + e(R)$. Note further that $t_i - t < t_i - t_{i-1} = \epsilon a$ and $C_j(\pi^*) \geq b > a$ for each $j = k, \cdots, |A^*|$. Thus, we have $C_j(\pi) = C_j(\pi^*)$ for each $j = 1, \cdots, k-1$ and $C_j(\pi) = C_j(\pi^*) + (t_i - t) \leq C_j(\pi^*) + \epsilon a \leq C_j(\pi^*) + \epsilon C_j(\pi^*) = (1 + \epsilon) C_j(\pi^*)$ for each $j = k, \cdots, |A^*|$. It follows that

$$C_{max}(\pi) \leq (1 + \epsilon) C_{max}(\pi^*) \text{ and } \sum w_j C_\pi \leq (1 + \epsilon) \sum w_j C_j(\pi^*).$$

Notice that both of the total rejection cost in π and π^* are $e(R^*)$. Thus, we can conclude that the objective value is increased by at most $1 + \epsilon$ times. This completes the proof of Lemma 2. □

Next, based on Lemma 2, we propose an FPTAS for problems $1|ONA(a,b), \text{rej}|C_{\max} + e(R)$ and $1|ONA(a,b), \text{rej}|\sum w_j C_j + e(R)$.

Lemma 3. *If algorithm A_ϵ^{MNA} is an FPTAS for problem $1|MNA(a,b), \text{rej}|f + e(R)$ with the time complexity of $O(T)$, then Algorithm 2 is also an FPTAS for problem $1|ONA(a,b), \text{rej}|f + e(R)$ with the time complexity of $O(\frac{nT}{\epsilon})$.*

Algorithm 2 A_ϵ^{ONA}

Step 1: Let A_ϵ^{MNA} be an FPTAS for problem $1|MNA(a,b), \text{rej}|f + e(R)$, where $f \in \{C_{\max}, \sum w_j C_j\}$.

Step 2: Applying algorithm A_ϵ^{MNA} to the problem $1|MNA(a,b), \text{rej}|f + e(R)$, we can obtain a schedule π_0.

Step 3: For each job J_j with $j = 1, \cdots, n$ and each integer t_i with $0 \leq t_i \leq a$ and $t_i + p_j \geq b$, we first schedule J_j in time interval $[t_i, t_i + p_j]$. Set $a' = t_i$, $b' = t_i + p_j$ and instance $I' = \{J_1, \cdots, J_n\} \setminus J_j$. Applying algorithm A_ϵ^{MNA} to the instance I' of problem $1|MNA(a',b'), \text{rej}|f + e(R)$, we can obtain a schedule $\pi(j, t_i)$.

Step 4: Choose the schedule among all schedules in $\{\pi_0\} \cup \{\pi(j, t_i) : 1 \leq j \leq n, 0 \leq t_i \leq a \text{ and } t_i + p_j \geq b\}$ with the smallest objective value.

Proof. Firstly, we prove that Algorithm 2 is an FPTAS for $1|ONA(a,b), \text{rej}|f + e(R)$. Let π^* be an optimal schedule for problem $1|ONA(a,b), \text{rej}|f + e(R)$. Let Z and Z^* be the objective values obtained from Algorithm 2 and the optimal schedule π^*, respectively. We also distinguish two cases in our discussion.

Case 1: No crossover job exists in π^*.

In this case, interval (a, b) is completely forbidden. That is, an ONA interval is equivalent to an MNA interval. Thus, we have $Z \leq Z(\pi_0) \leq (1+\epsilon)Z(\pi^*) = (1+\epsilon)Z^*$.

Case 2: There is a crossover job in π^*.

Let J_j be the crossover job in π^* and let t be the starting time of J_j in π^*. Clearly, we have $0 \leq t \leq a$ and $t + p_j \geq b$. In this case, no job in $I' = \{J_1, \cdots, J_n\} \setminus J_j$ can be processed in the interval $(t, t + p_j)$. Set $a' = t$ and $b' = t + p_j$. Thus, the remaining problem is equivalent to the instance I' of the problem $1|MNA(a',b'), \text{rej}|f + e(R)$. Let $Z^*(j, t)$ be the optimal objective value for problem $1|ONA(a,b), \text{rej}|f + e(R)$ under the constraint in which J_j is the crossover job and J_j starts its processing at time t. Note that algorithm A_ϵ^{MNA} is an FPTAS for problem $1|MNA(a,b), \text{rej}|f + e(R)$. Thus, we have $Z(\pi(j, t_i)) \leq (1+\epsilon)Z^*(j, t_i)$. Furthermore, by Lemma 2, we also have $Z^*(j, t_i) \leq (1+\epsilon)Z^*(j, t) = (1+\epsilon)Z^*$. It follows that

$$Z \leq Z(\pi(j, t_i)) \leq (1+\epsilon)Z^*(j, t_i) \leq (1+\epsilon)^2 Z^*(j, t) = (1+\epsilon)^2 Z^*.$$

From the above discussions, Algorithm 2 is an FPTAS for $1|ONA(a,b), \text{rej}|f + e(R)$. Note that algorithm A_ϵ^{MNA} is called at most $O(\frac{n}{\epsilon})$ times. Thus, the time complexity of Algorithm 2 is exactly $O(\frac{nT}{\epsilon})$. □

Note that Zhong et al. [24] presented an FPTAS with the time complexity of $O(\frac{n}{\epsilon})$ for problem $1|MNA(a,b), \text{rej}|C_{\max} + e(R)$. Zhao and Tang [25] presented an FPTAS with the time complexity

$O(\frac{n^4L^5}{\epsilon^4})$ for problem $1|\text{MNA}(a,b),\text{rej}|\sum w_jC_j + e(R)$, where $L = \log(\max\{n, \frac{1}{\epsilon}, b, \max e_j, (\max w_j) \cdot (\max p_j)\})$. Thus, we have the following corollaries.

Corollary 2. *Algorithm 1 is an FPTAS for problems $1|\text{ONA}(a,b),\text{rej}|C_{\max} + e(R)$ and $1|\text{ONA}(a,b),\text{rej}|\sum w_jC_j + e(R)$ with the time complexities of $O(\frac{n^2}{\epsilon^2})$ and $O(\frac{n^5L^5}{\epsilon^5})$, respectively.*

5. Conclusions and Future Work

In this paper, we are the first to consider scheduling with rejection and an operator non-availability interval simultaneously. The objective is to minimize the sum of the makespan (or the total weighted completion time) of the accepted jobs and the total rejection cost of the rejected jobs. Firstly, we build the relation between problem $1|\text{MNA}(a,b),\text{rej}|f + e(R)$ and problem $1|\text{ONA}(a,b),\text{rej}|f + e(R)$, where $f \in \{C_{\max}, \sum w_jC_j\}$. That is, by enumerating the crossover job and its starting time, problem $1|\text{ONA}(a,b),\text{rej}|f + e(R)$ can be decomposed into many subproblems of $1|\text{MNA}(a,b),\text{rej}|f + e(R)$. Consequently, by borrowing the previous algorithms for problem $1|\text{MNA}(a,b),\text{rej}|f + e(R)$, we provide a pseudo-polynomial-time algorithm and an FPTAS for problem $1|\text{ONA}(a,b),\text{rej}|f + e(R)$, respectively.

When there are $k \geq 2$ MNA or ONA intervals $(a_1, b_1), (a_2, b_2), \cdots, (a_k, b_k)$ on the machine, the corresponding problem is strongly NP-hard (see [3]) when k is arbitrary. However, when k is a fixed constant, it is possible to propose a pseudo-polynomial-time algorithm with a larger time complexity. Breit et al. [5] showed that, for any $\rho \geq 1$ and $k \geq 2$, there is no ρ-approximation algorithm for problem $1|\text{MNA}(a_i, b_i)|C_{\max}$ unless P = NP. It is easy to verify that the inapproximability result still holds when there are $k \geq 2$ MNA or ONA intervals on the machine and b_k is sufficiently large.

Note that, for problem $1|\text{ONA}(a,b),\text{rej}|\sum w_jC_j + e(R)$, the time complexity of the proposed FPTAS is $O(\frac{n^5L^5}{\epsilon^5})$. That is, the running time is very large and it is not strongly polynomial. Thus, an interesting problem is to design a faster (strongly polynomial) FPTAS for problem $1|\text{ONA}(a,b),\text{rej}|\sum w_jC_j + e(R)$. Note further that, when there are $k \geq 2$ MNA or ONA intervals and each MNA or ONA interval has a bounded length, it is possible to design an effective approximation algorithm. Thus, another interesting problem is to propose some approximation algorithms for the problems with multiple MNA or ONA intervals. Finally, it is also interesting to consider other objective functions such as $L_{\max} + e(R)$ and other machine setting such as parallel machines or shop machines.

Author Contributions: Conceptualization, methodology and writing-original manuscript, L.Z. (Lili Zuo) and Z.S.; project management, supervision and writing-review, L.L.; investigation, formal analysis and editing, L.Z. (Liqi Zhang).

Funding: This research was funded by the National Natural Science Foundation of China under grant number 11771406 and 11571321.

Acknowledgments: We are grateful to the Associate Editor and three anonymous reviewers for their valuable comments, which helped us significantly improve the quality of our paper.

Conflicts of Interest: The authors declare no conflict of interest.

References

1. Schmidt, G. Scheduling independent tasks with deadlines on semi-identical processors. *J. Oper. Res. Soc.* **1984**, *39*, 271–277. [CrossRef]
2. Adiri, I.; Bruno, J.; Frostig, E.; Kan, A.H.G.R. Single machine flowtime scheduling with a single breakdown. *Acta Inform.* **1989**, *26*, 679–696. [CrossRef]
3. Lee, C.Y. Machine scheduling with an availability constraints. *J. Glob. Optim.* **1996**, *9*, 363–382. [CrossRef]
4. Kacem, I. Approximation algorithms for the makespan minimization with positive tails on a single machine with a fixed nonavailability interval. *J. Comb. Optim.* **2009**, *17*, 117–133. [CrossRef]
5. Breit, J.; Schmidt, G.; Strusevich, V.A. Non-preemptive two-machine open shop scheduling with non-availability constraints. *Math. Methods Oper. Res.* **2003**, *57*, 217–234. [CrossRef]

6. Lee, C.Y. Parallel machine scheduling with non-simultaneous machine available time. *Discret. Appl. Math.* **1991**, *30*, 53–61. [CrossRef]
7. Kellerer, H. Algorithms for multiprocessor scheduling with machine release times. *IIE Trans.* **1998**, *30*, 991–999. [CrossRef]
8. Schmidt, G. Scheduling with limited machine availability. *Eur. J. Oper. Res.* **2000**, *121*, 1–15. [CrossRef]
9. Liao, C.J.; Shyur, D.L.; Lin, C.H. Makespan minimization for two parallel machines with an availability constraint. *Eur. J. Oper.* **2005**, *160*, 445–456. [CrossRef]
10. Aggoune, R. Minimising the makespan for the flow shop scheduling problem with availability constraints. *Eur. J. Oper. Res.* **2004**, *153*, 534–543. [CrossRef]
11. Burdett, R.L.; Kozan, E. Techniques for inserting additional trains into exist- ing timetables. *Transp. Res. Part B* **2009**, *43*, 821–836. [CrossRef]
12. Ma, Y.; Chu, C.; Zuo, C. A survey of scheduling with deterministic machine availability constraints. *Comput. Ind. Eng.* **2010**, *58*, 199–211. [CrossRef]
13. Brauner, N.; Frinke, G.; Lebacque, V.; Rapine, C.; Potts, C.; Struservich, V. Operator nonavailability periods. *4OR Q. J. Oper. Res.* **2009**, *7*, 239–253. [CrossRef]
14. Kacem, I.; Kellerer, H.; Seifaddini, M. Efficient approximation schemes for the maximum lateness minimization on a single machine with a fixed operator or machine non-availability interval. *J. Comb. Optim.* **2016**, *32*, 970–981. [CrossRef]
15. Chen, Y.; Zhang, A.; Tan, Z.Y. Complexity and approximation of single machine scheduling with an operator non-availability period to minimize total completion time. *Inf. Sci.* **2013**, *251*, 150–163. [CrossRef]
16. Wan, L.; Yuan, J.J. Single-machine scheduling with operator non-availability to minimize total weighted completion time. *Inf. Sci.* **2018**, *445*, 1–5. [CrossRef]
17. Burdett, R.L.; Corry, P.; Yarlagadda, P.K.D.V.; Eustace, C.; Smith, S. A flexible job shop scheduling approach with operators for coal export terminals. *Comput. Oper. Res.* **2019**, *104*, 15–36. [CrossRef]
18. Bartal, Y.; Leonardi, S.; Marchetti-Spaccamela, A.; Sgall, J.; Stougie, L. Multiprocessor scheduling with rejection. *Siam J. Discret. Math.* **2000**, *13*, 64–78. [CrossRef]
19. Cao, Z.G.; Zhang, Y.Z. Scheduling with rejection and nonidentical job arrivals. *J. Syst. Sci. Complex.* **2007**, *20*, 529–535. [CrossRef]
20. Zhang, L.Q.; Lu, L.F.; Yuan, J.J. Single machine scheduling with release dates and rejection. *Eur. J. Oper. Res.* **2009**, *198*, 975–978. [CrossRef]
21. Sengupta, S. Algorithms and approximation schemes for minimum lateness/tardiness scheduling with rejection. *Lect. Notes Comput. Sci.* **2003**, *2748*, 79–90.
22. Engels, D.W.; Karger, D.R.; Kolliopoulos, S.G.; Sengupta, S.; Uma, R.N.; Wein, J. Techniques for scheduling with rejection. *J. Algorithms* **2003**, *49*, 175–191. [CrossRef]
23. Shabtay, D.; Gaspar, N.; Kaspi, M. A survey on offline scheduling with rejection. *J. Sched.* **2013**, *16*, 3–28. [CrossRef]
24. Zhong, X.L.; Ou, J.W.; Wang, G.W. Order acceptance and scheduling with machine availability constraints. *Eur. J. Oper. Res.* **2014**, *232*, 435-442. [CrossRef]
25. Zhao, C.L.; Tang, H.Y. Single machine scheduling with an availability constraint and rejection. *Asia-Pac. J. Oper. Res.* **2014**, *31*, 1450037. [CrossRef]
26. Li, S.S.; Chen, R.X. Scheduling with rejection and a deteriorating maintenance activity on a single machine. *Asia-Pac. J. Oper. Res.* **2014**, *34*, 1750010. [CrossRef]

© 2019 by the authors. Licensee MDPI, Basel, Switzerland. This article is an open access article distributed under the terms and conditions of the Creative Commons Attribution (CC BY) license (http://creativecommons.org/licenses/by/4.0/).

Article

Transportation and Batching Scheduling for Minimizing Total Weighted Completion Time

Hongjun Wei *, Jinjiang Yuan and Yuan Gao

School of Mathematics and Statistics, Zhengzhou University, Zhengzhou 450001, China
* Correspondence: weihongjun@zzu.edu.cn

Received: 20 August 2019; Accepted: 3 September 2019; Published: 5 September 2019

Abstract: We consider the coordination of transportation and batching scheduling with one single vehicle for minimizing total weighted completion time. The computational complexity of the problem with batch capacity of at least 2 was posed as open in the literature. For this problem, we show the unary NP-hardness for every batch capacity at least 3 and present a polynomial-time 3-approximation algorithm when the batch capacity is at least 2.

Keywords: Transportation; batching scheduling; total weighted completion time; unary NP-hard; approximation algorithm

1. Introduction

Tang and Gong [1] first raised and studied the problem of transportation and batching scheduling (TBS). This model, which combines transportation and scheduling together, is motivated by a production environment in which a set of semi-finished jobs are transported from a holding area to a manufacturing facility for further processing by the available transporters and is used in many manufacturing systems. This is particularly true in the iron and steel industry.

Formally, we can describe the TBS problem in the following way. We are given a set of jobs, a set of vehicles (transporters), and a single batching machine that can handle batch jobs at the same time. Initially, all jobs and all vehicles are located at a holding area and available from time zero onward. When the production process begins, all jobs have to be transported by the vehicles to the batching machine and further processing is then carried out, where each vehicle can deliver one job at a time. The transportation time of a job is job-dependent, the empty moving times of the vehicles from the batching machine back to the holding area are identical, and the processing times of the batches on the batching machine are identical. The following notations are used in this scheduling model.

- $\mathcal{J} = \{J_1, J_2, \ldots, J_n\}$ is a set of n jobs to be processed.
- $M = \{1, 2, \ldots, m\}$ is a set of m vehicles used to transport the jobs.
- τ_j is the transportation time of job J_j, $j = 1, 2, \ldots, n$, from the holding area to the batching machine.
- τ is the empty moving time of each vehicle from the machine back to the holding area. In the sequel, we simply call τ the vehicle return time.
- c is the capacity of the batching machine. We require that every batch consists at most c jobs.
- p is the processing time of each batch, which is independent of the jobs composing the batch.
- C_j is the completion time of jobs J_j in a schedule, $j = 1, 2, \ldots, n$.
- $\alpha(b)$, which is an increasing function in b, is the processing (or batching) cost if a total of b batches are generated in of a schedule. The more batches there are, the more processing costs there will be.
- f is the scheduling cost which depends on the completion times $C_1, C_2, \ldots, C_n$ of the jobs.

The goal of the TBS problem is to find a feasible schedule that minimizes the scheduling cost plus the processing cost. We will denote this problem by $(m, c)|\tau|f + \alpha(b)$, where "$(m, c)$" means that we have m vehicles in the transportation stage and the batching machine has a capacity c for forming a batch and "τ" indicates the vehicle return time.

Production–transportation problems, which have some similarities as the TBS problems, have also been extensively studied in the literature. Hall and Potts [2] introduced and studied various single-machine and parallel-machine scheduling problems in which the various cost functions being considered are based on the delivery times and delivery cost. Chen [3] surveyed the existing models of integrated production and outbound distribution scheduling (IPODS) and presented a unified model representation method. The author also classified the existing models into several different classes and provided an overview for the optimality properties, computational tractability, and solution algorithms. As mentioned by Tang and Gong [1], the TBS problem differs from the IPODS problem. In fact, in the TBS setting, apart from the schedule of the semi-finished jobs in the transportation stage, we also consider the schedule of these jobs on the batching machine in the production stage.

Recall that a combinatorial optimization problem is binary (unary) NP-hard if it is NP-hard in the binary (unary) encoding.

Tang and Gong [1] studied the TBS problem which aims to minimize the sum of the total completion time of the jobs and the processing cost of the batching machine. For this problem, the authors proved the binary NP-hardness and further established a pseudo-polynomial-time algorithm and an FPTAS for any fixed m.

For the more "classic" scheduling objectives that exclude the processing cost, Zhu et al. [4] showed that the complexity result in Tang and Gong [1] is still valid, that is, the TBS problem with fixed m for minimizing the total completion time of the jobs is binary NP-hard. When m is arbitrary, Zhu et al. [4] showed that the TBS problem for minimizing the total completion time of the jobs is unary NP-hard. Moreover, they proved that the TBS problem for minimizing the sum of the total weighted completion time of the jobs and the processing cost of the batching machine is unary NP-hard even if $m = 1$ and $c = 3$. The computational complexity of the TBS problem with $m = 1$ for just minimizing the total weighted completion time of the jobs was posed as an open problem in Zhu et al. [4]. It should be noticed that, in the case where $\tau = 0$, i.e., the vehicle return time is given by 0, the model of transportation times in the TBS problems can be considered as a special case of setup times studied in Allahverdi [5] and Ciavotta et al., [6]

In this paper, we consider the TBS problem $(1,c)|\tau|\sum w_j C_j$, in which we have one single vehicle in the transportation stage, the scheduling criterion is to minimize the total weighted completion time of the jobs, and the processing cost is given by 0.

Note that when $c = 1$ and $\tau = 0$, problem $(1,c)|\tau|\sum w_j C_j$ degenerates to the classical two-machine flow-shop scheduling problem $F2|p_{2j} = p|\sum w_j C_j$. Recently, Wei and Yuan [7] showed that problem $F2|p_{2j} = p|\sum w_j C_j$ is unary NP-hard and admits a 2-approximation algorithm. More research of problem $F2||\sum w_j C_j$ can be found in Choi et al. [8] and Hoogeveen and Kawaguchi [9].

The unary NP-hardness of problem $F2|p_{2j} = p|\sum w_j C_j$, established in the work by the authors of [7], implies that problem $(1,1)|\tau = 0|\sum w_j C_j$ is unary NP-hard. However, in general, the computational complexity of problem $(1,c)|\tau|\sum w_j C_j$ for $c \geq 2$ is unaddressed.

In this paper, first, we show that for every $c \geq 3$ (including the possibility $c = n$), problem $(1,c)|\tau = 0|\sum w_j C_j$ is unary NP-hard. Then, for the general problem $(1,c)|\tau|\sum w_j C_j$ with the batch capacity $c \geq 2$, we present a polynomial-time approximation algorithm, which has a worst-case performance ratio of less than 3. The complexity of problem $(1,2)|\tau|\sum w_j C_j$ is still open.

2. Unary NP-Hardness Proof

To show the unary NP-hardness of problem $(1,c)|\tau = 0|\sum w_j C_j$ with $c \geq 3$, we will use the following decision problem "3-Partition" as the source problem. As outlined by Garey and Johnson [10], a 3-partition is unary NP-complete.

3-Partition: In an instance of the problem, we are given a set $\{a_1, a_2, \ldots, a_{3t}, B\}$ of $3t+1$ positive integers satisfying $\frac{1}{4}B < a_j < \frac{1}{2}B$ for $j = 1, 2, \ldots, 3t$ and $\sum_{j=1}^{3t} a_j = tB$. The decision asks, is there a partition of the index set $I = \{1, 2, \ldots, 3t\}$ into t parts $I_1, I_2, \ldots, I_t$ such that $|I_i| = 3$ and $\sum_{j \in I_i} a_j = B$ for all $i = 1, 2, \ldots, t$?

The following useful lemma states a basic algebraic result.

Lemma 1. *Let $x_1, x_2, \ldots, x_k$ be k positive numbers. Then $\sum_{i=1}^{k} x_i^2 \geq k \cdot \left(\frac{x_1 + x_2 + \cdots + x_k}{k}\right)^2$, and moreover the equality holds if and only if $x_1 = x_2 = \cdots = x_k = \frac{x_1 + x_2 + \cdots + x_k}{k}$.*

Theorem 1. *For every $c \geq 3$, problem $(1, c)|\tau = 0| \sum w_j C_j$ is unary NP-hard.*

Proof. For a given instance $(a_1, a_2, \ldots, a_{3t}; B)$ of 3-Partition, we first define

$$\Delta = t^2 B^2 + 1 = O(t^2 B^2) \tag{1}$$

and

$$M = t(t+3)(3\Delta + B)^2 = O(t^6 B^4). \tag{2}$$

Then we construct a scheduling instance of problem $(1, c)|\tau = 0| \sum w_j C_j$ as follows.

- There are $n = 3t + 1$ jobs $J_0, J_1, \ldots, J_{3t}$ of two types:
  (i) J_0, called the 0-job, has a transportation time $\tau_0 = 0$ and a weights $w_0 = M$, and
  (ii) $J_1, J_2, \ldots, J_{3t}$, called partition jobs, have transportation times $\tau_j = \Delta + a_j$ and weights $w_j = \Delta + a_j$ for $j = 1, 2, \ldots, 3t$.
- The number of vehicles is given by $m = 1$.
- The vehicle return time is given by $\tau = 0$.
- The batch machine capacity $c \geq 3$ is arbitrary, where $c = n$ is allowed.
- The batch processing time is given by $p = 3\Delta + B = O(t^2 B^2)$.
- The threshold value is given by

$$Q = M(3\Delta + B) + \frac{1}{2}t(t+3)(3\Delta + B)^2 = Mp + \frac{1}{2}t(t+3)p^2. \tag{3}$$

The above scheduling instance has $6t + 6$ parameters: τ_j, w_j ($j = 0, 1, \ldots, 3t$), τ, p, c, and Q, with Q being the largest one. Since $M = O(t^6 B^4)$ and $p = O(t^2 B^2)$, from Equation (3), we have $Q = Mp + \frac{1}{2}t(t+3)p^2 = O(t^8 B^6)$. This implies that the size of the above scheduling instance under the unary encoding is upper bounded by $O(t^9 B^6)$. Note that the size of the 3-partition instance under the unary encoding is given by $O(tB)$. Then, our scheduling instance can be constructed from the 3-partition instance in a polynomial time under the unary encoding. From the general principle of NP-hardness proof, we need to show in the following that the 3-Partition instance has a solution if and only if the scheduling instance has a feasible schedule π such that $\sum w_j C_j(\pi) \leq Q$.

Let us first suppose that the 3-Partition instance has a solution, which means that there is a partition of the index set $I = \{1, 2, \ldots, 3t\}$ into t parts $I_1, I_2, \ldots, I_t$ such that $|I_i| = 3$ and $\sum_{j \in I_i} a_j = B$ for all $i = 1, 2, \ldots, t$. Let $\mathcal{J}_0 = \{J_0\}$ and $\mathcal{J}_i = \{J_j : j \in I_i\}$ for all $i = 1, 2, \ldots, t$. We define a schedule π for the scheduling instance in the following way.

- The vehicle consecutively transports the $3t + 1$ jobs in the order

$$\mathcal{J}_0 \prec \mathcal{J}_1 \prec \cdots \prec \mathcal{J}_t \tag{4}$$

one by one, where the transportation order of the three jobs in each $\mathcal{J}_i$, $i = 1, 2, \ldots, t$, does not matter.
- The batching machine takes each $\mathcal{J}_i$, $i = 0, 1, 2, \ldots, t$, as a single batch and processes the $t + 1$ batches in the order described in Equation (4) as they are transported. Then we have totally $t + 1$ processing batches.

Note that the transportation time of the 0-job in $\mathcal{J}_0$ is 0, and the total transportation time of the three partition-jobs in $\mathcal{J}_i$, $i \in \{1, 2, \ldots, t\}$, is given by $\sum_{J_j \in \mathcal{J}_i} \tau_j = \sum_{j \in I_i} (\Delta + a_i) = 3\Delta + B$. From $p = 3\Delta + B$, the completion times of the $t + 1$ batches $\mathcal{J}_0 \prec \mathcal{J}_1 \prec \cdots \prec \mathcal{J}_t$ are given by $p, 2p, \ldots, (t+1)p$, respectively. Moreover, the weight of the 0-job in $\mathcal{J}_0$ is M and the total weight of the three partition-jobs in $\mathcal{J}_i$, $i \in \{1, 2, \ldots, t\}$, is given by $\sum_{J_j \in \mathcal{J}_i} w_j = \sum_{j \in I_i} (\Delta + a_i) = 3\Delta + B = p$. Then we have

$$\sum w_j C_j(\pi) = M \cdot p + p^2 \cdot (2 + 3 + \cdots + (t+1)) = Mp + \frac{1}{2} t(t+3) p^2.$$

From Equation (3), we have $Q = Mp + \frac{1}{2} t(t+3) p^2$, which leads to the relation $\sum w_j C_j(\pi) = Q$. Therefore, π is a required schedule. This proves the necessity.

We next prove the sufficiency. To this end, we suppose that π is a feasible schedule of the scheduling instance, such that $\sum w_j C_j(\pi) \leq Q$. Recall that $Q = Mp + \frac{1}{2} t(t+3) p^2$. Let $\mathcal{B}_0, \mathcal{B}_1, \ldots, \mathcal{B}_K$ be the batch sequence processed by the batching machine in π in this order.

If the 0-job J_0 completes after time p, we have $\sum w_j C_j(\pi) > w_0 C_0(\pi) \geq M(p+1) = Mp + M$. From the definition of M in Equation (2), we have $M = t(t+3)(3\Delta + B)^2 = t(t+3) p^2$. Thus, $\sum w_j C_j(\pi) > Mp + t(t+3) p^2 > Q$, contradicting the choice of π. Consequently, we have

$$C_0(\pi) = p, \; \mathcal{B}_0 = \{J_0\}, \text{ and } w_0 C_0(\pi) = Mp. \tag{5}$$

From Equation (5) and from the fact that $\sum w_j C_j(\pi) \leq Q = Mp + \frac{1}{2} t(t+3) p^2$, we have

$$\sum_{j=1}^{3t} w_j C_j(\pi) \leq \frac{1}{2} t(t+3) p^2. \tag{6}$$

From the above discussion, we know that the $3t$ partition-jobs $J_1, J_2, \ldots, J_{3t}$ are distributed into the K batches $\mathcal{B}_i$, $i = 1, 2, \ldots, K$. Then we define

$$I_i = \{j : J_j \in \mathcal{B}_i\} \text{ and } A_i = \sum_{j \in I_i} a_j, \; i = 1, 2, \ldots, K. \tag{7}$$

For each $i \in \{1, 2, \ldots, K\}$, we define $w^{(i)} = \sum_{j \in I_i} w_j$ and $\tau^{(i)} = \sum_{j \in I_i} \tau_j$. Since $w_j = \tau_j = \Delta + a_j$ for $j \in \{1, 2, \ldots, 3t\}$, we have

$$w^{(i)} = \tau^{(i)} = |I_i| \Delta + A_i, \; i = 1, 2, \ldots, K. \tag{8}$$

Since each batch $\mathcal{B}_i$ cannot be processed before all the jobs in $\mathcal{B}_1 \cup \mathcal{B}_2 \cup \cdots \cup \mathcal{B}_i$ are transported, we have

$$C_{\mathcal{B}_i}(\pi) \geq \tau^{(1)} + \tau^{(2)} + \cdots + \tau^{(i)} + p, \; i = 1, 2, \ldots, K. \tag{9}$$

Since the batches are processed in the order $\mathcal{B}_0 \prec \mathcal{B}_1 \prec \cdots \prec \mathcal{B}_K$ on the batching machine in π, we further have

$$C_{\mathcal{B}_i}(\pi) \geq (i+1) p, \; i = 1, 2, \ldots, K. \tag{10}$$

Note that $\sum_{i=1}^{K} \tau^{(i)} = \sum_{j=1}^{3t} \tau_j = \sum_{j=1}^{3t} (\Delta + a_j) = 3t\Delta + tB = tp$. We show in the following that $K = t$.

If $K \geq t+1$, let $\tau^* = \tau^{(t+1)} + \tau^{(t+2)} + \cdots + \tau^{(K)}$. Then $\Delta < \tau^* < tp - t\Delta$. From Equations (9) and (10), we have

$$\begin{aligned}
\sum_{j=1}^{3t} w_j C_j(\pi) &= \sum_{i=1}^{K} w^{(i)} C_{B_i}(\pi) \\
&= \sum_{i=1}^{t} \tau^{(i)} C_{B_i}(\pi) + \sum_{i=t+1}^{K} \tau^{(i)} C_{B_i}(\pi) \\
&\geq \sum_{i=1}^{t} \tau^{(i)} (\tau^{(1)} + \tau^{(2)} + \cdots + \tau^{(i)} + p) + \tau^*(t+2)p \\
&= \tfrac{1}{2} ((\sum_{i=1}^{t} (\tau^{(i)})^2 + (\sum_{i=1}^{t} \tau^{(i)})^2) + \sum_{i=1}^{t} \tau^{(i)} p + \tau^*(t+2)p \\
&= \tfrac{1}{2} \sum_{i=1}^{t} (\tau^{(i)})^2 + \tfrac{1}{2}(tp - \tau^*)^2 + (tp - \tau^*)p + \tau^*(t+2)p \\
&\geq \tfrac{1}{2} \cdot \tfrac{(tp-\tau^*)^2}{t} + \tfrac{1}{2}(tp - \tau^*)^2 + (tp - \tau^*)p + \tau^*(t+2)p \\
&> \tfrac{1}{2} t(t+3) p^2,
\end{aligned}$$

where the second inequality follows from Lemma 1 and the last inequality follows by a simple calculation. This contradicts the relation in Equation (6).

If $K \leq t - 1$, from Lemma 1 and Equation (9), we have

$$\begin{aligned}
\sum_{j=1}^{3t} w_j C_j(\pi) &= \sum_{i=1}^{K} w^{(i)} C_{B_i}(\pi) \\
&\geq \sum_{i=1}^{K} \tau^{(i)} (\tau^{(1)} + \tau^{(2)} + \cdots + \tau^{(i)} + p) \\
&= \tfrac{1}{2} ((\sum_{i=1}^{K} (\tau^{(i)})^2 + (\sum_{i=1}^{K} \tau^{(i)})^2) + \sum_{i=1}^{K} \tau^{(i)} p \\
&= \tfrac{1}{2} (\sum_{i=1}^{K} (\tau^{(i)})^2 + t^2 p^2) + t p^2 \\
&\geq \tfrac{1}{2} (t^2 p^2 / K + t^2 p^2) + t p^2 \\
&> \tfrac{1}{2} (t p^2 + t^2 p^2) + t p^2 \\
&= \tfrac{1}{2} t(t+3) p^2,
\end{aligned}$$

contradicting the relation in Equation (6) again.

The above discussion shows that $K = t$. Thus, from Lemma 1 and Equation (9) again, we have

$$\begin{aligned}
\sum_{j=1}^{3t} w_j C_j(\pi) &= \sum_{i=1}^{t} w^{(i)} C_{B_i}(\pi) \\
&\geq \sum_{i=1}^{t} \tau^{(i)} (\tau^{(1)} + \tau^{(2)} + \cdots + \tau^{(i)} + p) \\
&= \tfrac{1}{2} ((\sum_{i=1}^{t} (\tau^{(i)})^2 + (\sum_{i=1}^{t} \tau^{(i)})^2) + \sum_{i=1}^{t} \tau^{(i)} p \\
&= \tfrac{1}{2} (\sum_{i=1}^{t} (\tau^{(i)})^2 + t^2 p^2) + t p^2 \\
&\geq \tfrac{1}{2} (t p^2 + t^2 p^2) + t p^2 \\
&= \tfrac{1}{2} t(t+3) p^2 \\
&\geq \sum_{j=1}^{3t} w_j C_j(\pi),
\end{aligned}$$

where the last inequality follows from Equation (6). This means that all the inequalities in the above deduction must hold with equalities. In particular, we have $\sum_{i=1}^{t}(\tau^{(i)})^2 = tp^2$. From Lemma 1 again, it holds that $\tau^{(1)} = \tau^{(2)} = \cdots = \tau^{(t)} = p = 3\Delta + B$. From Equation (8), we conclude that

$$|I_1|\Delta + A_1 = |I_2|\Delta + A_2 = \cdots = |I_t|\Delta + A_t = 3\Delta + B. \qquad (11)$$

Since $A_i = \sum_{j \in I_i} a_j \leq tB$ for all $i = 1, 2, \ldots, t$ and the value $\Delta = t^2 B^2 + 1$ defined in (1) is sufficiently large, from Equation (11), we can easily deduce that $|I_i| = 3$ and $A_i = B$ for all $i = 1, 2, \ldots, t$. Consequently, the 3-partition instance has a solution. The result follows. □

3. Approximation

We assume in this section that $c \geq 2$. Given a job instance $\mathcal{J} = \{J_1, J_2, \ldots, J_n\}$ of problem $(1,c)|\tau|\sum w_j C_j$, we define

$$p_{1,j} = \tau_j + \tau, \ j = 1, 2, \ldots, n. \tag{12}$$

In $O(n \log n)$ time, we can renumber the n jobs in the nondecreasing order of the ratios $(p_{1j} + p/c)/w_j$ such that

$$(p_{1,1} + p/c)/w_1 \leq (p_{1,2} + p/c)/w_2 \leq \cdots \leq (p_{1,n} + p/c)/w_n. \tag{13}$$

In $O(n \log n)$ time, we can also obtain a permutation $(1', 2', \ldots, n')$ of $\{1, 2, \ldots, n\}$ such that $\tau_{1'} \geq \tau_{2'} \geq \cdots \geq \tau_{n'}$. Then, we define

$$\alpha = \min\{\tau_1, \tau_2, \cdots, \tau_n, \lfloor p/c \rfloor\} \tag{14}$$

and

$$\beta = \max\{p, \tau_{1'} + \tau_{2'} + \cdots + \tau_{c'} + c\tau\}. \tag{15}$$

For a schedule π of the job instance $\mathcal{J}$, we use $C_{1,j}(\pi)$ to denote the completion time of job J_j on the vehicle. Since the batch containing J_j must start at or after time $C_{1,j}(\pi)$, we have

$$C_j(\pi) \geq C_{1,j}(\pi) + p, \ j = 1, 2, \ldots, n. \tag{16}$$

By the job-exchanging argument, we can show that there must be an optimal schedule π of problem $(1,c)|\tau|\sum w_j C_j$ such that for the two jobs J_i and J_j,

$$C_{1,i}(\pi) < C_{1,j}(\pi) \Rightarrow C_i(\pi) \leq C_j(\pi). \tag{17}$$

Then we only consider schedules with the property in (17) in the sequel.

In the following we present an approximation algorithm for problem $(1,c)|\tau|\sum w_j C_j$ with the worst-case performance ratio at most $3\beta/(\beta + \alpha) \leq 3$, where $c \geq 2$. Our approximation algorithm can be described in the following way.

Algorithm 1. *For problem $(1,c)|\tau|\sum w_j C_j$ on instance $\mathcal{J}$.*

Step 1. Schedule the jobs in the order $J_1, J_2, \ldots, J_n$ on the vehicle without idle time.
Step 2. Form batches and process them on the batching machine by using the following strategy: When the batching machine is idle at time t and some jobs are available for processing at time t, it forms and process a new batch, which contains as many jobs as possible subject to the batch capacity c, by the rule that jobs with small subscriptions have the priority to be processed.

Clearly, Algorithm 1 runs in $O(n)$ time. To analyze the worst-case performance ratio of Algorithm 1, we first establish a lower bound of the optimal cost of problem $(1,c)|\tau|\sum w_j C_j$ on instance $\mathcal{J}$.

Lemma 2. *Let π^* be an optimal schedule of instance $\mathcal{J}$ and, for each $j \in \{1, 2, \ldots, n\}$, let $J_{[j]}$ denote the job that occupies the jth position on the vehicle in π^*. Then*

$$\sum_{j=1}^n w_{[j]} C_{[j]}(\pi^*) \geq \frac{1}{3} \Big(\sum_{j=1}^n \sum_{i=1}^j w_{[j]}(p_{1,[i]} + p/c) + (3\alpha + 2p - \tau) \sum_{j=1}^n w_{[j]} \Big). \tag{18}$$

Proof. Since π^* satisfies the property in (17), we may assume that $C_{[1]}(\pi^*) \leq C_{[2]}(\pi^*) \leq \cdots \leq C_{[n]}(\pi^*)$. Note that $c \geq 2$ is the batch capacity and each batch has a processing time p. Moreover, the first batch on the batching machine starts at a time greater than $\tau_{[1]} = p_{1,[1]} - \tau$. For each $j = 1, 2, \ldots, n$, at least $\lceil j/c \rceil$ batches are completed by time $C_{[j]}(\pi^*)$ on the batch machine. Then we have $C_{[j]}(\pi^*) \geq p_{1,[1]} - \tau + \lceil j/c \rceil p$ for $j = 1, 2, \ldots, n$. Consequently, we have

$$\sum_{j=1}^{n} w_{[j]} C_{[j]}(\pi^*) \geq (p_{1,[1]} - \tau) \sum_{j=1}^{n} w_{[j]} + p \sum_{j=1}^{n} \lceil j/c \rceil w_{[j]}. \tag{19}$$

From (16), for each $j = 1, 2, \cdots, n$, we also have $C_{[j]}(\pi^*) \geq C_{1,[j]}(\pi^*) + p$, which implies that $C_{[j]}(\pi^*) \geq \sum_{i=1}^{j} p_{1,[i]} + p - \tau$. Consequently, we have

$$\sum_{j=1}^{n} w_{[j]} C_{[j]}(\pi^*) \geq \sum_{j=1}^{n} \sum_{i=1}^{j} w_{[j]} p_{1,[i]} + (p - \tau) \sum_{j=1}^{n} w_{[j]}. \tag{20}$$

From the above two inequalities, (19) and (20), we obtain

$$
\begin{aligned}
3 \sum_{j=1}^{n} w_{[j]} C_{[j]}(\pi^*) &\geq \sum_{j=1}^{n} w_{[j]} \left(p_{1,[1]} - \tau + \lceil j/c \rceil p + 2 \sum_{i=1}^{j} p_{1,[i]} + 2(p - \tau) \right) \\
&\geq \sum_{j=1}^{n} w_{[j]} \left(\sum_{i=1}^{j} (p_{1,[i]} + p/c) + 3\alpha + 2p - \tau \right) \\
&= \sum_{j=1}^{n} \sum_{i=1}^{j} w_{[j]} (p_{1,[i]} + p/c) + (3\alpha + 2p - \tau) \sum_{j=1}^{n} w_{[j]}.
\end{aligned}
$$

Then the lemma follows immediately. □

The following lemma is also useful in our discussion.

Lemma 3. *For any c indices $i_1, i_2, \cdots, i_c \in \{1, 2, \cdots, n\}$, we have*

$$|p_{1,i_1} + p_{1,i_2} + \cdots + p_{1,i_c} - p| \leq \frac{\beta - \alpha}{\beta + \alpha} (p_{1,i_1} + p_{1,i_2} + \cdots + p_{1,i_c} + p).$$

Proof. Let $x = \min\{p_{1,i_1} + p_{1,i_2} + \cdots + p_{1,i_c}, p\}$ and $y = \max\{p_{1,i_1} + p_{1,i_2} + \cdots + p_{1,i_c}, p\}$. Then $|p_{1,i_1} + p_{1,i_2} + \cdots + p_{1,i_c} - p| = y - x$ and $p_{1,i_1} + p_{1,i_2} + \cdots + p_{1,i_c} + p = x + y$. From the definitions of α and β in Equations (14) and (15), we further have $\alpha \leq x \leq y \leq \beta$. This implies that $y\alpha \leq \beta\alpha \leq x\beta$. Now

$$
\begin{aligned}
|p_{1,i_1} + p_{1,i_2} + \cdots + p_{1,i_c} - p|(\beta + \alpha) \\
= (y - x)(\beta + \alpha) &= y\beta + y\alpha - x\beta - x\alpha \\
&\leq y\beta - y\alpha + x\beta - x\alpha \\
&= (x + y)(\beta - \alpha) \\
&= (\beta - \alpha)(p_{1,i_1} + p_{1,i_2} + \cdots + p_{1,i_c} + p).
\end{aligned}
$$

It follows that $|p_{1,i_1} + p_{1,i_2} + \cdots + p_{1,i_c} - p| \leq (\beta - \alpha)(p_{1,i_1} + p_{1,i_2} + \cdots + p_{1,i_c} + p)/(\beta + \alpha)$. □

Now we are ready to establish our final result. Recall that $c \geq 2$.

Theorem 2. *Algorithm 1 yields a schedule with cost no more than $3\beta/(\alpha + \beta)$ times the cost of an optimal schedule.*

Proof. Let π be the schedule of instance $\mathcal{J}$ generated by Algorithm 1. Since the jobs in $\mathcal{J}$ are scheduled in the order $J_1 \prec J_2 \prec \cdots \prec J_n$ on the vehicle without idle time, the implementation of Step 2 implies that, for $j = 1, 2, \ldots, c$, we have

$$\begin{aligned}
C_j(\pi) &\leq \sum_{i=1}^{j} p_{1,i} + 2p - \tau \\
&= \beta(\sum_{i=1}^{j} p_{1,i} + 2p - \tau)/(\beta + \alpha) + \alpha(\sum_{i=1}^{j} p_{1,i} + 2p - \tau)/(\beta + \alpha) \\
&\leq \beta(\sum_{i=1}^{j} p_{1,i} + 2p - \tau)/(\beta + \alpha) + 3\alpha\beta/(\beta + \alpha) \\
&\leq \beta(\sum_{i=1}^{j} (p_{1,i} + p/c) + 3\alpha + 2p - \tau)/(\beta + \alpha).
\end{aligned}$$

For each $j = c+1, c+2, \ldots, n$, we define $k_j = \lceil j/c \rceil - 1$ and $l_j = j - k_j c$. Then, we have

$$\begin{aligned}
C_j(\pi) &\leq p + \max\{C_{1,j}(\pi) + p, C_{j-c}(\pi)\} \\
&\leq p + \max\{C_{1,j-c}(\pi) + p + \sum_{h=1}^{c} p_{1,j-c+h}, \max\{C_{1,j-c}(\pi) + p, C_{j-2c}(\pi)\} + p\} \\
&\leq p + \max\{C_{1,j-c}(\pi) + p, C_{j-2c}(\pi)\} + \max\{\sum_{h=1}^{c} p_{1,j-c+h}, p\} \\
&\leq p + \max\{C_{1,l_j+c}(\pi) + p, C_{l_j}(\pi)\} + \sum_{i=1}^{k_j-1} \max\{\sum_{h=1}^{c} p_{1,l_j+ic+h}, p\} \\
&\leq p + \max\{C_{1,l_j+c}(\pi) + p, C_{1,l_j}(\pi) + 2p\} + \sum_{i=1}^{k_j-1} \max\{\sum_{h=1}^{c} p_{1,l_j+ic+h}, p\} \\
&\leq 2p + C_{1,l_j}(\pi) + \sum_{i=0}^{k_j-1} \max\{\sum_{h=1}^{c} p_{1,l_j+ic+h}, p\} \\
&= \sum_{i=1}^{l_j} p_{1,i} + 2p - \tau + \sum_{i=0}^{k_j-1} \max\{\sum_{h=1}^{c} p_{1,l_j+ic+h}, p\}.
\end{aligned}$$

By Lemma 3 and using the algebraic equality

$$2 \cdot \max\{x, y\} = x + y + |x - y|, \text{ for every two real numbers } x \text{ and } y,$$

we can obtain that

$$\begin{aligned}
&\max\{\sum_{h=1}^{c} p_{1,l_j+ic+h}, p\} \\
&= \tfrac{1}{2}(\sum_{h=1}^{c} p_{1,l_j+ic+h} + p + |\sum_{h=1}^{c} p_{1,l_j+ic+h} - p|) \\
&\leq \tfrac{1}{2}(\sum_{h=1}^{c} p_{1,l_j+ic+h} + p + (\sum_{h=1}^{c} p_{1,l_j+ic+h} + p)(\beta - \alpha)/(\beta + \alpha)) \\
&= \beta(\sum_{h=1}^{c} p_{1,l_j+ic+h} + p)/(\beta + \alpha).
\end{aligned}$$

Then, we have

$$\begin{aligned}
C_j(\pi) &\leq (\sum_{i=1}^{l_j} p_{1,i} + 2p - \tau) + \tfrac{\beta}{\beta+\alpha} \sum_{i=0}^{k_j-1} (\sum_{h=1}^{c} p_{1,l_j+ic+h} + p) \\
&= \tfrac{\alpha}{\beta+\alpha}(\sum_{i=1}^{l_j} p_{1,i} + 2p - \tau) + \tfrac{\beta}{\beta+\alpha}(\sum_{i=0}^{k-1}(\sum_{h=1}^{c} p_{1,l+ic+h} + p) + (\sum_{i=1}^{l_j} p_{1i} + 2p - \tau)) \\
&\leq \tfrac{\alpha}{\beta+\alpha} \cdot 3\beta + \tfrac{\beta}{\beta+\alpha}(\sum_{i=1}^{j} p_{1,i} + jp/c + 2p - \tau) \\
&= \tfrac{\beta}{\beta+\alpha}(\sum_{i=1}^{j}(p_{1,i} + p/c) + 3\alpha + 2p - \tau),
\end{aligned}$$

where $\sum_{i=1}^{l_j} p_{1,i} + 2p - \tau \leq 3\beta$ follows from the definition of β. Consequently, we have

$$\sum_{j=1}^{n} w_j C_j(\pi) \leq \frac{\beta}{\beta + \alpha} \left(\sum_{j=1}^{n} \sum_{i=1}^{j} w_j(p_{1,i} + p/c) + (3\alpha + 2p - \tau) \sum_{j=1}^{n} w_j \right). \tag{21}$$

Now let π^* be an optimal schedule of instance $\mathcal{J}$, and for each $j \in \{1, 2, \ldots, n\}$, let $J_{[j]}$ denote the job that occupies the jth position on the vehicle in π^*. Moreover, we consider the classical scheduling problem $1||\sum w_j C_j$ on job instance $\mathcal{J}' = \{J'_1, J'_2, \ldots, J'_n\}$, where each job J'_j has a processing time $p_{1,j} + p/c$ and a weight w_j. We consider the two schedules $\sigma = (J'_1, J'_2, \ldots, J'_n)$ and $\sigma^* = (J'_{[1]}, J'_{[2]}, \ldots, J'_{[n]})$ of problem $1||\sum w_j C_j$ on job instance $\mathcal{J}'$. From Smith [11], the well-known WSPT rule solves problem $1||\sum w_j C_j$ optimally. Thus, from the relations in (13), σ is an optimal schedule of problem $1||\sum w_j C_j$ on job instance $\mathcal{J}'$. It follows that $\sum_{j=1}^{n} w_j C_j(\sigma) \le \sum_{j=1}^{n} w_j C_j(\sigma^*)$. Note that $\sum_{j=1}^{n} w_j C_j(\sigma) = \sum_{j=1}^{n} \sum_{i=1}^{j} w_j(p_{1,i} + p/c)$ and $\sum_{j=1}^{n} w_j C_j(\sigma^*) = \sum_{j=1}^{n} \sum_{i=1}^{j} w_{[j]}(p_{1,[i]} + p/c)$. Then, we have

$$\sum_{j=1}^{n}\sum_{i=1}^{j} w_j(p_{1,i}+p/c) \le \sum_{j=1}^{n}\sum_{i=1}^{j} w_{[j]}(p_{1,[i]}+p/c). \tag{22}$$

By applying the inequality in Equation (22) to Equations (18) and (21), we obtain

$$\sum w_j C_j(\pi) \le \frac{3\beta}{\beta+\alpha} \sum w_{[j]} C_{[j]}(\pi^*)$$

This completes the proof. □

4. Conclusions

We studied the coordination of transportation and batching scheduling with one single vehicle for minimizing the total weighted completion time of the jobs without considering the processing cost of the batching machine. For this problem, we showed a unary NP-hardness of at least 3 for each batch capacity and presented a polynomial-time 3-approximation algorithm when the batch capacity is at least 2.

Future research may consider to include the processing cost in the objective function. In particular, approximation behavior of the problem $(1,c)|\tau|\sum w_j C_j + \alpha(b)$ with $\alpha(b) = \lambda b$ being a linear function in b is worthy of study. Moreover, when the batch capacity is given by $c = 2$, the computational complexity of problem $(1,2)|\tau|\sum w_j C_j + \alpha(b)$ is still open. A polynomial-time approximation scheme for solving problem $(1,c)|\tau|\sum w_j C_j$ is also expected.

Author Contributions: Conceptualization, methodology, and writing—original manuscript: H.W.; project management, supervision, and writing—review: J.Y.; data curation and formal analysis: Y.G.

Funding: This research was funded by the National Natural Science Foundation of China under grant numbers 11671368, 11771406, and 11901539. The third author was also supported by the China Postdoctoral Science Foundation under grant number 2019M652555.

Acknowledgments: The authors would like to thank the Associate Editor and the anonymous referees for their constructive comments and helpful suggestions.

Conflicts of Interest: The authors declare no conflict of interest.

References

1. Tang, L.X.; Gong, H. The coordination of transportation and batching scheduling. *Appl. Math. Mod.* **2009**, *33*, 3854–3862. [CrossRef]
2. Hall, N.G.; Potts, C.N. The coordination of scheduling and batch deliveries. *Ann. Oper. Res.* **2005**, *135*, 41–64. [CrossRef]
3. Chen, Z.L. Integrated production and outbound distribution scheduling: review and extensions. *Oper. Res.* **2010**, *58*, 130–148. [CrossRef]
4. Zhu, H.L.; Leus, R.; Zhou, H. New results on the coordination of transportation and batching scheduling. *Appl. Math. Mod.* **2016**, *40*, 4016–4022. [CrossRef]
5. Allahverdi, A. The third comprehensive survey on scheduling problems with setup times/costs. *Eur. J. Oper. Res.* **2015**, *246*, 345–378. [CrossRef]
6. Ciavotta, M.; Meloni, C.; Pranzo, M. Speeding up a Rollout algorithm for complex parallel machine scheduling. *Internat. J. Prod. Res.* **2016**, *54*, 4993–5009. [CrossRef]

7. Wei, H.J.; Yuan, J.J. Two-machine flow-shop scheduling with equal processing time on the second machine for minimizing total weighted completion time. *Oper. Res. Letters.* **2019**, *47*, 41–46. [CrossRef]
8. Choi, B.C.; Yoon, S.H.; Chung, S.J. Minimizing the total weighted completion time in a two-machine proportionate flow shop with different machine speeds. *Internat. J. Prod. Res.* **2006**, *44*, 715–728. [CrossRef]
9. Hoogeveen, J.A.; Kawaguchi, T. Minimizing total completion time in a two-machine flowshop: analysis of special cases. *Math. Oper. Res.* **1999**, *24*, 887–913. [CrossRef]
10. Garey, M.R.; Johnson, D.S. *Computers and Intractability: A Guide to the Theory of NP-Completeness*; Freeman: San Francisco, CA, USA, 1979.
11. Smith, W.E. Various optimizers for single stage production. *Naval Res. Logist. Quart.* **1956**, *3*, 59–66. [CrossRef]

© 2019 by the authors. Licensee MDPI, Basel, Switzerland. This article is an open access article distributed under the terms and conditions of the Creative Commons Attribution (CC BY) license (http://creativecommons.org/licenses/by/4.0/).

Article

Online Bi-Criteria Scheduling on Batch Machines with Machine Costs

Wenhua Li *, Weina Zhai and Xing Chai

School of Mathematics and Statistics, Zhengzhou University, Zhengzhou 450001, China;
zhaiweinalucky@163.com (W.Z.); chaixingstudy@163.com (X.C.)
* Correspondence: liwenhua@zzu.edu.cn

Received: 26 August 2019; Accepted: 8 October 2019; Published: 13 October 2019

Abstract: We consider online scheduling with bi-criteria on parallel batch machines, where the batch capacity is unbounded. In this paper, online means that jobs' arrival is over time. The objective is to minimize the maximum machine cost subject to the makespan being at its minimum. In unbounded parallel batch scheduling, a machine can process several jobs simultaneously as a batch. The processing time of a job and a batch is equal to 1. When job J_j is processed on machine M_i, it results cost c_{ij}. We only consider two types of cost functions: $c_{ij} = a + c_j$ and $c_{ij} = a \cdot c_j$, where a is the fixed cost of machines and c_j is the cost of job J_j. The number of jobs is n and the number of machines is m. For this problem, we provide two online algorithms, which are showed to be the best possible with a competitive ratio of $(1 + \beta_m, \lceil \frac{n}{m} \rceil)$, where β_m is the positive root of the equation $(1 + \beta_m)^{m+1} = \beta_m + 2$.

Keywords: bi-criteria scheduling; online algorithm; makespan; maximum machine cost; competitive ratio

1. Introduction

In this article, we consider an online bi-criteria scheduling problem to minimize the maximum machine cost subject to the makespan achieving its minimum. Online means that jobs' arrival is over time. It means, until when a job arrives, all information about it, including its arrival time, processing time and processing cost, is not known by us. For a minimization problem that is relevant to a single objective function, the competitive ratio of an online algorithm A is defined to be

$$\rho_A = \sup\{\frac{f(A,I)}{OPT(A,I)} : I \text{ is any job instance and } OPT(A,I) > 0\}.$$

Here, $f(A, I)$ is the objective value in algorithm A for any input instance I, $OPT(A, I)$ is the optimal objective value in the offline circumstance, respectively. We say algorithm A is the best possible if there doesn't exist any algorithm A' such that $\rho_{A'} < \rho_A$.

Parallel-batch was first studied by Uzsoy et al. [1,2]. There are two classes of parallel-batch models that have been widely considered in the literature, the unbounded version $b = \infty$ and the bounded version $b < \infty$, where b is the batch capacity. That is, at most b jobs can be processed simultaneously in one batch. The processing time of one batch is defined as the longest job in it. Since in this paper we consider that the jobs are with identical processing time, the processing time of the batches is 1.

In traditional scheduling theory, most problems are concerned with the minimization of one certain function. There have been many achievements such as, for minimizing maximum completion time when jobs have the same processing times, Zhang et al. [3] provided two best possible online algorithmss with $1 + \beta_m$ and $1 + \alpha$-competitive ratio for the unbounded model $b = \infty$ and bounded model $b < \infty$, respectively, and β_m satisfies $(1 + \beta_m)^{m+1} = \beta_m + 2$, $\alpha = \frac{\sqrt{5}-1}{2}$. When jobs have diverse processing times, Tian et al. [4] and Liu et al. [5] independently gave two best possible online algorithms with competitive ratio of $1 + \alpha_m$, and α_m is the positive solution of the equation

$\alpha_m^2 + m \cdot \alpha_m - 1 = 0$. Fang et al. [6] presented a best possible online algorithm with a competitive ratio of $1 + \phi$ for a set of processing time scheduling problems, where $\phi = \frac{\sqrt{5}-1}{2}$. For minimizing a maximum weighted completion time problem, Li et al. [7] established a best possible online algorithm with a competitive ratio of $\frac{\sqrt{5}+1}{2}$. For minimizing total weighted completion time problem, Cao et al. [8] gave a best possible online algorithm with a competitive ratio of ρ_m, where ρ_m is the positive solution of $\rho_m^{m+1} - \rho_m - 1 = 0$. Some reviews for parallel-batch scheduling research can be found in [9–14].

Today, with the rapid development of science and technology, minimization of one certain function doesn't satisfy the needs of things. In addition, jobs' objective functions may have certain kinds of aspects to minimize. In recent years, there have been some results about minimizing bi-criteria objective functions such as Ma et al. [15], who considered an online trade-off scheduling problem that minimize makespan and total weighted completion time on a single machine, presenting a nondominated $(1 + \alpha, 1 + \frac{1}{\alpha})$−competitive online algorithm for each α with $0 < \alpha \leq 1$. Liu et al. [16] considered the single machine online trade-off scheduling problem, which minimizes the makespan and maximum lateness. They established a nondominated $(\rho, 1 + \frac{1}{\rho})$−competitive online algorithm with $1 \leq \rho \leq \frac{\sqrt{5}+1}{2}$. Here, a (ρ_1, ρ_2)-competitive online algorithm is called nondominated if there is no other (ρ_1', ρ_2')-competitive online algorithm A' such that $(\rho_1', \rho_2') \leq (\rho_1, \rho_2)$ and either $\rho_1' < \rho_1$ or $\rho_2' < \rho_2$. In addition, Lee et al. [17] considered two bi-criteria scheduling problems: one is minimizing the maximum machine cost subject to the total completion time achieving its minimum, another is minimizing the total machine cost subject to the makespan achieves its minimum. As these two problems are strongly NP-hard, they proposed fast heuristics and found their worst-case performance bounds.

Another class of scheduling problems with bi-criteria is to minimize a secondary objective function f_2 subject to a primary objective function f_1 being at its minimum, and the objective is denoted by $Lex(f_1, f_2)$. In practical production, the producer wants to reduce the cost of the machine as soon as it is finished. Given m parallel batch machines M_i, $1 \leq i \leq m$, and n jobs J_j, $1 \leq j \leq n$. Every machine has a fixed cost a_i, job J_j has cost c_j, and $1 \leq j \leq n$. When job J_j is processed on machine M_i, this will result in different costs c_{ij}, $1 \leq i \leq m$, $1 \leq j \leq n$. Suppose that $x_{ij} = 1$ if job j is processed on machine i, otherwise $x_{ij} = 0$. Thus, the total machine cost, is named TMC, and $TMC = \sum_{i=1}^{m} \sum_{j=1}^{n} c_{ij} x_{ij}$, and the maximum machine cost is named MMC, and $MMC = \max_{i=1}^{m} \{\sum_{j=1}^{n} c_{ij} x_{ij}\}$. In Lee et al. [17], they studied the offline bi-criteria scheduling problems, for which the objective functions are minimizing MMC subject to the constraint that $\sum C_j$ is minimized and minimizing TMC subject to the constraint that $C_{\max}$ is minimized, where C_j is the completion time of job J_j, $\sum C_j$ is total completion time of jobs, and $C_{\max}$ is the maximum completion time of jobs. They considered three kinds of cost functions: $c_{ij} = c_j$, $c_{ij} = a_i + c_j$, and $c_{ij} = a_i \cdot c_j$. In our article, we consider online algorithms to minimize the maximum machine cost subject to the makespan achieving its minimum, and the objective function is denoted as $Lex(C_{\max}, MMC)$. Here, we assume that all machines have the same fixed cost a, and we consider two kinds of costs: $c_{ij} = a + c_j$ and $c_{ij} = a \cdot c_j$, $1 \leq i \leq m$, $1 \leq j \leq n$. Since the jobs are processed in batches in our model, the cost of a batch processed on some machine is defined as the maximum cost of the jobs in it. Then, the cost of one machine is the total cost of the batches on it. This problem can be written in the three-field notation as $Pm|online, p - batch, b = \infty, p_j = 1|Lex(C_{\max}, MMC)$, when $c_{ij} = a + c_j$ or $c_{ij} = a \cdot c_j$, $1 \leq i \leq m$, $1 \leq j \leq n$.

For the online scheduling problem to minimize a primary objective function f_1 and a secondary objective function f_2, we say that an online algorithm A is $(\rho_{A,1}, \rho_{A,2})$-competitive if it is $\rho_{A,1}$-competitive when minimizing f_1 and $\rho_{A,2}$-competitive when minimizing f_2. In the case that $\rho_{A,1}$ is the competitive ratio of A for minimizing f_1 and $\rho_{A,2}$ is the competitive ratio of A for minimizing f_2, we also say that the online algorithm A has a competitive ratio of $(\rho_{A,1}, \rho_{A,2})$. Suppose that the best possible competitive ratio is ρ when minimizing f_1. We say that the online algorithm A is the best possible, if $\rho_{A,1} = \rho$ and there is no other online algorithm A' such that $\rho_{A',1} = \rho$ and $\rho_{A',2} < \rho_{A,2}$.

This paper is organized as four sections as follows. In Section 2, the parameters and notations are introduced. In Section 3, the lower bounds of the competitive ratio are presented. In Section 4, two best possible online algorithms with a competitive ratio of $(1 + \beta_m, \lceil \frac{n}{m} \rceil)$ are showed, where β_m is the positive root of the equation $(1 + \beta_m)^{m+1} = \beta_m + 2$.

The objective considered in this paper is to minimize the maximum machine cost subject to the makespan being at its minimum. In addition, the algorithms studied in this paper are extensions of the results about makespan in the literature.

2. Preliminaries and Notations

Some preliminaries and notations that will be used in the paper are shown in the following:

- $c_{\max} = \max\{c_1, c_2, \cdots, c_n\}$: The maximum cost of the jobs;
- B_{l_i}: The lth batch on machine M_i;
- c_{B_l}: The cost of batch B_l, denoted as c_{B_l}, is the maximum cost of jobs belonging to batch B_l;
- c_{M_i}: The cost of machine M_i, i.e., the total cost of all batches on machine M_i;
- $U(t)$: The set of the unscheduled available jobs at time t;
- $r_{\max}(t)$: The last release time of jobs in $U(t)$;
- r_j: The release time of job J_j;
- $r_{\max}$: The last release time of all jobs;
- s_l: The starting time of batch B_l by an online algorithm;
- σ and π: The schedules generated by an online algorithm and an offline optimal algorithm, respectively;
- $C_{\max}(\sigma)$ and $C_{\max}(\pi)$: The maximum completion time in σ and the maximum completion time in π, respectively;
- $MMC(\sigma)$ and $MMC(\pi)$: The maximum machine cost in σ and the maximum machine cost in π, respectively.

For minimizing a single criterion, there have been many results about minimizing makespan. For example, the following lemma shows one result to minimize $C_{\max}$. From Theorem 3 of Zhang et al. [3], we have

Lemma 1. *For problem $Pm|online, p-batch, b = \infty, p_j = 1|C_{\max}$, the competitive ratio of the best possible online algorithm is $1 + \beta_m$, where β_m is the positive root of the equation $(1 + \beta_m)^{m+1} = \beta_m + 2$.*

3. The Lower Bound

Theorem 1. *For problem $Pm|online, p-batch, b = \infty, p_j = 1|Lex(C_{\max}, MMC)$, when $c_{ij} = a + c_j$ or $c_{ij} = a \cdot c_j$, $1 \leq i \leq m$, $1 \leq j \leq n$, there exists no online algorithm with a competitive ratio less than $(1 + \beta_m, \lceil \frac{n}{m} \rceil)$.*

Proof. Supposing that the fixed cost of each machine is a, the cost of each job is 1, which means $c_{\max} = 1$. Let ψ be the set of the best possible solutions of the objective function $C_{\max}$. Then, for $\forall \sigma \in \psi$, we prove that there exists no online algorithm that satisfies $\frac{MMC(\sigma)}{MMC(\pi)} < \lceil \frac{n}{m} \rceil$, subjected to the constraint

$$\frac{C_{\max}(\sigma)}{C_{\max}(\pi)} \leq 1 + \beta_m. \quad (1)$$

We use adversary strategy to prove this conclusion. Let $\mathcal{A}$ be an arbitrary online algorithm, and ϵ is an arbitrarily small positive number. Suppose the first job J_1 arrives at 0 and starts at s_1. From (1), we can know that $s_1 \leq \beta_m$. Otherwise, we have

$$\frac{C_{\max}(\sigma)}{C_{\max}(\pi)} = \frac{s_1 + 1}{r_1 + 1} > 1 + \beta_m,$$

a contradiction. Job J_{i+1} arrives at $s_i + \epsilon$ and starts at s_{i+1}, $1 \leq i \leq n-1$. We claim that $s_i \leq (1+\beta_m)r_i + \beta_m$. Otherwise,

$$\frac{C_{\max}(\sigma)}{C_{\max}(\pi)} = \frac{s_i + 1}{r_i + 1} > \frac{(1+\beta_m)r_i + \beta_m + 1}{r_i + 1} = 1 + \beta_m,$$

Then, $C_{\max}(\sigma) > (1+\beta_m)C_{\max}(\pi)$, a contradiction.

Hence, n jobs are processed as n batches on m machines.

When $c_{ij} = a + c_j$, after n jobs are processed, there must be not less than $\lceil \frac{n}{m} \rceil$ jobs on one machine. Because the cost of each job is 1, the maximum machine cost is $MMC(\sigma) \geq \lceil \frac{n}{m} \rceil \times (a+1)$. In π, all jobs can form one batch starting at the last time when the job arrives, so $MMC(\pi) = a + 1$. Then, we get $\frac{MMC(\sigma)}{MMC(\pi)} \geq \lceil \frac{n}{m} \rceil$.

When $c_{ij} = a \cdot c_j$, similarly after n jobs are finished, there must be one machine that does not have less than $\lceil \frac{n}{m} \rceil$ jobs. Since each job's cost is 1, the maximum machine cost is $MMC(\sigma) \geq \lceil \frac{n}{m} \rceil \times (a \cdot 1)$. In π, all jobs can form one batch starting at the time which the last job arrives, so $MMC(\pi) = a \cdot 1$. Then, we get $\frac{MMC(\sigma)}{MMC(\pi)} \geq \lceil \frac{n}{m} \rceil$.

Therefore, for problem $Pm|online, p-batch, b=\infty, p_j=1|.Lex(C_{\max}, MMC)$, when $c_{ij} = a + c_j$ or $c_{ij} = a \cdot c_j$, $1 \leq i \leq m, 1 \leq j \leq n$, and there exists no online algorithm in which the competitive ratio is less than $(1+\beta_m, \lceil \frac{n}{m} \rceil)$. □

4. Best Possible Online Algorithms

Here, there are two online algorithms for this problem.

Algorithm H^1

At current time t, if some machine is idle, $U(t) \neq \emptyset$, when $t \geq (1+\beta_m)r_{\max}(t) + \beta_m$; then, start the jobs in $U(t)$ as a batch on the idle machine that has the minimum machine cost at the moment. Otherwise, do nothing but wait.

Algorithm H^2

At current time t, if some machine is idle, $U(t) \neq \emptyset$, when $t \geq (1+\beta_m)r_{\max}(t) + \beta_m$; then, start the jobs in $U(t)$ as a batch on the idle machine that has the minimum number of batches at the moment. Otherwise, do nothing but wait.

Following the notation in Zhang et al. [3], we also call batches that start at $(1+\beta_m)r_{\max}(t) + \beta_m$ regular batches. From Lemma 1 of Zhang et al. [3], we have

Lemma 2. *All batches generated by algorithm H^1 and H^2 are regular batches.*

Lemma 3. *When $c_{ij} = a + c_j$, Then, $MMC(\pi) = a + c_{\max}$; When $c_{ij} = a \cdot c_j$, then $MMC(\pi) = a \cdot c_{\max}$.*

Proof. The offline optimal objective case of the maximum machine cost MMC is: all jobs can form one batch starting at the last arrival time on an arbitrary machine. Thus, when $c_{ij} = a + c_j$, the maximum machine cost is $MMC(\pi) = a + c_{\max}$; when $c_{ij} = a \cdot c_j$, the maximum machine cost is $MMC(\pi) = a \cdot c_{\max}$. □

Theorem 2. *For problem $Pm|online, p-batch, b=\infty, p_j=1|Lex(C_{\max}, MMC)$, when $c_{ij} = a + c_j$ or $c_{ij} = a \cdot c_j$, $1 \leq i \leq m, 1 \leq j \leq n$, algorithm H^1 is a best possible online algorithm with a competitive ratio of $(1+\beta_m, \lceil \frac{n}{m} \rceil)$.*

Proof. When $c_{ij} = a + c_j$, suppose that the schedule generated by algorithm H_1 is σ'_1. From Lemmas 1 and 2, we can know that

$$C_{\max}(\sigma'_1) = (1 + \beta_m)r_{\max} + \beta_m + 1 = (1 + \beta_m)(r_{\max} + 1) \leq (1 + \beta_m)C_{\max}(\pi).$$

In the following, we prove $MMC(\sigma'_1) \leq \lceil \frac{n}{m} \rceil \cdot MMC(\pi)$. Suppose, in σ'_1, that the machine M_x has the maximum machine cost, Then,

$$MMC(\sigma'_1) = c_{M_x}. \qquad (2)$$

We distinguish the following cases:

Case 1 The number of batches on machine M_x is no more than $\lceil \frac{n}{m} \rceil$. Thus,

$$\begin{aligned} c_{M_x} &\leq \lceil \frac{n}{m} \rceil \cdot a + \sum_{1 \leq l \leq \lceil \frac{n}{m} \rceil} c_{B_{l_x}} \\ &\leq \lceil \frac{n}{m} \rceil \cdot (a + c_{\max}). \end{aligned}$$

In addition, by (2), we have

$$MMC(\sigma'_1) = c_{M_x} \leq \lceil \frac{n}{m} \rceil \cdot (a + c_{\max}).$$

From Lemma 3, we get

$$MMC(\sigma'_1) \leq \lceil \frac{n}{m} \rceil \cdot MMC(\pi).$$

Case 2 The number of batches on machine M_x is more than $\lceil \frac{n}{m} \rceil$. Thus, there must be one machine that has less than $\lceil \frac{n}{m} \rceil$ batches. Suppose machine $M_{x'}$ is the machine that has less than $\lceil \frac{n}{m} \rceil$ batches; let B_y be the last batch to process on machine M_x.

Firstly, if machine $M_{x'}$ is idle directly before s_y, let the total cost of batches that start before s_y on $M_{x'}$ be V_1. From algorithm H^1, because the number of batches on machine $M_{x'}$ is no more than $\lceil \frac{n}{m} \rceil - 1$, so

$$V_1 \leq (\lceil \frac{n}{m} \rceil - 1) \cdot a + \sum_{1 \leq l \leq \lceil \frac{n}{m} \rceil - 1} c_{B_{l_{x'}}}. \qquad (3)$$

Moreover, from algorithm H^1, the total cost of batches that start before s_y on machine M_x is not greater than the total cost of batches start before s_y on machine $M_{x'}$, that is

$$c_{M_x} - (a + c_{B_y}) \leq V_1,$$

then from (3), we have

$$\begin{aligned} c_{M_x} &\leq V_1 + (a + c_{B_y}) \\ &\leq (\lceil \frac{n}{m} \rceil - 1) \cdot a + \sum_{1 \leq l \leq \lceil \frac{n}{m} \rceil - 1} c_{B_{l_{x'}}} + (a + c_{B_y}) \\ &\leq \lceil \frac{n}{m} \rceil \cdot a + (\lceil \frac{n}{m} \rceil - 1) \cdot c_{\max} + c_{\max} \\ &= \lceil \frac{n}{m} \rceil \cdot (a + c_{\max}). \end{aligned}$$

In addition, by (2), we have

$$MMC(\sigma'_1) = c_{M_x} \leq \lceil \frac{n}{m} \rceil \cdot (a + c_{\max}).$$

Lemma 3 shows that

$$MMC(\sigma'_1) \leq \lceil \frac{n}{m} \rceil \cdot MMC(\pi).$$

Secondly, if machine $M_{x'}$ is busy directly before s_y, let $B_{y'}$ be the last batch that is on machine M_x such that machine $M_{x'}$ is idle directly before $s_{y'}$. Supposing that there are k batches between B_y and $B_{y'}$, we denote them as $B'_1, B'_2, \cdots, B'_k$.

Claim $B_{y'}$ must exist and is not the first batch on M_x.

Otherwise, $B_{y'}$ does not exist or it is the first batch on M_x. This means that machine $M_{x'}$ is busy when $B_{2_x}, B_{3_x}, \cdots$ start on machine M_x. Since the number of batches on machine M_x is more than $\lceil \frac{n}{m} \rceil$, the number of batches on machine $M_{x'}$ must be not less than $\lceil \frac{n}{m} \rceil$, contradicting the assumption that the number of batches on machine $M_{x'}$ is less than $\lceil \frac{n}{m} \rceil$. Thus, the claim holds.

Let the total cost of batches start before $s_{y'}$ on machine $M_{x'}$ be V_2. Then, from the definition of $B_{y'}$ and $M_{x'}$, the number of batches that start before $s_{y'}$ on machine $M_{x'}$ is no more than $\lceil \frac{n}{m} \rceil - (k+2)$. Thus,

$$\begin{aligned} V_2 &\leq [\lceil \frac{n}{m} \rceil - (k+2)] \cdot a + \sum_{1 \leq l \leq \lceil \frac{n}{m} \rceil - (k+2)} c_{B_{l_{x'}}} \\ &\leq [\lceil \frac{n}{m} \rceil - (k+2)] \cdot a + [\lceil \frac{n}{m} \rceil - (k+2)] \cdot c_{\max} \\ &= [\lceil \frac{n}{m} \rceil - (k+2)] \cdot (a + c_{\max}). \end{aligned}$$

That is,

$$V_2 \leq [\lceil \frac{n}{m} \rceil - (k+2)] \cdot (a + c_{\max}). \tag{4}$$

Furthermore, by algorithm H^1, the total cost of batches starting before $s_{y'}$ on machine M_x is not greater than the total cost of batches starting before $s_{y'}$ on machine $M_{x'}$, then

$$c_{M_x} - (a + c_{B_y}) - (k \cdot a + \sum_{1 \leq l \leq k} c_{B'_l}) - (a + c_{B_{y'}}) \leq V_2.$$

Thus, from (4), we know that

$$\begin{aligned} c_{M_x} &\leq V_2 + (a + c_{B_{y'}}) + (k \cdot a + \sum_{1 \leq l \leq k} c_{B'_l}) + (a + c_{B_y}) \\ &\leq [\lceil \frac{n}{m} \rceil - (k+2)] \cdot (a + c_{\max}) + (a + c_{\max}) + (k \cdot a + k \cdot c_{\max}) + (a + c_{\max}) \\ &= \lceil \frac{n}{m} \rceil \cdot (a + c_{\max}). \end{aligned}$$

In addition, by (2), we have

$$MMC(\sigma'_1) = c_{M_x} \leq \lceil \frac{n}{m} \rceil \cdot (a + c_{\max}).$$

Lemma 3 shows that

$$MMC(\sigma'_1) \leq \lceil \frac{n}{m} \rceil \cdot MMC(\pi).$$

We know that $k \geq 1$. When $k = 0$, similar to the above discussion, the conclusion also holds.

When $c_{ij} = a \cdot c_j$, suppose the schedule produced by algorithm H^1 is σ_1''. From Lemmas 1 and 2, we obtain that $C_{\max}(\sigma_1'') \leq (1 + \beta_m) C_{\max}(\pi)$. In the following, we want to prove that $MMC(\sigma_1'') \leq \lceil \frac{n}{m} \rceil \cdot MMC(\pi)$. Supposing that machine M_w has maximum machine cost in σ_1'', Then,

$$MMC(\sigma_1'') = c_{M_w}. \tag{5}$$

We distinguish the following cases:

Case 3 The number of batches is no more than $\lceil \frac{n}{m} \rceil$ on machine M_w. Then,

$$\begin{aligned} c_{M_w} &\leq \sum_{1 \leq l \leq \lceil \frac{n}{m} \rceil} (a \cdot c_{B_{l_w}}) \\ &= a \cdot \sum_{1 \leq l \leq \lceil \frac{n}{m} \rceil} c_{B_{l_w}} \\ &\leq \lceil \frac{n}{m} \rceil \cdot a \cdot c_{\max}. \end{aligned}$$

Thus,

$$c_{M_w} \leq \lceil \frac{n}{m} \rceil \cdot a \cdot c_{\max}. \tag{6}$$

From (5) and (6), we have

$$MMC(\sigma_1'') = c_{M_w} \leq \lceil \frac{n}{m} \rceil \cdot (a \cdot c_{\max}).$$

Lemma 3 shows that

$$MMC(\sigma_1'') \leq \lceil \frac{n}{m} \rceil \cdot MMC(\pi).$$

Case 4 The number of batches is more than $\lceil \frac{n}{m} \rceil$ on machine M_w. Thus, there must be one machine that has fewer than $\lceil \frac{n}{m} \rceil$ batches. Supposing that machine $M_{w'}$ is the machine for which the number of batches on it is less than $\lceil \frac{n}{m} \rceil$, let B_z be the last batch to process on machine M_w.

If machine $M_{w'}$ is idle directly before s_z, we denote the total cost of batches start before s_z on machine $M_{w'}$ as V_1', by algorithm H^1 because the number of batches on machine $M_{w'}$ is no more than $\lceil \frac{n}{m} \rceil - 1$, hence

$$V_1' \leq a \cdot \sum_{1 \leq l \leq \lceil \frac{n}{m} \rceil - 1} c_{B_{l_{w'}}}. \tag{7}$$

Furthermore, by algorithm H^1, the total cost of batches starting before s_z on machine M_w is not greater than the total cost of batches starting before s_z on machine $M_{w'}$; then,

$$c_{M_w} - (a \cdot c_{B_z}) \leq V_1'.$$

From (7), we have

$$\begin{aligned} c_{M_w} &\leq V_1' + a \cdot c_{B_z} \\ &\leq a \cdot \sum_{1 \leq l \leq \lceil \frac{n}{m} \rceil - 1} c_{B_{l_{w'}}} + a \cdot c_{B_z} \\ &\leq a \cdot (\lceil \frac{n}{m} \rceil - 1) \cdot c_{\max} + a \cdot c_{\max} \\ &= a \cdot \lceil \frac{n}{m} \rceil \cdot c_{\max}. \end{aligned}$$

In addition, by (6), we have

$$MMC(\sigma_1'') = c_{M_w} \leq \lceil \frac{n}{m} \rceil \cdot a \cdot c_{\max}.$$

Lemma 3 shows that
$$MMC(\sigma_1'') \leq \lceil \frac{n}{m} \rceil \cdot MMC(\pi).$$

If machine $M_{w'}$ is busy directly before s_z, let $B_{z'}$ be the last batch on machine M_w such that machine $M_{w'}$ is idle directly before $s_{z'}$. Similarly, suppose there are k batches between $B_{z'}$ and B_z, we denote them as $B_1', B_2', \cdots, B_k'$. From the discussion of case 2 in $c_{ij} = a + c_j$ situation, such $B_{z'}$ must exist and is not the first batch on M_w. Denoting the total cost of batches starting before $s_{z'}$ on machine $M_{w'}$ is V_2'; thus, by the definition of $B_{z'}$ and $M_{w'}$, the number of batches starting before $s_{z'}$ on machine $M_{w'}$ is no more than $\lceil \frac{n}{m} \rceil - (k+2)$. Then,

$$\begin{aligned} V_2' &\leq \sum_{1 \leq l \leq \lceil \frac{n}{m} \rceil - (k+2)} a \cdot c_{B_{l_{w'}}} \\ &= a \cdot \sum_{1 \leq l \leq \lceil \frac{n}{m} \rceil - (k+2)} c_{B_{l_{w'}}} \\ &\leq a \cdot [\lceil \frac{n}{m} \rceil - (k+2)] \cdot c_{\max}. \end{aligned}$$

Thus,
$$V_2' \leq a \cdot [\lceil \frac{n}{m} \rceil - (k+2)] \cdot c_{\max}. \tag{8}$$

Moreover, by algorithm H^1, the total cost of batches starting before $s_{z'}$ on machine M_w is not greater than the total cost of batches starting before $s_{z'}$ on machine $M_{w'}$, so

$$c_{M_w} - (a \cdot c_{B_z}) - (a \cdot \sum_{1 \leq l \leq k} c_{B_l'}) - (a \cdot c_{B_{z'}}) \leq V_2'.$$

Therefore, from (8), we get

$$\begin{aligned} c_{M_w} &\leq V_2' + a \cdot c_{B_{z'}} + a \cdot \sum_{1 \leq l \leq k} c_{B_l'} + a \cdot c_{B_z} \\ &\leq a \cdot [\lceil \frac{n}{m} \rceil - (k+2)] \cdot c_{\max} + a \cdot c_{\max} + a \cdot k \cdot c_{\max} + a \cdot c_{\max} \\ &= \lceil \frac{n}{m} \rceil \cdot a \cdot c_{\max}. \end{aligned}$$

In addition, by (6), we have
$$MMC(\sigma_1'') = c_{M_w} \leq \lceil \frac{n}{m} \rceil \cdot (a \cdot c_{\max}).$$

Lemma 3 shows that
$$MMC(\sigma_1'') \leq \lceil \frac{n}{m} \rceil \cdot MMC(\pi).$$

We know that $k \geq 1$. When $k = 0$, similar to the above discussion, the conclusion also holds.

Overall, for problem $Pm|online, p-batch, b = \infty, p_j = 1|Lex(C_{\max}, MMC)$, when $c_{ij} = a + c_j$ or $c_{ij} = a \cdot c_j$, $1 \leq i \leq m, 1 \leq j \leq n$, algorithm H^1 is a $(1 + \beta_m, \lceil \frac{n}{m} \rceil)$-competitive online algorithm.

Combining theorem 1, we obtain that algorithm H^1 is a best possible online algorithm. □

Lemma 4. *In algorithm H^2, there are at most $\lceil \frac{n}{m} \rceil$ batches on each machine.*

Proof. When $n \leq m$, there is at most one batch on each machine, so the conclusion holds naturally. When $n > m$, suppose, after the kth batch has been processed, that there are at most $\lceil \frac{k}{m} \rceil$ batches on each machine, and $m \leq k \leq n - 1$. In the following, we have an induction on k, to prove that, after the

$(k+1)$th batch has been processed, there are at most $\lceil \frac{k+1}{m} \rceil$ batches on each machine. The batches are denoted by $B_1, B_2, \cdots$, such that $s_1 < s_2 < \cdots$.

Case 1 $k = qm + l$, and $1 \leq q \leq \lceil \frac{n}{m} \rceil - 1$, $1 \leq l \leq m - 1$, where q, l are integers—as after the kth batch has been processed, there are $k - qm = l$ machines in which their batch numbers are $\lceil \frac{k}{m} \rceil$, and other $m - (k - qm) = m - l$ machines in which their batch numbers are $\lceil \frac{k}{m} \rceil - 1$. Let s_k be the release time of kth batch, and r_{k+1} be the latest release time of jobs in $(k+1)$th batch. Then, we get $r_{k+1} > s_k$; otherwise, jobs in the $(k+1)$th batch will process with jobs in the kth batch, a contradiction. Furthermore, by algorithm H^1, we can get that the starting time of the $(k+1)$th batch is $(1 + \beta_m)r_{k+1} + \beta_m$. In addition, because

$$
\begin{aligned}
(1 + \beta_m)r_{k+1} + \beta_m &> (1 + \beta_m)s_k + \beta_m \\
&= (1 + \beta_m)^2 r_k + (1 + \beta_m)\beta_m + \beta_m \\
&> (1 + \beta_m)^2 s_{k-1} + (1 + \beta_m)\beta_m + \beta_m \\
&\vdots \\
&> (1 + \beta_m)^m s_{k+1-m} + \sum_{i=0}^{m-1} \beta_m (1 + \beta_m)^i \\
&= (1 + \beta_m)^m s_{k+1-m} + (1 + \beta_m)^m - 1 \\
&= \frac{\beta_m + 2}{\beta_m + 1}(s_{k+1-m} + 1) - 1 \\
&= s_{k+1-m} + \frac{s_{k+1-m} + 1}{\beta_m + 1} \\
&\geq s_{k+1-m} + 1,
\end{aligned}
$$

$$(1 + \beta_m)r_{k+1} + \beta_m > s_{k+1-m} + 1. \tag{9}$$

We use s_{k+1-m} to represent the starting time of the $(k+1-m)$th batch. From (9), it shows that, when the $(k+1)$th batch starts, the $(k+1-m)$th batch has been completed. Then, l batches that start before the $(k+1-m)$th batch also have been completed. We define these l batches as $B_{k+1-m-l}, B_{k+1-m-(l-1)}, \cdots, B_{k-m}$. Then, when the $(k+1)$th batch starts, $l + 1$ batches $B_{k+1-m-l}, B_{k+1-m-(l-1)}, \cdots, B_{k-m}, B_{k+1-m}$ have been completed. In addition, because of $k = qm + l$, when the $(k+1)$th batch starts, there is at least one machine that is idle. In addition, it can be known, by algorithm H^2, that the number of batches on this idle machine is $\lceil \frac{k}{m} \rceil - 1$. Hence, after the $(k+1)$th batch is completed, the number of batches on this idle machine is $\lceil \frac{k}{m} \rceil - 1 + 1 = \lceil \frac{k}{m} \rceil$. Moreover, because $\lceil \frac{k}{m} \rceil = \lceil \frac{k+1}{m} \rceil$ when $k = qm + l$, there are $k - qm + 1 = l + 1$ machines whose batch numbers are $\lceil \frac{k+1}{m} \rceil$, and other $m - (k - qm + 1) = m - l - 1$ machines that have $\lceil \frac{k+1}{m} \rceil - 1$ batches. This means that there are at most $\lceil \frac{k+1}{m} \rceil$ batches on each machine. The result follows.

Case 2 $k = qm$ and $1 \leq q \leq \lceil \frac{n}{m} \rceil - 1$, where q is an integer. After the $(k+1)$th batch is processed, one machine has $\lceil \frac{k}{m} \rceil + 1$ batches, and the number of batches on other machines is still $\lceil \frac{k}{m} \rceil$. Furthermore, $\lceil \frac{k}{m} \rceil + 1 = \lceil \frac{k+1}{m} \rceil$ when $k = qm$. Thus, every machine has at most $\lceil \frac{k+1}{m} \rceil$ batches. The results follow. □

Theorem 3. *For problem $Pm|online, p - batch, b = \infty, p_j = 1|Lex(C_{max}, MMC)$, when $c_{ij} = a + c_j$ or $c_{ij} = a \cdot c_j$, $1 \leq i \leq m$, $1 \leq j \leq n$, algorithm H^2 is the best possible online algorithm with a competitive ratio of $(1 + \beta_m, \lceil \frac{n}{m} \rceil)$.*

Proof. When $c_{ij} = a + c_j$, suppose the schedule generated by algorithm H^2 is σ_2'. From Lemma 1 and Lemma 2, we have

$$C_{max}(\sigma_2') = (1 + \beta_m)r_{max} + \beta_m + 1 = (1 + \beta_m)(r_{max} + 1) \leq (1 + \beta_m) \cdot C_{max}(\pi).$$

In the following, we want to prove $MMC(\sigma_2') \leq \lceil \frac{n}{m} \rceil \cdot MMC(\pi)$.
Suppose machine M_x has the maximum machine cost, Then,

$$MMC(\sigma_2') = c_{M_x}. \tag{10}$$

From Lemma 4, we know that the number of batches on machine M_x is no more than $\lceil \frac{n}{m} \rceil$; then, from (10), we get

$$\begin{aligned}
MMC(\sigma_2') &= c_{M_x} \\
&\leq \lceil \frac{n}{m} \rceil \cdot a + \sum_{1 \leq l \leq \lceil \frac{n}{m} \rceil} c_{B_{l_x}} \\
&\leq \lceil \frac{n}{m} \rceil \cdot a + \lceil \frac{n}{m} \rceil \cdot c_{\max} \\
&= \lceil \frac{n}{m} \rceil \cdot (a + c_{\max}).
\end{aligned}$$

In addition, Lemma 3 shows that

$$MMC(\sigma_2') \leq \lceil \frac{n}{m} \rceil \cdot MMC(\pi).$$

When $c_{ij} = a \cdot c_j$, suppose the schedule produced by algorithm H^2 is σ_2''. From Lemmas 1 and 2, we can get $C_{\max}(\sigma_2'') \leq (1 + \beta_m) \cdot C_{\max}(\pi)$—the following to prove $MMC(\sigma_2'') \leq \lceil \frac{n}{m} \rceil \cdot MMC(\pi)$.
Assume that machine M_w is the machine with the maximum cost, Then,

$$MMC(\sigma_2'') = c_{M_w}. \tag{11}$$

From Lemma 4, we know that the number of batches on machine M_w is no more than $\lceil \frac{n}{m} \rceil$; then, from (11), we get

$$\begin{aligned}
MMC(\sigma_2'') &= c_{M_w} \\
&\leq \sum_{1 \leq l \leq \lceil \frac{n}{m} \rceil} a \cdot c_{B_{l_w}} \\
&\leq \lceil \frac{n}{m} \rceil \cdot (a \cdot c_{\max}).
\end{aligned}$$

From Lemma 3, we have

$$MMC(\sigma_2'') \leq \lceil \frac{n}{m} \rceil \cdot MMC(\pi).$$

To sum up, for problem $Pm|online, p-batch, b = \infty, p_j = 1|Lex(C_{\max}, MMC)$, when $c_{ij} = a + c_j$ or $c_{ij} = a \cdot c_j, 1 \leq i \leq m, 1 \leq j \leq n$, algorithm H^2 is an online algorithm with a competitive ratio of $(1 + \beta_m, \lceil \frac{n}{m} \rceil)$.

Combining Theorem 1, it implies that algorithm H^2 is a best possible online algorithm. □

5. Conclusions

In this paper, we established two best possible online algorithms for problem $Pm|online, p-batch, b = \infty, p_j = 1|Lex(C_{\max}, MMC)$, when $c_{ij} = a + c_j$ or $c_{ij} = a \cdot c_j, 1 \leq i \leq m, 1 \leq j \leq n$. The algorithms provided in this paper are to minimize the maximum machine cost subject to the makespan being at its minimum. They are extensions of the algorithm in [3], which is only minimizing the makespan. Here, we suppose that all machines have the same fixed cost a; for further research, extending this problem to different machine costs a_i is still an important research topic that needs to be studied.

Author Contributions: Conceptualization and methodology, W.L.; formal analysis and writing—original manuscript, W.Z. and X.C.; project management, supervision and writing-review, W.L. and X.C.

Funding: This Research was funded by the National Natural Science Foundation of China (Nos. 11571321, 11971443 and 11771406) and Postgraduate Education Reform Project of Henan Province (No. 2019SJGLX051Y).

Acknowledgments: We are grateful to the Associate Editor and four anonymous reviewers for their valuable comments, which helped us significantly improve the quality of our paper.

Conflicts of Interest: The authors declare no conflict of interest.

References

1. Uzsoy, R.; Lee, C.Y.; Martin-Vega, L.A. A review of production planning and scheduling models in the semiconductor industry, part I: System characteristics, performance evaluation and production planning. *IIE Trans.* **1992**, *24*, 47–61. [CrossRef]
2. Uzsoy, R.; Lee, C.Y.; Martin-Vega, L.A. A survey of production planning and scheduling models in the semiconductor industry, part II: Shop-floor control. *IIE Trans.* **1994**, *26*, 44–55. [CrossRef]
3. Zhang, G.; Cai, X.; Wong, C.K. Optimal online algorithms for scheduling on parallel batch processing machines. *IIE Trans.* **2003**, *35*, 175–181. [CrossRef]
4. Tian, J.; Cheng, T.C.E.; Ng, C.T.; Yuan, J. Online scheduling on unbounded parallel-batch machines to minimize the makespan. *Inf. Process. Lett.* **2009**, *109*, 1211–1215. [CrossRef]
5. Liu, P.; Lu, X.; Fang, Y. A best possible deterministic online algorithm for minimizing makespan on parallel batch machines. *J. Sched.* **2012**, *15*, 77–81. [CrossRef]
6. Fang, Y.; Liu, P.; Lu, X. Optimal online algorithms for one batch machine with grouped processing times. *J. Comb. Optim.* **2011**, *22*, 509–516. [CrossRef]
7. Li, W. A best possible online algorithm for the parallel-machine scheduling to minimize the maximum weighted completion time. *Asia-Pac. J. Oper. Res.* **2015**. [CrossRef]
8. Cao, J.; Yuan, J.; Li, W.; Bu, H. Online scheduling on batching machines to minimise the total weighted completion time of jobs with precedence constraints and identical processing times. *Int. J. Syst. Sci.* **2011**, *42*, 51–55. [CrossRef]
9. Brucker, P.; Gladky, A.; Hoogeveen, H.; Kovalyov, M.Y.; Potts, C.N.; Tautenhahn, T.; van de Velde, S.L. Scheduling a batching machine. *J. Sched.* **1998**, *1*, 31–54. [CrossRef]
10. Lee, C.Y.; Uzsoy, R. Minimizing makespan on a single batch processing machine with dynamic job arrivals. *Int. J. Prod. Res.* **1999**, *37*, 219–236. [CrossRef]
11. Liu, Z.H.; Yu, W.C. Scheduling one batch process or subject to job release dates. *Discrete Appl. Math.* **2000**, *105*, 129–136. [CrossRef]
12. Li, W.; Yuan, J.; Lin, Y. A note on special optimal batching structures to minimize total weighted completion time. *J. Comb. Optim.* **2007**, *14*, 475–480. [CrossRef]
13. Tian, J.; Fu, R.; Yuan, J. Online over time scheduling on parallel-batch machines: A survey. *J. Oper. Res. Soc. China* **2014**, *2*, 445–454. [CrossRef]
14. Li, W.; Chai, X. The medical laboratory scheduling for weighted flow-time. *J. Comb. Optim.* **2019**, *37*, 83–94. [CrossRef]
15. Ma, R.; Yuan, J. Online trade-off scheduling on a single machine to minimize makespan and total weighted completion time. *Int. J. Prod. Econ.* **2014**, *158*, 114–119. [CrossRef]
16. Liu, Q.; Yuan, J. Online trade-off scheduling on a single machine to minimize makespan and maximum lateness. *J. Comb. Optim.* **2016**, *32*, 385–395. [CrossRef]
17. Lee, K.; Leung, J.Y.T.; Jia, Z.; Li, W.; Pinedo, M.L.; Lin, B.M.T. Fast approximation algorithms for bi-criteria scheduling with machine assignment costs. *Eur. J. Oper. Res.* **2014**, *238*, 54–64. [CrossRef]

© 2019 by the authors. Licensee MDPI, Basel, Switzerland. This article is an open access article distributed under the terms and conditions of the Creative Commons Attribution (CC BY) license (http://creativecommons.org/licenses/by/4.0/).

Article

Dynamic Restructuring Framework for Scheduling with Release Times and Due-Dates

Nodari Vakhania

Centro de Investigación en Ciencias, Universidad Autónoma del Estado de Morelos, Morelos 62209, Mexico; nodari@uaem.mx

Received: 8 October 2019; Accepted: 12 November 2019; Published: 14 November 2019

Abstract: Scheduling jobs with release and due dates on a single machine is a classical strongly NP-hard combination optimization problem. It has not only immediate real-life applications but also it is effectively used for the solution of more complex multiprocessor and shop scheduling problems. Here, we propose a general method that can be applied to the scheduling problems with job release times and due-dates. Based on this method, we carry out a detailed study of the single-machine scheduling problem, disclosing its useful structural properties. These properties give us more insight into the complex nature of the problem and its bottleneck feature that makes it intractable. This method also helps us to expose explicit conditions when the problem can be solved in polynomial time. In particular, we establish the complexity status of the special case of the problem in which job processing times are mutually divisible by constructing a polynomial-time algorithm that solves this setting. Apparently, this setting is a maximal polynomially solvable special case of the single-machine scheduling problem with non-arbitrary job processing times.

Keywords: scheduling algorithm; release-time; due-date; divisible numbers; lateness; bin packing; time complexity

1. Introduction

Scheduling jobs with release and due-dates on single machine is a classical strongly NP-hard combination optimization problem according to Garey and Johnson [1]. In many practical scheduling problems, jobs are released non-simultaneously and they have individual due-dates by which they ideally have to complete. Since the problem is NP-hard, the existing exact solution algorithms have an exponential worst-case behavior. The problem is important not only because of its immediate real-life applications, but also because it is effectively used as an auxiliary component for the solution of more complex multiprocessor and shop scheduling problems.

Here, we propose a method that can, in general, be applied to the scheduling problems with job release times and due-dates. Based on this method, we carry out a detailed study of the single-machine scheduling problem disclosing its useful structural properties. These properties give us more insight into the complex nature of the problem and its bottleneck feature that makes it intractable. At the same time, the method also helps us to expose explicit conditions when the problem can be solved in polynomial time. Using the method, we establish the complexity status of the special case of the problem in which job processing times are mutually divisible by constructing a polynomial-time algorithm that solves this setting. This setting is a most general polynomially solvable special case of the single-machine scheduling problem when jobs have restricted processing times but job parameters are not bounded: if job processing times are allowed to take arbitrary values from set $\{p, 2p, 3p, \dots\}$, for an integer p, the problem remains strongly NP-hard [2]. At the same time, the restricted setting may potentially have practical applications in operating systems (we address this issue in more detail in Section 12).

Problem description. Our problem, commonly abbreviated in the scheduling literature as $1|r_j|L_{\max}$ (the notation suggested by Graham et al. [3]), can be stated as follows. There are given n jobs $\{1, 2, \ldots, n\}$ and a single machine. Each job j has (uninterruptible) *processing time* p_j, *release time* r_j and *due-date* d_j: p_j is the time required by job j on the machine; r_j is the time moment by which job j becomes available for scheduling on the machine; and d_j is the time moment, by which it is desirable to complete job j on the machine (informally, the smaller is job due-date, the more urgent it is, and the late completion is penalized by the objective function).

The problem restrictions are as follows. The first basic restriction is that the machine can handle at most one job at a time.

A *feasible schedule* S is a mapping that assigns to every job j its starting time $t_j(S)$ on the machine, such that

$$t_j(S) \geq r_j \tag{1}$$

and

$$t_j(S) \geq t_k(S) + p_k, \tag{2}$$

for any job k included earlier in S (for notational simplicity, we use S also for the corresponding job-set).

The inequality in Equation (1) ensures that no job is started before its release time, and the inequality in Equation (2) ensures that no two jobs overlap in time on the machine.

$$c_j(S) = t_j(S) + p_j$$

is the *completion time* of job j in schedule S.

The *delay* of job j in schedule S is

$$t_j(S) - r_j.$$

An optimal schedule is a feasible schedule minimizing the maximum job *lateness*

$$L_{\max} = \max\{j | c_j - d_j\}$$

(besides the lateness, there exist other due-date oriented objective functions). $L_{\max}(S)$ ($L_j(S)$, respectively) stands for the maximum job lateness in schedule S (the lateness of job j in S, respectively). The objective is to find an optimal schedule.

Adopting to the standard three-field scheduling notation, we abbreviate the special case of problem $1|r_j|L_{\max}$ with divisible job processing times by $1|p_j : divisible, r_j|L_{\max}$. In that setting, we restrict job processing times to the mutually divisible ones: given any two neighboring elements in a sequence of job processing times ordered non-decreasingly, the first one exactly divides the second one (this ratio may be 1). A typical such sequence is formed by the integers each of which is (an integer) power of 2 multiplied by an integer $p \geq 1$.

A brief introduction to our method. Job release times and due-dates with due-date orientated objective functions compose a sloppy combination for most of the scheduling problems in the sense that it basically contributes to their intractability. In such problems, the whole scheduling horizon can be partitioned, roughly, into two types of intervals, the rigid one and the flexible ones. In an optimal schedule, every rigid interval (that potentially may contribute to the optimal objective value) is occupied by a specific set of (urgent) jobs, whereas the flexible intervals can be filled out by the rest of the (non-urgent) jobs in different ways. Intuitively, the "urgency" of a job is determined by its due-date and the due-dates of close-by released jobs; a group of such jobs may form a rigid sequence in a feasible schedule if the differences between their due-dates are "small enough". The remaining jobs are to be "dispelled" in between the rigid sequences.

This kind of division of the scheduling horizon, which naturally arises in different machine environments, reveals an inherent relationship of the scheduling problems with a version of bin packing problem and gives some insight into a complicated nature of the scheduling problems with job release times and due-dates. As shown below, this relationship naturally yields a general algorithmic framework based on the binary search.

A bridge between the scheduling and the bin packing problems is constructed by a procedure that partitions the scheduling horizon into the rigid and the flexible intervals. Exploring a recurrent nature of the scheduling problem, we develop a polynomial-time recursive procedure that partitions the scheduling horizon into the rigid and flexible intervals. After this partition, the scheduling of the rigid intervals is easy but scheduling of the flexible intervals remains non-trivial. Optimal scheduling of the flexible intervals, despite the fact that these intervals are formed by non-urgent jobs, remains NP-hard. To this end, we establish further structural properties of the problem, which yield a general algorithmic framework that may require exponential time. Nevertheless, we derive a condition when the framework will find an optimal solution in polynomial time. This condition reveals a basic difficulty that would face any polynomial-time algorithm to create an optimal solution.

Some kind of compactness property for the flexible segments may be guaranteed if they are scheduled in some special way. In particular, we show that the compactness property can be achieved by an underlying algorithm that works for the mutually divisible job processing times. The algorithm employs some nice properties of a set of mutually divisible numbers.

In terms of time complexity, our algorithmic framework solves problem $1|r_j|L_{max}$ in time $O(n^2 \log n \log p_{max})$ if our optimality condition is satisfied. Whenever during the execution of the framework the condition is not satisfied, an additional implicit enumeration procedure can be incorporated (to maintain this work within a reasonable size, here we focus solely to exact polynomial-time algorithms). Our algorithm for problem $1|p_j : divisible, r_j|L_{max}$ yields an additional factor of $O(n \log p_{max})$, so its time complexity is $O(n^3 \log n \log p_{max}^2)$.

Some previous related work. Coffman, Garey and Johnson [4] previously showed that some special cases of a number of weakly NP-hard bin packing problems with divisible item sizes can be solved in polynomial time (note that our algorithm implies a similar result for a strongly NP-hard scheduling problem). We mention briefly some earlier results concerning our scheduling problem. As to the exponential-time algorithms, the performance of venerable implicit enumeration algorithms by McMahon and Florian [5] and Carlier [6] has not yet been surpassed. There is an easily seen polynomial special case of the problem when all job release times or due-dates are equal (Jackson [7]), or all jobs have unit processing times (Horn [8]). If all jobs have equal integer length p, the problem $1|p_j = p, r_j|L_{max}$ can also be solved in polynomial time $O(n^2 \log n)$. Garey et al. [9] described how the union and find tree with path compression can be used to reduce the time complexity to $O(n \log n)$. The problem $1|p_j \in \{p, 2p\}, r_j|L_{max}$, in which job processing times are restricted to p and $2p$, for an integer p, can also be solved in polynomial $O(n^2 \log n \log p)$ time [10]. If we bound the maximum job processing time p_{max} by a polynomial function in n, $P(n) = O(n^k)$, and the maximal difference between the job release times by a constant R, then the problem $1/p_{max} < P, |r_j - r_i| < R/L_{max}$ remains polynomially solvable [2]. When $P(n)$ is a constant or it is $O(n)$, the time complexity of the algorithm by [2] is $O(n^2 \log n \log p_{max})$; for $k \geq 2$, it is $O(n^{k+1} \log n \log p_{max})$. The algorithm becomes pseudo-polynomial without the restriction on p_{max} and it becomes exponential without the restriction on job release times. In another polynomially solvable special case the jobs can be ordered so that $d_1 \leq \cdots \leq d_n$ and $d_1 - \alpha r_1 - \beta p_1 \geq \cdots \geq d_n - \alpha r_n - \beta p_n$, for some $\alpha \in [0, +\infty)$ and $\beta \in [0, 1]$ Lazarev and Arkhipov [11]. The problem allows fast $O(n \log n)$ solution if for any pair of jobs j, i with $r_i > r_j$ and $d_i < d_j$, $d_j - r_j - p_j \leq d_i - r_i - p_i$, and if $r_i + p_i \geq r_j + p_j$ then $d_i \geq d_j$ [12].

The structure of this work. This paper consists of two major parts. In Part 1, an algorithmic framework for a single machine environment and a common due-date oriented objective function, the maximum job lateness, is presented, whereas, in Part 2, the framework is finished to a polynomial-time algorithm for the special case of the problem with mutually divisible job processing times. In Section 2, we give a brief informal introduction to our method. Section 3 contains a brief overview of the basic concepts and some basic structural properties that posses the schedules enumerated in the framework. In Section 4, we study recurrent structural properties of our schedules, which permit the partitioning of the scheduling horizon into the two types of intervals. In Section 5, we describe how our general framework is incorporated into a binary search procedure. In Section 6, we give an aggregated description of our main framework based on the partitioning of the scheduling horizon into the flexible and the rigid segments, and show how the rigid segments are scheduled in an optimal solution. In Section 7, we describe a procedure which is in charge of the scheduling of the non-urgent segments, and formulate our condition when the main procedure will deliver an optimal solution. This completes Part 1. Part 2 consists of Sections 8–11, and is devoted to the version of the general single-machine scheduling problem with mutually divisible job processing times (under the assumption that the optimality condition of Section 7 is not satisfied). In Section 8, we study the properties of a set of mutually divisible numbers that we use to reduce the search space. Using these properties, we refine our search in Section 9. In Section 10, we give the final examples illustrating the algorithm for divisible job processing times. In Section 11, we complete the correctness proof of that algorithm. The conclusions in Section 12 contain final analysis, possible impact, extensions and practical applications of the proposed method and the algorithm for the divisible job processing times.

2. An Informal Description of the General Framework

In this section, we give a brief informal introduction to our method (the reader may choose to skip it and go to formal definitions of the next section). We mention above the ties of our scheduling problem with a version of bin packing problem, in which there is a fixed number of bins of different capacities and the objective is to find out if there is a feasible solution respecting all the bin capacities. To see the relationship between the bin packing and the scheduling problems, we analyze the structure of the schedules that we enumerate. In particular, the scheduling horizon will contain two types of sequences formed by the "urgent" jobs (that we call kernels) and the remaining sequences formed by the "non-urgent" jobs (that we call bins). A key observation is that a kernel may occupy a quite restricted time interval in any optimal schedule, whereas the bin intervals can be filled out by the non-urgent jobs in different ways. In other words, the urgent jobs are to be scheduled within the rigid time intervals, whereas non-urgent ones are to be dispelled within the flexible intervals. Furthermore, the time interval within which each kernel is to be scheduled can be "adjusted" in terms of the delay of its earliest scheduled job. In particular, it suffices to consider the feasible schedules in which the earliest job of a kernel K is delayed by at most some magnitude, e.g., δ_K; $\delta_K \in [0, \Delta_K]$, where Δ_K is the initial delay of the earliest scheduled job of that kernel (intuitively, Δ_K can be seen as an upper bound on the possible delay for kernel K, a magnitude, by which the earliest scheduled job of kernel K can be delayed without surpassing the minimal so far achieved maximum job lateness). As shown below, for any kernel K, $\Delta_K < p_{\max} = \max_j\{p_j\}$. Observe that, if $\delta_K = 0$, i.e., when we restrict our attention to the feasible schedules in which kernel K has no delay, the lateness of the latest scheduled job of that kernel is a lower bound on the optimal objective value. In this way, we can calculate the time intervals which are to be assigned to every kernel relatively easily. The bins are formed by the remaining time intervals. The length of a bin, i.e., that of the corresponding time interval, will not be prior fixed until the scheduling of that bin is complete (roughly, because there might be some valid range for the "correct" Δ_Ks).

Then, roughly, the scheduling problem reduces to finding out if all the non-kernel jobs can "fit" feasibly (with respect to their release times) into the bins without surpassing the currently allowable lateness for the kernel following that bin; recall that the "allowable lateness" of kernel K is determined by δ_K. We "unify" all the δ_Ks to a single δ (common for all the kernels), and carry out binary search to find an optimal δ within the interval $[0, \max_K \Delta_K$ (the minimum δ such that all the non-kernel jobs fit into the bins; the less is δ, the less is the imposed lateness for the kernel jobs).

Thus, there is a fixed number of bins of different capacities (which are the lengths of the corresponding intervals in our setting), and the items which are to be assigned to these bins are non-kernel jobs. We aim to find out if these items can feasibly be packed into these bins. A simplified version of this problem, in which no specified time interval with each bin is associated and the items can be packed in any bin, is NP-hard. In our version, whether a job can be assigned to a bin depends, in a straightforward way, on the interval of that bin and on the release time of that job (a feasible packing is determined according to these two parameters).

If the reader is not yet too confused, we finally note that the partition of jobs into kernel and non-kernel ones is somewhat non-permanent: during the execution of our framework, a non-kernel job may be "converted" into a kernel one. This kind of situation essentially complicates the solution process and needs an extra treatment. Informally, this causes the strong NP-hardness of the scheduling problem: our framework will find an optimal solution if no non-kernel job converts to a kernel one during its execution (the so-called instance of Alternative (b2)). We observe this important issue in later sections, starting from Section 7.

3. Basic Definitions

This subsection contains definitions which consequently gain in structural insight of problem $1|r_j|L_{\max}$ (see, for instance, [2,13]). First, we describe our main schedule generation tool. Jackson's extended heuristics (Jackson [7] and Schrage [14]), also referred to as the *Earliest Due-date* heuristics (ED-heuristics), is commonly used for scheduling problems with job release times and due-dates. ED-heuristics is characterized by n scheduling times: these are the time moments at which a job is assigned to the machine. Initially, the earliest scheduling time is set to the minimum job release time. Among all jobs released by a given scheduling time (the jobs available by that time moment), one with the minimum due-date is assigned to the machine (ties can be broken by selecting a longest job). Iteratively, the next scheduling time is the maximum between the completion time of the latest assigned so far job to the machine and the minimum release time of a yet unassigned job (note that no job can be started before the machine gets idle, and no job can be started before its release time). Among all jobs available by each scheduling time, a job with the minimum due-date is determined and is scheduled on the machine at that time. Thus, whenever the machine becomes idle, ED-heuristics schedules an available job giving the priority to a most urgent one. In this way, it creates no gap that can be avoided (by scheduling some already released job).

3.1. Structural Components in an ED-Schedule

While constructing an ED-schedule, a *gap* (an idle machine-time) may be created (a maximal consecutive time interval during which the machine is idle; by our convention, there occurs a 0-length gap (c_j, t_i) if job i is started at its release time immediately after the completion of job j.

An ED-schedule can be seen as a sequence of somewhat independent parts, the so-called *blocks*; each block is a consecutive part in that schedule that consists of a sequence of jobs successively scheduled on the machine without any gap in between any neighboring pair of them; a block is preceded and succeeded by a (possibly a 0-length) gap.

As shown below in this subsection, by modifying the release times of some jobs, ED-heuristics can be used to create different feasible solutions to problem $1|r_j|L_{\max}$. All feasible schedules that we consider are created by ED-heuristics, which we call ED-schedules. We construct our initial ED-schedule, denoted by σ, by applying ED-heuristics to the originally given problem instance. Then, we slightly modify the original problem instance to generate other feasible ED-schedules.

Kernels. Now, we define our kernels and the corresponding bins formally. Recall that kernel jobs may only occupy restricted intervals in an optimal schedule, whereas the remaining bin intervals are to be filled in by the rest of the jobs (the latter jobs are more flexible because they may be "moved freely" within the schedule, without affecting the objective value to a certain degree, as we show below).

Let $B(S)$ be a block in an ED-schedule S containing job o that realizes the maximum job lateness in that schedule, i.e.,

$$L_o(S) = \max_j\{L_j(S)\} = L_{\max}(S). \tag{3}$$

Among all jobs in block $B(S)$ satisfying Equation (3), the latest scheduled one is called an *overflow job* in schedule S.

A *kernel* in schedule S is a longest continuous job sequence ending with an overflow job o, such that no job from this sequence has a due-date greater than d_o (for notational simplicity, we use K also for the corresponding job-set). For a kernel K, we let $r(K) = \min_{i \in K}\{r_i\}$. We may observe that the number of kernels in schedule S equals to the number of the overflow jobs in it. Besides, since every kernel is contained within a single block, it may include no gap. We denote by $K(S)$ the earliest kernel in schedule S. The following proposition states an earlier known fact from [13]. Nevertheless, we also give its proof as it gains some intuition on the used here techniques.

Proposition 1. *The maximum lateness of a job of kernel K in ED-schedule S is the minimum possible if the earliest scheduled job of that kernel starts at time $r(K)$. Hence, if schedule S contains a kernel with this property, then it is optimal.*

Proof. By the definition, for any job $j \in K$, $d_j \leq d_o$ (job j is no-less urgent than the overflow job o), whereas note that the maximum lateness of a job of kernel K in schedule S is $L_o(S)$. At the same time, the jobs in kernel K form a tight (continuous) sequence without any gap. Let S' be a complete schedule in which the order of jobs of kernel K differs to that in schedule S and let job o' realizes the maximum lateness of a job of kernel K in schedule S'. Then, from the above observations and the fact that the earliest job of kernel K starts at its release time in schedule S, it follows that

$$L_o(S) \leq L_{o'}(S').$$

Hence,

$$L_{\max}(S') \geq L_o(S) = L_{\max}(S) \tag{4}$$

and schedule S is optimal. □

Emerging jobs. In the rest of this section, let S be an ED-schedule with kernel $K = K(S)$ and with the overflow job $o \in K$ such that the condition in Proposition 1 does not hold. That is, there exists job e with $d_e > d_o$ scheduled before all jobs of kernel K that imposes a forced delay (right-shift) for the jobs of that kernel. By creating an alternative feasible schedule in which job e is rescheduled after kernel K, this kernel may be (re)started earlier, i.e., the earliest scheduled job of kernel K may be restarted earlier than the earliest scheduled job of that kernel has started in schedule S. We need some extra definitions before we define the so-obtained alternative schedule formally.

Suppose job i precedes job j in ED-schedule S. We say that i *pushes* j in S if ED-heuristics may reschedule job j earlier if job i is forced to be scheduled after job j.

If (by the made assumption immediately behind Proposition 1) the earliest scheduled job of kernel K does not start at its release time, then it is immediately preceded and pushed by Job l with $d_l > d_o$, the so-called *delaying emerging* job for kernel K (we use l exclusively for the delaying emerging job).

Besides the delaying emerging job, there may exist job e with $d_e > d_o$ scheduled before kernel K (hence before Job l) in schedule S pushing jobs of kernel K in schedule S. Any such job as well as Job l is referred to as an *emerging job* for K.

We denote the set of emerging jobs for kernel K in schedule S by $E(K)$. Note that $l \in E(K)$ and since S is an ED-schedule, $r_e < r(K)$, for any $e \in E(K)$, as otherwise a job of kernel K with release time $r(K)$ would have been included at the starting time of job e in schedule S.

Besides jobs of set $E(K)$, schedule S may contain job j satisfying the same parametric conditions as an emerging job from set $E(K)$, i.e., $d_j > d_o$ and $r_j < r(K)$, but scheduled after kernel K. We call such a job a *passive emerging* job for kernel K (or for the overflow job o) in schedule S. We denote the set of all the passive emerging jobs for kernel $K = K(S)$ by $EP(K)$.

Note that any $j \in EP(K)$ is included in block $B(S)$ (the block in schedule S containing kernel K) in schedule S. Note also that, potentially, any job $j \in EP(K)$ can be feasibly scheduled before kernel K as well. A job not from set $E(K) \cup EP(K)$ is a *non-emerging* job in schedule S.

In summary, all jobs in $E(K) \cup EP(K)$ are less urgent than all jobs of kernel K and any of them may be included before or after that kernel within block $B(S)$. The following proposition is not difficult to prove (e.g., see [13]).

Proposition 2. *Let S' be a feasible schedule obtained from schedule S by the rescheduling a non-emerging job of schedule S after kernel K. Then, The inequality in Equation (4) holds.*

Activation of an emerging job. Because of the above proposition, it suffices to consider only the rearrangements in schedule S that involve the jobs from set $E(K) \cup EP(K)$. As the first pass, to restart kernel K earlier, we may create a new ED-schedule S_e obtained from schedule S by the rescheduling an emerging job $e \in E(K)$ after kernel K (we call this operation the *activation* of job e for kernel K). In ED-schedule S_e, besides job e, all jobs in $EP(K)$ are also scheduled (remain to be scheduled) after kernel K. Technically, we create schedule S_e by increasing the release times of job e and jobs in $EP(K)$ to a sufficiently large magnitude (e.g., the maximum job release time in kernel K), so that, when ED-heuristics is newly applied, neither job e nor any of the jobs in set $EP(K)$ will be scheduled before any job of kernel K.

It is easily seen that kernel K (regarded as a job-set) restarts earlier in ED-schedule S_e than it has started in schedule S. In particular, the earliest job of kernel K is immediately preceded by a gap and starts at time $r(K)$ in schedule S_l, whereas the earliest scheduled job of kernel K in schedule S starts after time $r(K)$ (the reader may have a look at the work of Vakhania, N. [13] for more details on the relevant issues).

L-schedules. We call a complete feasible schedule S^L in which the lateness of no job is more than threshold L, an *L-schedule*. In schedule S, job i is said to *surpass the L-boundary* if $L_i(S) > L$.

The magnitude
$$\lambda_i(S) = L_i(S) - L \tag{5}$$

is called the *L-delay* of job i in schedule S.

3.2. Examples

We illustrate the above introduced notions in the following two examples.

Example 1. *We have a problem instance with four jobs $\{l, 1, 2, 3\}$, defined as follows:*
$r_l = 0$, $p_l = 16$, $d_l = 100$,
$r_1 = 5$, $p_1 = 2$, $d_1 = 8$,
$r_2 = 4$, $p_2 = 4$, $d_2 = 10$,
$r_3 = 3$, $p_3 = 8$, $d_3 = 12$.

The initial ED-schedule σ is illustrated in Figure 1. There is a single emerging job in that schedule, which is the delaying emerging Job l pushing the following scheduled Jobs 1–3, which constitute the kernel in σ; Job 3 is the overflow job o in schedule σ, which consists of a single block. $L_{max}(\sigma) = L_3(\sigma) = 30 - 12 = 18$.

ED-schedule σ_l, depicted in Figure 2, is obtained by activating the delaying emerging Job l in schedule σ (the release time of Job l is set to that of job 1 and ED-heuristics is newly applied). Kernel in that schedule is formed by Jobs 1 and 2, Job 2 is the overflow job with $L_{max}(\sigma_l) = L_2(\sigma_l) = 17 - 10 = 7$, whereas Job 3 becomes the delaying emerging job in schedule σ_l.

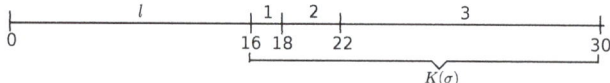

Figure 1. The initial ED-schedule σ for Example 1.

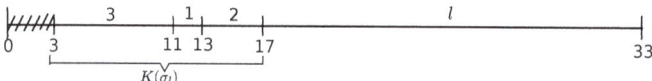

Figure 2. The ED-schedule σ_l for Example 1.

Example 2. *In our second (larger) problem instance, we have eight jobs $\{l, 1, 2, \ldots, 7\}$, defined as follows:*
$r_l = 0$, $p_l = 32$, $d_l = 50$,
$r_1 = 3$, $p_1 = 4$, $d_1 = 23$,
$r_2 = 10$, $p_2 = 2$, $d_2 = 22$,
$r_3 = 11$, $p_3 = 8$, $d_3 = 20$,
$r_4 = 0$, $p_4 = 8$, $d_4 = 67$,
$r_5 = 54$, $p_5 = 4$, $d_5 = 58$,
$r_6 = 54$, $p_6 = 4$, $d_6 = 58$,
$r_7 = 0$, $p_7 = 8$, $d_7 = 69$.

The initial ED-schedule σ is illustrated in Figure 3. Job l is the delaying emerging job, and Jobs 4 and 7 are passive emerging jobs. The kernel $K^1 = K(\sigma)$ is formed by Jobs 3, 2, and 1 (Job 1 being the overflow job).

ED-schedule σ_l is depicted in Figure 4. There arises a (new) kernel $K^2 = K(\sigma_l)$ formed by Jobs 5 and 6, whereas Job 4 is the delaying emerging job (Job 7 is the passive emerging job for both, kernels K^1 and K^2). Job 6 is the overflow job, with $L_{\max}(\sigma_l) = L_6(\sigma_l) = 68 - 58 = 10$.

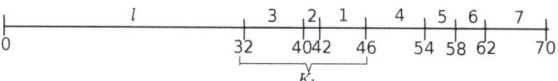

Figure 3. The initial ED-schedule σ for Example 2.

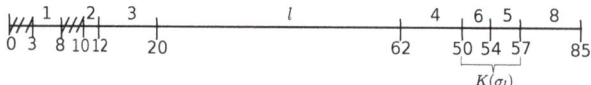

Figure 4. ED-schedule σ_l for Example 2.

4. Recurrent Substructures for Kernel Jobs

In this section, we describe a recursive procedure that permits us to determine the rigid intervals of a potentially optimal schedule (as we show below, these intervals not necessarily coincide with kernel intervals detected in ED-schedules). The procedure relies on an important recurrent substructure property, which is also helpful for the establishment of the ties of the scheduling problem with bin packing problems.

We explore the recurrent structure of our scheduling problem by analyzing ED-schedules. To start with, we observe that in ED-schedule S_l (where l is the delaying emerging job for kernel $K = K(S)$), the processing order of the jobs in kernel K can be altered compared to that in schedule S. Since the time interval that was occupied by Job l in schedule S gets released in schedule S_l, some jobs of kernel K may be scheduled within that interval (recall that by the construction, no job from $EP(K)$ may occupy that interval). In fact, the processing order of jobs of kernel K in schedules S and S_l might be different: recall from Section 3 that a job $j \in K$ with $r_j = r(K)$ will be included the first within the above interval in schedule S_l (whereas kernel K in schedule S is not necessarily initiated by job j; the reader may compare ED-schedules of Figures 1 and 2 and those of Figures 3 and 4 of Examples 1 and 2, respectively).

We call job $j \in K$ *anticipated* in schedule S_l if it is rescheduled to an earlier position in that schedule compared to its position in schedule S (in ED-schedules of Figures 2 and 4, Job 3 and Jobs 1 and 2, respectively, are the anticipated ones). In other words, job j surpasses at least one job i in schedule S_l such that i has surpassed j in schedule S (we may easily observe that, due to ED-heuristics, this may only happen if $q_j < q_i$, as otherwise job j would have been included before job i already in schedule S). Recall from Section 3 that the earliest scheduled job of kernel K is immediately preceded by a newly arisen gap in schedule S_l (in ED-schedules of Figures 2 and 4 it is the gap $[0,3)$). Besides, a new gap in between the jobs of kernel K may also arise in schedule S_l if there exists an anticipated job since, while rescheduling the jobs of kernel K, there may occur a time moment at which some job of that kernel completes but no other job is available in schedule S_l. Such a time moment in ED-schedule of Figure 4 is 7, which is extending up to the release Time 10 of Job 2 resulting in a new gap $[7,10)$ arising within the jobs of kernel K^1.

It is apparent now that jobs of kernel K (kernel K^1 in the above example) may be redistributed into several continuous parts separated by the gaps in schedule S_l (the first such part in ED-schedule of Figure 4. consists of the anticipated Job 1 and the second part consists of Jobs 2 and 3, where Job 2 is another anticipated job).

If there arises an anticipated job so that the jobs of kernel K are redistributed into one or more continuous parts in schedule S_l, then kernel K is said to *collapse*; if kernel K collapses into a single continuous part, then this continuous part and kernel K, considered as job-sets, are the same, but the corresponding job sequences are different because of an anticipated job. It follows that, if kernel K collapses, then there is at least one anticipated job in schedule S_l that converts to the delaying emerging job in that schedule (recall from Proposition 1 that schedule S is optimal if it possesses no delaying emerging job).

Throughout this section, we concentrate our attention to the part of schedule S_l initiating at the starting time of Job l in schedule S and containing all the newly arisen continuous parts of kernel K in that schedule that we denote by $S_l[K]$. We treat this part as an independent ED-schedule consisting of solely the jobs of the collapsed kernel K (recall that no job distinct from a job of kernel K may be included in schedule S_l until all jobs of kernel K are scheduled, by the definition of that schedule). For the instance of Example 1 with $S = \sigma$, schedule $\sigma_l[K]$ is the part of the ED-schedule of Figure 2 that initiates at at Time 0 and ends at Time 17. For the instance of Example 2, schedule $\sigma_l[K_1]$ starts at Time 0 and ends at Time 20 (see Figure 4).

We distinguish three different types of continuous parts in schedule $S_l[K]$. A continuous part that consists of only anticipated jobs (contains no anticipated job, respectively) is called an *anticipated* (*uniform*, respectively) continuous part. A continuous part which is neither anticipated nor uniform is called *mixed* (hence, mixed continuous part contains at least one anticipated and one non-anticipated job).

We observe that in ED-schedule of Figure 2 schedule $\sigma_l[K]$ consists of a single mixed continuous part with the anticipated Job 3, which becomes the new delaying emerging job in that schedule. Schedule $\sigma_l[K_1]$ of Example 2 (Figure 4) consists of two continuous parts, the first of which is anticipated with a single anticipated Job 1, and the second one is mixed with the anticipated Job 2. The latter job becomes the delaying emerging job in schedule $\sigma_l[K_1]$ and is followed by Job 3, which constitutes the unique kernel in schedule $\sigma_l[K_1]$.

Substructure Components

The decomposition of kernel K into the continues parts has the recurrent nature. Indeed, we easily observe that schedule $S_l[K]$ has its own kernel $K_1 = K((S_l)[K])$. If kernels K and K_1 (considered as sequences) are different, then the decomposition process naturally continues with kernel K_1 (otherwise, it ends by Point (4) of Proposition 3). For instance, in Example 1, kernel K_1 is constituted by Jobs 1 and 2 (Figure 2) and, in Example 2, it is constituted by Job 3 (see Figure 4) (in Lemma 4, we show that schedule $S_l[K]$ may contain only one kernel, which is from the last continuous part of that schedule). In turn, if kernel K_1 possesses the delaying emerging job, it may also collapse, and this process may recurrently be repeated. This gives the rise to a recurrent substructure decomposition of kernel K. The process continues as long as the next arisen kernel may again collapse, i.e., it possesses the delaying emerging job. Suppose there is the delaying emerging job l_1 for kernel K_1 in schedule $S_l[K]$. We recurrently define a (sub)schedule $S_{l,l_1}[K, K_1]$ of schedule $S_l[K]$ containing only jobs of kernel K_1 and in which the delaying emerging job l_1 is activated for that kernel, similarly to what is done for schedule $S_l[K]$. This substructure definition applies recursively as long as every newly derived (sub)schedule contains a kernel that may collapse, i.e., it possesses the delaying emerging job (this kernel belongs to the last continuous part of the (sub)schedule, as we prove in Lemma 4). This delaying emerging job is activated and the next (sub)schedule is similarly created.

We refer to the created is this (sub)schedules as the substructure *components* arisen as a result of the collapsing of kernel K and the following arisen kernels during the decomposition process. As already specified, the first component in the decomposition is $S_l[K]$ with kernel $K_1 = K(S_l[K])$, the second one is $S_{l,l_1}[K, K_1]$ with kernel $K_2 = K(S_{l,l_1}[K, K_1])$, the third one is $S_{l,l_1,l_2}[K, K_1, K_2]$, with kernel $K_3 = K(S_{l,l_1,l_2}[K, K_1, K_2])$, where l_2 is the delaying emerging job of kernel K_2, and so on, with the last *atomic component* being $S_{l,l_1,...,l_k}[K, K_1, ..., K_k]$ such that the kernel $K^* = K(S_{l,l_1,...,l_k}[K, K_1, ..., K_k])$ of that component has no delaying emerging job (here, l_k is the delaying emerging job of kernel K_k). Note that the successively created components during the decomposition form an embedded substructure in the sense that the set of jobs that contains each next generated component is a proper subset of that of the previously created one: substructure component $S_{l,l_1,...,l_j}[K, K_1, ..., K_j]$, for any $j \leq k$, contains only jobs of kernel K_j, whereas clearly $|K_j| < |S_{l,l_1,...,l_{j-1}}[K, K_1, ..., K_{j-1}]|$ (as kernel K_j does not contain, at least, job l_j, i.e., no activated delaying emerging job pertains to the next generated substructure component).

Below, we give a formal description of the procedure that generates the complete decomposition of kernel K, i.e., it creates all the substructure components of that kernel.

PROCEDURE Decomposition(S, K, l)
{S is an ED-schedule with kernel K and delaying emerging Job l}
   WHILE $S_l[K]$ is not atomic DO
   BEGIN
      $S := S_l[K]$; $K :=$ the kernel in component $S_l[K]$;
      $l :=$ the delaying emerging job of component $S_l[K]$;
      CALL PROCEDURE Decomposition(S, K, l)
   END.

We illustrate the decomposition procedure on our two problem instances.

Example 1 (continuation). In the decomposition of kernel $K(\sigma)$ of Example 1, in ED-schedule of Figure 2, kernel K_1 of substructure component $S_l[K]$ consists of Jobs 1 and 2, and Job 3 is the corresponding delaying emerging job. Figure 5 illustrates schedule $\sigma_{l,3}$ obtained from schedule σ_l of Figure 2 by the activation of the (second) emerging Job 3 (which, in fact, is optimal for the instance of Example 1, with $L_{max}(\sigma_{l,3}) = L_3(\sigma_{l,3}) = 18 - 12 = 6$). A new substructure component $S_{l,3}[K, K_1]$ consisting of jobs of kernel K_1 is a mixed continuous part with the anticipated Job 2. Kernel K_2 of that component consists of Job 1, whereas Job 2 is the delaying emerging job for that sub-kernel ($L_1(\sigma_{l,3}) = 10 - 8 = 2$). Figure 6 illustrates ED-schedule $\sigma_{l,3,2}$ that contains the next substructure component $S_{l,3,2}[K, K_1, K_2]$ consisting of Job 1. Substructure component $S_{l,3,2}[K, K_1, K_2]$ is uniform and is the last atomic component in the decomposition, as it possesses no delaying emerging job and forms the last (atomic) kernel K_3 in the decomposition (with no delaying emerging job). $L_{max}(S_{l,3,2}[K, K_1, K_2]) = L_1(\sigma_{l,3,2}) = 7 - 8 = -1$. Note that the kernel in component $S_{l,3,2}[K, K_1, K_2]$

coincides with that component and is not a kernel in ED-schedule $\sigma_{l,3,2}$ (the overflow job in that schedule is Job 3 with $L_{max}(\sigma_{l,3,2}) = L_3(\sigma_{l,3,2}) = 19 - 12 = 7$).

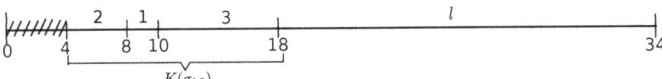

Figure 5. ED-schedule representing the second substructure component in the decomposition of kernel K for Example 1.

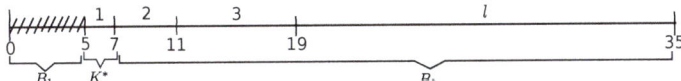

Figure 6. ED-schedule representing the third (atomic) substructure component in the decomposition of kernel K for Example 1.

Example 2 (continuation). Using this example, we illustrate the decomposition of two different kernels, which are denoted by K^1 and K^2 abvoe. In the decomposition of kernel K^1, in ED-schedule σ_l of Figure 4, we have two continuous parts in substructure component $S_l[K^1]$, the second of which contains kernel K_1^1 consisting of Job 3; the corresponding delaying emerging job is Job 2. The next substructure component $S_{l,2}[K^1, K_1^1]$ consisting of Job 3 (with the lateness $19 - 20 = -1$) is uniform and it is an atomic component that completes the decomposition of kernel K^1. This component can be seen in Figure 7 representing ED-schedule $\sigma_{l,2}$ obtained from schedule σ_l of Figure 4 by the activation of the emerging Job 2 for kernel K_1^1.

Once the decomposition of kernel K^1 is complete, we detect a new kernel K^2 consisting of Jobs 5 and 6 in the ED-schedule $\sigma_{l,2}$ depicted in Figure 7 (the same kernel is also represented in the ED-schedule σ_l of Figure 4). Kernel K^2 possesses the delaying emerging Job 4. The first substructure component $S_4[K^2]$ in the decomposition of kernel K^2 consists of a single uniform continuous part, which forms also the corresponding kernel K_1^2. The latter kernel has no delaying emerging job and the component $S_4[K^2]$ is atomic (see Figure 8).

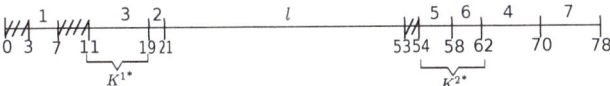

Figure 7. ED-schedule representing the second (atomic) substructure component in the decomposition of kernel K^1 and kernel K^2 for Example 2.

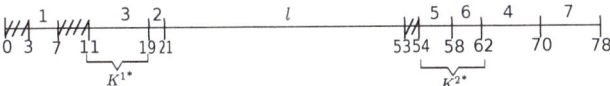

Figure 8. ED-schedule representing the atomic substructure components for kernels K^1 and K^2 for Example 2.

We need a few auxiliary lemmas to prove the validity of the decomposition procedure. For notational simplicity, we state them in terms of schedule S with kernel K and the component $S_l[K]$ (instead of referring to an arbitrary component $S_{l,l_1,...,l_j}[K, K_1, \ldots, K_j]$ with kernel K_{j+1} and the following substructure component $S_{l,l_1,...,l_j,l_{j+1}}[K, K_1, \ldots, K_j, K_{j+1}]$). The next proposition immediately follows from the definitions.

Proposition 3. *Suppose kernel K collapses. Then:*

(1) *The anticipated jobs from kernel K are non-kernel jobs in schedule $S_l[K]$.*
(2) *Any continuous part in schedule $S_l[K]$ is either anticipated or uniform or mixed.*

(3) If schedule $S_l[K]$ consists of a single continuous part then it is mixed.
(4) If $K((S_l)[K]) = K$ (considering the kernels as job sequences), then schedule $S_l[K]$ consists of (a unique) uniform part that forms its kernel $K((S_l)[K])$. This kernel has no delaying emerging job and hence cannot be further decomposed.

Lemma 1. *Let A be an anticipated continuous part in component $S_l[K]$. Then for any job $j \in A$,*

$$L_j(S_l[K]) < L_{\max}(S_l[K]),$$

i.e., an anticipated continuous part may not contain kernel $K(S_l[K])$.

Proof. Let G be the set of all jobs which have surpassed job j in schedule S and were surpassed by j in $S_l[K]$ (recall the definition of an anticipated part). For any job $i \in G$, $d_j \geq d_i$ since job j is released before jobs in set G and it is included after these jobs in ED-schedule S. This implies that $L_j(S_l[K]) < L_i(S_l[K]) \leq L_{\max}(S_l[K])$. The lemma is proved. □

Lemma 2. *A uniform continuous part U in component $S_l[K]$ (considered as an independent ED-schedule), may contain no delaying emerging job.*

Proof. Schedule U has no anticipated job, i.e., the processing order of jobs in U in both schedules S and $S_l[K]$ is the same. Observe that U constitutes a sub-sequence of kernel K in schedule S. However, kernel K has a single delaying emerging Job l that does not belong to schedule $S_l[K]$. Since U is part of $S_l[K]$ and it respects the same processing order as schedule S, it cannot contain the delaying emerging job. □

Lemma 3. *Suppose a uniform continuous part $U \in S_l[K]$ contains a job realizing the maximum lateness in component $S_l[K]$. Then,*

$$L_{\max}(U) \leq L_{\max}(S^{\text{opt}}), \tag{6}$$

i.e., the lateness of the corresponding overflow job is a lower bound on the optimal objective value.

Proof. Considering part U as an independent schedule, it may contain no emerging job (Lemma 2). At the same time, the earliest scheduled job in U starts at its release time since it is immediately preceded by a gap, and the lemma follows from Proposition 1. □

Lemma 4. *Only a job from the last continuous part $C \in S_l[K]$ may realize the maximum job lateness in schedule $S_l[K]$.*

Proof. The jobs in the continuous part C: (i) either were the latest scheduled ones from kernel K in schedule S; or (ii) the latest scheduled ones of schedule S have anticipated the corresponding jobs in C in schedule $S_l[K]$. In Case (ii), these anticipated jobs may form part of C or be part of a preceding continuous part P. In the latter sub-case, due to a gap in between the continuous parts in $S_l[K]$, the jobs of continuous part P should have been left-shifted in schedule $S_l[K]$ no less than the jobs in continuous part C and our claim follows. The former sub-case of Case (ii) is obviously trivial. In Case (i), similar to in the earlier sub-case, the jobs from the continuous parts preceding C in $S_l[K]$ should have been left-shifted in $S_l[K]$ no less than the jobs in C (again, due to the gap in between the continuous parts). Hence, none of them may have the lateness more than that of a job in continuous part C. □

Proposition 4. *PROCEDURE Decomposition(S, K, l) finds the atomic component of kernel K in less than $\kappa/2$ iterations, where κ is the number of jobs in kernel K. The kernel of that atomic component is formed by a uniform continuous part, which is the last continuous part of that component.*

Proof. With every newly created substructure component during the decomposition of a kernel with κ jobs, the corresponding delaying emerging job is associated. At every iteration of the procedure, the delaying emerging job is activated, and that job does not belong to the next generated component. Then, the first claim follows as every kernel contains at least one job. Hence, the total number of the created components during all calls of the collapsing stage is bounded above by $\kappa/2$.

Now, we show the second claim. From Lemma 4, the last continuous part of the atomic component contains the overflow job of that component. Clearly, the last continuous part of any component cannot be anticipated, whereas any mixed continuous part (seen as an independent schedule) contains an emerging job, hence a component with the last mixed continuous part cannot be atomic. Then, the last continuous part of the atomic component is uniform (see Point (2) in Proposition 3), and since it possesses no delaying emerging job (Lemma 2), it wholly constitutes the kernel of that component. □

From here on, let $K^* = K(S_{l,l_1,\ldots,l_k}[K, K_1, \ldots, K_k])$, where $S_{l,l_1,\ldots,l_k}[K, K_1, \ldots, K_k]$ is the atomic component in the decomposition of kernel K, and let ω^* be the overflow job in kernel K^*. By Proposition 4, K^* (the *atomic kernel* in the decomposition) is the only kernel in the atomic component $S_{l,l_1,\ldots,l_k}[K, K_1, \ldots, K_k]$ and is also the last uniform continuous part of that component.

Corollary 1. *There exists no L-schedule if*

$$L_{\max}(K^*) = L_{\omega^*}(S_{l,l_1,\ldots,l_k}[K, K_1, \ldots, K_k]) > L.$$

In particular, $L_{\max}(K^*)$ is a lower bound on the optimum objective value.

Proof. By Lemma 4 and Proposition 4, kernel K^* is the last continuous uniform part of the atomic component $S_{l,l_1,\ldots,l_k}[K, K_1, \ldots, K_k]$). Then, by Proposition 4 and the inequality in Equation (6),

$$L_{\max}(K^*) = L_{\max}(S_{l,l_1,\ldots,l_k}[K, K_1, \ldots, K_k]) \leq L_{\max}(S^{\mathrm{opt}}).$$

□

Theorem 1. *PROCEDURE Decomposition(S, K, l) forms all substructure components of kernel K with the last atomic component and atomic kernel K^* in time $O(\kappa^2 \log \kappa)$ (where κ is the number of jobs in kernel K).*

Proof. First, observe that, for any non-atomic component $S_{l,l_1,\ldots,l_j}[K, K_1, \ldots, K_j]$ ($j < k$) created by the procedure, the kernel $K_{j+1} = K(S_{l,l_1,\ldots,l_j}[K, K_1, \ldots, K_j])$ of that component is within its last continuous part (Lemma 4). This part cannot be anticipated or uniform (otherwise, it would not have been non-atomic). Thus, the last continuous part M in that component is mixed and hence it contains an anticipated job. The latest scheduled anticipated job in M is the delaying emerging job l_{j+1} for kernel K_{j+1} in the continuous part M. Then, the decomposition procedure creates the next component $S_{l,l_1,\ldots,l_j,l_{j+1}}[K, K_1, \ldots, K_j, K_{j+1}]$ in the decomposition (consisting of the jobs of kernel K_{j+1}) by activating job l_{j+1} for kernel K_{j+1}.

Consider now the last atomic component $S_{l,l_1,\ldots,l_k}[K, K_1, \ldots, K_k]$. By Proposition 4, atomic kernel K^* of component $S_{l,l_1,\ldots,l_k}[K, K_1, \ldots, K_k]$ is the last uniform continuous part in that component. By the inequality in Equation (6), $L_{\max}(S_{l,l_1,\ldots,l_k}[K, K_1, \ldots, K_k]) = L_{\max}(K^*)$ is a lower bound on the optimal objective value and hence the decomposition procedure may halt: the atomic kernel K^* cannot be decomposed and the maximum job completion time in that kernel cannot be further reduced. Furthermore, if $L_{\max}(K^*) > L$, then there exists no L-schedule (Corollary 1).

As to the time complexity, the total number of iterations (recursive calls of PROCEDURE Decomposition(S, K, l)) is bounded by $\kappa/2$ (where κ is the number of jobs in kernel K, see Proposition 4). At every iteration i, kernel K_{i+1} and job l_{i+1} can be detected in time linear in the number of jobs in component $S_{l,l_1,\ldots,l_i}[K, K_1, \ldots, K_i]$, and hence the condition in WHILE can be verified with the same cost. Besides, at iteration i, ED-heuristics with cost $O(\kappa \log \kappa)$ is applied, which yields the overall time complexity $O(\kappa^2 \log \kappa)$ of PROCEDURE Decomposition(S, K, l). □

Corollary 2. *The total cost of the calls of the decomposition procedure for all the arisen kernels in the framework is $O(n^2 \log n)$.*

Proof. Let $K_1, \ldots, K_k$ be all the kernels that arise in the framework. For the purpose of this estimation, assume κ, $k\kappa \leq n$, is the number of jobs in every kernel (this will give an amortized estimation). Since every kernel is processed only once, the total cost of the calls of the decomposition procedure for kernels $K_1, \ldots, K_k$ is then

$$kO(\kappa^2 \log \kappa) \leq \frac{n}{\kappa} O(\kappa^2 \log \kappa) < O(n^2 \log n).$$

□

5. Binary Search

In this section, we describe how binary search can be beneficially used to solve problem $1|r_j|L_{\max}$. Recall from the previous section that PROCEDURE Decomposition(S, K, l) extracts the atomic kernel K^* from kernel K (recall that l is the corresponding delaying emerging job—without loss of generality, assume that it exists, as otherwise the schedule S with $K(S) = K$ is optimal by Proposition 1). Notice that, since the kernel of every created component in the decomposition is from its last continuous part (Lemma 4), there is no intersection between the continuous parts of different components excluding the last continuous part of each component. All the continuous parts of all the created components in the decomposition of kernel K except the last continuous part of each component are merged in time axes resulting in a partial ED-schedule which initiates at time $r(K)$ and has the number of gaps equal to the number of its continuous parts minus one (as every two neighboring continuous parts are separated by a gap). It includes (feasibly) all the jobs of kernel K except ones from the atomic kernel K^* (that constitutes the last continuous part of the atomic component, see Proposition 4). By merging this partial schedule with the atomic kernel K^*, we obtain another feasible partial ED-schedule consisting of all the jobs of kernel K, which we denote by $S^*[K]$. We extend PROCEDURE Decomposition(S, K, l) with this construction. It is easy to see that the time complexity of the procedure remains the same. Thus, from here on, we let the output of PROCEDURE Decomposition(S, K, l) be schedule $S^*[K]$.

Within the gaps in partial schedule $S^*[K]$, some *external* jobs for kernel K, ones not in schedule $S^*[K]$, will be included. During such an expansion of schedule $S^*[K]$ with the external jobs, the right-shift (a forced delay) of the jobs from that schedule by some constant units of time, which is determined by the current trial δ in the binary search procedure, will be allowed (in this section, we define the interval from which trial δs are taken).

At an iteration h of the binary search procedure with trial δ_h, one or more kernels may arise. Iteration h starts by determining the earliest arisen kernel, which, as we show below, depends on the value of trial δ_h. This kernel determines the initial partition of the scheduling horizon into one kernel and two non-kernel (bin) intervals. Repeatedly, during the scheduling of a non-kernel interval, a new kernel may arise, which is added to the current set of kernels at iteration h. Every newly arisen kernel is treated similarly in a recurrent fashion. We denote by $\mathcal{K}$ the set of kernels detected by the current state of computation at iteration h (omitting parameter h for notational simplicity). For every newly arisen kernel $K \in \mathcal{K}$, PROCEDURE Decomposition(S, K, l) is invoked and partial schedule $S^*[K]$ is expanded by external jobs. Destiny feasible schedule of iteration h contains all the extended schedules $S^*[K]$, $K \in \mathcal{K}$.

The next proposition easily follows from the construction of schedule $S^*[K]$, Lemma 4 and Corollary 1:

Proposition 5. $K^* = K(S^*[K])$ (K^* is the only kernel in schedule $S^*[K]$) and

$$L_{\max}(S^*[K]) = L_{\max}(K^*) \leq L_{\max}(S^{\text{opt}}),$$

i.e., $L_{\max}(S^*[K])$ is a lower bound on the optimum objective value.

$$L^*_{\max} = \max_{K \in \mathcal{K}}\{L_{\max}(K^*)\}$$

is a stronger lower bound on the objective value.

Now, we define an important kernel parameter used in the binary search. Given kernel $K \in \mathcal{K}$, let

$$\delta(K^*) = L^*_{\max} - L_{\max}(K^*) \geq 0, \qquad (7)$$

i.e., $\delta(K^*)$ is the amount of time by which the starting time of the earliest scheduled job of kernel K^* can be right-shifted (increased) without increasing lower bound $L^*_{\max}$. Note that for every $K \in \mathcal{K}$, $\delta(K^*) + L_{\max}(K^*)$ is the same magnitude.

Example 2 (continuation). For the problem instance of Example 2, $L_{\max}(K^{1*}) = L_3(\sigma_{l,2}) = 19 - 20 = -1$, $L_{\max}(K^{2*}) = L_6(\sigma_{l,2,4}) = 62 - 58 = 4$; hence, $\delta(K^{1*}) = 5$ and $\delta(K^{2*}) = 0$ (recall that atomic kernel K^{1*} consists of a single Job 3, and atomic kernel K^{2*} consists of Jobs 5 and 6; hence, the lower bound $L^*_{\max} = 4$ is realized by atomic kernel K^{2*}).

Proposition 6. Let S be a complete schedule and $\mathcal{K}$ be the set of the kernels detected prior to the creation of schedule S. The starting time of every atomic kernel K^*, $K \in \mathcal{K}$, can be increased by $\delta(K^*)$ time units (compared to its starting time in schedule $S^*[K]$) without increasing the maximum lateness $L_{\max}(S)$.

Proof. Let $(K')^*$, $K' \in \mathcal{K}$, be an atomic kernel that achieves lower bound $L^*_{\max}$, i.e., $L_{\max}((K')^*) = L^*_{\max}$ (equivalently, $\delta((K')^*) = 0$). By Equation (7), if the completion time of every job in atomic kernel $K^* \neq (K')^*$ is increased by $\delta(K^*)$, the lateness of none of these jobs may become greater than that of the overflow job from kernel $(K')^*$, which proves the proposition as $L_{\max}(S) \geq L^*_{\max}$. □

We immediately obtain the following corollary:

Corollary 3. In an optimal schedule S_{opt}, every atomic kernel K^*, $K \in \mathcal{K}$, starts either no later than at time $r(K^*) + \delta(K^*)$ or no later than at time $r(K^*) + \delta(K^*) + \delta$, for some $\delta \geq 0$.

An extra delay δ might be unavoidable for a proper accommodation of the non-kernel jobs. Informally, δ is the maximum extra delay that we will allow for every atomic kernel in the iteration of the binary search procedure with trial value δ. For a given iteration in the binary search procedure with trial δ, the corresponding threshold, an upper limit on the currently allowable maximum job lateness, L_δ-boundary (or L-boundary) is

$$L_\delta = L^*_{\max} + \delta = L_{\max}(K^*) + \delta(K^*) + \delta \ (K \in \mathcal{K}). \qquad (8)$$

We call L_δ − schedule a feasible schedule in which the maximum lateness of any job is at most $L_\delta = L^*_{\max} + \delta$ (see Equation (8)).

Note that, since to every iteration a particular δ corresponds, the maximum allowable lateness at different iterations is different. The concept of the overflow job at a given iteration is consequently redefined: such a job *must* have the lateness greater than L_δ. Note that this implicitly redefines also the notation of a kernel at that iteration of the binary search procedure.

It is not difficult to determine the time interval from which the trial δs can be derived. Let Δ be the delay of kernel $K(\sigma)$ imposed by the delaying emerging Job l in initial ED-schedule σ, i.e.,

$$\Delta = c_l(\sigma) - r(K(\sigma)). \tag{9}$$

Example 1 (continuation). For the problem instance of Example 1, for instance, $\Delta = 16 - 3 = 13$ (see Figure 1).

Proposition 7.
$$L_{max}(\sigma) - L^*_{max} \leq \Delta. \tag{10}$$

Proof. This is a known property that easily follows from the fact that no job of kernel $K(\sigma)$ could have been released by the time $t_l(\sigma)$, as otherwise ED-heuristics would have been included the former job instead of Job l in schedule σ. □

Assume, for now, that we have a procedure that, for a given L-boundary (see Equation (8)), finds an L-schedule S^L if it exists, otherwise, it outputs a "no" answer.

Then, the binary search procedure incorporates the above verification procedure as follows. Initially, for $\delta = \Delta$, $L^*_{max} + \Delta$-schedule σ already exists. For $\delta = 0$ with $L = L^*_{max}$, if there exists no L^*_{max}-schedule then the next value of δ is $[\Delta/2]$. Iteratively, if an L-schedule with $L = L^*_{max} + \delta$ for the current δ exists, the δ is increased correspondingly, otherwise it is decreased correspondingly in the binary search mode.

Proposition 8. *The L-schedule S^L corresponding to the minimum $L = L^*_{max} + \delta$ found in the binary search procedure is optimal.*

Proof. First, we show that trial δs can be derived from the interval $[0, \Delta]$. Indeed, the left endpoint of this interval can clearly be 0 (potentially yielding a solution with the objective value L^*_{max}). By the inequality in Equation (10), the maximum job lateness in any feasible ED-schedule in which the delay of some kernel is more than Δ would be no less than $L_{max}(\sigma)$, which obviously proves the above claim.

Now note that the minimum L-boundary yields the minimal possible lateness for the kernel jobs subject to the condition that no non-kernel job surpasses L-boundary. This obviously proves the proposition. □

By Proposition 8, the problem $1|r_j|L_{max}$ can be solved, given that there is a verification procedure that, for a given L-boundary, either constructs L_δ-schedule S^{L_δ} or answers correctly that it does not exist. The number of iterations in the binary search procedure is bounded by $\log p_{max}$ as clearly, $\Delta < p_{max}$. Then, note that the running time of our basic framework is $\log p_{max}$ multiplied by the running time of the verification procedure. The rest of this paper is devoted to the construction of the verification procedure, invoked in the binary search procedure for trial δs.

6. The General Framework for Problem $1|r_j|L_{max}$

In this section, we describe our main algorithmic framework which basic components form the binary search and the verification procedures. The framework is for the general setting $1|r_j|L_{max}$ (in the next section, we give an explicit condition when the framework guarantees the optimal solution of the problem). At every iteration in the binary search procedure, we intend to keep the delay of jobs from each partial schedule $S^*[K]$, $K \in \mathcal{K}$ within the allowable margin determined by the current L_δ-boundary.

For a given threshold L_δ, we are concerned with the existence of a partial L_δ-schedule that includes all the jobs of schedule $S^*[K]$ and probably some external jobs. We refer to such partial schedule as an *augmented L_δ-schedule* for kernel K and denote it by $S^{L_\delta}[K]$ (we specify the scope of that schedule more accurately later in this section).

Due to the allowable maximum job lateness of $L_\delta \geq L_{opt}$ in schedule $S^{L_\delta}[K]$, in the case that the earliest scheduled job of kernel K^* gets pushed by some (external) job l^* in schedule $S^{L_\delta}[K]$, that job will be considered as the delaying emerging job iff

$$c_{l^*}(S^L[K]) \geq r(K^*) + \delta(K^*) + \delta.$$

For a given threshold $L = L_\delta$, the allowable L-*bias* for jobs of kernel K^* in schedule $S^L[K]$

$$\beta_L(K^*) = L - L_{\max}(K^*). \tag{11}$$

The intuition behind this definition is that the jobs of kernel K^* in schedule $S^*[K]$ can be right-shifted by $\beta_L(K^*)$ time units without surpassing the L boundary (see Proposition 9 below).

Proposition 9. *In an L-schedule S^L, all the jobs of schedule $S^*[K]$ are included in the interval of schedule $S^*[K]$. Furthermore, any job in $S^*[K] \setminus K^*$ can be right-shifted provided that it remains scheduled before the jobs of kernel K^*, whereas the jobs from kernel K^* can be right-shifted by at most $\beta_L(K^*)$.*

Proof. Let j be the earliest scheduled job of atomic kernel K^* in schedule $S^*[K]$. By right-shifting job j by $\beta_L(K^*)$ time units (Equation (11)) we get a new (partial) schedule S' in which all the jobs are delayed by $\beta_L(K^*)$ time units with respect to schedule $S^*[K]$ (note that the processing order of the jobs of atomic kernel K^* need not be altered in schedule S' as the jobs of kernel K^* are scheduled in ED-order in schedule $S^*[K]$). Hence,

$$\max_{i \in S^*[K]} L_i(S') \leq \max_{i \in S^*[K]} \{L_i(S^*[K]) + \beta_L(K^*)\} = \max_{i \in S^*[K]} \{L_i(S^*[K])\} + \beta_L(K^*).$$

By substituting for $\beta_L(K^*)$ using Equation (11) and applying that $\max_{i \in S^*[K]}\{L_i(S^*[K])\} = L_{\max}(K^*)$, we obtain

$$\max_{i \in S^*[K]} L_i(S') \leq L.$$

Hence, the lateness of any job of atomic kernel K^* is no more than L. Likewise, any other job from schedule $S^*[K]$ can be right-shifted within the interval of $S^*[K]$ without surpassing the magnitude $L_{\max}(K^*) \leq L$ given that it is included before the jobs of kernel K^* (see the proof of Lemma 4). □

6.1. Partitioning the Scheduling Horizon into the Bin and Kernel Segments

By Proposition 9, all jobs from the atomic kernel K^* are to be included with a possible delay (right-shift) of at most $\beta_L(K^*)$ in L-schedule S^L. The rest of the jobs from schedule $S^*[K]$ are to "dispelled" before the jobs of K^* within the interval of that schedule. Since schedule $S^*[K]$ contains the gaps, some additional external jobs may also be included within the same time interval. According to this observation, we partition every complete feasible L-schedule into two types of segments, rigid and flexible ones. The rigid segments are to be occupied by the atomic kernels, and the rest of the (flexible) segments, which are called *bin* segments or intervals, are left for the rest of the jobs (we use term bin for both, the corresponding time interval and for the corresponding schedule portion interchangeably). For simplicity, we refer to the segments corresponding to the atomic kernels as kernel segments or intervals.

In general, we have a bin between two adjacent kernel intervals, and a bin before the first and after the last kernel interval. Because of the allowable right-shift $\beta_L(K^*)$ for the jobs of an atomic kernel K^*, the starting and completion times of the corresponding kernel and bin intervals are not priory fixed. We denote by $B^-(K)$ ($B^+(K)$, respectively) the bin before (after, respectively) the kernel interval corresponding to the atomic kernel K^* of kernel K. There are two bins in schedule $\sigma_{l,3,2}$, surrounding the atomic kernel consisting of Job 1 in Figure 6. We have three bins in schedules depicted in

Figures 8 and 9 for the problem instance of Example 2, $B_1 = B^-(K^1)$, $B_2 = B^+(K^1) = B^-(K^2)$ and $B_3 = B^+(K^2)$ (schedule of Figure 9 incorporates an optimal arrangement of jobs in these bins).

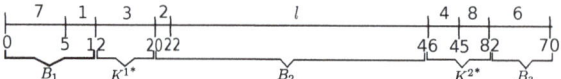

Figure 9. An optimal L-schedule for Example 2 with three bins ($L = L_{\max}^* = 4$).

The scope of augmented L-schedule $S^L[K]$ for kernel K includes that of bin $B^-(K)$ and that of the atomic kernel K^*. These two parts are scheduled independently. The construction of second part relies on the next proposition that easily follows from Proposition 9:

Proposition 10. *No job of the atomic kernel K^* will surpass the L-boundary if the latest scheduled job of bin $B^-(K)$ completes no later than at time moment*

$$\psi_L(K) = r(K^*) + \beta_L(K^*) \tag{12}$$

(the latest time moment when atomic kernel K^* may start in an L-schedule) and the jobs of that kernel are scheduled by ED-heuristics from time moment $\psi_L(K)$.

We easily arrange the second part of augmented schedule $S^L[K]$, i.e., one including the atomic kernel K^*, as specified in Proposition 10. Hence, from here on, we are solely concerned with the construction of the the first part, i.e., that of bin $B^-(K)$, which is a complicated task and basically contributes to the complexity status of problem $1|r_j|L_{\max}$.

We refer to a partial feasible L-schedule for the first part of schedule $S^L[K]$ (with its latest job completion time not exceeding the magnitude $\psi_L(K)$, at which the second part initiates) as a *preschedule* of kernel K and denote it by $PreS(K)$. Note that the time interval of preschedule $PreS(K)$ coincides with that of bin $B^-(K)$; in this sense, $PreS(K)$ is a schedule for bin $B^-(K)$.

Kernel preschedules are generated in Phase 1, described in Section 7. If Phase 1 fails to construct an L-preschedule for some kernel, then Phase 2 described in Section 9 is invoked (see Proposition 12 in Section 5). Phase 2 basically uses the construction procedure of Phase 1 for the new problem instances that it derives.

6.1.1. The Main Partitioning Procedure

Now, we describe the main procedure (PROCEDURE MAIN) of our algorithm, that is in charge of the partitioning of the scheduling horizon into the kernel and the corresponding bin intervals. This partition is dynamically changed and is updated in a recurrent fashion each time a new kernel arises. The occurrence of each new kernel K during the construction of a bin, the split of this bin into smaller bins and the collapsing of kernel K induce the recurrent nature in our method (not surprising, the recurrence is a common feature in the most common algorithmic frameworks such are dynamic programming and branch-and-bound).

Invoked for kernel K (K is a global variable), PROCEDURE MAIN first calls PROCEDURE Decomposition(S, K, l) that forms schedule $S^*[K]$ ending with the atomic kernel K^* (see the beginning of Section 5 and Propositions 5 and 9).

PROCEDURE MAIN incorporates properly kernel K into the current partition updating respectively the current *configuration* $\mathcal{C}(\delta, K)$ defined by a trial δ, the current set of kernels $\mathcal{K}$ together with the corresponding $\delta(M^*)$s (see Equation (7)) and the augmented schedules $S^{L_\delta}[M]$, for $M \in \mathcal{K}$, constructed so far.

Given trial δ and kernel K, the configuration $\mathcal{C}(\delta, K)$ is unique, and there is a unique corresponding schedule $\Sigma_{\mathcal{C}(\delta,K)}$ with $K = K(\Sigma_{\mathcal{C}(\delta,K)})$ that includes the latest generated (so far) augmented schedules $S^{L_\delta}[M]$, $M \in \mathcal{K}$.

PROCEDURE MAIN starts with the initial configuration $\mathcal{C}(\Delta, K)$ with $\delta = \Delta$, $K = K(\sigma)$, $\Sigma_{\mathcal{C}(\Delta,K)} = \sigma$ and $\mathcal{K} = \emptyset$ (no bin exists yet in that configuration).

Iteratively, PROCEDURE MAIN, invoked for kernel K, creates a new configuration $\mathcal{C}(\Delta, K)$ with two new surrounding bins $B^-(K)$ and $B^+(K)$ and the atomic kernel K^* in between these bins. These bins arise within a bin of the previous configuration (the later bin disappears in the updated configuration). Initially, atomic kernel $(K(\sigma))^*$ splits schedule σ in two bins $B^-(K(\sigma))$ and $B^+(K(\sigma))$.

Two (atomic) kernels in schedule $\Sigma_{\mathcal{C}(\delta,K)}$ are *tied* if they belong to the same block in that schedule.

Given configuration $\mathcal{C}(\delta, K)$, the longest sequence of the augmented L-schedules of the pairwise tied kernels in schedule $\Sigma_{\mathcal{C}(\delta,K)}$ is called a *secondary block*.

We basically deal with the secondary block containing kernel K and denote it by $\mathcal{B}_K$ (we may omit argument K when this is not important). An essential characteristic of a secondary block is that every job that pushes a job from that secondary block belongs to the same secondary block. Therefore, the configuration update in PROCEDURE MAIN can be carried out solely within the current secondary block $\mathcal{B}_K$.

As we show below, PROCEDURE MAIN will create an L-schedule for an instance of $1|p_j : divisible, r_j|L_{\max}$ whenever it exists (otherwise, it affirms that no L-schedule for that instance exists). The same outcome is not guaranteed for an instance of $1|r_j|L_{\max}$, in general. In Theorem 3, we give an explicit condition under which an L-schedule for an instance of $1|r_j|L_{\max}$ will always be created, yielding a polynomial-time solution for the general setting. Unfortunately, if the above condition is not satisfied, we cannot, in general, affirm that there exists no feasible L-augmented schedule, even if our framework fails to find it for an instance of problem $1|r_j|L_{\max}$.

6.1.2. PROCEDURE AUGMENTED(K, δ), Rise of New Kernels and Bin Split

PROCEDURE MAIN uses PROCEDURE AUGMENTED(K, δ) as a subroutine. PROCEDURE AUGMENTED(K, δ), called for kernel K with threshold L_δ, is in charge of the creation of an L_δ-augmented schedule $S^{L_\delta}[K]$ respecting the current configuration $\mathcal{C}(\delta, K)$. PROCEDURE AUGMENTED(K, δ) constructs the second part of schedule $S^{L_\delta}[K]$ (one including the atomic kernel K^*) directly as specified in Proposition 10. The most time consuming part of PROCEDURE AUGMENTED(K, δ) is that of the construction of the preschedule $PreS(K)$ of schedule $S^{L_\delta}[K]$. This construction is carried out at Phase 1 described in Section 7.

After a call of PROCEDURE AUGMENTED(K, δ), during the construction of an L-preschedule $PreS(K)$ at Phase 1, a new kernel K' may arise (the reader may have a look at Proposition 12 and Lemma 5 from the next section). Then, PROCEDURE AUGMENTED(K, δ) returns the newly arisen kernel K' and PROCEDURE MAIN, invoked for that kernel, updates the current configuration. Since the rise of kernel K' splits the earlier bin $B^-(K)$ into two new surrounding bins $B^-(K')$ and $B^+(K') = B^-(K)$ of the new configuration, the bin $B^-(K)$ of the previous configuration disappears and is "replaced" by a new bin $B^-(K) = B^+(K')$ of the new configuration. Correspondingly, the scope of a preschedule for kernel K is narrowed (the former bin $B^-(K)$ is "reduced" to the newly arisen bin $B^+(K') = B^-(K)$).

In this way, as a result of the rise of a new kernel within the (current) bin $B^-(K)$ and the resultant bin split, PROCEDURE AUGMENTED(K, δ) may be called more than once for different (gradually decreasing in size) bins: The initial bin $B^-(K)$ splits into two bins, the resultant new smaller bin $B^-(K)$ may again be split, and so on. Thus, to the first call of PROCEDURE AUGMENTED(K, δ) the largest bin $B^-(K)$ corresponds, and the interval of the new arisen bin for every next call of the procedure is a proper sub-interval of that of the bin corresponding to the previous call of the procedure. Note that each next created preschedule is composed of the jobs from the corresponding bin.

PROCEDURE AUGMENTED(K, δ) has three outcomes. If no new kernel during the construction of preschedule $PreS(K)$ respecting the current configuration arises, the procedure completes with the successful outcome generating an L-augmented schedule $S^{L_\delta}[K]$ respecting the current configuration (in this case, schedule $S^{L_\delta}[K]$ may form part of the complete L-augmented schedule if the later schedule exists). PROCEDURE MAIN incorporates L_δ-augmented schedule $S^{L_\delta}[K]$ into the current configuration (the first IF statement in the iterative step in the description of the next subsection).

With the second outcome, a new kernel K' during the construction of preschedule $PreS(K)$ within bin $B^-(K)$ arises (Proposition 12 and Lemma 5). Then, PROCEDURE AUGMENTED(K, δ) returns kernel K' and PROCEDURE MAIN is invoked for this newly arisen kernel and it updates the current configuration, respectively (see the iterative step in the description). Then, PROCEDURE MAIN calls recursively PROCEDURE AUGMENTED(K', δ) for kernel K' and the corresponding newly arisen bin $B^-(K')$ (this call is now in charge of the generation of an L-preschedule $PreS(K')$ for kernel K', see the second IF statement in the iterative step of the description in the next subsection).

With the third (failure) outcome, Phase 1 (invoked by PROCEDURE AUGMENTED(K, δ) for the creation of an L-preschedule $PreS(K)$) fails to create an L-preschedule respecting the current configuration (an IA(b2), defined in the next section, occurs (see Proposition 12). In this case, PROCEDURE MAIN invokes Phase 2. Phase 2 is described in Section 9. Nevertheless, the reader can see a brief description of that phase below:

Phase 2 uses two subroutines, PROCEDURE sl-SUBSTITUTION(K) and PROCEDURE ACTIVATE(s), where s is an emerging job. PROCEDURE sl-SUBSTITUTION(K) generates modified configurations with an attempt to create an L-preschedule $PreS(K)$ respecting a newly created configuration, in which some preschedules of the kernels, preceding kernel K in the secondary block $\mathcal{B}_K$ are reconstructed. These preschedules are reconstructed by the procedure of Phase 1, which is called by PROCEDURE ACTIVATE(s). PROCEDURE ACTIVATE(s), in turn, is repeatedly called by PROCEDURE sl-SUBSTITUTION(K) for different emerging jobs in the search of a proper configuration (each call of PROCEDURE ACTIVATE(s) creates a new configuration by a call of Phase 1). If at Phase 2 a configuration is generated for which Phase 1 succeeds to create an L-preschedule $PreS(K)$ respecting that configuration (the successful outcome), the augmented L-schedules corresponding to the reconstructed preschedules remain incorporated into the current schedule $\Sigma_{\mathcal{C}(\delta,K)}$.

6.1.3. Formal Description of PROCEDURE MAIN

The formal description of PROCEDURE MAIN below is completed by the descriptions of Phases 1 and 2 in the following sections. For notation simplicity, in set operations, we use schedule notation for the corresponding set of jobs. Given a set of jobs A, we denote by $ED(A)$ the ED-schedule obtained by the application of ED-heuristics to the jobs of set A.

Whenever a call of PROCEDURE MAIN for kernel K creates an augmented L-schedule $S^{L_\delta}[K]$, the procedure completes secondary block $\mathcal{B}_K$ by merely applying ED-heuristics to the remaining available jobs, ones to be included in that secondary block; i.e., partial ED-schedule $ED(\mathcal{B}_K \setminus \cup_{M \in \mathcal{B}_K} \{S^{L_\delta}[M]\})$ is generated and is merged with the already created part of block $\mathcal{B}_K$ to complete the block (the rest of the secondary blocks are left untouched in the updated schedule $\Sigma_{\mathcal{C}(\delta,K)}$).

PROCEDURE MAIN returns L_δ-schedule with the minimal δ, which is optimal by Lemma 8.

PROCEDURE MAIN
Initial step: {Determine the initial configuration $\mathcal{C}(\Delta, K), K = K(\sigma)$}
Start the binary search with trial $\delta = \Delta$
{see Equation (9) and the inequality in Equation (10)}
$\Sigma_{\mathcal{C}(\Delta,K)} := \sigma$
{initialize the set of kernels}
$K := K(\sigma); \mathcal{K} := K$
{set the initial lower bound and the initial allowable delay for kernel K}
$L^*_{max} := L_{max}(K^*); \delta(K^*) := 0$
IF schedule σ contains no kernel with the delaying emerging job, output σ and halt
{σ is optimal by Proposition 1}

Iterative step:

{Update the current configuration $\mathcal{C}(\delta, K)$ with schedule $\Sigma_{\mathcal{C}(\delta,K)}$ as follows:}
{update the current set of kernels}
$\mathcal{K} := \mathcal{K} \cup K$;
{update the current lower bound}
$L^*_{\max} := \max\{L^*_{\max}, L_{\max}(K^*)\}$;
{update the corresponding allowable kernel delays (see Equation (7))}
$\delta(M^*) := L^*_{\max} - L_{\max}(M^*)$, for every kernel $M \in \mathcal{K}$
Call PROCEDURE AUGMENTED(K, δ) {construct an L_δ-augmented schedule $S^{L_\delta}[K]$}
IF during the execution of PROCEDURE AUGMENTED(K, δ) a new kernel K' arises
{update the current configuration according to the newly arisen kernel}
　　THEN $K := K'$; repeat *Iterative step*
　　IF the outcome of PROCEDURE AUGMENTED(K, δ) is failure THEN call Phase 2
{at Phase 2 new configuration is looked for such that there exist preschedule $PreS(K)$ respecting that configuration, see Section 9}
　　IF L_δ-augmented schedule $S^{L_\delta}[K]$ is successfully created
{the outcome of PROCEDURE AUGMENTED(K, δ) and that of Phase 2 is successful, hence complete secondary block $\mathcal{B}_K$ by ED-heuristics if there are available jobs which were not included in any of the constructed augmented schedules, i.e., $\mathcal{B}_K \setminus \cup_{M \in \mathcal{B}_K}\{S^{L_\delta}[M]\} \neq \emptyset$}
　　THEN update block $\mathcal{B}_K$ and schedule $\Sigma_{\mathcal{C}(\delta,K)}$ by merging it with partial ED-schedule
　　　　$ED(\mathcal{B}_K \setminus \cup_{M \in \mathcal{B}_K}\{S^{L_\delta}[M]\})$
　　　　(leave in the updated schedule $\Sigma_{\mathcal{C}(\delta,K)}$ the rest of the secondary blocks as they are)
　　IF (the so updated) schedule $\Sigma_{\mathcal{C}(\delta,K)}$ is an L_δ-schedule
{continue the binary search with the next trial δ}
　　THEN $\delta :=$ the next trial value and repeat *Iterative step*; return the generated L_δ-schedule with the
　　　　minimum δ and halt if all the trial δs were already considered
　　ELSE {there is a kernel with the delaying emerging job in schedule $\Sigma_{\mathcal{C}(\delta,K)}$}
　　　　$K := K(\Sigma_{\mathcal{C}(\delta,K)})$; repeat *Iterative step*
　　IF L_δ-augmented schedule $S^{L_\delta}[K]$ could not been created
{the outcome of Phase 2 is failure and hence there exists no L_δ-schedule; continue the binary search with the next trial δ}
　　THEN $\delta :=$ the next trial value and repeat *Iterative step*; return the generated L_δ-schedule with the
　　　　minimum δ and halt if all the trial δs were already considered

7. Construction of Kernel Preschedules at Phase 1

At Phase 1, we distinguish two basic types of the available (yet unscheduled) jobs which can feasibly be included in bin $B^-(K)$, for every $K \in \mathcal{K}$. Given a current configuration, we call jobs that can only be scheduled within bin $B^-(K)$ y-jobs; we call jobs which can also be scheduled within some succeeding bin(s) the x-jobs for bin $B^-(K)$ or for kernel K. In this context, y-jobs have higher priority.

We have two different types of the y-jobs for bin $B^-(K)$. The set of the *Type (a)* y-jobs is formed by the jobs in set $K \setminus K^*$ and yet unscheduled jobs not from kernel K released within the interval of bin $B^-(K)$. The rest of the y-jobs are ones released before the interval of bin $B^-(K)$, and they are referred to as the *Type (b)* y-jobs.

Recall that the interval of bin $B^-(K)$ begins right after the atomic kernel of the preceding bin (or at $\min_i r_i$ if K is the earliest kernel in $\mathcal{K}$) and ends with the interval of schedule $S^*[K]$. The following proposition immediately follows:

Proposition 11. *Every x-job for bin $B^-(K)$ is an external job for kernel K, and there may also exist the external y-jobs for that kernel. A Type (a) y-job can feasibly be scheduled only within bin $B^-(K)$, whereas Type (b) y-jobs can potentially be scheduled within a preceding bin (as they are released before the interval of bin $B^-(K)$).*

Phase 1 for the construction of preschedule $PreS(K)$ of kernel K consists of two passes. In Pass 1 y-jobs of bin $B^-(K)$ are scheduled. In Pass 2, x-jobs of bin $B^-(K)$ are distributed within that bin. We know that all Type (a) y-jobs can be feasibly scheduled within bin $B^-(K)$ without surpassing the L-boundary (since they were so scheduled in that bin), and these jobs may only be feasibly scheduled within that bin. Note that, respecting the current configuration with the already created augmented schedules for the kernels in set $\mathcal{K}$), we are forced to include, besides Type (a) y-jobs, also all the Type (b) y-jobs into bin $B^-(K)$. If this does not work at Phase 1 in the current configuration, we try to reschedule some Type (b) y-jobs to the earlier bins in Phase 2 by changing the configuration.

7.1. Pass 1

Pass 1 consists of two steps. In Step 1, ED-heuristics is merely applied to all the y-jobs for scheduling bin $B^-(K)$.

If the resultant ED-schedule $PreS(K,y)$ is a feasible L-schedule (i.e., no job in it surpasses the current L-boundary and/or finishes after time $\psi_L(K)$), Step 1 completes with the successful outcome and Pass 1 outputs $PreS(K,y)$ (in this case, there is no need in Step 2), and Phase 1 continues with Pass 2 that augments $PreS(K,y)$ with x-jobs, as described in the next subsection.

If schedule $PreS(K,y)$ is not an L-schedule (there is a y-job in that schedule surpassing the L-boundary), *Pass 1* continues with Step 2.

Proposition 12 specifies two possible cases when preschedule $PreS(K,y)$ does not contain all the y-jobs for bin $B^-(K)$, and Step 1 fails to create an L-preschedule for kernel K at the current configuration.

Proposition 12. *Suppose $PreS(K,y)$ is not a feasible L-schedule, i.e., there arises a y-job surpassing the current L-boundary and/or completing after time $\psi_L(K)$.*

(1) If there is a Type (b) y-job surpassing the L-boundary, then there exists no feasible partial L-preschedule for kernel K containing all the Type (b) y-jobs for this kernel (hence there is no complete feasible L-schedule respecting the current configuration).

(2) If there is a Type (a) y-job y surpassing the L-boundary and there exists a feasible partial L-preschedule for kernel K containing all the y-jobs, it contains a new kernel consisting of some Type (a) y-jobs including job y.

Proof. We first show Case (2). As already mentioned, all Type (a) y-jobs may potentially be included in bin $B^-(K)$ without surpassing the L-boundary and be completed by time $\psi_L(K)$ (recall Equation (12)). Hence, since y is a Type (a) y-job, it should have been pushed by at least one y-job i with $d_i > d_y$ in preschedule $PreS(K,y)$. Then, there exists the corresponding kernel with the delaying emerging y-job (containing job y and possibly other Type (a) y-jobs).

Now, we prove Case (1). Let y be a Type (b) y-job that was forced to surpass the L-boundary and/or could not be completed by time moment $\psi_L(K)$. In the latter case, ED-heuristics could create no gap in preschedule $PreS(K,y)$ as all the Type (b) y-jobs were released from the beginning of the construction, and Case (1) obviously follows. In the former case, job y is clearly pushed by either another Type (b) y-job or a Type (a) y-job. Let k be a job pushing job y. Independently of whether k is a Type (a) or Type (b) y-job, since job y is released from the beginning of the construction and job k was included ahead job y, by ED-heuristics, $d_k \leq d_y$. Then, no emerging job for job y may exist in preschedule $PreS(K,y)$ and Case (1) again follows as all the Type (a) y-jobs must be included before time $\psi_L(K)$. □

For convenience, we refer to Case (1) in Proposition 12 as an *instance of Alternative (b2)* (IA(b2) for short) with Type (b) y-job y (we let y be the latest Type (b) y-job surpassing the L-boundary and/or completing after time $\psi_L(K)$). (The behavior alternatives were introduced in a wider context earlier in [13].) If an IA(b2) in bin $B^-(K)$ arises and there exists a complete L-schedule, then, in that schedule, some Type (b) y-job(s) from bin $B^-(K)$ is (are) included within the interval of some bin(s) preceding bin $B^-(K)$ in the current secondary block $\mathcal{B}_K$ (we prove this in Proposition 16 in Section 7).

In *Step 2*, Cases (1) and (2) are dealt with as follows. For Case (1) (an IA(b2)), Step 2 invokes PROCEDURE sl-SUBSTITUTION(K) of Phase 2. PROCEDURE sl-SUBSTITUTION(K) creates one or more new (temporary) configurations, as described in Section 7. For every created configuration, it reconstructs some bins, preceding bin $B^-(K)$ in the secondary block $\mathcal{B}_K$ incorporating some Type (b) y-jobs for bin $B^($K$)$ into the reconstructed preschedules. The purpose of this is to find out if there exists an L-preschedule $PreS(K)$ respecting the current configuration and construct it if it exists.

For Case (2) in Proposition 12, Step 2 returns the newly arisen kernel K' and PROCEDURE MAIN is invoked with that kernel, which updates the current configuration respectively. PROCEDURE MAIN then returns the call to PROCEDURE AUGMENTED(K', δ) (see the description of Section 4) (note that, since PROCEDURE AUGMENTED(K', δ) invokes *Phase 1* now, for kernel K', Case (2) yields recursive calls of Phase 1).

7.2. Pass 2: DEF-Heuristics

If Pass 1 successfully completes, i.e., creates a feasible L-preschedule $PreS(\bar{K}, y)$, Pass 2, described in this subsection, is invoked (otherwise, IA(b2) with a Type (b) y-job from bin $B^-(K)$ arises and Phase 2 is invoked). Throughout this section, $PreS(K, y)$ stands for the output of Pass 1 containing all the y-jobs for bin $B^-(K)$. At Pass 2, the x-jobs released within the remaining available room in preschedule $PreS(K, y)$ are included by a variation of the *Next Fit Decreasing* heuristics, adopted for our scheduling problem with job release times. We call this variation *Decreasing Earliest Fit* heuristics, DEF-heuristics for short. It works with a *list* of x-jobs for kernel K sorted in non-increasing order of their processing times, the ties being broken by sorting jobs with the same processing time in the non-decreasing order of their due-dates.

DEF-heuristics, iteratively, selects next job x from the list and initially appends this job to the current schedule $PreS(K, y)$ by scheduling it at the earliest idle-time moment t' before time $\psi_L(K)$ (any unoccupied time interval in bin $B^-(K)$ before time $\psi_L(K)$ is an idle-time interval in that bin). Let $PreS(K, y, +x)$ be the resultant partial schedule, that is obtained by the application of ED-heuristics from time moment t' to job x and to the following y-jobs from schedule $PreS(K, y)$ which may possibly right-shifted in schedule $PreS(K, y, +x)$) (compared to their positions in schedule $PreS(K, y)$). In the description below, the assignment $PreS(K, y) := PreS(K, y, +x)$ updates the current partial schedule $PreS(K, y)$ according to the rearrangement in schedule $PreS(K, y, +x)$, removes job x from the list and assigns to variable x the next x-job from the list.

PROCEDURE DEF($PreS(K, y), x$)
IF job x completes before or at time $\psi_L(K)$ in schedule $PreS(K, y, +x)$ {i.e., $t' + p_x$ falls within the current bin}
THEN GO TO Step (A) {verify the conditions in Steps (A) and (B)}
ELSE remove job x from the list {job x is ignored for bin $B^-(K)$}; set x to the next job from the list;
 CALL PROCEDURE DEF($PreS(K, y), x$)
(A) IF job x does not push any y-job in schedule $PreS(K, y, +x)$ {x can be scheduled at time moment t' without the interference with any y-job, i.e., $t' + p_x$ is no greater than the starting time of the next y-job in preschedule $PreS(K, y)$} and it completes by time moment $\psi_L(K)$ in schedule $PreS(K, y, +x)$
 THEN $PreS(K, y) := PreS(K, y, +x)$; CALL PROCEDURE DEF($PreS(K, y), x$)
(B) IF job x pushes some y-job in schedule $PreS(K, y, +x)$
 THEN {verify the conditions in Steps (B.1)–(B.3)}
 (B.1) IF in schedule $PreS(K, y, +x)$ no (right-shifted) y-job surpasses L-boundary and
  all the jobs are completed by time moment $\psi_L(K)$
  THEN $PreS(K, y) := PreS(K, y, +x)$; CALL PROCEDURE DEF($PreS(K, y), x$)
 (B.2) IF in schedule $PreS(K, y, +x)$ some y-job completes after time moment $\psi_L(K)$
 THEN set x to the next x-job from the list and CALL PROCEDURE DEF($PreS(K, y), x$).
 We need the following auxiliary lemma before we describe Step (B.3):

Lemma 5. *If a (right-shifted) y-job surpasses L-boundary in schedule $PreS(K, y, +x)$, then there arises a new kernel in that schedule (in bin $B^-(K)$) consisting of solely Type (a) y-jobs, and x is the delaying emerging job of that kernel.*

Proof. Obviously, by the condition in the lemma, there arises a new kernel in schedule $PreS(K, y, +x)$, call it K', and it consists of y-jobs following job x in schedule $PreS(K, y, +x)$. Clearly, x is the delaying emerging job of kernel K'. Such a right-shifted job y cannot be of Type (b) as otherwise it would have been included within the idle-time interval (occupied by job x) at Pass 1. Hence, kernel K' consists of only Type (a) y-jobs. □

Due to the above lemma, PROCEDURE DEF continues as follows:
(B.3) IF in schedule $PreS(K, y, +x)$ the lateness of some (right-shifted) y-job exceeds L THEN return the newly arisen kernel K' and invoke PROCEDURE MAIN with kernel K' {this updates the current configuration respectively and makes a recursive call of Phase 1 now for kernel K'}
IF the list is empty THEN OUTPUT($PreS(K, y)$) and halt.

This completes the description of Pass 2 and that of Phase 1.

From here on, we let $PreS(K) = PreS(K, y, x)$ be the output of Phase 1 (a feasible L-preschedule for kernel K containing all the y-jobs for bin $B^-(K)$). An easily seen property of PROCEDURE DEF and preschedule $PreS(K, y, x)$ is summarized in the following proposition.

Proposition 13. *An L-preschedule cannot be obtained by replacing any x-job $x \in PreS(K, y, x)$ with a longer available x-job in preschedule $PreS(K, y, x)$. Hence, the omission of job x from preschedule $PreS(K, y, x)$ will create a new gap which may only be filled in by including job(s) with the same or smaller processing time.*

Let v and χ be the number of y-jobs and x-jobs of bin $B^-(K)$, respectively, $\nu = v + \chi$ is the total number of jobs in that bin, and let v_1 be the number of Type (b) y-jobs. The next theorem gives a valid upper bound on the cost of a call of PROCEDURE AUGMENTED(K, δ) at Phase 1 (including all the recursive calls that the initial call may yield).

Theorem 2. *The total cost of a call of Phase 1 for a kernel K is $O(\nu^2 \log \nu)$. Hence, the cost of a call of PROCEDURE AUGMENTED(K, δ) is the same.*

Proof. At Step 1 of Pass 1, during the construction of preschedule $PreS(K)$ ED-heuristics with an upper bound on its running time $O(v \log v)$ for scheduling up to v y-jobs is used, whereas at less than v_1 scheduling times a new kernel may arise (as the delaying emerging job may only be a Type (b) y-job). Phase 1 invokes PROCEDURE MAIN which, in turn, calls the decomposition procedure for each of these kernels. By Lemma 2, the total cost of all the calls of the decomposition procedure can be estimated as $O(\kappa_1^2 \log \kappa_1 + \kappa_2^2 \log \kappa_2 + \cdots + \kappa_{v_1}^2 \log \kappa_{v_1})$, where $\kappa_1, \ldots, \kappa_{v_1}$ is the number of jobs in each of the v_1 arisen kernels, correspondingly. Let m be the mean arithmetic of all these κs. Since any newly arisen kernel may contain only y-jobs for bin $B^-(K)$ and no two kernels may have a common job, $v_1 m \leq v$. The maximum in the sum is reached when all the κs are equal to m, and from the above sum another no-smaller magnitude $O(v_1 m^2 \log m) \leq O(v_1 (v/v_1)^2 \log(v/v_1)) \leq O(v^2 \log v)$ is obtained (in the first and second inequalities, $v_1 m \leq v$ and $v_1 \geq 1$, respectively, are applied).

Then, the total cost of Pass 1 for kernel K (including that of Step 2, Case (2)) is $O(v_1 v \log v + v^2 \log v) = O(v^2 \log v)$. The cost of Steps (A), (B.1) and (B.2) of Pass 2 is that of ED-heuristics, i.e., $O(\chi \log \chi)$. At Step (B.3), since the delaying emerging job for every newly arisen kernel is a distinct x-job for bin $B^-(K)$, the number of the calls of PROCEDURE MAIN for all the newly arisen kernels after the initial call of PROCEDURE AUGMENTED(K, δ), and hence the number of the recursive calls of Phase 1 for kernel K, is bounded by χ. Similar to what is done for Pass 1, we let $\kappa_1, \ldots, \kappa_{v_1}$ be the number of jobs in each of the χ arisen kernels, respectively. Again, by Lemma 2, the total cost of all the calls of PROCEDURE MAIN to the decomposition procedure is

$O(\kappa_1^2 \log \kappa_1 + \kappa_2^2 \log \kappa_2 + \cdots + \kappa_\chi^2 \log \kappa_\chi)$. We let again m be the mean arithmetic of all these κs, $\chi m \leq v$ and obtain an upper bound $O(\chi^2 \log \chi + \chi m^2 \log m) \leq O(\chi^2 \log \chi + \chi(v/\chi)^2 \log(v/\chi)) \leq O(v^2 \log v)$ on the cost of Pass 2 and hence the total cost of Phase 1 is $O(v^2 \log v)$.

The second claim in theorem follows as the cost of the generation of the second part of an augmented L_δ-schedule is absorbed by that of the first part. Indeed, recall that for a call of PROCEDURE AUGMENTED(K, δ), the second part of schedule $S^{L_\delta}[K]$ consisting of the jobs of the atomic kernel K^*, is constructed by ED-heuristics in time $O(\kappa' \log \kappa')$, where $\kappa' \leq \kappa$ is the total number of jobs in atomic kernel K^*, and κ is the number of jobs in kernel K (Proposition 10). Similar to above in this proof, we can show that the construction of the second part of the augmented schedules for the calls of PROCEDURE AUGMENTED for all the arisen kernels (for the same δ) is $O(n \log n)$. □

At this stage, we can give an optimality (sufficient) condition for problem $1|r_j|L_{\max}$, that is helpful also in that it exhibits where the complex nature of the problem is "hidden". Dealing with an IA(b2) is a complicated task as it implies the solution of NP-hard set/numerical problems such as 3-PARTITION yet with additional restrictions that impose job release times. As to the solution provided by PROCEDURE MAIN, as we have seen above, the recurrences at Step 2, Case (2) in Pass 1, and at Step (B.3) at Pass 2 do not, in fact, cause an exponential behavior.

Theorem 3. *PROCEDURE MAIN finds an optimal solution to problem $1|r_j|L_{\max}$ in time $O(n^2 \log n \log p_{\max})$ if no IA(b2) at Phase 1 arises.*

Proof. The proof is quite straightforward, we give a scratch. The initial step takes time $O(n \log n)$ (the cost of ED-heuristics). At iterative step, the cost of updates of $L_{\max}^*$ and $\delta(M^*)$, $M \in \mathcal{K}$ and that of the detection of every newly arisen kernel is bounded by the same magnitude. It is easy to see that an L-preschedule for every kernel in $\mathcal{K}$ will be generated at Phase 1 if no Type (b) y-job is forced to surpasses the L-boundary, or, equivalently, no IA(b2) arises (only a y-job may be forced to surpass the L-boundary, whereas, if a Type (a) y-job surpasses it, PROCEDURE MAIN proceeds with the newly arisen kernel). Hence, PROCEDURE AUGMENTED(K, δ) will create a feasible L-augmented schedule for every kernel (since no IA(b2) at Pass 1 arises). Then, it remains to estimate the calls of PROCEDURE AUGMENTED(K, δ) in the iterative step. The cost of a call of PROCEDURE AUGMENTED(K, δ) for a given kernel K including all the embedded recursive calls is $O(v^2 \log v)$ (Theorem 2). These recursive calls include the calls for all the kernels which may arise within bin $B^-(K)$. Hence, for the purpose of our estimation, it suffices to distinguish the calls PROCEDURE AUGMENTED(K, δ) and PROCEDURE AUGMENTED(M, δ) for two distinct kernels K and M such that bins $B^-(K)$ and $B^-(M)$ have no jobs in common. Then, similar to what is done to estimate the cost of Pass 1 in the proof of Theorem 2, we easily get an overall (amortized) cost of $O(n^2 \log n)$ for PROCEDURE MAIN for a given trial δ. Then, we obtain the overall cost of $O(n^2 \log n \log p_{\max})$ for PROCEDURE MAIN taking into account that there are no more than $\log p_{\max}$ trial δs. □

8. Construction of Compact Preschedules for Problem $1|p_j : divisible, r_j|L_{\max}$

This section starts Part 2, in which our basic task is to develop an auxiliary algorithm that deals with an IA(b2) occurred at Phase 1 (recall that if no IA(b2) occurs, PROCEDURE MAIN with PROCEDURE AUGMENTED(K, δ) using Phase 1 already solves problem $1|r_j|L_{\max}$). A compact feasible schedule, one without any redundant gap, has properties that are helpful for the establishment of the existence or the non-existence of a complete L-schedule whenever during the construction of a kernel preschedule at Phase 1 an instance of Alternative (b2) arises. In this section, we study the compactness properties for instances of problem $1|p_j : divisible, r_j|L_{\max}$.

Since the basic construction components of a complete feasible schedule are the secondary blocks, it suffices to deal with compact secondary blocks. A secondary block $\mathcal{B}$ is *compact* if there is no feasible L-schedule containing all the jobs of that block with the total length of all the gaps in it no-less than that in block $\mathcal{B}$.

We can keep the secondary blocks compact if the processing times of some non-kernel jobs are mutually divisible. For the commodity and without loss of generality, we assume that the processing times of the non-kernel jobs are powers of 2 (precisely, we identify the specific non-kernel jobs for which mutual divisibility is required on the fly). Below, we give a basic property of a set of divisible numbers and then we give another useful property of a kernel preschedule with divisible job processing times, which are used afterwards.

Lemma 6. *For a given job x, let $J^-(x)$ be the set of jobs $J^-(x) = \{i|p_i < p_x\}$ such that $p(J^-(x)) > p_x$ and the processing times of jobs in set $J^-(x) \cup \{x\}$ are mutually divisible. Then, there exists a proper subset J' of set $J^-(x)$ with $p(J') = p_x$ (that can be found in an almost liner time).*

Proof. The following simple procedure finds subset J'. Sort the jobs in set $J^-(x)$ in non-increasing order of their processing times, say $\{x_1, \ldots, x_k\}$. It is straightforward to see that, because of the divisibility of the processing times of the jobs in set $J^-(x) \cup \{x\}$, there exists integer $l < k$ such that $\sum_{i=1}^{l} x_i = p_x$, i.e., $J' = \{x_1, \ldots, x_k\}$. □

Lemma 7. *Preschedule $PreS(K, y, x)$, constructed at Pass 2 of Phase 1 for an instance of $1|p_j :$ divisible, $r_j|L_{\max}$, contains no gap except one that may possibly arise immediately before time moment $\psi_L(K)$.*

Proof. By the way of contradiction, suppose I is an internal gap in schedule $PreS(K, y)$ of Pass 1. Note that initially, gap I was completely occupied in bin $B^-(K)$ in schedule σ, and that it is succeeded by at least one y-job in preschedule $PreS(K, y)$. That is, the x-jobs with the total length of at least $|I|$ should have been available while scheduling the interval of gap I in PROCEDURE DEF at Pass 2. Then, an idle time interval within the interval of gap I in preschedule $PreS(K, y, x)$ of Pass 2 may potentially occur only at the end of that interval, say at time moment τ, due to the non-permitted interference in schedule $PreS(K, y, +x)$ of an available (and not yet discarded) x-job with a succeeding y-job, say y (Step (B)). Note that job y is a Type (a) y-job (if it were of Type (b), then it would have been included ahead any x-job in bin $B^-(K)$) and that the lateness of that job did not exceed L before kernel K was detected in schedule $\Sigma(\mathcal{C}(\delta, K))$. Let X be the set of the x-jobs preceding job y in the interval of gap I in the latter schedule, and X' be the corresponding set of the x-jobs in preschedule $PreS(K, y, x)$ (by our construction, $P(X) > P(X')$). In PROCEDURE DEF, during the construction of schedule $PreS(K, y, x)$, at time moment τ there must have been no job with processing time $p(X) - p(X')$ or less available. However, this is not possible since, because of the divisibility of job processing times, set X must contain such a job (and that job must have been available and yet unscheduled). The existence of a gap from time moment τ in the interval of gap I in schedule $PreS(K, y, x)$ has led to a contradiction and hence it cannot exist. □

In the rest of this section, we assume that preschedule $PreS(K)$ contains a gap; i.e., it ends with a gap (Lemma 7). Our objective is to verify if that gap can be reduced. To this end, we define two kinds of jobs such that their interchange may possibly be beneficial.

The first type of jobs are formed from set $EP(K, L)$, the set of the passive emerging jobs for kernel K in the current configuration with threshold $L = L_\delta$. Recall that a job from set $EP(K, L)$ is included after kernel K in schedule $\Sigma_{\mathcal{C}(\delta,K)}$ but it may feasibly be included (as an x-job) in a preschedule of kernel K (in bin $B^-(K)$).

Recall at the same time, that a job from preschedule $PreS(K)$ which may be rescheduled after all jobs of kernel K without surpassing the L-boundary is one from set $E(K, L)$, the set of emerging jobs for kernel K at the current configuration (such a job was included as an x-job in preschedule $PreS(K)$).

A vulnerable component of a secondary block is a preschedule in it, in the sense that we can maintain a secondary block compact if every preschedule that it contains is also *compact*, i.e., there exists no other preschedule (for the same kernel) with the total length of the gaps less than that in the former preschedule (see Corollary 4 at the end of this section). A key informal observation here is that, if a preschedule for kernel K is not compact, then a compact one can only be obtained from the former preschedule by replacing some jobs from set $E(K, L)$ with some jobs from set $EP(K, L)$, whereas nothing is to be gained by substituting any jobs from a compact preschedule by any jobs from set $EP(K, L)$ (Proposition 14 below).

Let $A \subseteq E(K, L)$ and $B \subseteq EP(K, L)$. Consider A and B as potential "swap" subsets and denote by $PreS(K, -A, +B)$ the preschedule for kernel K obtained by interchanging the roles of jobs from sets A and B while reconstructing the current preschedule $PreS(K)$ by the procedure of Phase 1. Technically, preschedule $PreS(K, -A, +B)$ can be constructed at Phase 1 for the restricted problem instance $PI(PreS(K), -A, +B)$ that contains all jobs from preschedule $PreS(K)$ and set B but does not contain ones in set A (so jobs from set A are activated for kernel K). Note that a job from set A belongs to $PreS(K, -A, +B)$, and, along with the remaining jobs from preschedule $PreS(K)$, some job(s) from set B may also be included in $PreS(K, -A, +B)$.

Proposition 14. *If an L-preschedule $PreS(K)$ is not compact then there exist sets A and B such that an L-preschedule $PreS(K, -A, +B)$ is compact.*

Proof. Among the jobs included in schedule $\Sigma_{\mathcal{C}(\delta,K)}$ after preschedule $PreS(K)$, the available room (the gap) from preschedule $PreS(K)$ may only potentially be used by job(s) from set $EP(K, L)$. By the construction of Phase 1, this will not be possible unless some emerging job(s) from preschedule $PreS(K)$ is (are) rescheduled after kernel K. Then, this kind of the interchange of the jobs from set $E(K, L)$ with the jobs from set $EP(K, L)$ yields the only potentially improving rearrangement of the jobs in preschedule $PreS(K)$, and the proposition follows. □

Let us say that set A *covers* set B if preschedule $PreS(K, -A, +B)$ includes all jobs from problem instance $PI(PreS(K), -A, +B)$. Since we wish to reduce the total gap length in preschedule $PreS(K)$, $p(A) < p(B)$ must hold, which is our assumption from now on (we use $p(A)$ for the total processing time in job-set A; below, we use $p_{min}\{A\}$ for the minimum job processing time in A).

Let $\gamma(K)$ the total gap length in preschedule $PreS(K) \in \Sigma_{\mathcal{C}(\delta,K)}$. We call

$$ST(K) = \gamma(K) + \beta_L(K^*) \tag{13}$$

the *store* of kernel K in the current configuration $\mathcal{C}(\delta, K)$. It is easily seen that $ST(K)$ is the maximum available vacant room in preschedule $PreS(K) \in \Sigma_{\mathcal{C}(\delta,K)}$:

Proposition 15. *The total length of the jobs (the gaps, respectively) in preschedule $PreS(K) \in \Sigma_{\mathcal{C}(\delta,K)}$ might be increased (decreased, respectively) by at most $ST(K)$ time units in any L-preschedule for kernel K. If set A covers set B, then the store of kernel K in an updated configuration with preschedule $PreS(K, -A, +B)$ is*

$$ST(K) - (P(B) - P(A)).$$

Lemma 8. *If $ST(K) < p_{min}\{E(K, L)\}$, then preschedule $PreS(K)$ is compact. If preschedule $PreS(K)$ is not compact, then $ST(K) \geq p_{min}\{A\}$, for any $A \subseteq E(K, L)$.*

Proof. By the condition in lemma, the gap in preschedule $PreS(K)$ (see Lemma 7) can potentially be occupied only by a job j with $p_j \leq p_{min}\{E(K, L)\}/2$ (see Proposition 15). There may exist no such job in set $EP(K, L)$ as otherwise it would have been included in preschedule $PreS(K)$ as an x-job at Pass 2. Now, it can be straightforwardly seen that no interchange of jobs in set $E(K, L)$ from preschedule $PreS(K)$ with jobs from set $EP(K, L)$ may reduce the gap, because of the divisibility of the processing times of the jobs in sets $E(K, L)$ and $EP(K, L)$, and the first claim in lemma follows from Proposition 14.

Now, we show the second claim. Suppose preschedule $PreS(K)$ is not compact. Then, there exist sets A and B such that A covers B and preschedule $PreS(K, -A, +B)$ results in the reduction of the store of kernel K by $p(B) - p(A)$ (see Equation (13) and Propositions 14 and 15). Because of the divisibility of job processing times in sets A and B, $p(B) - p(A)$ is a multiple of $p_{min}\{A \cup B\}$. Hence, if $ST(K) < p_{min}\{A \cup B\}$, then preschedule $PreS(K)$ is compact; $ST(K) \geq p_{min}\{A \cup B\}$ must hold if $PreS(K)$ is not compact. $ST(K) \geq p_{min}\{B\}$ is not possible, as otherwise a job from set B with processing time $p_{min}\{B\}$ would have been included in preschedule $PreS(K)$ at Pass 2 of Phase 1. It follows that $ST(K) \geq p_{min}\{A\}$. □

Due to Lemma 8, from here on, assume that $ST(K) \geq p_{min}\{A\}$. It is not difficult to see that not all $ST(K)$ time units may potentially be useful. In particular, let $\nu \geq 1$ be the maximum integer such that $ST(K) \geq \nu p_{min}\{A\}$, and let $p' = \nu p_{min}\{A\}$.

Lemma 9. *A feasible L-preschedule $PreS(K, -A, +B)$ contains gap(s) with the total length of at least $ST(K) - p'$; hence, $p(B) \leq p(A) + p'$ when set A covers set B. Furthermore, $p_{min}(EP(K, L)) = 2^\kappa p_{min}\{A\}$, for some integer $\kappa \geq 1$, and $p' \leq 2^\kappa p_{min}\{A\}$.*

Proof. The first claim easily follows from the definitions and the mutual divisibility of the processing times of jobs in sets A and B, and inequality $p(B) \leq p(A) + p'$ immediately follows. As to the second claim, first we note that, for any $\pi \in EP(K, L)$, $p_\pi > ST(K)$, as otherwise job π would have been included in preschedule $PreS(K)$ at Pass 2 of Phase 1. Then, $p_{min}(EP(K, L)) > p'$, whereas $p' \geq p_{min}\{A\}$. Hence, $p_{min}(EP(K, L)) > p_{min}\{A\}$. Now, the second claim follows from the fact that the processing times of jobs in sets $EP(K, L)$ and A are powers of 2. □

Example 3. *Suppose $p_{min}\{A\} = 4$ and $ST(K) = 23$. Then, $p' = 5p_{min}\{A\} = 20$, hence a gap of length 3 is unavoidable. Let $p_{min}(EP(K, L)) = 2^3 p_{min}\{A\} = 32$. Since the shortest job that set B may contain has processing time 32, the most we may expect is to form set A of three jobs of (the minimal) length 4, set B being formed by a single job with the length 32. Then, after swapping sets A and B, we have a residue $32 - 3 \times 4 = 20$. Because of these extra 20 units, the available idle space of length 23 is reduced to 3 in schedule $S^L(K, -A, +B)$ in which set A covers set B. In that schedule, a gap of (the minimal possible) length $23 - 20 = 3$ occurs.*

We may restrict our attention to sets A and B which do not contain equal-length jobs, as otherwise we may simply discount the corresponding jobs from both sets. In particular, for given A and B with $i \in A$ and $j \in B$ with $p_i = p_j$, we obtain sets $A(-i)$ and $B(-j)$ by eliminating job i and job j, respectively, from sets A and B, respectively. Let $A(-all\_equal, B)$ and $B(-all\_equal, A)$ be the reduced sets A and B, respectively, obtained by the repeated application of the above operation for all equal-length jobs. Sets $A(-all\_equal, B)$ and $B(-all\_equal, A)$ contain no equal-length jobs. We have proved the following lemma.

Lemma 10. *If set A covers set B, then set $A(-all\_equal, B)$ covers set $B(-all\_equal, A)$, where $p(B) - p(A) = p(B(-all\_equal, A)) - p(A(-all\_equal, B))$.*

Theorem 4. *If set A covers set B, then there are also (reduced) sets $A' \subseteq A$ and $B' \subseteq B$, where set B' contains a single element $\pi \in EP(K, L)$ with the minimum processing time in set B and with $P(B') - p' \leq P(A') < P(B')$ such that set A' covers set B' and $P(B') - P(A') = P(B) - P(A)$.*

Proof. Let A and B be the reduced sets that contain no equal-length jobs and such that A covers B (see Lemma 10). We can further reduce sets A and B by discounting, similarly, for each job $j \in B$, jobs from set A, for which processing times sum up to p_j. In particular, take a longest job $j \in B$ and longest jobs from set A that sum up to p_j. Due to the divisibility of job processing times and the inequalities $p(B) > p(A)$ and $p_{min}(EP(K, L)) = 2^\kappa p_{min}\{A\}$ (see Lemma 9), this will be possible as long as the total processing time in A is no smaller than p_j. The sets A and B are reduced respectively, and the same operation for these reduced sets is repeated until the total processing time of the remaining jobs in the reduced set A is less than p_j. Then, we are left with a single job $j \in B$ (one with the minimum processing time in B) and the jobs in set A with the total processing time less than p_j, and such that $p_j - p(A) \leq p'$ (see Lemma 9).

Let A' and B' be the reduced sets obtained from sets A and B, respectively. By the construction of set A' and B' and the fact that set A covers set B, it immediately follows that $P(B') - P(A') = P(B) - P(A)$ and that set A' covers set B'. □

Now, we show that the current secondary block $\mathcal{B}_K$ will be kept compact if we merely unify the compact preschedules in schedule $\Sigma_{C(\delta,K)}$.

Theorem 5. *A secondary block $\mathcal{B}$ consisting of compact L-preschedules is compact.*

Proof. If the time interval of every preschedule $PreS(K)$ from block $\mathcal{B}$ extends up to time $\psi_L(K)$ and it contains no gap then the secondary block $\mathcal{B}$ is clearly compact. Suppose there is preschedule $PreS(K)$ from block $\mathcal{B}$ that contains a gap and/or completes before time $\psi_L(K)$. First, we observe that no extra job can be included within preschedule $PreS(K)$ to obtain another L-preschedule with an extended time interval and/or with less total gap length. Indeed, let x', $p_{x'} < p_x$, be a shortest available x-job from set $\in J^-(x)$. By PROCEDURE DEF, schedule $PreS(K, +x')$ is not a feasible L-preschedule for kernel K (as otherwise PROCEDURE DEF would include job x' in preschedule $PreS(K)$ at Pass 2). Thus, job x' may only feasibly be included in preschedule $PreS(K)$ by removing a longer job x from that preschedule. However, such a rearrangement may, at most, fill in the former execution interval of job x due to the above made observation and Lemma 6.

To prove the lemma, now it clearly suffices to show that nothing is to be gained by a job rearrangement in preschedule $PreS(K)$ that involves, besides the jobs from sets $E(K, L)$ and $EP(K, L)$, the jobs from a preschedule preceding preschedule $PreS(K)$.

Let $PreS'(K)$ be an arbitrary L-preschedule for kernel K (one respecting the current threshold L_δ). Without loss of generality, assume preschedules $PreS(K)$ and $PreS'(K)$ start at the same time, whereas none of them may complete after time $\psi_L(K)$ (Equation (12)). Let W and Z, respectively, be the sets of integer numbers, the processing times of jobs in the current preschedule $PreS(K) \in \Sigma_{C(\delta,K)}$ and in preschedule $PreS'(K)$, respectively (here, we assume that sets W and Z consist of mutually divisible integer numbers, possibly with some repetitions).

Similar to what is done in Lemma 10 and Theorem 4, we discount the same numbers from sets W and Z and the numbers from one set which sum up to another number from the other set (taking a combination with the longest possible jobs). Note that both sets are reduced by the same amount (a sum of powers of 2). Denote by W' and Z' the resultant sets.

If $p(W') \geq p(Z')$ then the total gap length in preschedule $PreS(K)$ cannot be more than that in preschedule $PreS'(K)$, and the theorem follows if the condition holds for all preschedules in block $\mathcal{B}$.

Otherwise, suppose $p(W') < p(Z')$. By the definition of the sets W' and Z' and the store of kernel K (Equation (13)), $p(Z') - p(W') = p(Z) - p(W) \leq ST(K)$ (see Theorem 4) and the preschedule for kernel K consisting of the jobs associated with the set of processing times $\{W \setminus W'\} \cup Z'$ will have the same total gap length as preschedule $PreS'(K)$ (the substitution of the jobs corresponding to set W' by those from set Z' would result in a preschedule with the same total gap length as that in preschedule $PreS'(K)$). By the construction of preschedule $PreS(K)$ at Phase 1, no job x with processing time from set Z' which could have been feasibly included within preschedule $PreS(K)$ was available during the construction of that preschedule. Hence, every such job x should have been already scheduled in a preschedule $PreS(K')$ preceding preschedule $PreS(K)$ in block $\mathcal{B}$. By rescheduling job x from preschedule $PreS(K')$ to preschedule $PreS(K)$, the total gap length in the newly created preschedule of kernel K will be reduced by p_x, but a new gap of the same length will occur in the resultant new preschedule of kernel K' as there is no other suitable job available (otherwise, it would have been included in preschedule $PreS(K)$). Hence, the total gap length in block $\mathcal{B}$ will remain the same. Thus, no matter how the jobs are redistributed among the preschedules from block $\mathcal{B}$, the total length of the remaining gaps in that block will remain the same. The lemma is proved. □

Corollary 4. *If a secondary block $\mathcal{B}$ is constituted by the preschedules created at Phase 1, then it is compact.*

Proof. For every kernel $K \in \mathcal{B}$, if an L-preschedule $PreS(K, x, y)$ of Phase 1 is not compact then there exist sets $A \subseteq E(K, L)$ and $B \subseteq EP(K, L)$ such that an L-preschedule $PreS(K, -A, +B)$ is compact (Proposition 14). By Theorem 4, $B = \{\pi\}$, for some job $\pi \in EP(K, L)$. However, since for every job $j \in A$, $p_j < p_\pi$ (see Lemma 9), set A cannot cover set B in preschedule $PreS(K, -A, +B)$, as otherwise job π would have been included in preschedule $PreS(K, x, y)$ at Pass 2 instead of the shorter jobs from set A. It follows that every preschedule from block $\mathcal{B}$ is compact, and the corollary follows from Theorem 5. □

9. Phase 2: Search for an L-preschedule When an IA(b2) at Phase 1 Arises

Throughout this section, we consider the scenario when a compact preschedule for a newly arisen kernel K cannot be constructed at Phase 1, i.e., an IA(b2) with a Type (b) y-job y at Pass 1 arises. Recall that this happens when Pass 1 is unable to include job y in preschedule $PreS(K, y)$ in the current configuration (see Proposition 12). Phase 2, invoked from Phase 1, generates one or more new problem instances and calls back Phase 1 to create the corresponding new configurations. Thus, Phase 2 has no proper algorithmic features except that it generates new problem instances.

We refer to the earliest occurrence of IA(b2) in secondary block $\mathcal{B}_K$ at Phase 1 as the basic case. In the inductive case (abbreviated IA(b2-I)), IA(b2) repeatedly arises in the current secondary block (roughly, we "stay" in the current secondary block for IA(b2-I) in the inductive case, whereas we are brought to a new secondary block with every newly occurred IA(b2) in the basic case). In general, different occurrences of an IA(b2-I) in the inductive case may occur for different kernels, where all of them pertain to the current secondary block $\mathcal{B}$.

Throughout this section, let K^- be the kernel immediately preceding kernel K in block $\mathcal{B}_K$. We let y be an incoming job in bin $B^-(K) = B^+(K^-)$ at Phase 1; y is an incoming job in the first bin of block $\mathcal{B}_K$ if there exists no K^-. Note that r_y is no smaller than the starting time of block $\mathcal{B}_K$, and, since it can feasibly be scheduled within every bin that initiates at or after time r_y up to (and including) bin $B^-(K)$, y is a former x-job for any such a bin (except that it is a Type (b) job for bin $B^-(K)$), i.e., it may potentially be included in any of these bins. We explore such possibility and seek for a suitable distribution of all the x-jobs and Type (b) y-jobs into these bins at Phase 2.

Proposition 16. *Suppose during the construction of preschedule $PreS(K, y)$ an IA(b2)/IA(b2-I) with job y occurs and there exists schedule S^L. Then, job y or a Type (b) y-job included between kernel K^- and job y in bin $B^-(K)$ is scheduled before kernel K^- in schedule S^L.*

Proof. Note that the critical block in schedule $\Sigma_{C(\delta,K)}$ coincides with the secondary block $\mathcal{B}_{K^-}$, and it is compact when the above IA(b2)/IA(b2-I) occurs by Corollary 4. Then, job y cannot be restarted earlier in any feasible L-schedule in which the same jobs (which were included in preschedule $PreS(K, y)$ at Pass 1) are left scheduled before job y. The lemma obviously follows if y is the earliest considered job to be scheduled in bin $B^-(K)$. Otherwise, job y may potentially be started earlier either by scheduling it before kernel K^- or by decreasing (left-shifting) its current early starting time. The latter will only be possible if some job included in bin $B^-(K)$ ahead of job y is rescheduled behind job y. By the construction at Phase 1, any job included in bin $B^-(K)$ ahead of job y is a no less urgent than job y y-job and it cannot be rescheduled after job y without surpassing the L-boundary. Then, job y may be left-shifted only if one of the latter jobs is rescheduled before kernel K^-. However, this is not possible for a Type (a) y-job and the lemma is proved. □

By the above proposition, either job y or a Type (b) y-job included between kernel K^- and job y in bin $B^-(K)$ is to be rescheduled before kernel K^-. In particular, the following observations are evident:

- (1) If job y, is the first scheduled job in bin $B^-(K)$ or is preceded only by Type (a) y-jobs in that bin, then job y is to be entirely rescheduled before kernel K^-.

- (2) If job y is preceded by some Type (b) y-job(s), then either job y or some of these Type (b) y-job(s) is (are) to be rescheduled before kernel K^-. Since in any L-schedule job y needs to be left-shifted by at least λ_y amount of time (the L-delay of job y (see Equation (5))), the total processing time of these Type (b) y-jobs to be rescheduled before kernel K^- must be no-less than λ_y.

Let us denote by Λ_y the set of the y-jobs to be rescheduled before kernel K^- as defined in Cases (1) and (2) above. Set Λ_y will not be explicitly defined; it will be formed implicitly during the activation procedure that we describe in this section. In Case (1) above, set Λ_y will contain a single job y, hence $p_s \geq p_y$ must clearly hold, whereas, in Case (2), p_s must clearly be no-less than the minimum processing time of a y-job in set Λ_y. Let $\bar{p}_{\min\{y\}}$ be the minimum processing time among these y-jobs. The next proposition follows:

Proposition 17. $p_s \geq \bar{p}_{\min\{y\}}$.

9.1. The Activation of a Substitution Job

Given that an IA(b2)/IA(b2-I) with job y after kernel K^- arises, $s \in \mathcal{B}_K$ is called a *substitution* job if $d_s > d_y$. Intuitively, job s is an emerging job for job y (the latter job surpasses the current L-boundary, and in this sense, it is a potential overflow job).

PROCEDURE ACTIVATE(s) that activates substitution job s has some additional features compared to the basic definition of Section 2, as we describe in this subsection (in the next subsection, we complete the description of Phase 2 by a subroutine that tries different substitution jobs to determine a "right" one).

Let $B\{(s)\}$ be the bin from secondary block $\mathcal{B}_K$ containing substitution job s (it follows that s was included as an x-job in bin $B\{(s)\}$). PROCEDURE ACTIVATE(s) reconstructs preschedules for the kernels in the current schedule $\Sigma_{\mathcal{C}(\delta,K)}$ between the kernel K' with $B^-(K') = B\{(s)\}$ (the kernel with its first surrounding bin $B\{(s)\}$) and kernel K^-, including these two kernels, calling Phase 1 for each of these kernels (the kernel preschedules are reconstructed in their precedence order). This reconstruction leads to a new temporal configuration. PROCEDURE ACTIVATE(s) aims to verify if there exists a feasible L-preschedule for kernel K respecting this configuration. If it does not exist, PROCEDURE sl-SUBSTITUTION(K), described in the next subsection, tries another substitution job for kernel K, calling again Phase 1 for kernel K; each call creates a new temporary configuration and is carried out for a specially derived problem instance that depends on the selected substitution job.

For notational simplicity, we denote every newly constructed preschedule of kernel K by $PreS(K)$; we distinguish preschedules constructed at different calls of Phase 1 just by referring to the call with the corresponding substitution job, and will normally use $PreS(K)$ for the latest so far created preschedule for kernel K.

In the inductive case, the activation procedure for a substitution job s calls Phase 1 with a non-empty set $\mathcal{S}_\mathcal{B}$ of the substitution jobs, ones in the state of activation in the secondary block $\mathcal{B}$ by the corresponding call of Phase 1 (note that $s \notin \mathcal{S}_\mathcal{B}$). As already noted, the activation procedure may be called for different kernels which belong to the current secondary block, so that this block may contain a preschedule, already reconstructed by an earlier call of the activation procedure for another kernel from that block (set $\mathcal{S}_\mathcal{B}$ contains all the corresponding substitution jobs).

Problem instances for the basic and inductive cases. The problem instances for the basic and inductive cases are different, as we specify now. The problem instance PI($current, +y, [s]$) of the *basic* case contains the jobs in schedule $\Sigma_{\mathcal{C}(\delta,K)}$ from all the bins between bin $B\{(s)\}$ and bin $B^-(K^-)$, including the jobs of bins $B^-(K^-)$ and $B\{(s)\}$ except job s, job y and all the y-jobs included before job y in preschedule $PreS(K,y)$ of Pass 1 (the latter y-jobs are ones which were already included in bin $B^-(K)$ at Pass 1 when the IA(b2) with job y has occurred; note that no x-job for bin $B^-(K)$ is included in instance PI($current, +y, [s]$)).

The problem instance of the *inductive* case contains the same set of jobs as that in the basic case, and it also contains the substitution jobs from set S_B. For the sake of simplicity, we denote that problem instance also by PI(*current*, $+y$, [s]).

Successful and failure outcomes. As already specified, the activation of job s consists of the rescheduling of preschedules of bins $B\{(s)\}, \ldots, B^-(K^-)$ by a call of Phase 1 for instance PI(*current*, $+y$, [s]) in this precedence order (note that while rescheduling these bins only the jobs from that instance are considered). As we show at the end of this subsection in Lemma 11, all these bins will be successfully reconstructed at Phase 1.

PROCEDURE ACTIVATE(s) halts either with the successful outcome or with the failure outcome. For every successful outcome, the current call of Phase 2 (invoked for the IA(b2) with job y) completes and Phase 1 is repeatedly invoked from PROCEDURE MAIN for the construction of a new preschedule $PreS(K)$ for kernel K. Intuitively, the difference between the configurations after this new and the previous calls of Phase 1 for kernel K is that, as a result of the new call, no job from problem instance PI(*current*, $+y$, [s]) may again surpass the L-boundary, and job y is already included in current secondary block in the new configuration. We omit a straightforward proof of the next proposition.

Proposition 18. *If there is a job from instance PI(current, $+y$, [s]) that the activation procedure could not include in any of the reconstructed bins $B\{(s)\}, \ldots, B^-(K^-)$, this job is a y-job for bin $B^-(K)$ (or it is a job from set S_B in the inductive case). If a former y-job is of Type (a), then all such Type (a) y-jobs can be included in bin $B^-(K)$ during the construction of a new preschedule PreS(K) for kernel K at Phase 1.*

Note that, independently of the outcome, the activation procedure cannot include job s before any of the Type (b) y-jobs for bin $B^-(K)$ from instance PI(*current*, $+y$, [s]) in the basic case. However, as shown below, job s may be included ahead some of these Type (b) y-jobs at a later call of the activation procedure for a substitution job, different from job s, in the inductive case.

Extension of Phase 1 for a call from the inductive case. The activation procedure for the inductive case takes a special care on the jobs from set S_B while invoking Phase 1 for instance PI(*current*, $+y$, [s]) (or instance PI(*current*, $+y$, [$\emptyset$]) which we define below). In particular, when Phase 1 is called from the inductive case, two types of the x-jobs are distinguished during the (re)construction of a preschedule $PreS(\bar{K})$, $\bar{K} \neq K$ (one of the bins $B\{(s)\}, \ldots, B^-(K^-)$). The *Type (b)* x-jobs are ones which are also x-jobs for bin $B^-(K)$, and the rest of the x-jobs are *Type (a)* x-jobs. We observe that a Type (a) x-job for bin $B^-(\bar{K})$ will transform to a Type (b) y-job for bin $B^-(K)$ unless it is included in one of the preceding reconstructed bins $B^-(\bar{K})$, and that a substitution job from set S_B is a Type (b) x-job for any bin $B^-(\bar{K})$.

Phase 1, when invoked from the inductive case, is extended with an additional, Pass 3, designed for scheduling the substitution jobs from set S_B. Pass 3 uses the algorithm of Pass 2, DEF-heuristics, but with a different input, restricted solely to Type (b) x-jobs (hence, a former substitution job from S_B may potentially be included at Pass 3). There is a respective modification in the input of Pass 2, which consists now of only Type (a) x-jobs (hence no substitution job from set S_B will be included at Pass 2). Pass 3 is invoked after Pass 2, and Pass 2 is invoked after Pass 1, which remains unmodified while rescheduling each of the bins $B\{(s)\}, \ldots, B^-(K^-)$.

Once (in both basic and inductive cases) preschedules of bins $B\{(s)\}, \ldots, B^-(K^-)$ are reconstructed (Lemma 11), Phase 1 continues with the reconstruction of preschedule $PreS(K)$ as follows.

- (A) If there remains no unscheduled job from instance PI(*current*, $+y$, [s]) (except possibly jobs from set S_B in the inductive case), i.e., all these jobs are included in one of the reconstructed bins $B\{(s)\}, \ldots, B^-(K^-)$, the activation procedure halts with the successful outcome.

If there is a job from instance PI(*current*, $+y$, [s]) that could not have been included in any of the reconstructed bins $B\{(s)\}, \ldots, B^-(K^-)$ (excluding jobs from set S_B in the inductive case), then it is a y-job for bin $B^-(K)$ (and it might also be a job from set S_B in the inductive case). PROCEDURE ACTIVATE(s) proceeds as described below.

- (B) If every job from instance PI(*current*, $+y$, $[s]$) that could not have been included in any of the reconstructed bins $B\{(s)\}, \ldots, B^-(K^-)$ is a Type (a) y-job for bin $B^-(K)$ (or a job from set S_B in the inductive case), the outcome of the activation of job s is again successful (see Proposition 18). {all the Type (a) y-jobs for bin $B^-(K)$ will fit in that bin}.
- If there is a Type (b) y-job for bin $B^-(K)$ from instance PI(*current*, $+y$, $[s]$) that could not have been included in any of the reconstructed bins $B\{(s)\}, \ldots, B^-(K^-)$, the outcome of the activation procedure depends on whether Phase 1 will succeed to construct L-preschedule $PreS(K)$ including all such Type (b) y-jobs.

(C1) If during the construction of preschedule $PreS(K)$ at Pass 1 an iteration is reached at which all the Type (b) y-jobs from instance PI(*current*, $+y$, $[s]$) are included, then the outcome of the activation of job s is again successful and Phase 1 continues with the construction of (a new) preschedule $PreS(K)$ for kernel K by considering all the available jobs (including job s) without any further restriction.

(C2) If the above iteration during the construction of preschedule $PreS(K)$ does not occur, then either (C2.1) a new kernel K' including the corresponding type (a) y-job(s) arises or (C2.2) an IA(b2) with a Type (b) y-job occurs (see Proposition 12).

In Case (C2.1), Step 2 of Pass 1 returns kernel K' and calls PROCEDURE MAIN to update the current configuration (see the description of Pass 1 in Section 7.1).

In Case (C2.2), PROCEDURE ACTIVATE(s) completes with the failure outcome (then PROCEDURE sl-SUBSTITUTION(K), described in the next subsection, looks for another substitution job s' and calls repeatedly PROCEDURE ACTIVATE(s')).

This completes the description of PROCEDURE ACTIVATE(s). In the next subsection, we describe how we select a substitution job in the basic and inductive cases completing the description of Phase 2.

Lemma 11. *PROCEDURE ACTIVATE(s) creates an L-preschedule, for every reconstructed bin $B\{(s)\}, \ldots, B^-(K^-)$ with the cost of Phase 1.*

Proof. In this proof, we refer to a call of PROCEDURE ACTIVATE(s) from the condition of the lemma as the current call of that procedure; note that, for the inductive case, there should have been performed earlier calls of the same procedure within the current secondary block. In particular, prior to the current call of PROCEDURE ACTIVATE(s), every bin $B^-(\bar{K}) \in \{B\{(s)\}, \ldots, B^-(K^-)\}$ was (re)constructed directly at Phase 1 (one or more times). The current call reconstructs bin $B^-(\bar{K})$ (preschedule $PreS(\bar{K})$) once again. Recall also that problem instance PI(*current*, $+y$, $[s]$) contains additional job y and the Type (b) y-jobs preceding that job by the construction of the preschedule for kernel K at Pass 1 (these jobs were included prior to the occurrence of an IA(b2) with job y). All these Type (b) y-jobs for bin $B^-(K)$ become x-jobs for a bin $B^-(\bar{K})$ after the current call of PROCEDURE ACTIVATE(s).

Again, the activation procedure calls Phase 1, and by the construction of Phase 1, it will suffice to show that during the reconstruction of any of the bins $B^-(\bar{K})$, there will occur no Type (b) y-job that cannot be included in the newly created preschedule $PreS(\bar{K})$ (note that no such Type (a) y-job may arise). Let us now distinguish two kinds of Type (b) y-jobs for bin $B^-(\bar{K})$: a Type (b) y-job that was also a Type (b) y-job during the previous (re)construction of preschedule $PreS(\bar{K})$, and a newly arisen Type (b) y-job for bin $B^-(\bar{K})$, i.e., one that was earlier included as an x-job in a preceding preschedule $PreS(K')$ but which turned out to be a Type (b) y-job during the current construction of preschedule $PreS(\bar{K})$.

The lemma is obviously true if there exists no latter kind of a y-job for bin $B^-(\bar{K})$. To the contrary, suppose job x was scheduled in bin $B^-(K')$ (preceding bin $B^-(\bar{K})$) as an x-job, but it was forced to be rescheduled to (a later) bin $B^-(\bar{K})$ as an y-job during the current call of PROCEDURE

ACTIVATE(s). Then, during the current construction of preschedule $PreS(K')$ (the last call of PROCEDURE ACTIVATE(s) that has invoked Phase 1) a new x-job z was included before job x was considered at Pass 2 of Phase 1. By DEF-heuristics (Pass 2), this may only occur if a job scheduled in bin $B^-(K')$ at the previous call is not considered at the current call during the construction of that bin (the preschedule $PreS(K')$). Let N be the set consisting of all such jobs. By the definition of instance PI($current, +y, [s]$) and the activation procedure, a job in set N may be job s or a job which was left-shifted within the time intervals liberated by job s or by other left-shifted job(s).

Thus, job z has now occupied the time intervals within which job x and job(s) in set N were scheduled. $p_z \geq 2p_x$, as otherwise job x would have been considered and included in bin $B^-(K')$ ahead of job z by DEF-heuristics (recall that the smallest job processing time, larger than p_x is $2p_x$). Then, $p(N) < p_x$ is not possible, since otherwise $p(N) + p_x < 2p_x \leq p_z$ and the length of the released time intervals would not be sufficient to include job z in bin $B^-(K')$ (hence, job z would not push out job x). If $p(N) = p_x$, because of the divisibility of job processing times and by DEF-heuristics, job z may only push out job x if $p_z = 2p_x = 2p(N)$. Then, p_z is greater than the processing time of any job in set N. However, in this case, job z would have been included at the previous call in bin $B^-(\bar{K})$ ahead of job x and the jobs in set N since it is longer than any of these jobs, a contradiction.

If now at the current call $p(N) > p_x$, a job can be included ahead of job z in preschedule $PreS(K')$ within the time intervals earlier occupied by the jobs in set N. Let p', $p' \leq p(N)$, be the length of the remaining total idle-time intervals. If $p_z \leq p'$, then job z cannot push out job z since it fits within the remaining idle-time interval. If $p_z > p'$, then p_z must be no smaller than the smallest power of 2 greater than $p_x + p'$. Hence, job z cannot fit within the intervals of the total length of $p_x + p'$, and, again, it cannot pull out job x.

We showed that job z cannot exist, hence job x does not exist and PROCEDURE ACTIVATE(s) creates an L-preschedule for the bins $B\{(s)\}, \ldots, B^-(K^-)$. The cost of the procedure is the same as that of Phase 1 since the cost of the creation of problem instance PI($current, +y, [s]$) is obviously absorbed by the cost of Phase 1. □

9.2. Selecting a Substitution Job

Now, we describe PROCEDURE sl-SUBSTITUTION(K) that repeatedly activates different substitution jobs for an IA(b2) occurred at Phase 1 (using PROCEDURE ACTIVATE(s)) to determine one for which PROCEDURE ACTIVATE(s) completes with the successful outcome (whenever there exists such a substitution job). From here on, we refer to the original precedence order of the substitution jobs in the current secondary block $\mathcal{B}_K$ (their precedence order corresponding to the last configuration in which none of them were activated).

Lemma 12. *Suppose an IA(b2)/IA(b2-I) with job y arises and s' and s'' are the substitution jobs such that job s'' preceded job s'. Then, if the outcome of activation of job s' is the failure then outcome of activation of job s'' will also be the failure.*

Proof. Let j be any candidate job to be rescheduled before kernel K^-, i.e., $j = y$ or j is any of the Type (b) y-jobs included after kernel K^- before the above IA(b2)/IA(b2-I) with job y has occurred (see Proposition 16). Job j is released either: (1) before the (former) execution interval of job s'; or (2) within or after that interval. In Case (1), job j can immediately be included in bin $B\{(s')\}$. Moreover, as $p_{s'} > p_j$, if j cannot be included in bin $B\{(s')\}$, it can also not be included in any other bin before kernel K^- (one preceding bin $B\{(s')\}$). In Case (2), job j cannot be included before kernel K^- unless some jobs from bin $B\{(s')\}$ and the following bins are left-shifted within the idle-time interval released by job s' (releasing, in turn, the idle-time within which job j may be included). Again, since $p_{s'} > p_j$, job j will fit within the idle-time interval released by job s', given that all the intermediate jobs are "sufficiently" left-shifted. Since job s' succeeds job s'', the activation of job s' will left-shift these jobs no-less than the activation of job s'' (being a substitution job, s' is "long enough"). The lemma now obviously follows. □

Determining the sl-substitution job. We use the above lemma for the selection of a right substitution job. Let us call the shortest latest scheduled substitution job such that the outcome of its activation is successful, the *sl-substitution* job for job y. We show in Lemma 16 that, if there exists no sl-substitution job, there exists no L-schedule.

Our procedure for determining the sl-substitution job is easy to describe. PROCEDURE sl-SUBSTITUTION(K) (invoked for an IA(b2) with a Type (b) y-job from Phase 1 during the construction of preschedule $PreS(K)$) finds the sl-substitution job or otherwise returns the failure outcome. Iteratively, it calls PROCEDURE ACTIVATE(s) for the next substitution job s (a candidate for the sl-substitution job) until PROCEDURE ACTIVATE(s) delivers a successful outcome or all the candidate jobs (which may potentially be the sl-substitution job) are considered.

The order in which the candidate substitution jobs are considered is dictated by Lemma 12. Recall from Proposition 17 that a substitution job is at least as long as $\bar{p}_{\min\{y\}}$. Let $\bar{p} \geq \bar{p}_{\min\{y\}}$ be the minimum processing time no smaller than $\bar{p}_{\min\{y\}}$ of any yet unconsidered substitution job. PROCEDURE sl-SUBSTITUTION(K), iteratively, among all yet unconsidered substitution jobs with processing time $\bar{p}$ determines the latest scheduled substitution job s and calls PROCEDURE ACTIVATE(s) (see Lemma 12). If the outcome of PROCEDURE ACTIVATE(s) is successful, the outcome of PROCEDURE sl-SUBSTITUTION(K) is also successful and it returns job s (s is the sl-substitution job). Otherwise, if there exits the sl-substitution job, it is longer than job s. $\bar{p}$ is set to the next smallest processing time larger than the current $\bar{p}$, s becomes the latest scheduled substitution job with the processing time $\bar{p}$ and PROCEDURE ACTIVATE(s) is called again. The procedure continues in this fashion as long as the latest outcome is the failure and $\bar{p}$ can be increased (i.e., a substitution job with the processing time greater than that of the latest considered one exists). Otherwise, PROCEDURE sl-SUBSTITUTION(K) halts with the failure outcome.

Let μ be the number of non-kernel jobs in the current secondary block $\mathcal{B}_K$.

Lemma 13. *PROCEDURE sl-SUBSTITUTION finds the sl-substitution job or establishes that it does not exist by verifying at most $\log p_{\max}$ substitution jobs in time $O(\log p_{\max} \mu^2 \log \mu)$.*

Proof. The preprocessing step of PROCEDURE sl-SUBSTITUTION creates a list in which the substitution jobs are sorted in non-decreasing order of their processing times, whereas the jobs of the same processing time are included into the inverse precedence order of these jobs in that list. The preprocessing step takes time $O(\mu \log \mu)$.

Since the processing time of every next tried substitution job is larger than that of the previous one, the procedure works on $\log p_{\max}$ iterations (assuming that the processing times of the substitution jobs are powers of 2). By Lemma 12, among all the candidate substitution jobs with the same processing time, it suffices to consider only the latest scheduled one. For the failure outcome, by the same lemma, it suffices to consider the latest scheduled substitution job with the next smallest processing time (given that the procedure starts with the latest scheduled substitution job with the smallest processing time).

At every iteration, the corresponding bins from the current secondary block $\mathcal{B}_K$ are rebuilt at Phase 1. Applying Theorem 2 and the fact that different bins have no common jobs, we easily obtain that the cost of the reconstruction of all the bins at that iteration is $O(\mu^2 \log \mu)$ and hence the total cost is $O(\mu \log \mu + \log p_{\max} \mu^2 \log \mu) = O(\log p_{\max} \mu^2 \log \mu)$. □

10. More Examples

Before we prove the correctness of our algorithm for problem $1|p_j : divisible, r_j|L_{\max}$, we give final illustrations using the problem instances of Examples 1 and 2 and one additional problem instance, for which an IA(b2) arises. Recall that Figures 5 and 9 represent optimal solutions for the former two problem instances.

For the problem instance of Example 1, in the schedule of Figure 6 the collapsing of kernel K is complete and the decomposition procedure identifies the atomic kernel K^*; hence, the corresponding two bins are determined. The atomic kernel K^* consists of Job 1 with the lateness $-1 = L^*_{max}$. The binary search is carried out within the interval $[0, 13)$ ($\Delta = 16 - 3 = 13$). For $\delta = 7$, the L_δ-boundary is $-1 + 7 = 6$. At Phase 1, bins B_1 and B_2 are scheduled as depicted in the schedule of Figure 5 (in bin 1 only a single x-Job 2 at Pass 2 can be included, whereas bin B_2 is packed at Pass 1 with two y-Jobs 3 and l). Hence, the L-schedule of Figure 5 for $L = 6$ is successfully created. For the next $\delta = 4$, the L_δ-boundary is $-1 + 4 = 3$. Bin B_1 is scheduled similarly at the iteration with $\delta = 7$; while scheduling bin B_2 at Phase 1, an IA(b2) with y-Job 3 occurs (since its lateness results to be greater than 3), but there exists no substitution job. Hence, there exists no L_δ-schedule for $\delta = 4, L = 3$. Phase 1 will complete with the similar outcome for the iteration in the binary search with $\delta = 6$, and the algorithm halts with the earlier obtained feasible solution for $\delta = 7$.

For the problem instance of Example 2, the schedule of Figure 8 represents the result of the decomposition of both arisen kernels K^1 and K^2 (kernel K^2 arises once the decomposition of kernel K^1 is complete and bin B_1 gets scheduled). We have $L_{max}(K^{1*}) = L_3(\sigma_{l,2}) = 19 - 20 = -1$, whereas $\Delta = 32 - 3 = 29$. For $\delta = 0$, bin B_1 may contain only Job 1. Once bin B_1 is scheduled, the second kernel K^2 arises. The result of its collapsing is reflected in Figure 8. We have $L^*_{max} = L_{max}(K^{2*}) = L_6(\sigma_{l,2,4}) = 62 - 58 = 4$. Then, $\delta(K^{1*}) = 5$ (while $\delta(K^{2*}) = 0$), and an extra delay of 5 is now allowed for kernel K^1. Note that the current secondary block $\mathcal{B}_{K^2}$ includes all the three bins. For $\delta = 0$, bin B_1 is newly rescheduled and at Pass 2 of Phase 1 an x-Job 7 is now included in that bin (due to the allowable extra delay for kernel K^{1*}). No other job besides Job l can be included in bin B_2, and the last bin B_3 is formed by Job 4. A complete L-schedule (with $L_\delta = L^*_{max} + \delta = 4 + 0 = 4$) with the objective value equal to a lower bound 4 is successfully generated (see Figure 9).

Example 4. *In this example, we modify the problem instance of Example 2. The set of jobs is augmented with one additional Job 8, and the parameters of Jobs 4 and 7 are modified as follows:*
$r_4 = 0, p_4 = 8, d_4 = 66$,
$r_7 = 0, p_7 = 4, d_7 = 60$,
$r_8 = 0, p_8 = 4, d_8 = 63$.

Figure 10 represents the last step in the decomposition of kernel K^1, which is the same as for the problem instance of Example 2 (the schedules represented in Figures 10–13 have different scaling due to the differences in their lengths). This decomposition defines the two bins surrounding atomic kernel K^{1*}. The binary search is invoked for $\delta = 0$; since K^{1*} is the only detected kernel so far, $\delta(K^{1*}) = 0$ and $L_\delta = L_3(S^*[K^1]) = -1$. The first bin is successfully packed with an additional external x-Job 7 at Pass 2 of Phase 1 (since there exists no y-job, Pass 1 is not invoked). PROCEDURE MAIN proceeds by applying ED-heuristics from Time 21 during which the second kernel K^2 arises. Figure 11 represents the resultant partial schedule with the first packing of bin B^1 and kernel K^2. Figure 12 represents the result of the full decomposition of kernel K^2 (which is again the same as for the problem instance of Example 2). Now, $\delta(K^{2*}) = 0$ and $\delta(K^{1*}) = 5$. Bin B_1 is repacked, in which a longer x-Job 4 can now be included, and bin B_2 at Phase 1 is packed (at Pass 1 an y-Job 2, and at Pass 2 an x-Job l is included in that bin, see Figure 12). PROCEDURE MAIN is resumed to expand the current partial schedule, but now an IA(b2) with the earliest included Job 7 arises (as its resultant lateness is $66 - 60 = 6 > 4$). Job 4 from bin B_1 is the sl-substitution job. The result of its activation is reflected in Figure 13: bin B_1 is repacked now with x-Jobs 7 and 8, bin B_2 remains the same, and the last Job 4 is included in bin B_3 at Phase 0, yielding a complete S^{L_0}-schedule with the optimal objective value 4 (both Jobs 6 and 4 realize this optimal value).

Figure 10. Full decomposition of kernel K^1.

Figure 11. Extended schedule $S^L(current, K^1)$ with kernel K^2.

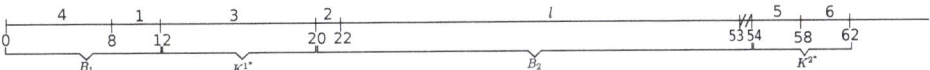

Figure 12. In schedule $S^L(current, K^2)$ Job 7 cannot be included.

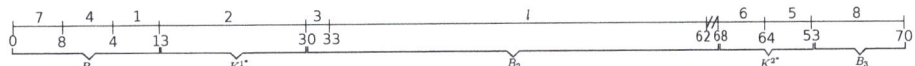

Figure 13. An optimal schedule S^L in which bins B_1, B_2 and $\bar{B}_3$ are successfully repacked.

11. Correctness of the Framework for Jobs with Divisible Processing Times

In Section 8, we show that, for an instance of $1|p_j : divisible, r_j|L_{\max}$, the current secondary block is kept active at Phase 1 (Corollary 4). Now, we generalize this result proving a similar statement for a secondary block that gets reconstructed at Phase 2 (we cannot affirm a similar property for an instance of $1|r_j|L_{\max}$, a reason PROCEDURE MAIN may not provide an optimal solution to the general problem). For the commodity in the proof that we present here, we introduce a few new definitions.

First, we observe that a call of PROCEDURE ACTIVATE(s) may create the new, so-called *critical gap(s)* in a reconstructed preschedule in the current secondary block $\mathcal{B}_K$. To every critical gap in secondary block $\mathcal{B}_K$, the substitution job from set $\mathcal{S}_B$ which activation has yielded that gap, corresponds. Denote by $CG(s)$ the set of all the currently remaining (yet unused at that configuration) critical gaps yielded by the activation of a substitution job s; let $|CG(s)|$ be the total length of these gaps.

A substitution job $s \in \mathcal{S}_B$ is *stable* if $|CG(s)| = 0$. When a substitution job s is activated, the total length of the critical gaps arisen after its activation depends, in particular, on p_s. For example, in the basic case, or in the inductive case if substitution jobs in $\mathcal{S}_B$ are stable, the new critical gaps with the total length $p_s - p_y$ will arise, where y is the y-job for which s was activated.

If an activated substitution job s is non-stable and during a later call of PROCEDURE ACTIVATE(s'), $s' \ne s$, some y-job within the interval of the gaps in $CG(s)$ is included, $|CG(s)|$ will be reduced. In this way, job s may eventually become stable.

For a substitution job $s' \in \mathcal{S}_B$, we let $Y(s')$ be the set of all the newly included y-jobs in the reconstructed bins after a call of PROCEDURE ACTIVATE(s') (see Lemma 11).

Suppose a call of PROCEDURE ACTIVATE(s) (succeeding an earlier call of PROCEDURE ACTIVATE(s')) includes job s' before all the y-jobs from set $Y(s')$. Then, job s' is said to be *inactivated*. The intuition behind this definition is that job s' will not necessity remain in the state of activation for all jobs from set $Y(s')$ in the case that the activation of a new substitution job s gives a sufficient room for a proper accommodation of the jobs in set $Y(s')$ (this is rectified in more details in the proof below). At the same time, we may note that job s' may not be included in any of the newly reconstructed bins and neither in bin $B^-(K)$ (then it eventually will be included within a succeeding bin of the secondary block $\mathcal{B}_K$).

Lemma 14. *At Phase 2, the current secondary block is kept compact given that for every occurrence of an IA(b2-I) the corresponding sl-substitution job exists.*

Proof. For the basic case, before the activation of the corresponding sl-substitution job, say s_1, the critical block $\mathcal{B}_K$ is compact by Corollary 4. Since s_1 is the sl-substitution job, block $\mathcal{B}_K$ will remain compact after the above activation. We are brought to the inductive case if an IA(b2) repeatedly arises in the above block, the first occurrence of an IA(b2-I) with the number of the activated substitution jobs $k = 1$.

We proceed with the proof using the induction on the number of the activated substitution jobs. We now prove our claim for $k = 2$, in the case that the second substitution job s_2 is activated in the current secondary block. Consider the following possibilities. Originally, job s_2 either: (i) succeeded job s_1; or (ii) preceded job s_1.

In Case (i), if $p_{s_2} \geq 2p_{s_1}$, all the y-jobs already included within $CG(s_1)$ together with jobs in set Λ_{y_2} can be feasibly scheduled within $CG(s_2)$ as $p_{s_2} \geq 2p(\Lambda_{y_2})$. Hence, after a call of PROCEDURE ACTIVATE(s_2) job s_1 will be inactivated at Pass 3 (see Lemma 11). Hence, job s_1 becomes stable and we are left with a single substitution job s_2 in the state of activation.

In Case (ii), note that no job from set Λ_{y_2} was included within $CG(y_1)$ after a call of PROCEDURE ACTIVATE(s_1). Hence, $|CG(s_1)| < p(\Lambda_{y_2})$. If job s_2 is long enough and all jobs in $Y(s_1)$ are released early enough and can fit within the newly released space by job s_2 after a call of PROCEDURE ACTIVATE(s_2), once Pass 3 of the activation procedure completes, job s_1 will again become stable and we are again left with a single substitution job s_2 in the state of activation.

Since in the above considered cases, the only non-stable substitution job is s_2, our claim follows from case $k = 1$ and the fact that s_2 is the sl-substitution job. It only remains to consider the cases when job s_1 remains in the state of the activation after a call of PROCEDURE ACTIVATE(s_2), i.e., both substitution jobs s_1 and s_2 remain in the state of activation. This happens in Case (i) if $p_{s_2} \leq p_{s_1}$ (note that in this case $p_{s_2} \leq p(\Lambda_{y_2})$ also holds as otherwise job s_2, instead of job s_1, would have been selected as the sl-substitution job for job y_1). Either jobs in set Λ_{y_2} are not released early enough to be included within $CG(s_1)$ or $|CG(s_1)|$ is not large enough. Hence, another substitution job needs to be activated to include jobs in Λ_{y_2} (see Lemma 15 below). Since s_2 is the sl-substitution job, $|CG(s_2)|$ is the minimal possible. The lemma holds if job s_1 again becomes stable. Otherwise, note that, since both s_1 and s_2 are the sl-substitution jobs, the only remaining possibility to be considered is when a single substitution job s with $p_s < p_{s_1} + p_{s_2}$ (instead of jobs s_1 and s_2) is activated.

Consider the following sub-cases: (1) $|CG(s_1)| \geq p(\Lambda_{y_2})$; and (2) $|CG(s_1)| < p(\Lambda_{y_2})$. In Case (1). jobs in Λ_{y_2} are not released early enough to be included within $CG(s_1)$ as otherwise they would have been included by an earlier call of PROCEDURE ACTIVATE(s_1). Hence, no job preceding originally job s_1 can be beneficially activated. At the same time, any substitution job succeeded originally job s_1 is longer than s_1 (by the definition job s_1 and PROCEDURE sl-SUBSTITUTION). Then, $p_s \geq 2p_{s_1}$ because of the divisibility of job processing times. In Case (2), $p_s \geq 2p_{s_1}$ must also hold as otherwise all jobs in set $Y(s_1)$ together with jobs in set Λ_{y_2} would not fit within the time intervals that potentially might be liberated by a call of PROCEDURE ACTIVATE(s).

Hence, in both Cases (1) and (2) above, $p_s < p_{s_1} + p_{s_2}$ is not possible and hence the activation of job s will yield the critical gaps with a total length no less than our procedure yields, and the lemma follows. The proof for the Case (ii) when the jobs in $Y(s_1)$ do not fit within $CG(s_2)$ or they are not released early enough is quite similar to case (i) above (the roles of jobs s_1 and s_2 being interchanged).

For the inductive pass with $k \geq 3$, let s_k be the next activated sl-substitution job and let $S_B = \{s_1, \ldots, s_{k-1}\}$ be the substitution jobs in the state of activation in the current critical block B). By the inductive assumption, block B was compact before job s_k is activated. Now, we show that the block remains compact once job s_k is activated. This follows if s_k, as before, remains the only (non-stable) substitution job in the state of activation after a call of PROCEDURE ACTIVATE(s_k). Otherwise, originally, job s_k: (i) succeeded all the jobs $\{s_1, \ldots, s_{k-1}\}$; (ii) preceded these jobs; or (iii) was scheduled in between their original positions. We use similar arguments as for $k = 2$. We give a scratch.

In Case (ii), note that the time intervals released by a call of PROCEDURE ACTIVATE(s_k) will be available for the jobs from set $Y(s_1) \cup \cdots \cup Y(s_{k-1})$ during the execution of the procedure at Pass 2 of Phase 1, and they may potentially be left-shifted to these intervals. Because of the mutual divisibility of processing times of these jobs and by the construction of Pass 2, the total length of the remaining idle-time intervals, if any, will be the minimal possible (this can be straightforwardly seen). It follows that, at Pass 3, the corresponding jobs from $\mathcal{S}_\mathcal{B}$ will become inactivated and hence stable, whereas the rest of them are to stay in the state of activation, and our claim follows from the inductive assumption.

In Case (i), all jobs from $Y(s_1) \cup \cdots \cup Y(s_{k-1})$ are released early enough to be included within the intervals newly released by a call of PROCEDURE ACTIVATE(s_k). Again, because of the mutual divisibility of processing times of these jobs and by the construction of Pass 2, the remaining idle-time intervals, if any, will be the minimal possible, and at Pass 3 the corresponding substitution jobs will be inactivated.

The proof of Case (iii) merely combines those for Cases (i) and (ii): at Pass 2, the intervals released by a call of PROCEDURE ACTIVATE(s_k) might be used by jobs from $Y(s_1) \cup \cdots \cup Y(s_{k-1})$ preceding and also succeeding these intervals, and the corresponding jobs from $\{s_1, \ldots, s_{k-1}\}$ will again become stable. □

Lemma 15. *Suppose an IA(b2)/IA(b2-I) with job y during the construction of preschedule PreS(K) arises and there exists an L-schedule S^L. Then, a substitution job is scheduled after kernel K^- in schedule S^L. That is, there exists no L-schedule if there exists no substitution job.*

Proof. The lemma is a kind of reformulation of Proposition 16. For the basic case, before the activation of the sl-substitution job s_1, the secondary block $\mathcal{B}_K$ is compact by Corollary 4. Similar to in the proof of Proposition 16, we can see that the current starting time of job y cannot be reduced by any job rearrangement that leaves the same set of jobs scheduled before job y. Hence, some emerging x-job s from one of the bins from the secondary block $\mathcal{B}_K$ pushing job y is included behind job y in schedule S^L (recall that $d_s > d_y$ must hold as, otherwise, once rescheduled after kernel K^-, job s will surpass the L-boundary or will force another y-job to surpass it). Job s cannot be from bin $B^-(K)$ since no x-job can be included ahead of job y during the construction of $PreS(K)$ as job y is released from the beginning of that construction (and it would have been included at Pass 1 of Phase 1 before any x-job is considered at Pass 2). Therefore, job s belongs to one of the bins preceding bin $B^-(K)$ in block $\mathcal{B}_K$. The proof for the inductive case is similar except that it uses Lemma 14 instead of Corollary 4. □

Lemma 16. *If there exists no sl-substitution job, then no L-schedule exists.*

Proof. If there exists no substitution job at all, then the statement follows from Lemma 15. Otherwise, the outcome of the activation of every tried substitution is the failure. We claim that there exists no L-preschedule that contains the jobs from problem instance PI(current, +y, [s]) together with all the jobs from all the (intermediate) kernels between the bins $B\{(s)\}$ and $B^-(K^-)$. Let s be the earliest tried substitution job by PROCEDURE sl-SUBSTITUTION(K). If job s becomes non-stable after a call of PROCEDURE ACTIVATE(s), then, due to the failure outcome, it must be the case that the corresponding y-job(s) (see Proposition 16) cannot be left-shifted within the time intervals liberated by job s (because of their release times). Hence, neither they can be left-shifted by activation of any substitution job preceding job s (Lemma 12). Otherwise, it must have been stable once activated, but the interval released by job s is not long enough (again, due to the failure outcome). Hence, only another longer substitution job may be of a potential benefit, whereas the latest scheduled one, again, provides the maximum potential left-shift for the above y-job(s). We continue applying this argument to every next tried substitution job. Our claim and hence the lemma follow due to the failure outcome for the latest tried (the longest) substitution job. □

Now, we immediately obtain the following corollary that already shows the correctness of PROCEDURE MAIN for divisible processing times:

Corollary 5. *For every trial δ, PROCEDURE MAIN generates an L_δ-schedule if the outcome of every call of PROCEDURE sl-SUBSTITUTION(K) for an IA(b2) is successful (or no IA(b2) arises at all); otherwise (there exists no sl-substitution job for some IA(b2)), no L_δ-schedule exists.*

Theorem 6. *PROCEDURE MAIN optimally solves problem $1|p_j\ :\ divisible, r_j|L_{\max}$ in time $O(n^3 \log n \log^2 p_{\max})$.*

Proof. The soundness part immediately follows from Corollary 5 and the definition of the binary search in Section 5 (see Proposition 8). We show the time complexity. Due to Theorem 3, it remains to estimate an additional cost yielded by Phase 2 for instances of alternative (b2). Recall from Theorem 2 that, for every arisen kernel K, the cost of the generation of L_δ-augmented schedule $S^{L_\delta}[K]$ for a given δ is $O(\nu^2 \log \nu)$, where ν is the total number of jobs in bin $B^-(K)$. Recall also that this cost includes the cost of all the embedded recursive calls for all the kernels which may arise within bin $B^-(K)$. Similar to in the proof of Theorem 3, it suffices to distinguish the calls of PROCEDURE AUGMENTED(K,δ) and PROCEDURE AUGMENTED(M,δ) for two distinct kernels K and M such that bins $B^-(K)$ and $B^-(M)$ have no jobs in common. Now, we count the number of such calls of PROCEDURE AUGMENTED(K,δ) from Phase 2 by PROCEDURE sl-SUBSTITUTION(K). The number of times, an IA(b2) at Phase 1 may arise is bounded by ν_1, the number of Type (b) y-jobs (note that any Type (b) y-job may yield at most one IA(b2)). Hence, for any bin $B^-(K)$, PROCEDURE AUGMENTED(K,δ) may be called less than ν_1 times for different instances of Alternative (b2), whereas for the same IA(b2) no more than $p_{\max}$ different substitution jobs might be tried (Lemma 13). Hence, the total number of calls of PROCEDURE AUGMENTED(K,δ) is bounded above by $O(\nu_1 + p_{\max})$, which easily yields the overall bound $O(n^3 \log n \log^2 p_{\max})$. □

12. Possible Extensions and Applications

We describe our framework for the single-machine environment and with a due-date oriented objective function $L_{\max}$. It might be a subject of a future research to adopt and extend the proposed framework for other machine environments with this or another due-date oriented objective function. Both the recurrence substructure properties and the schedule partitioning into kernel and bin intervals can be extended for the identical machine environment and shop scheduling problems with job due-dates. Less straightforward would be its adaptation for the uniform machine environment, and, unlikely, the approach can be extended to the unrelated machine environment.

The framework can obviously be converted to a powerful heuristic algorithm, as well as to an exact implicit enumeration scheme for a general setting with arbitrary job processing times. For both heuristic and enumerative approaches, it will clearly suffice to augment the framework with an additional search procedure invoked for the case when the condition of Theorem 3 is not satisfied.

Based on the constructed framework, we have obtained an exact polynomial-time algorithm for problem $1|p_j : divisible, r_j|L_{\max}$. A natural question is whether, besides the scheduling and bin packing problems ([4]), there are other NP-hard combinatorial optimization problems for which restrictions with divisible item sizes are polynomially solvable (the properties of mutually divisible numbers exploited in reference [4] and here could obviously be helpful).

Finally, we argue that scheduling problems with divisible job processing times may naturally arise in practice. As an example, consider the problem of distribution of the CPU time and the computer memory, the basic functions of the operating systems. In Linux operating systems buddy memory allocation is used, in which memory blocks of sizes of powers of 2 are allocated. To a request for memory of size K, the system allocates a block of size 2^k where $2^{k-1} < K \leq 2^k$ (if currently there is no available block of size 2^k, it splits the shortest available block of size 2^{k+1} or more). In buddy systems, memory allocation and deallocation operations are naturally simplified, as an $O(n)$ time search is reduced to $O(\log n)$ time using binary tree representation for blocks.

A similar "buddy" approach for the CPU time sharing in operating systems would assume the "rounding" of the arriving requests with arbitrary processing times within the allowable patterns of processing times—the powers of 2. In the CPU time sharing, the system must decide which of the arriving requests to assign to the processor and when. The request may arrive over time or, in the case of the scheduled maintenance and other scheduled computer services (for example, operating system updates), the arrival time of the requests and their processing times are known in advance. The latter scenario fits into our model. One may think on the rounding of a processing time of a request up or down to a closer power of 2. Alternatively, to avoid unnecessary waste of the processor time, one may always round down and process the remaining small part in a parallel or sequential manner immediately upon the completion of the main part or later on. Possible efficient and practical strategies for "completing" the solution with divisible processing times in a single-processor or multiprocessor environment deserves an independent study.

The "buddy" approach for the CPU time sharing in operating systems is justified by our results, as we show that the scheduling problems with mutually divisible processing times can be solved essentially more efficiently than with arbitrary job processing times. The degree of the "waste" during the rounding of the memory blocks and processing requirements is somewhat similar and comparable in both the memory allocation and the CPU time sharing methods. In the case of the memory allocation we may waste an extra memory, and in the case of the time sharing we waste an extra time (which would influence on the quality of the solution of course). It is important and not trivial how an input with arbitrary job processing times can be converted to an input with divisible processing times, and how close the obtained optimal solution for the instance with divisible times will be from an optimal solution for the original instance. This interesting topic can be a subject of a future independent study.

Funding: This research received no external funding.

Conflicts of Interest: The author declares no conflict of interest.

References

1. Garey, M.R.; Johnson, D.S. *Computers and Intractability: A Guide to the Theory of NP-Completeness*; Freeman: San Francisco, CA, USA, 1979.
2. Vakhania, N.; Werner, F. Minimizing maximum lateness of jobs with naturally bounded job data on a single machine in polynomial time. *Theor. Comput. Sci.* **2013**, *501*, 72–81, doi:10.1016/j.tcs.2013.07.001. [CrossRef]
3. Graham, R.L.; Lawler, E.L.; Lenstra, J.K.; Kan, A.H.G.R. Optimization and approximation in deterministic sequencing and scheduling: A survey. *Ann. Discret. Math.* **1979**, *5*, 287–328.
4. Coffman, E.G., Jr.; Garey, M.R.; Johnson, D.S. Bin packing with divisible item sizes. *J. Complex.* **1987**, *3*, 406–428. [CrossRef]
5. McMahon, G.; Florian, M. On scheduling with ready times and due-dates to minimize maximum lateness. *Oper. Res.* **1975**, *23*, 475–482. [CrossRef]
6. Carlier, J. The one-machine sequencing problem. *Eur. J. Oper. Res.* **1982**, *11*, 42–47. [CrossRef]
7. Jackson, J.R. Scheduling a production line to minimize the maximum lateness. In *Management Science Research Report 43*; University of California: Los Angeles, CA, USA, 1955.
8. Horn, W.A. Some simple scheduling algorithms. *Naval Res. Logist. Q.* **1974**, *21*, 177–185. [CrossRef]
9. Garey, M.R.; Johnson, D.S.; Simons, B.B.; Tarjan, R.E. Scheduling unit-time tasks with arbitrary release times and deadlines. *SIAM J. Comput.* **1981**, *10*, 256–269. [CrossRef]
10. Vakhania, N. Single-Machine Scheduling with Release Times and Tails. *Ann. Oper. Res.* **2004**, *129*, 253–271. [CrossRef]
11. Lazarev, A.A.; Arkhipov, D.I. Minimization of the Maximal Lateness for a Single Machine. *Autom. Remote Control.* **2016**, *77*, 656–671. [CrossRef]
12. Vakhania, N. Fast solution of single-machine scheduling problem with embedded jobs. *Theor. Comput. Sci.* **2019**. [CrossRef]

13. Vakhania, N. A better algorithm for sequencing with release and delivery times on identical processors. *J. Algorithms* **2003**, *48*, 273–293. [CrossRef]
14. Schrage, L. Obtaining optimal solutions to resource constrained network scheduling problems, March 1971. Unpublished manuscript.

© 2019 by the author. Licensee MDPI, Basel, Switzerland. This article is an open access article distributed under the terms and conditions of the Creative Commons Attribution (CC BY) license (http://creativecommons.org/licenses/by/4.0/).

Article

Online Batch Scheduling of Simple Linear Deteriorating Jobs with Incompatible Families

Wenhua Li [1,*], Libo Wang [1], Xing Chai [2] and Hang Yuan [3]

1. School of Mathematics and Statistics, Zhengzhou University, Zhengzhou 450001, China; wanglibo0224@163.com
2. College of Science, Henan University of Technology, Zhengzhou 450001, China; chaixingstudy@163.com
3. Department of Economics, State University of New York at Binghamton, Binghamton, NY 13902, USA; hyuan3@binghamton.edu
* Correspondence: liwenhua@zzu.edu.cn

Received: 31 December 2019; Accepted: 24 January 2020; Published: 1 February 2020

Abstract: We considered the online scheduling problem of simple linear deteriorating job families on m parallel batch machines to minimize the makespan, where the batch capacity is unbounded. In this paper, simple linear deteriorating jobs mean that the actual processing time p_j of job J_j is assumed to be a linear function of its starting time s_j, i.e., $p_j = \alpha_j s_j$, where $\alpha_j > 0$ is the deterioration rate. Job families mean that one job must belong to some job family, and jobs of different families cannot be processed in the same batch. When $m = 1$, we provide the best possible online algorithm with the competitive ratio of $(1 + \alpha_{\max})^f$, where f is the number of job families and $\alpha_{\max}$ is the maximum deterioration rate of all jobs. When $m \geq 1$ and $m = f$, we provide the best possible online algorithm with the competitive ratio of $1 + \alpha_{\max}$.

Keywords: online algorithm; batch scheduling; linear deterioration; job families; competitive ratio

1. Introduction

1.1. Background

In this paper, all jobs arrive over time, i.e., each job has an arrival time. Before the jobs arrive, we do not know any information, including arrival time, processing time, deterioration rate, etc. Due to the unknown information of the jobs, the online algorithm is not guaranteed to be optimal. Borodin and El-Yaniv [1] used the competitive ratio to measure the quality of an online algorithm. For a minimization scheduling problem, we define the competitive ratio of the online algorithm A as:

$$\rho = \sup\{A(\mathcal{I})/OPT(\mathcal{I}) : \mathcal{I} \text{ is any instance such that } OPT(\mathcal{I}) > 0 \}.$$

where $\mathcal{I}$ is any job instance and $A(\mathcal{I})$ and $OPT(\mathcal{I})$ are the objective values obtained from the algorithm A and an optimal offline scheduling OPT, respectively. In this study, the objective was to minimize the makespan. An online algorithm A is called the best possible if no other online algorithms A^* produce a smaller competitive ratio.

Parallel-batch means that one batch processing machine can process b jobs simultaneously as a batch. The processing time of a batch is the maximum processing time of all jobs in this batch. All jobs in a batch have the same starting time, processing time, and completion time. According to the number of jobs contained in a batch, Brucker et al. [2] divided the model into two cases: the unbounded model ($b = \infty$) and the bounded model ($b < \infty$).

Job families mean that one job must belong to some job family, and jobs of different families cannot be processed in the same batch. Online scheduling problems on parallel batch machines with incompatible job families have been studied extensively. Fu et al. [3] studied the online algorithm on a single machine to minimize the makespan. Li et al. [4] examined the online scheduling of incompatible unit-length job families with lookahead on a single machine. Tian et al. [5] analyzed the problem on m parallel machines. However, research is lacking on the parallel-batch online scheduling with incompatible deteriorating job families.

Traditional scheduling problems assume that the processing time of a job is fixed. However, in real life, one job will take longer when it has a later starting time. For example, in steel production and financial management [6,7], the processing time is longer when it starts later. In the steel-making process, strict requirements are placed on temperature. If the waiting time is too long, the temperature of molten steel will drop. So, it will take time to heat up again before further processing. Other examples are provided in cleaning and fire fighting. The scheduling problem of deteriorating jobs was first introduced by Browne and Yechiali [8] and Gupta and Gupta [9], independently. Both considered minimizing makespan on a single machine. Since then, this topic has attracted considerable attentions. Gawiejnowicz and Kononov [10] considered the general properties of scheduling with fixed job processing time and scheduling with job processing time as proportional linear functions of the job starting time. The relevant research includes [11–18], among many others. Recently, some works have been published about online algorithms for linear deteriorating jobs [19–23].

To minimize the makespan of the online scheduling problem on m parallel machines with linear deteriorating jobs, Cheng et al. [19] constructed an algorithm and proved that the bound of the competitive ratio of the algorithm is tight, where $m = 2$ and the largest deterioration rate of jobs is known in advance. Yu et al. [22] proved that no deterministic online algorithm is better than $(1 + \alpha_{max})$-competitive when $m = 2$, where α_{max} is the maximum deterioration rate of all jobs.

1.2. Research Problem

Our contribution is to extend the online scheduling problem on m parallel batch machines with simple linear deteriorating job families to minimize the makespan. Here, batch capacity is unbounded, i.e., $b = \infty$. We use f-family to denote there are f job families. We constructed the best possible online algorithm with the competitive ratio of $(1 + \alpha_{max})^f$ when $m = 1$, where f is the number of job families and α_{max} is the maximum deterioration rate of all jobs. When $m \geq 1$ and $m = f$, we created the best possible online algorithm with the competitive ratio of $1 + \alpha_{max}$.

We examined the online batch scheduling of simple linear deteriorating job families. The actual processing time p_j of job J_j is assumed to be a linear function of its starting time s_j, i.e., $p_j = \alpha_j s_j$, where $\alpha_j > 0$ is the deterioration rate, which is unknown until it arrives. The objective was to minimize makespan. Assume that the arrival time of all jobs is greater than or equal to $t_0 > 0$; otherwise, jobs arriving at time 0 can be completed at time 0. We used the three-field notation $\alpha|\beta|\gamma$ [24] to represent one scheduling problem.

This paper is organized as follows. In Section 2, we consider the problem $1|\text{online},r_j,\text{p-batch}, b = \infty,\text{f-family}, p_j = \alpha_j t|C_{max}$, where f is the number of job families. We prove the lower bound and provide the best possible online algorithm with the competitive ratio of $(1 + \alpha_{max})^f$. In Section 3, we consider the problem $Pm|\text{online},r_j,\text{p-batch}, b = \infty,\text{m-family}, p_j = \alpha_j t|C_{max}$, where m is the number of machines. We prove the lower bound and provide the best possible online algorithm with the competitive ratio of $1 + \alpha_{max}$, where α_{max} is the maximum deterioration rate of all jobs.

Throughout this paper, we use σ and π to denote the schedules obtained from an online algorithm and an optimal offline schedule, respectively. Let $C_{max}(\sigma)$ and $C_{max}(\pi)$ be the objective values of σ and π, respectively, and α_{max} be the maximum deterioration rate of all jobs. Let ϵ be an arbitrary small positive number.

2. Single Batch Machine ($m = 1$)

In this section, we consider the online scheduling on an unbounded batch machine and the jobs belong to f incompatible deteriorating job families. The number of job families, f, is known in advance.

We prove the lower bound and provide the best possible online algorithm with the competitive ratio of $(1+\alpha_{max})^f$.

Theorem 1. *For problem $1|online, r_j, p\text{-}batch, b=\infty, f\text{-}family, p_j = \alpha_j t|C_{max}$, the competitive ratio of any online algorithm is not less than $(1+\alpha_{max})^f$.*

Proof. Let H be any online algorithm and I be a job instance provided by the adversary. In instance I, all the jobs have the same deterioration rate of α.

At time t_0, f jobs from the f different job families arrive by the adversary. If a job is scheduled by H to process at time t and t is in the time interval $[t_0, (1+\alpha)^f t_0)$, then at time $t+\epsilon$, the adversary releases a copy of this job, which belongs to the same job family. Let s be the starting time of the first job whose completion time is at least $(1+\alpha)^f t_0$. $(1+\alpha)s \geq (1+\alpha)^f t_0$. so,

$$s \geq (1+\alpha)^{f-1} t_0. \tag{1}$$

Case 1 $(1+\alpha)^{f-1} t_0 \leq s < (1+\alpha)^f t_0$.

In this case, there are still f jobs from f distinct job families that are not processed at time $(1+\alpha)s$. So,

$$C_{max}(\sigma) \geq (1+\alpha)^f (1+\alpha)s = (1+\alpha)^{f+1} s. \tag{2}$$

We assume that the jobs processed in the time interval $[t_0, (1+\alpha)s)$ belong to k distinct job families, say, $\mathcal{F}_1, \mathcal{F}_2, \cdots, \mathcal{F}_k$, where $1 \leq k \leq f$. The other $f-k$ job families are defined by $\mathcal{F}_{k+1}, \mathcal{F}_{k+2}, \cdots, \mathcal{F}_f$. Let s_i be the last starting time of the jobs in $\mathcal{F}_i$ that start before or at time s for $1 \leq i \leq k$, and satisfy $s_1 < s_2 < \cdots < s_k$. Clearly, $s_k = s$. From the construction of instance I, we know that the last arrival time of the jobs in $\mathcal{F}_i$ ($1 \leq i \leq k$) is $s_i + \epsilon$ and the arrival time of the jobs in $\mathcal{F}_i$ ($k+1 \leq i \leq f$) is t_0.

Construct a schedule π' below: the jobs in $\mathcal{F}_i$ ($1 \leq i \leq f$) form a batch starting at time s_i', where:

$$s_i' = \begin{cases} (1+\alpha)^{i-(k+1)} t_0, & k+1 \leq i \leq f \\ \max\{s_i + \epsilon, (1+\alpha)^{(f+i)-(k+1)} t_0\}, & 1 \leq i \leq k. \end{cases}$$

We can see that π' is feasible, and the maximum completion time of the jobs in π' is the completion time of the jobs in $\mathcal{F}_k$. So,

$$C_{max}(\pi') = (1+\alpha) \max\{s_k + \epsilon, (1+\alpha)^{f-1} t_0\}. \tag{3}$$

Since $s_k = s$, we have $C_{max}(\pi') = \max\{(1+\alpha)(s+\epsilon), (1+\alpha)^f t_0\}$. and $C_{max}(\pi) \leq C_{max}(\pi')$. If $C_{max}(\pi') = (1+\alpha)(s+\epsilon)$, then by Equation (2) we know that:

$$\frac{C_{max}(\sigma)}{C_{max}(\pi)} \geq \frac{(1+\alpha)^{f+1} s}{(1+\alpha)(s+\epsilon)} \to (1+\alpha)^f = (1+\alpha_{max})^f, \epsilon \to 0.$$

If $C_{max}(\pi') = (1+\alpha)^f t_0$, then by Equations (1) and (2), we know that:

$$\frac{C_{max}(\sigma)}{C_{max}(\pi)} \geq \frac{(1+\alpha)^{f+1} s}{(1+\alpha)^f t_0} = \frac{(1+\alpha)s}{t_0} \geq (1+\alpha)^f = (1+\alpha_{max})^f.$$

Case 2 $s \geq (1+\alpha)^f t_0$.

According to the constructing of I, s_k is the last starting time of the jobs in time interval $[t_0, (1+\alpha)^f t_0)$, and $s_k + \epsilon$ is the last arrival time of all jobs.

Since s is the starting time of the first job whose completion time is at least $(1+\alpha)^f t_0$, we obtain $(1+\alpha)s_k < (1+\alpha)^f t_0$, i.e., $s_k < (1+\alpha)^{f-1} t_0$. From Equation (3), we have:

$$\begin{aligned} C_{\max}(\pi') &= \max\{(1+\alpha)(s_k+\epsilon), (1+\alpha)^f t_0\} \\ &\to \max\{(1+\alpha)s_k, (1+\alpha)^f t_0\} \\ &= (1+\alpha)^f t_0, \end{aligned}$$

as $\epsilon \to 0$.

By the definition of s, f jobs from f distinct job families have not been processed at time s. So,

$$C_{\max}(\sigma) \geq (1+\alpha)^f s.$$

Thus,

$$\frac{C_{\max}(\sigma)}{C_{\max}(\pi)} \geq \frac{(1+\alpha)^f s}{(1+\alpha)^f t_0} = \frac{s}{t_0} \geq (1+\alpha)^f = (1+\alpha_{\max})^f.$$

The result follows. □

Before introducing the online algorithm, given a batch B, we define some notations in the following:

$J(B)$: the last job with maximum deterioration rate in B.
$r(B)$: the arrival time of $J(B)$.
$s(B)$: the starting time of $J(B)$ in σ.
$\alpha(B)$: the deterioration rate of $J(B)$.
$U(t)$: the set of the unprocessed jobs at time t.
$B_i(t)$: the set of the unprocessed jobs of the same family at time t, $1 \leq i \leq f$, which is a waiting batch at time t if $B_i(t) \neq \emptyset$.
$\mathcal{B}(t)$: the set of the waiting batches at time t.
$|\mathcal{B}(t)|$: the number of all waiting batches at time t.
$r(\mathcal{B}(t)) = \min\{r(B) : B \in \mathcal{B}(t)\}$.

The online algorithm, called A_1 (Algorithm 1), can be stated as follows. Without causing confusion, assume that $B_i(t) = B_i$ in the following.

Algorithm 1: A_1

Input: Job instance I. **do**
Step 0: Set $t = t_0$.
Step 1: If $\mathcal{B}(t) = \emptyset$, then go to Step 5.
Step 2: Let $\mathcal{B}(t) = \{B_1, B_2, \cdots, B_k\}$ such that $\alpha(B_1) \geq \alpha(B_2) \geq \cdots \geq \alpha(B_k)$, where $k \leq f$.
Step 3: If $t \geq (1+\alpha(B_1))^k t_0$, then process the batch B_1 at time t. Reset $t = (1+\alpha(B_1))t$. Return to Step 1.
Step 4: If $t < (1+\alpha(B_1))^k t_0$, then reset $t = \min\{(1+\alpha(B_1))^k t_0, t^*\}$, where t^* is the arrival time of the next job. Go to Step 2.
Step 5: If new jobs arrive after t, then reset t as the arrival time of the first new job. Go to Step 1.
Output: Job schedule σ.

Example 1. *To make the algorithm more intuitive, we present an instance $\mathcal{I}_1$ in Table 1, where $\mathcal{F}_1$ and $\mathcal{F}_2$ are two different families. As shown in Figure 1, σ is the schedule generated by A_1 and π is an optimal offline schedule for $\mathcal{I}_1$, where $B_1 = \{J_1, J_3\}$ and $B_2 = \{J_2\}$.*

We have $C_{\max}(\sigma) = 81$ and $C_{\max}(\pi) = 9$.

Table 1. Instance $\mathcal{I}_1$.

Job	Arrival Time	Deterioration Rate
$J_1 \in \mathcal{F}_1$	$r_1 = t_0 = 1$	2
$J_2 \in \mathcal{F}_2$	$r_2 = t_0 = 1$	2
$J_3 \in \mathcal{F}_1$	$r_3 = 2$	1

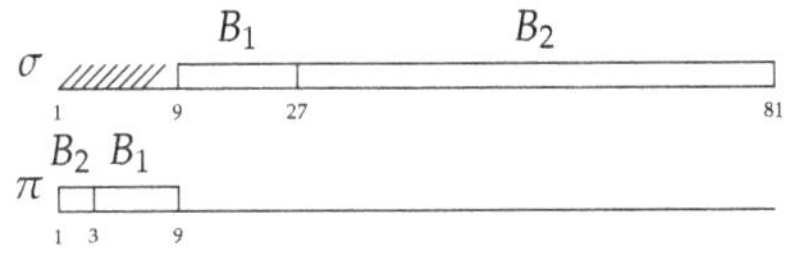

Figure 1. Schedule for Instance $\mathcal{I}_1$.

Suppose that r_l is the last arrival time. Let $s \geq r_l$ be the minimum time, such that s is the starting time of some batch and there is no idle time between s and $C_{\max}(\sigma)$ in σ. Let $\mathcal{B}$ be the set of the batches that process between s and $C_{\max}(\sigma)$ in σ and $s(\mathcal{B})$ be the start time of $\mathcal{B}$. Since $s(\mathcal{B}) = s \geq r_l$, each batch in $\mathcal{B}$ is from a different family and $\mathcal{B} = \mathcal{B}(s)$. From Algorithm 1, $\mathcal{B} = \{B_1, B_2, \cdots, B_k\}$ with $\alpha(B_1) \geq \alpha(B_2) \geq \cdots \geq \alpha(B_k)$, where $k \leq f$. Then,

$$C_{\max}(\sigma) = \prod_{i=1}^{k}(1 + \alpha(B_i))s(\mathcal{B}). \qquad (4)$$

The following two lemmas are the competition ratio analyses of Algorithm 1.

Lemma 1. *Suppose the machine has an idle time immediately before $s(\mathcal{B})$ in σ. Then, $C_{\max}(\sigma)/C_{\max}(\pi) \leq (1 + \alpha_{\max})^f$.*

Proof. Since an idle time occurs immediately before $s(\mathcal{B})$ in σ, from Algorithm 1, we have: $s(\mathcal{B}) = \max\{r_l, (1 + \alpha(B_1))^k t_0\}$.

If $s(\mathcal{B}) = r_l$, then for each $B_i \in \mathcal{B}$ with $1 \leq i \leq k$, $J(B_i)$ arrives at time r_l. From Equation (4), we know that $C_{\max}(\sigma) = \prod_{i=1}^{k}(1 + \alpha(B_i))r_l \leq C_{\max}(\pi)$.

If $s(\mathcal{B}) = (1 + \alpha(B_1))^k t_0$, then from Equation (4) we have $C_{\max}(\sigma) = \prod_{i=1}^{k}(1 + \alpha(B_i))(1 + \alpha(B_1))^k t_0$. Since $C_{\max}(\pi) \geq \prod_{i=1}^{k}(1 + \alpha(B_i))t_0$, so $C_{\max}(\sigma)/C_{\max}(\pi) \leq (1 + \alpha(B_1))^k \leq (1 + \alpha_{\max})^f$. □

Lemma 2. *Suppose the machine has no idle time immediately before $s(\mathcal{B})$ in σ. Then, $C_{\max}(\sigma)/C_{\max}(\pi) \leq (1 + \alpha_{\max})^f$.*

Proof. Since the machine has no idle time immediately before $s(\mathcal{B})$ in σ, $s(\mathcal{B})$ is the completion time of some batch, say B^*, in σ. We have $s(\mathcal{B}) = (1 + \alpha(B^*))s(B^*)$. From the definition of s, we know $s(B^*) < r_l$.

We suppose that $\mathcal{B}$ is divided into two sets, $\mathcal{B}_1$ and $\mathcal{B}_2$, such that:

$$\mathcal{B}_1 = \{B_i \in \mathcal{B} : r(B_i) > s(B^*)\},$$
$$\mathcal{B}_2 = \{B_j \in \mathcal{B} : r(B_j) \leq s(B^*)\}.$$

Since $\mathcal{B} = \mathcal{B}_1 \cup \mathcal{B}_2$, $s(\mathcal{B}) = (1 + \alpha(B^*))s(B^*)$, then from Equation (4) we have:

$$\begin{aligned}
C_{\max}(\sigma) &= \prod_{i=1}^{k}(1+\alpha(B_i))s(\mathcal{B})\\
&= \prod_{B_i\in\mathcal{B}}(1+\alpha(B_i))s(\mathcal{B})\\
&= \prod_{B_i\in\mathcal{B}_1}(1+\alpha(B_i))\prod_{B_j\in\mathcal{B}_2}(1+\alpha(B_j))(1+\alpha(B^*))s(B^*).
\end{aligned}$$

From the definition of $\mathcal{B}_1$, we know $C_{\max}(\pi) \geq \prod_{B_i\in\mathcal{B}_1}(1+\alpha(B_i))s(B^*)$. Hence,

$$\begin{aligned}
C_{\max}(\sigma) &= \prod_{B_i\in\mathcal{B}_1}(1+\alpha(B_i))\prod_{B_j\in\mathcal{B}_2}(1+\alpha(B_j))(1+\alpha(B^*))s(B^*)\\
&\leq \prod_{B_j\in\mathcal{B}_2}(1+\alpha(B_j))(1+\alpha(B^*))C_{\max}(\pi).
\end{aligned}$$

According to the definition of $\mathcal{B}_2$ and Algorithm 1, each batch in set $\mathcal{B}_2$ and B^* belongs to the different family. Then, at most $f-1$ batches exist in $\mathcal{B}_2$. Hence,

$$C_{\max}(\sigma)/C_{\max}(\pi) \leq \prod_{B_j\in\mathcal{B}_2}(1+\alpha(B_j))(1+\alpha(B^*)) \leq (1+\alpha_{\max})^f.$$

□

By Lemmas 1 and 2, and Theorem 1, we can reach the final conclusion.

Theorem 2. *For problem $1|online,r_j,p\text{-}batch, b=\infty, f\text{-}family, p_j=\alpha_j t|C_{\max}$, Algorithm 1 has a competitive ratio of $(1+\alpha_{\max})^f$ and is the best possible.*

3. Parallel Batch Machines ($m \geq 1$)

In this section, we consider the online scheduling on m parallel batch machines and the jobs belong to m incompatible deteriorating job families. We prove the lower bound and construct the best possible online algorithm with a competitive ratio of $1+\alpha_{\max}$.

Theorem 3. *For problem $Pm|online,r_j,p\text{-}batch, b=\infty, m\text{-}family, p_j=\alpha_j t|C_{\max}$, no online algorithm exists with a competitive ratio less than $1+\alpha_{\max}$.*

Proof. Let H be any online algorithm and I be a job instance provided by the adversary. In the instance I, all the jobs have a deterioration rate of α.

At time t_0, m jobs $J_1, J_2, \cdots, J_m$ from different families arrive. Suppose that job J_j starts processing at time s_j in σ, $j=1,2,\cdots,m$.

If a job J_k exists such that $s_k \geq (1+\alpha)t_0$, where $1 \leq k \leq m$, then the adversary does not release other jobs. Hence,

$$C_{\max}(\sigma) \geq (1+\alpha)s_k \geq (1+\alpha)^2 t_0 \text{ and } C_{\max}(\pi) = (1+\alpha)t_0.$$

We have $C_{\max}(\sigma)/C_{\max}(\pi) \geq 1+\alpha = 1+\alpha_{\max}$.

If for each job J_j with $1 \leq j \leq m$, $s_j < (1+\alpha)t_0$, let $J_l \in \{J_1, J_2, \cdots, J_m\}$ is the last starting job. At time $s_l + \epsilon$, a copy of the job $J_j (j=1,2,\cdots,m)$ arrives. We have:

$$C_{\max}(\sigma) \geq (1+\alpha)^2 s_l \text{ and } C_{\max}(\pi) = (1+\alpha)(s_l+\epsilon).$$

Hence, $C_{\max}(\sigma)/C_{\max}(\pi) \geq (1+\alpha)s_l/(s_l+\epsilon) \to 1+\alpha = 1+\alpha_{\max}$, as $\epsilon \to 0$.
The result follows. □

Before providing the online algorithm, we define some notations used in the following:

$U(t)$: the set of the unprocessed jobs at time t.
$\alpha_{\max}(t)$: the maximum deterioration rate of the jobs arrived at t or before t.
$m(t)$: the number of the idle machines at time t.
$f(t)$: the number of job families in $U(t)$ at time t.
$B_i(t)$: the nonempty set of the unprocessed jobs of the same family at time t, where $1 \leq i \leq f(t)$.
$J^i(t)$: the job with the maximum deterioration rate in $B_i(t)$, where $1 \leq i \leq f(t)$.
$\alpha^i(t)$: the deterioration rate of the job $J^i(t)$, where $1 \leq i \leq f(t)$.
Without loss of generality, assume that $\alpha^1(t) \geq \alpha^2(t) \geq \cdots \geq \alpha^{f(t)}(t)$. The online algorithm, called A_2 (Algorithm 2), can be stated as follows.

Algorithm 2: A_2

Input: Job instance I. **do**
Step 0: Set $t = t_0$.
Step 1: If $U(t) = \emptyset$, then go to Step 6.
Step 2: If $m(t) = m$, and $t \geq (1 + \alpha_{\max}(t))t_0$, then at time t, start $B_i(t)$ as a single batch on the idle machine for any $i = 1, 2, \cdots, f(t)$. Reset $t = (1 + \alpha^{f(t)}(t))t$. Go to Step 1.
Step 3: If $m(t) = m$, and $t < (1 + \alpha_{\max}(t))t_0$, then reset $t = t^*$, such that t^* is either the arrival time of the next job or $(1 + \alpha_{\max}(t))t_0$. Go to Step 1.
Step 4: If $m(t) < m$, and $t \geq (1 + \alpha_{\max}(t))(1 + \alpha^1(t))t_0$, then at time t, start $B_i(t)$ as a single batch on the idle machine for any $i = 1, 2, \cdots, \min\{m(t), f(t)\}$. Reset $t = t^*$, such that t^* is either the arrival time of the next job or $(1 + \alpha^{\min\{m(t), f(t)\}}(t))t$. Go to Step 1.
STEP 5: If $m(t) < m$, and $t < (1 + \alpha_{\max}(t))(1 + \alpha^1(t))t_0$, then reset $t = t^*$, such that either t^* is the arrival time of the next job or $m(t^*) > m(t)$, or $t^* = (1 + \alpha_{\max}(t))(1 + \alpha^1(t))t_0$.
Step 6: If new jobs arrive after t, then reset t as the arrival time of the first new job. Go to Step 1.
Output: Job schedule σ.

Example 2. *To make the algorithm more intuitive, we present an instance $\mathcal{I}_2$ in Table 2. Figure 2 depicts the schedule generated by Algorithm 2 and Figure 3 is an optimal offline schedule for $\mathcal{I}_2$.*
We have $C_{\max}(\sigma) = 12$ and $C_{\max}(\pi) = 8$.

Table 2. Instance $\mathcal{I}_2$.

Job	Arrival Time	Deterioration Rate
$J_1 \in \mathcal{F}_1$	$r_1 = t_0 = 1$	2
$J_2 \in \mathcal{F}_2$	$r_2 = t_0 = 1$	1
$J_3 \in \mathcal{F}_3$	$r_3 = t_0 = 1$	1
$J_4 \in \mathcal{F}_1$	$r_4 = 4$	1

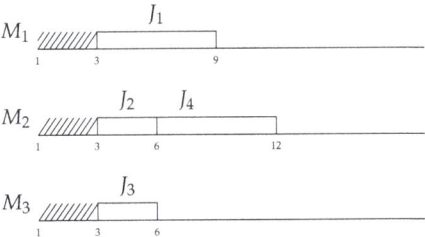

Figure 2. Schedule generated by A_2 for Instance $\mathcal{I}_2$.

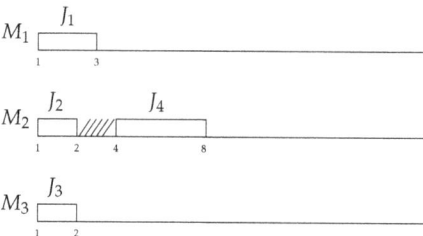

Figure 3. Optimal offline schedule for Instance $\mathcal{I}_2$.

Suppose that Algorithm 2 generates n batches $B_1, B_2, \cdots B_n$. For batch B_i, we define some notations in the following:

J_i: the job with the maximum deterioration rate in B_i.
α_i: the deterioration rate of J_i or the deterioration rate of B_i.
r_i: the arrival time of J_i.
s_i: the starting time of B_i in σ, suppose that $s_1 \leq s_2 \leq \cdots \leq s_n$.

The following is the competition ratio analysis of the Algorithm 2.

Let B_l be the first batch in σ assuming the objective value. B_l is a regular batch if $s_l = \max\{(1 + \alpha_{\max}(s_l))t_0, r_l\}$ or $\max\{(1 + \alpha_{\max}(s_l))t_0, r_l\} < s_l \leq (1 + \alpha_{\max}(s_l))(1 + \alpha^1(s_l))t_0$.

Lemma 3. *Suppose that only one job J_i exists in batch B_i of σ, $i = 1, 2, \cdots, n$, then the value of $C_{\max}(\sigma)/C_{\max}(\pi)$ does not decrease.*

Proof. From Algorithm 2, the start time of batch B_i is only related to the maximum deterioration rate of the jobs in this batch and the maximum deterioration rate of all jobs that have arrived. So, the value of $C_{\max}(\sigma)$ does not change when we assume each batch B_i has only one job J_i. The reduction in the number of jobs may decrease the value of $C_{\max}(\pi)$, so the value of $C_{\max}(\sigma)/C_{\max}(\pi)$ does not decrease. □

In the following, we assume that only one job J_i exists in batch B_i of σ, $i = 1, 2, \cdots, n$. Per Lemma 3, this does not influence the competition ratio analysis of Algorithm 2.

Lemma 4. $\alpha_l = \alpha^1(s_l)$.

Proof. Obviously, $\alpha_l \leq \alpha^1(s_l)$. If $\alpha_l < \alpha^1(s_l)$, since $J^1(s_l) \in U(s_l)$, then

$$C_{\max}(\sigma) \geq (1 + \alpha^1(s_l))s_l > (1 + \alpha_l)s_l.$$

This contradicts the completion time of B_l being the maximum completion time. Hence, $\alpha_l = \alpha^1(s_l)$. □

Lemma 5. *If the batch B_l is a regular batch, then $C_{\max}(\sigma)/C_{\max}(\pi) \leq 1 + \alpha_{\max}$.*

Proof. Since B_l is a regular batch, then

$$s_l = \max\{(1 + \alpha_{\max}(s_l))t_0, r_l\},$$

or

$$\max\{(1 + \alpha_{\max}(s_l))t_0, r_l\} < s_l \leq (1 + \alpha_{\max}(s_l))(1 + \alpha^1(s_l))t_0.$$

Case 1 $s_l = \max\{(1 + \alpha_{\max}(s_l))t_0, r_l\}$.

Then

$$\begin{aligned}
C_{\max}(\sigma) &= (1+\alpha_l)s_l \\
&= \max\{(1+\alpha_l)(1+\alpha_{\max}(s_l))t_0, (1+\alpha_l)r_l\} \\
&\leq (1+\alpha_{\max})\max\{(1+\alpha_l)t_0, r_l\} \\
&\leq (1+\alpha_{\max})C_{\max}(\pi).
\end{aligned}$$

Case 2 $\max\{(1+\alpha_{\max}(s_l))t_0, r_l\} < s_l \leq (1+\alpha_{\max}(s_l))(1+\alpha^1(s_l))t_0$.

In this case, at time s_l, some batches must have a start time less than s_l being processed. Let B_a be the last such batch to start, then $s_a < s_l < (1+\alpha_a)s_a$. Hence,

$$C_{\max}(\sigma) = (1+\alpha_l)s_l \leq (1+\alpha_l)(1+\alpha_a)s_a. \tag{5}$$

Suppose that $r_l \leq s_a$. Since $s_l > s_a$, the batch with a larger deterioration rate has higher priority in σ, then $\alpha_l \leq \alpha_a \leq \alpha^1(s_a)$ per Algorithm 2. At time s_a, J_l does not start processing. This indicates that there is no machine that can process J_l at time s_a, i.e., $m(s_a) < f(s_a) \leq m$. From Algorithm 2, we know that $s_a \geq (1+\alpha_{\max}(s_a))(1+\alpha^1(s_a))t_0$. By Lemma 4, we have $\alpha_l = \alpha^1(s_l)$. Hence,

$$\alpha^1(s_a) \geq \alpha_a \geq \alpha_l = \alpha^1(s_l).$$

By the definition of B_a, we have $\alpha_{\max}(s_a) = \alpha_{\max}(s_l)$, so

$$s_a \geq (1+\alpha_{\max}(s_a))(1+\alpha^1(s_a))t_0 \geq (1+\alpha_{\max}(s_l))(1+\alpha^1(s_l))t_0 \geq s_l.$$

This contradicts $s_a < s_l$. Hence $r_l > s_a$.
Thus, $C_{\max}(\pi) \geq (1+\alpha_l)r_l > (1+\alpha_l)s_a$. From Equation (5), we have

$$C_{\max}(\sigma)/C_{\max}(\pi) < 1+\alpha_a \leq 1+\alpha_{\max}.$$

□

In the following, we discuss the case where B_l is not a regular batch. This implies that no machine is idle immediately before time s_l, where $s_l > \max\{(1+\alpha_{\max}(s_l))(1+\alpha^1(s_l))t_0, r_l\}$. Renumber the m last batches starting on the m machines before time s_l to $B_{l,1}, B_{l,2}, \cdots, B_{l,m}$, such that $s_{l,1} \leq s_{l,2} \leq \cdots \leq s_{l,m}$. By Lemma 4, we have $\alpha_l = \alpha^1(s_l)$. So,

$$C_{\max}(\sigma) = (1+\alpha_l)s_l \leq (1+\alpha_l)\min_{1\leq j\leq m}\{(1+\alpha_{l,j})s_{l,j}\}. \tag{6}$$

If $s_{l,1} = s_{l,2} = \cdots = s_{l,k} = \cdots = s_{l,m} < s_l$, then $J_{l,1}, J_{l,2}, \cdots, J_{l,m}$ belong to m different job families and one of them belongs to the same family with J_l. Then $r_l > s_{l,1}$ and $C_{\max}(\pi) \geq (1+\alpha_l)r_l > (1+\alpha_l)s_{l,1}$. From Equation (6), we have:

$$C_{\max}(\sigma) \leq (1+\alpha_l)\min_{1\leq j\leq m}\{(1+\alpha_{l,j})s_{l,1}\} < \min_{1\leq j\leq m}(1+\alpha_{l,j}) \leq 1+\alpha_m.$$

In the following, we suppose that $s_{l,i} < s_{l,i+1}$ for some $i \in \{1, 2, \cdots, m-1\}$. Let k be the index that satisfies $s_{l,1} = s_{l,2} = \cdots = s_{l,k} < s_{l,k+1} \leq \cdots \leq s_{l,m} < s_l$, then $\alpha_{l,1} \geq \alpha_{l,2} \geq \cdots \geq \alpha_{l,k}$ and $J_{l,1} = J^1(s_{l,1})$. If $k \geq 2$, then we observe that any two jobs from $\{J_{l,1}, J_{l,2}, \cdots, J_{l,k}\}$ belong to different job families. Define

$$I_1 = \{J_{l,1}, J_{l,2}, \cdots, J_{l,k}\} \text{ and } I_2 = \{J_{l,k+1}, \cdots, J_{l,m}\}.$$

Lemma 6. *For any job $J_{l,j} \in I_2$, we have $s_{l,j} \geq (1+\alpha_{\max}(s_{l,j}))(1+\alpha_{l,j})t_0$.*

Proof. Since $s_{l,1} = s_{l,2} = \cdots = s_{l,k} < s_{l,k+1} \leq \cdots \leq s_{l,m} < s_l$, there is no idle machine immediately before time s_l, and $B_{l,1}, B_{l,2}, \cdots, B_{l,m}$ is the m last batches starting on the m machines before time s_l, then $m(t) < m$ for any time $t \in [s_{l,k+1}, s_{l,m}]$. From Algorithm 2, we have: $s_{l,j} \geq (1 + \alpha_{\max}(s_{l,j}))(1 + \alpha_{l,j})t_0$ for any job $J_{l,j} \in I_2$. □

Lemma 7. *If $r_l > s_{l,k+1}$, then $C_{\max}(\sigma)/C_{\max}(\pi) \leq 1 + \alpha_{\max}$.*

Proof. Since $r_l > s_{l,k+1}$, then $C_{\max}(\pi) \geq (1 + \alpha_l)r_l > (1 + \alpha_l)s_{l,k+1}$. From Equation (6), we have:

$$C_{\max}(\sigma) \leq (1 + \alpha_l) \min_{1 \leq j \leq m} \{(1 + \alpha_{l,j})s_{l,j}\} \leq (1 + \alpha_l)(1 + \alpha_{l,k+1})s_{l,k+1}.$$

Hence,

$$C_{\max}(\sigma)/C_{\max}(\pi) < 1 + \alpha_{l,k+1} \leq 1 + \alpha_{\max}.$$

□

Lemma 8. *If $r_l \leq s_{l,k+1}$, then $C_{\max}(\sigma)/C_{\max}(\pi) \leq 1 + \alpha_{\max}$.*

Proof. Since $r_l \leq s_{l,k+1}$, from Algorithm 2, we have:

$$\alpha_l \leq \min\{\alpha_{l,j} | J_{l,j} \in I_2\}. \tag{7}$$

If a job $J_{l,h} \in I_2 \setminus \{J_{l,k+1}\}$ exists such that $r_{l,h} > s_{l,k+1}$, then from Equation (7), we obtain:

$$C_{\max}(\pi) \geq (1 + \alpha_{l,h})r_{l,h} > (1 + \alpha_{l,h})s_{l,k+1} \geq (1 + \alpha_l)s_{l,k+1}.$$

From Equation (6), we have:

$$C_{\max}(\sigma) \leq (1 + \alpha_l) \min_{1 \leq j \leq m} \{(1 + \alpha_{l,j})s_{l,j}\} \leq (1 + \alpha_l)(1 + \alpha_{l,k+1})s_{l,k+1}.$$

Hence,

$$C_{\max}(\sigma)/C_{\max}(\pi) < 1 + \alpha_{l,k+1} \leq 1 + \alpha_{\max}.$$

Suppose that, for any job $J_{l,h} \in I_2 \setminus \{J_{l,k+1}\}$, $r_{l,h} \leq s_{l,k+1}$. Since $r_{l,k+1} \leq s_{l,k+1}$ and $r_l \leq s_{l,k+1}$, then the arrival time of all jobs from $I_2 \cup \{J_l\}$ is less than $s_{l,k+1}$. Thus, any two jobs from $I_2 \cup \{J_l\}$ belong to distinct job families.

Claim At least one job in $I_2 \cup \{J_l\}$ has an arrival time greater than $s_{l,k}$.

Otherwise, if the arrival time of all jobs is less than or equal to $s_{l,k}$, then all jobs in $I_1 \cup I_2 \cup \{J_l\}$ are available at time $s_{l,1}$, and each job independently forms a batch in σ. We obtain that every two jobs from $I_1 \cup I_2 \cup \{J_l\}$ belong to distinct job families. Since $I_1 \cup I_2 \cup \{J_l\} = \{J_{l,1}, J_{l,2}, \cdots, J_{l,m}\} \cup \{J_l\}$, then $f(s_{l,1}) = m + 1 > m$. This contradicts $f(s_{l,1}) \leq m$. The claim follows.

Since at least one job from $I_2 \cup \{J_l\}$ arrives after $s_{l,k}$, from Equation (7), we have:

$$C_{\max}(\pi) > (1 + \alpha_l)s_{l,k}.$$

From Equation (6), we have:

$$C_{\max}(\sigma) \leq (1 + \alpha_l) \min_{1 \leq j \leq m} \{(1 + \alpha_{l,j})s_{l,j}\} \leq (1 + \alpha_l)(1 + \alpha_{l,k})s_{l,k}.$$

Hence,

$$C_{\max}(\sigma)/C_{\max}(\pi) < 1 + \alpha_{l,k} \leq 1 + \alpha_{\max}.$$

From Lemmas 5, 7 and 8, and Theorem 3, we obtain the following theorem.

Theorem 4. *For problem $Pm|online,r_j,p\text{-}batch, b = \infty, m\text{-}family, p_j = \alpha_j t|C_{max}$, Algorithm 2 has a competitive ratio of $1 + \alpha_{max}$ and is the best possible.*

4. Conclusions and Future Research

In this paper, we outlined two best possible online algorithms. The first algorithm for problem $1|online,r_j,p\text{-}batch, b = \infty, f\text{-}family, p_j = \alpha_j t|C_{max}$ is a simple delay algorithm. We obtained the delay time by analyzing the properties of the unprocessed jobs, providing the best possible online algorithm with the competitive ratio of $(1 + \alpha_{max})^f$. The second algorithm for problem $Pm|online,r_j,p\text{-}batch, b = \infty, m\text{-}family, p_j = \alpha_j t|C_{max}$ is a more complex delay algorithm. We obtained the different delay times depending on the number of idle machines and provide the best possible online algorithm with the competitive ratio of $1 + \alpha_{max}$. The results are shown in Table 3.

Table 3. Summary of results.

Parallel Machine	Number of Families	Optimum Rate
$m = 1$	f	$(1 + \alpha_{max})^f$; best possible
$m \geq 1$	$f = m$	$1 + \alpha_{max}$; best possible

In future research, the general linear deterioration effect, such as $p_j = \alpha_j s_j + \beta_j$, is worthy of research. In additional, for the online scheduling problem on m parallel machines with linear deteriorating jobs to minimize the makespan, Yu et al. [22] only proved that no deterministic online algorithm is better than $(1 + \alpha_{max})$-competitive when $m = 2$, where α_{max} is the maximum deterioration rate of all jobs. However, no best possible online algorithm has been reported. This is also a topic for further study.

Author Contributions: Conceptualization, methodology and funding acquisition, W.L.; formal analysis and writing-original draft preparation, L.W.; writing—review, supervision and project administration, W.L. and X.C.; investigation, H.Y. All authors have read and agreed to the published version of the manuscript.

Funding: Research supported by NSFC (Nos. 11571321,11971443 and 11771406).

Conflicts of Interest: The authors declare no conflict of interest.

References

1. Borodin, A.; El-Yaniv, R. *Online Computation and Competitive Analysis*; Cambridge University Press: Cambridge, UK, 1998.
2. Brucker, P.; Gladky, A.; Hoogeveen, H.; Kovalyov, M.Y.; Potts, C.N.; Tautenhahn, T.; van de Velde, S.L. Scheduling a batching machine. *J. Sched.* **1998**, *1*, 31–54. [CrossRef]
3. Fu, R.Y.; Cheng, T.C.E.; Ng, C.T.; Yuan, J.J. An optimal online algorithm for single parallel-batch machine scheduling with incompatible job families to minimize makespan. *Oper. Res. Lett.* **2013**, *41*, 216–219. [CrossRef]
4. Li, W.H.; Yuan, J.J.; Yang, S.F. Online scheduling of incompatible unit-length job families with lookahead. *Theor. Comput. Sci.* **2014**, *543*, 120–125. [CrossRef]
5. Tian, J.; Cheng, T.C.E.; Ng, C.T.; Yuan, J.J. Online scheduling on unbounded parallel-batch machines with incompatible job families. *Theor. Comput. Sci.* **2011**, *412*, 2380–2386. [CrossRef]
6. Kunnathur, A.S.; Gupta, S.K. Minimizing the makespan with late start penalties added to processing times in a single facility scheduling problem. *Eur. J. Oper. Res.* **1990**, *47*, 56–64. [CrossRef]
7. Mosheiov, G. Schedulig jobs under simple linear deterioration. *Comput. Oper. Res.* **1994**, *21*, 653–659. [CrossRef]

8. Browne, S.; Yechiali, U. Scheduling deteriorating jobs on a single processor. *Oper. Res.* **1990**, *38*, 495–498. [CrossRef]
9. Gupta, J.N.D.; Gupta, S.K. Single facility scheduling with nonlinear processing times. *Comput. Ind. Eng.* **1988**, *14*, 387–393. [CrossRef]
10. Gawiejnowicz, S.; Kononov, A. Isomorphic scheduling problems. *Ann. Oper. Res.* **2014**, *213*, 131–145. [CrossRef]
11. Gawiejnowicz, S. Scheduling deteriorating jobs subject to job or machine availability constraints. *Eur. J. Oper. Res.* **2007**, *180*, 472–478. [CrossRef]
12. Ji, M.; Cheng, T.C.E. Batch scheduling of simple linear deteriorating jobs on a single machine to minimize makespan. *Eur. J. Oper. Res.* **2010**, *202*, 90–98. [CrossRef]
13. Lee, W.; Wu, C.; Chung, Y. Scheduling deteriorating jobs on a single machine with release times. *Comput. Ind. Eng.* **2008**, *54*, 441–452. [CrossRef]
14. Ng, C.T.; Li, S.S.; Cheng, T.C.E.; Yuan, J.J. Preemptive scheduling with simple linear deterioration on a single machine. *Theor. Comput. Sci.* **2010**, *411*, 3578–3586. [CrossRef]
15. Ji, M.; He, Y.; Cheng, T.C.E. Scheduling linear deteriorating jobs with an availability constraint on a single machine. *Theor. Comput. Sci.* **2006**, *362*, 115–126. [CrossRef]
16. Agnetis, A.; Billaut, J.; Gawiejnowicz, S.; Pacciarelli, D.; Soukhal, A. *Multiagent Scheduling*; Springer: Berlin/Heidelberg, Germany, 2014.
17. Gawiejnowicz, S. *Time-Dependent Scheduling*; Springer: Berlin/Heidelberg, Germany, 2008.
18. Strusevich, V.; Rustogi, K. *Scheduling with Time-Changing Effects and Rate-Modifying Activities*; Springer: Berlin/Heidelberg, Germany, 2017.
19. Cheng, M.B.; Sun, S.J. A heuristic MBLS algorithm for two semi-online parallel machine scheduling problems with deterioration jobs. *J. Shanghai Univ.* **2007**, *11*, 451–456. [CrossRef]
20. Liu, M.; Zheng, F.; Wang, S.; Huo, J. Optimal algorithms for online single machine scheduling with deteriorating jobs. *Theor. Comput. Sci.* **2012**, *445*, 75–81. [CrossRef]
21. Ma, R.; Tao, J.P.; Yuan, J.J. Online scheduling with linear deteriorating jobs to minimize the total weighted completion time. *Appl. Math. Comput.* **2016**, *273*, 570–583. [CrossRef]
22. Yu, S.; Ojiaku, J.T.; Wong, P.W.H.; Xu, Y.F. Online makespan scheduling of linear deteriorating jobs on parallel machines. In Proceedings of the International Conference on Theory and Applications of Models of Computation 2012, Beijing, China, 16–21 May 2012; pp. 260–272.
23. Yu, S.; Wong, P.W.H. Online scheduling of simple linear deteriorating jobs to minimize the total general completion time. *Theor. Comput. Sci.* **2013**, *487*, 95–102. [CrossRef]
24. Graham, R.L.; Lawler, E.L.; Lenstra, J.K.; Kan, A.H.G.R. Optimization and approximation in deterministic sequencing and scheduling: A survey. *Ann. Discret.* **1979**, *5*, 646–675.

© 2020 by the authors. Licensee MDPI, Basel, Switzerland. This article is an open access article distributed under the terms and conditions of the Creative Commons Attribution (CC BY) license (http://creativecommons.org/licenses/by/4.0/).

Article

On the Dual and Inverse Problems of Scheduling Jobs to Minimize the Maximum Penalty

Alexander A. Lazarev [1], Nikolay Pravdivets [1,*] and Frank Werner [2]

1. Institute of Control Sciences, 117997 Moscow, Russia; jobmath@mail.ru
2. Fakultät für Mathematik, Otto-von-Guericke-Universität Magdeburg, 39106 Magdeburg, Germany; frank.werner@ovgu.de
* Correspondence: pravdivets@ipu.ru; Tel.: +7-495-334-87-51

Received: 11 June 2020; Accepted: 6 July 2020; Published: 10 July 2020

Abstract: In this paper, we consider the single-machine scheduling problem with given release dates and the objective to minimize the maximum penalty which is NP-hard in the strong sense. For this problem, we introduce a dual and an inverse problem and show that both these problems can be solved in polynomial time. Since the dual problem gives a lower bound on the optimal objective function value of the original problem, we use the optimal function value of a sub-problem of the dual problem in a branch and bound algorithm for the original single-machine scheduling problem. We present some initial computational results for instances with up to 20 jobs.

Keywords: single-machine scheduling; minimization of maximum penalty; dual problem; inverse problem; branch and bound

MSC: 90B35; 90C57

---

1. Introduction

We consider single-machine scheduling problems, where a set of n jobs $N = \{1, 2, \ldots, n\}$ has to be processed on a single machine starting at time τ. For each job j, a release date r_j, a processing time p_j and a due date d_j are given. Many scheduling problems require the minimization of some maximum term. Denote by C_j the completion time of job j, then the minimization of the makespan (i.e., the maximum completion time of the jobs)

$$C_{max} = \max_{j=1,\ldots,n} C_j,$$

or the minimization of maximum lateness

$$L_{max} = \max_{j=1,\ldots,n} \{C_j - d_j\}$$

are well-known examples of such an optimization criterion.

In this paper, we consider two related problems of such a min-max problem, namely a dual problem as well as an inverse problem of the single-machine scheduling problem with given release dates and minimizing the maximum penalty. While the original problem $1|r_j|L_{max}$ is NP-hard in the strong sense [1], we prove that both the dual and inverse problems of this problem can be solved in polynomial time.

Due to the NP-hardness of the problem $1|r_j|L_{max}$, several branch and bound algorithms have been developed and special cases of the problem have been considered, see e.g., [2–9]. In [9], it has been shown that if the release dates for all jobs are from the interval $[d_j - p_j - A, d_j - A]$ for all jobs and some constant A, the problem can be solved in $O(n \log n)$ time if no machine idle times are allowed and in $O(n^2 \log n)$ time if machine idle times are allowed. Another important special case of this problem has been considered in [10]. In that paper, it is shown that for naturally bounded job data, the problem can be polynomially solved. More precisely, a polynomial time solution of the variant is given when the maximal job processing time and the differences between the job release dates are bounded by a constant. The binary search procedure presented in this work determines an optimal solution in $O(n^2 \log n \log p_{max})$ or $O(d_{max} n \log n \log p_{max})$ time, where p_{max} is the maximal processing time and d_{max} is the maximal due date.

In [11], some computational experiences and applications for the more general case with additional precedence constraints are reported. Their algorithm turned out to be superior to earlier algorithms. Some applications to job shop scheduling are also discussed. A more recent branch and bound algorithm for this single-machine problem with release dates and precedence constraints has been given in [12]. This algorithm uses four heuristics for finding initial upper bounds and a variable neighborhood search procedure. It solved most considered instances within one minute of CPU time. Several approximation schemes for four variants of the problem with additional non-availability or deadline constraints have been derived in [13]. An approximation algorithm for this single-machine problem with additional workload dependent maintenance duration has been presented in [14]. This algorithm is even optimal for some special cases of the problem. A hybrid metaheuristic search algorithm for the single-machine problem with given release dates and precedence constraints has been developed in [15]. Computational tests have been made using own instance sets with 100 jobs and instances from [12] with up to 1000 jobs. The hybridization of the elektromagnetism algorithm with tabu search leads to a tradeoff between diversification and intensification strategies. The metric approach is another recent possibility for solving the problem $1|r_j|L_{max}$ approximately with guaranteed maximal error, see e.g., [16]. The introduced metric delivers an upper bound on the absolute error of the objective function value. Taking a given instance of some problem and using the introduced metric, the nearest instance is determined for which a polynomial or pseudo-polynomial algorithm is known. Then a schedule is constructed for this instance and applied to the original instance.

There have been also considered problems with several optimization criteria. In [17], the single-machine problem with the primary criterion of minimizing maximum lateness and the secondary criterion of minimizing the maximum job completion time has been investigated. The author gives dominance properties and conditions when the Pareto-optimal set can be found in polynomial time. The derived properties allow extension of the basic framework to exponential implicit enumeration schemes and polynomial approximation algorithms. The problem of finding the Pareto-optimal set for two criteria in the case when there are constraints on the source data have been considered in [18,19]. In [18], the idea of the dual approach is considered in detail, but there is no sufficient experimental study of the effectiveness of this approach. Lazarev et al. [20] considered the problem of minimizing maximum lateness and the makespan in the case of equal processing times and proposed a polynomial time approach for finding the Pareto-optimal set of feasible solutions. They presented two approaches, the efficiency of which depends on the number of jobs and the accuracy of the input-output parameters.

The dual and inverse problems considered in this paper are maximization problems. In the literature, there exist some works on other single-machine maximization problems, usually under the assumption of no inserted idle times between the processing of jobs on the machine. Maximization problems in single-machine problems were considered e.g., in [21,22]. The complexity and some algorithms for single-machine total tardiness maximization problems have been discussed in [23]. In [24], a pseudo-polynomial algorithm for the single-machine total tardiness maximization problem has been transformed by a graphical algorithm into a polynomial one.

The remainder of this paper is as follows. In Section 2, we introduce the dual problem of the single-machine problem, where the maximum penalty term of a job should be minimized. Section 3 considers an inverse problem, where the minimum of the penalty terms should be maximized. The solution of this dual problem is embedded into a branch and bound algorithm for the original problem. Some computational results for this branch and bound algorithm are given in Section 4. The paper finishes with some concluding remarks.

2. The Dual Problem

Let us consider the general formulation of the NP-hard problem of minimizing the maximum penalty or cost $\varphi_{\max}$ for a set of jobs on a single machine, i.e., problem $1 \mid r_j \mid \varphi_{\max}$. The machine cannot process more than one job of the set $N = \{1, \ldots, n\}$ at a moment. Preemptions of the processing of a job are prohibited. Let $\varphi_{j_k}(C_{j_k}(\pi))$ denote the penalty for job $j \in N$ if it is processed as the k-th job in the sequence $\pi = (j_1, j_2, \ldots, j_k, \ldots, j_n)$. We assume that all $\varphi_{j_k}(C_{j_k}(\pi)), k = 1, 2, \ldots, n$, are arbitrary non-decreasing penalty functions.

By μ^* we denote the optimal value of the objective function:

$$\mu^* = \min_{\pi \in \Pi(N)} \max_{k=1,\ldots,n} \varphi_{j_k}(C_{j_k}(\pi)), \tag{1}$$

where $\Pi(N) = \{\pi_1, \pi_2, \ldots, \pi_{n!}\}$ denotes the set of all permutations (schedules) of the jobs of the set N.

In the scheduling literature, many special cases of the following general dual problem are considered. One wishes to find an optimal job sequence π^* and the corresponding objective function value ν^* such that

$$\nu^* = \max_{k=1,\ldots,n} \min_{\pi \in \Pi(N)} \varphi_{j_k}(C_{j_k}(\pi)). \tag{2}$$

For convenience, we introduce a notation that takes into account the position of the job in the schedule. Let a schedule (job sequence) $\pi \in \Pi(N)$ be given by $\pi = (j_1, j_2, \ldots, j_n)$. For the job, which is processed as the k-th job in the sequence, $k = 1, 2, \ldots, n$, under a schedule π, we denote:

$$\nu_k = \min_{\pi \in \Pi(N)} \varphi_{j_k}(C_{j_k}(\pi)), k = 1, 2, \ldots, n. \tag{3}$$

Obviously,

$$\nu^* = \max_{k=1,\ldots,n} \nu_k. \tag{4}$$

Lemma 1 ([25]). *Let $\varphi_j(t), j = 1, 2, \ldots, n$, be arbitrary non-decreasing penalty functions in the problem $1 \mid r_j \mid \varphi_{\max}$. Then we have $\nu_n \geq \nu_k$ for all $k = 1, 2, \ldots, n$, i.e., $\nu^* = \nu_n$.*

Proof. Suppose that there exists a number $k, k < n$, such that $\nu_k > \nu_n$. Assume that $\pi_n = (j_1, j_2, \ldots, j_n)$ is a schedule for which the value ν_n is obtained. Then we consider the schedule

$$\pi = (j_{n-k+1}, j_{n-k+2}, \ldots, j_n, j_1, j_2, \ldots, j_{n-k}).$$

Please note that under the schedule π, the job j_n will be carried out as the k-th job in the sequence. Since $C_{j_n}(\pi_n) \geq C_{j_n}(\pi)$, we have

$$\nu_n = \varphi_{j_n}(C_{j_n}(\pi_n)) \geq \varphi_{j_n}(C_{j_n}(\pi)) \geq \nu_k.$$

Due to the assumption $\nu_k > \nu_n$, we obtain the inequality

$$\nu_k > \nu_n \geq \nu_k$$

which is a contradiction. The lemma has been proved. □

Thus, the solution of the dual problem $1|r_j|\varphi_{max}$ is reduced to the problem of finding the value v_n. We enumerate all jobs in order of non-decreasing release dates: $r_{i_1} \leq r_{i_2} \leq \cdots \leq r_{i_n}$. Due to

$$v_n = \min_{\pi \in \Pi(N)} \varphi_{j_n}(C_{j_n}(\pi)),$$

we will put each of the jobs j of the set N onto the last (i.e., the n-th) position. The other $n-1$ jobs of the set $N \setminus \{j\}$ are arranged in their original order starting at time τ. This gives the earliest completion time of processing the jobs from the set $N \setminus \{j\}$. This procedure is formally summarized in Algorithm 1. The input data of Algorithm 1 is the pair (N, τ), where τ is used to calculate C_j.

Algorithm 1: Solution of the dual problem of the problem $1 \mid r_j \mid \varphi_{max}$

1. Construct the schedule $\pi^r = (i_1, i_2, \ldots, i_n)$, in which all jobs are sequenced according to non-decreasing release dates: $r_{i_1} \leq r_{i_2} \leq \cdots \leq r_{i_n}$.
2. For $k = 1, 2, \ldots, n$, find the value $\varphi_{i_k}(C_{i_k}(\pi_k))$ for the schedule $\pi_k = (\pi^r \setminus i_k, i_k)$.
3. Find the value $v^* = \min_{k=1,\ldots,n} \varphi_{i_k}(C_{i_k}(\pi_k))$ and the job i_k, which gives the value v^*.

The complexity of Algorithm 1 can be estimated as follows. We need $O(n \log n)$ operations to construct the schedule π^r. We need $O(n)$ operations to find each of the n values $\varphi_{i_k}(C_{i_k}(\pi_k))$. Therefore, to determine the value v^* and the corresponding job i_k, for which the value v^* is obtained, no more than $O(n^2)$ operations are required.

Theorem 1 ([25]). *Let $\varphi_j(t), j = 1, 2, \ldots, n$, be arbitrary non-decreasing penalty functions for the problem $1 \mid r_j \mid \varphi_{max}$. Then the inequality $\mu^* \geq v^*$ holds.*

Proof. Suppose the opposite, i.e., there exists an instance of the problem $1 \mid r_j \mid \varphi_{max}$, for which the inequality $\mu^* < v^*$ holds. Let $\pi^* = (j_1, j_2, \ldots, j_n)$ be an optimal schedule for this instance. Then we have

$$\varphi_{j_n}(C_{j_n}(\pi^*)) \leq \mu^* < v^*,$$

which contradicts equality (3):

$$v^* = v_n = \min_{\pi \in \Pi(N)} \varphi_{j_n}(C_{j_n}(\pi)).$$

The theorem has been proved. □

The obtained estimate can be efficiently used in constructing schemes of a branch and bound method for solving the problem $1 \mid r_j \mid \varphi_{max}$, and for estimating the error of approximate solutions when the branch and bound algorithm stops without finding an optimal solution.

We denote by $\{N', \tau', v', \pi', B'\}$ the sub-problem of processing the jobs of the set $N' \subseteq N$ from time $\tau' \geq \tau$ on according to some partial sequence π' for the jobs of the set $N \setminus N'$, where v' is the lower bound obtained by solving the dual problem of this instance, $\tau' = C_{max}(\pi', \tau)$ is the start time of the planning horizon for the jobs from the set B', which is equal to the makespan value for the sequence π', and τ is the time when the machine is ready to process the jobs from the set N. B' is the set of jobs that cannot be placed on the corresponding first position of the schedule.

The subsequent Algorithm 2 implements one of the possible schemes of the branch and bound method using the solution of the dual problem. The branching in Algorithm 2 is carried out as a result of dividing the current sub-instance into two instances: put the job f (the job with the smallest due date from the set of jobs ready for processing) at the next position in the schedule and prohibit the inclusion of the job f at the next position in the schedule (by increasing the possible start time of job f).

Denote

$$r_j(\tau) = \max\{r_j, \tau\}, \quad r(N, \tau) = \min_{j \in N} r_j(\tau).$$

Let $|N| > 1$. We choose a job $f = f(N, \tau)$ from the set N such that

$$f(N, \tau) = \arg \min_{j \in N \setminus B} \left\{ d_j \mid r_j(\tau) = r(N, \tau) \right\}.$$

If $N = \{i\}$, then $f(N, \tau) = i$ for all τ. Here B is a set of jobs that cannot be placed on the current first position. Denote by π^* the currently best schedule constructed for all jobs.

We note that a deeper comparative discussion of the characteristics of the particular search strategies can be found in [26,27].

Algorithm 2: Solution of the problem $1 \mid r_j \mid L_{\max}$ by the branch and bound method based on the solution of the dual problem

1. **Initial step**
   Let $\pi^* = \emptyset$. The list of instances contains the original instance $\{N, \tau, \nu, \emptyset, \emptyset\}$, where ν is a lower bound on the optimal objective function value obtained by solving the dual problem of this instance.

2. **Main step**
   (a) From the list of instances, select an instance $\{N', \tau', \nu', \pi', B'\}$ with a minimal lower bound ν'.
   (b) Find the job $f = f(N', \tau')$ from the set $N' \setminus B'$ with the smallest due date from the number of jobs ready for processing at time τ'.
   (c) Replace the instance $\{N', \tau', \nu', \pi', B'\}$ by the two instances $\{N_1, \tau_1, \nu_1, \pi_1, B_1\}$ and $\{N_2, \tau_2, \nu_2, \pi_2, B_2\}$ in the list of instances, where:

   - $N_1 = N' \setminus \{f\}$, $\tau_1 = \max\{r_f, \tau'\} + p_f$, $B_1 = \emptyset$, $\pi_1 = (\pi', f)$, ν_1 is a lower bound obtained by solving the dual problem of this instance by Algorithm 3;
   - $N_2 = N'$, $\tau_2 = \tau'$, $B_2 = B' \cup \{f\}$, $\pi_2 = \pi'$, ν_2 is a lower bound obtained by solving the dual problem of this instance by Algorithm 3.

   (d) If, after completing this step of the algorithm, we obtain $\{\pi_1\} = N$, that is, all jobs are ordered, then $\pi^* = \arg \min\{L_{\max}(\pi_1, \tau), L_{\max}(\pi^*, \tau)\}$.
   (e) Exclude all instances $\{N', \tau', \nu', \pi', B'\}$, for which $\nu' \geq L_{\max}(\pi^*, \tau)$.

3. **Termination step**
   If the list of instances is empty, STOP, otherwise repeat the main step 2.

To find the value ν in step 2(c), we need to modify Algorithm 1 taking into account a list B of jobs that cannot be placed on the current position. The input data of Algorithm 3 is the triplet (N, τ, B), where τ is used to calculate C_j.

Algorithm 3: Modification of the solution algorithm for the dual problem of the problem $1 \mid r_j \mid \varphi_{\max}$ with respect to a list of jobs that cannot be on the first position

1. Construct the schedule $\pi^r = (i_1, i_2, \ldots, i_n)$, in which all jobs are sequenced according to non-decreasing release dates: $r_{i_1} \leq r_{i_2} \leq \cdots \leq r_{i_n}$.
2. For $k = 1, 2, \ldots, n$, if $N \setminus (B \cup \{i_k\}) \neq \emptyset$, find a job

$$i_l = \arg \min_{j \in N \setminus (B \cup \{i_k\})} r_j$$

and construct the schedule $\pi_k = (i_l, \pi^r \setminus \{i_l, i_k\}, i_k)$.
Find the value $\varphi_{i_k}(C_{i_k}(\pi_k))$.
If $N \setminus (B \cup \{i_k\}) = \emptyset$, then we assume $C_{i_k} = +\infty$.
3. Find the value $v^* = \min_{k=1,\ldots,n} \varphi_{i_k}(C_{i_k}(\pi_k))$ and the job i_k, which gives the value v^*.

It is easy to see that this algorithm can be used to solve the more general problem $1 \mid r_j \mid \varphi_{\max}$. In addition, if the algorithm is stopped without an empty list of instances due to a time limit, the current schedule π^* can be taken as an approximate solution of the problem.

Hence, although the original problem $1 \mid r_j \mid \varphi_{\max}$ is NP-hard in the strong sense (recall that problem $1 \mid r_j \mid L_{\max}$ is NP-hard in the strong sense), the dual problem turned out to be polynomially solvable.

If precedence relations are specified between the jobs by an acyclic graph G, then the dual problem of the problem $1 \mid r_j, prec \mid \varphi_{\max}$ can also be solved in a similar way. Since the argumentation is similar, we skip the details. Here, the core is to solve the problem $1 \mid r_j, prec \mid C_{\max}$. Jobs without successors according to the precedence graph G will be put one-by-one to the last positions in the job sequence. Thus, the dual problem of the problem $1 \mid r_j, prec \mid \varphi_{\max}$ is also polynomially solvable.

For problems with $m > 1$ machines, e.g., problem $Pm \mid r_j, prec \mid \varphi_{\max}$, the core consists of solving the dual problem, which is the partition problem. This dual problem is NP-hard in the ordinary sense.

Thus, although in mathematical programming the original and dual problems have usually the same complexity status, it turned out that the dual problems of the scheduling problems considered in this paper have a lower complexity than the original problems. This interesting fact should be investigated further in more detail also for other scheduling problems.

3. The Inverse Problem of the Maximum Lateness Problem

The inverse problem of the NP-hard problem of minimizing maximum lateness $1 \mid r_j \mid L_{\max}$ consists of finding a schedule π, which reaches the maximum minimal lateness and finding the value

$$\lambda^* = \max_{\pi \in \Pi(N)} \min_{k=1,\ldots,n} L_{j_k}(C_{j_k}(\pi)). \tag{5}$$

Please note that for this problem, inserted idle times of the machine are prohibited.

This problem was solved only for the case of simultaneous availability of the set N for processing, i.e., $r_j = 0$, for all $j \in N$ in [28]. We consider the general case of the problem $1 \mid r_j \mid \max L_{\min}$.

Lemma 2. *There exists an optimal schedule $\pi = (i_1, \ldots, i_n)$ for the problem $1 \mid r_j \mid \max L_{\min}$, for which*

$$d_{i_k} - p_{i_k} \leq d_{i_{k+1}} - p_{i_{k+1}}, \quad k = 2, 3, \ldots, n-1, \tag{6}$$

and

$$\lambda^* = \min_{k=1,\ldots,n} L_{i_k}(C_{i_k}(\pi)).$$

Proof. Assume that at least one of inequalities (6) is not satisfied for an optimal schedule $\pi' = (j_1, \ldots, j_n)$ and let
$$\lambda^* = \min_{k=1,\ldots,n} L_{i_k}(C_{i_k}(\pi')).$$

In what follows, the proof will consist of two stages, which can be repeated several times.

Step 1. If there are no machine idle times in the schedule π', then go to step 2. Let there be idle times according to schedule π', and consider the last of them:
$$C_{j_k} < r_{j_{k+1}} \quad \text{and} \quad r_{j_m} \leq C_{j_{m-1}}, \quad m = k+2, \ldots, n.$$

Construct the schedule $\pi'' = (j_{k+1}, j_1, \ldots, j_k, j_{k+2}, \ldots, j_n)$. Since
$$C_j(\pi'') \geq C_j(\pi') \quad \text{for all } j \in N,$$

the value of the minimal lateness will not decrease. There will be no idle time under the schedule π'', and the optimal value λ^* will be saved. Set $\pi' := \pi''$ and go to step 2.

Step 2. If the schedule π' meets the conditions of Lemma 2, the proof is completed. If there exist two jobs j_l, j_{l+1}, for which
$$d_{j_l} - p_{j_l} > d_{j_{l+1}} - p_{j_{l+1}},$$

then exchange the jobs j_l, j_{l+1} which yields the schedule
$$\pi'' = (j_1, \ldots, j_{l-1}, j_{l+1}, j_l, j_{l+2}, \ldots, j_n).$$

As there are no machine idle times under the schedule π', we have
$$r_{j_l} \leq C_{j_{l-1}}(\pi').$$

There are the following possible cases:

(1) Let $r_{j_{l+1}} \leq C_{j_{l-1}}(\pi')$. Obviously, in this case we have
$$C_{j_k}(\pi') = C_{j_k}(\pi''), \quad k = 1, 2, \ldots, l-1, l+2, \ldots, n. \tag{7}$$

According to the assumptions, inequality
$$C_{j_{l-1}}(\pi') + p_{j_l} + p_{j_{l+1}} - d_{j_{l+1}} > C_{j_{l-1}}(\pi') + p_{j_l} - d_{j_l}. \tag{8}$$

holds. Moreover, we have
$$C_{j_{l-1}}(\pi') + p_{j_{l+1}} - d_{j_{l+1}} > C_{j_{l-1}}(\pi') + p_{j_l} - d_{j_l}; \tag{9}$$

$$C_{j_{l-1}}(\pi') + p_{j_{l+1}} - d_{j_l} > C_{j_{l-1}}(\pi') + p_{j_l} - d_{j_l}. \tag{10}$$

Formulas (7)–(10) show that the maximal lateness is not reduced. Set $\pi' := \pi''$ and repeat step 2.

(2) Let $r_{j_{l+1}} > C_{j_{l-1}}(\pi')$. In this case, we have
$$C_{j_k}(\pi'') = C_{j_k}(\pi'), \quad k = 1, 2, \ldots, l-1, \tag{11}$$

$$C_{j_k}(\pi'') > C_{j_k}(\pi'), \quad k = l+2, \ldots, n. \tag{12}$$

According to the assumptions, we have
$$C_{j_{l+1}}(\pi'') - d_{j_{l+1}} > C_{j_l}(\pi') - d_{j_l}; \tag{13}$$

$$C_{j_l}(\pi'') - d_{j_l} > C_{j_{l+1}}(\pi') - d_{j_{l+1}}. \tag{14}$$

Formulas (8) and (11)–(14) show that the maximal lateness is not reduced. Set $\pi' := \pi''$ and go to step 1.

In a finite number of steps, we construct an optimal schedule satisfying the conditions of the lemma. The lemma has been proved. □

Algorithm 4 constructs n schedules, one of which is satisfying the conditions of the lemma.

Algorithm 4: Solution of the inverse problem $1 \mid r_j \mid \max L_{\min}$

1. Enumerate all jobs of the set N according to $d_1 - p_1 \leq d_2 - p_2 \leq \cdots \leq d_n - p_n$.
2. For $k = 1, 2, \ldots, n$ do:
   (a) construct the schedule $\pi_k = (k, 1, \ldots, k-1, k+1, \ldots, n)$ and
   (b) determine $\lambda_k = \min\limits_{j=1,\ldots,n} L_j(C_j(\pi_k))$.
3. Calculate $\lambda^* = \max\limits_{k=1,\ldots,n} \lambda_k$.

$O(n \log n)$ operations are needed for renumbering the jobs of the set N. $O(n)$ operations are needed for constructing the schedule π_k and calculating the value λ_k, $k = 1, \ldots, n$. Thus, no more than $O(n^2)$ operations are needed to find the value λ^*.

The objective function value of a solution of the problem of maximizing minimal lateness $1 \mid r_j \mid \max L_{\min}$ is a lower bound on the optimal objective function value for the original problem $1 \mid r_j \mid L_{\max}$.

Theorem 2 ([18]). *For the optimal function values of the problem $1 \mid r_j \mid L_{\max}$ and the corresponding inverse problem $1 \mid r_j \mid \max L_{\min}$, the inequality $\mu^* \geq \lambda^*$ holds.*

Proof. Denote by π' and π'' optimal schedules for the problems $1 \mid r_j \mid L_{\max}$ and $1 \mid r_j \mid \max L_{\min}$, respectively. There exist jobs $k', k'' \in N$, for which the following inequalities hold:

$$\mu^* = C_{k'}(\pi') - d_{k'} \geq C_j(\pi') - d_j \quad \text{for all } j \in N; \tag{15}$$

$$C_j(\pi'') - d_j \geq C_{k''}(\pi'') - d_{k''} = \lambda^* \quad \text{for all } j \in N. \tag{16}$$

Please note that jobs k' and k'' can be identical. Let $\pi'' = (j_1, \ldots, j_n)$. Obviously, the following inequality is satisfied for job j_1:

$$C_{j_1}(\pi') - d_{j_1} \geq C_{j_1}(\pi'') - d_{j_1}. \tag{17}$$

According to inequalities (15)–(17), we get

$$\mu^* = C_{k'}(\pi') - d_{k'} \geq C_{j_1}(\pi') - d_{j_1} \geq C_{j_1}(\pi'') - d_{j_1} \geq C_{k''}(\pi'') - d_{k''} = \lambda^*,$$

i.e., $\mu^* \geq \lambda^*$. The theorem has been proved. □

4. Computational Results

In this section, we present some results of the numerical experiments carried out on randomly generated test instances. The numerical experiments were carried out on a PC Intel® Core™ i5-4210U CPU @ 1.70GHz, 4 cores; 8 GB DDR4 RAM.

Various methods of generating test instances for different types of scheduling problems are described in [29]. For the problem $1|r_j|L_{max}$ with n jobs, the authors suggest the following generation scheme:

- $r_0 = 0$ and $r_i = r_{i-1} + X_i$, where $X_i \sim \exp(\lambda)$, for $i = 1,...,n$;
- $p_i \sim N(\mu, \sigma)$ for $i = 1,...,n$, truncated below a known lower bound;
- $d_i = r_i + kE[p_i]$ for $i = 1,...,n$, where $k \geq 1$.

The authors suggest that λ, μ, σ and k are generation parameters that can be fixed by the user. Applying this generation scheme, we generate release dates which are independent from the processing times, while the due dates correlate with the processing times, as it usually happens in real problems. We set

$$\lambda = \frac{1}{100}, \quad \mu = 100, \quad \sigma = 40, \quad k = 1$$

and generated 15 instances for each $n \in \{3, 4, ..., 20\}$. The results are shown in Table 1, where the number of branching points are given for each of the 15 instances for any value of n.

Table 1. Number of branching points in Algorithm 2 for the test instances generated according to [29].

n	\multicolumn{15}{c}{Number of the Instance}
	1	2	3	4	5	6	7	8	9	10	11	12	13	14	15
3	4	2	2	2	2	2	2	2	2	2	2	2	4	2	2	2
4	3	3	3	3	3	3	3	3	3	5	4	3	3	3	3
5	9	4	4	25	4	4	31	4	4	8	4	4	4	4	18
6	5	5	64	227	161	13	7	5	5	5	5	5	84	15	5
7	6	25	16	6	6	6	31	147	310	34	35	6	172	6	222
8	7	234	33	7	7	23	7	391	7	7	7	7	194	7	7
9	1271	8	1847	8	3786	2934	8	8	62	8	8	721	8	208	8
10	9	690	9	9	66	114	9	9	9	9	9	9	3944	6584	9
11	10	13,098	10	5071	10	13,992	11	10	6081	773	591	303	10	3772	10
12	11	22,736	11	11	35,910	57,407	32,798	11	139	164,613	11	11	166	11	207,119
13	12	12	213,433	1109	*	12	183	227	794,567	12	*	12	30,946	*	12
14	13	13	458	*	5933	21,774	4394	13	13	13	*	32,246	*	13	*
15	2293	14	*	*	*	*	14	*	14	14	*	*	14	14	*
16	15	*	15	*	15	*	15	*	15	*	*	*	172,876	15	*
17	*	16	16	*	16	*	16	16	16	16	*	9128	*	16	16
18	17	*	*	17	*	*	17	17	17	17	*	*	*	*	*
19	*	18	18	*	*	*	*	*	4999	18	18	*	*	18	18
20	*	19	2103	*	*	19	20	19	19	19	19	19	19	19	*

An asterisk (*) in Table 1 means that the solution could not be found within 15 min. According to this table, a large part of the instances can be solved with several branching points not greater than the number of jobs. However, some instances appeared to be hard and required a much larger number of branching points. It can be observed that these hard instances generated according to [29] display a rather different solution behavior and need very different numbers of branching points for their exact solution. This interesting phenomenon deserves further detailed investigations which are planned as future work by the authors.

Next, we consider an instance of the problem as a point in the $3n$-dimensional space, where each value of r_i, p_i, d_i represents one of the dimensions. We consider the vector from the zero point to the point of the instance. The complexity of an instance is defined by the direction of the vector, but not by its length. Therefore, to explore the complexity of different instances, we can take points on the surface of the $3n$-dimensional cube. For the processing times and the release dates, we consider only non-negative values. As a result, we obtain points on the quarter of the surface of the $3n$-dimensional cube with $r_i \geq 0$ and $p_i \geq 0$. Let the size of the cube be 100. We generated 300,000 points on the surface for problems with 4, 5, 6, 7, 8 and 9 jobs (i.e., in the 12-, 15-, 18-, 21, 24, 27-dimensional spaces, respectively). All instances have been solved, and we counted the number of iterations (i.e., branching points) for each problem instance. The results are shown in Figure 1 for the instances with 4 jobs, in Figure 2 for the instances with 5 jobs, in Figure 3 for the instances with 6 jobs, in Figure 4 for the instances with 7 jobs, in Figure 5 for the instances with 8 jobs and in Figure 6 for the instances with 9 jobs. At the y-axis, the numbers are given how often a particular number of branching points (Figure 1) or an interval for the number of branching points (Figures 2–6) has occurred among the 300,000 instances for each number n of jobs. For example, in Figure 1, one can see that among the 300,000 solved instances with 5 jobs, there were 72,897 instances with 3 branching points, 16,131 instances with 4 branching points, 29,342 instances with 5 branching points, and so on. It can be observed that the maximal number of 20 branching points was reached for 1294 instances, which is approximately equal to 0.4% of all instances. In Figure 2, the numbers of branching points are grouped in intervals of 5 in each column, i.e., there were 86 026 instances with several branching points between 0 and 4 (actually 4, because it is the minimum possible number of branching points for instances with 5 jobs), 28,566 instances with several branching points between 5 and 9, etc. For instances with 5 jobs, the maximum number of branching points was 93, and it was reached for only 33 instances. As it can be seen in the figures, most of instances can be solved by a small number of branches. For a larger number of jobs, one can detect a smaller number of hard instances with a large number of branching points. Thus, for the instances with 9 jobs, among the 300,000 solved instances, there were only two instances with several branching points more than 180,000: one with 184,868, and the other with 191,887.

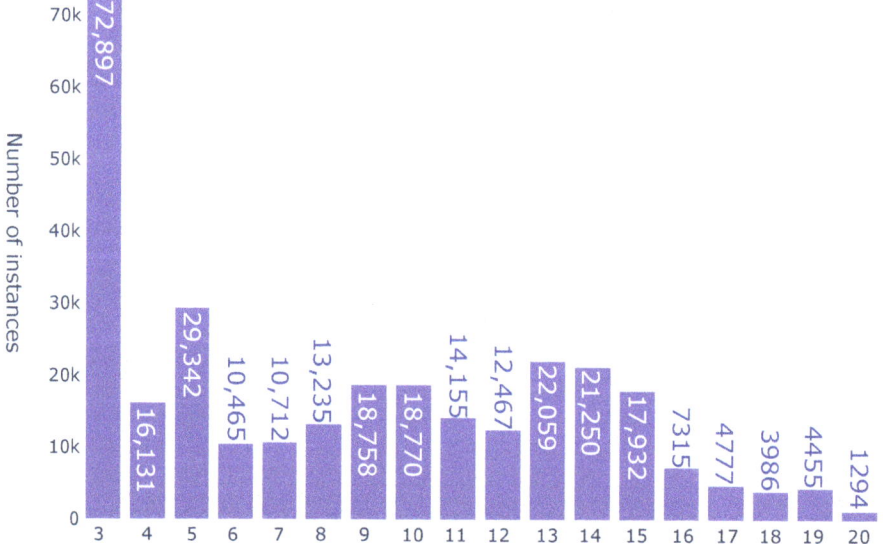

Figure 1. Number of branching points for the instances with 4 jobs.

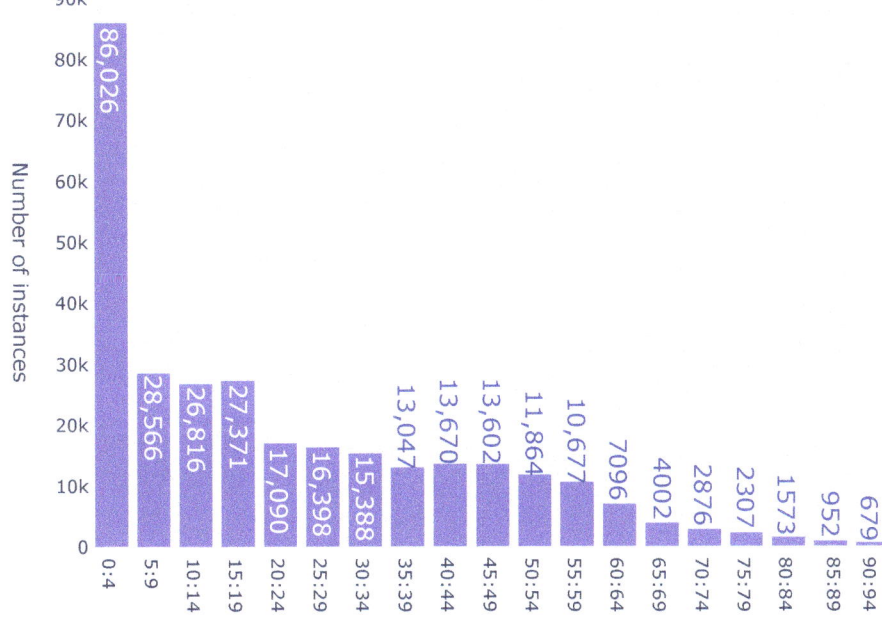

Figure 2. Number of branching points for the instances with 5 jobs.

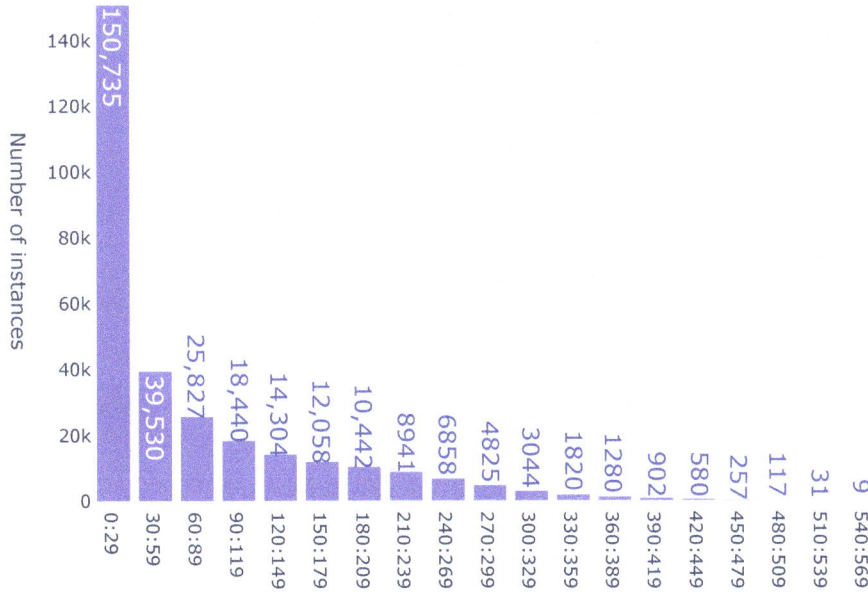

Figure 3. Number of branching points for the instances with 6 jobs.

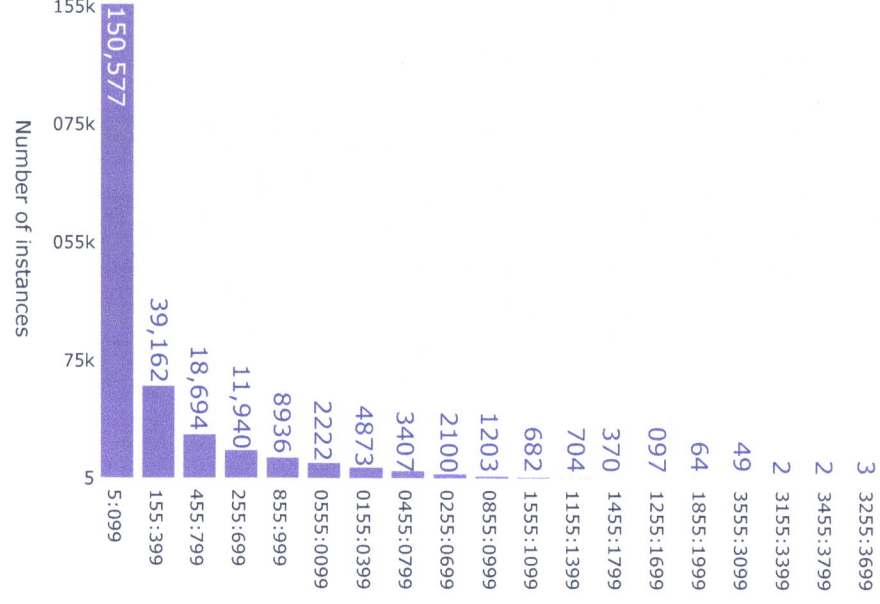

Figure 4. Number of branching points for the instances with 7 jobs.

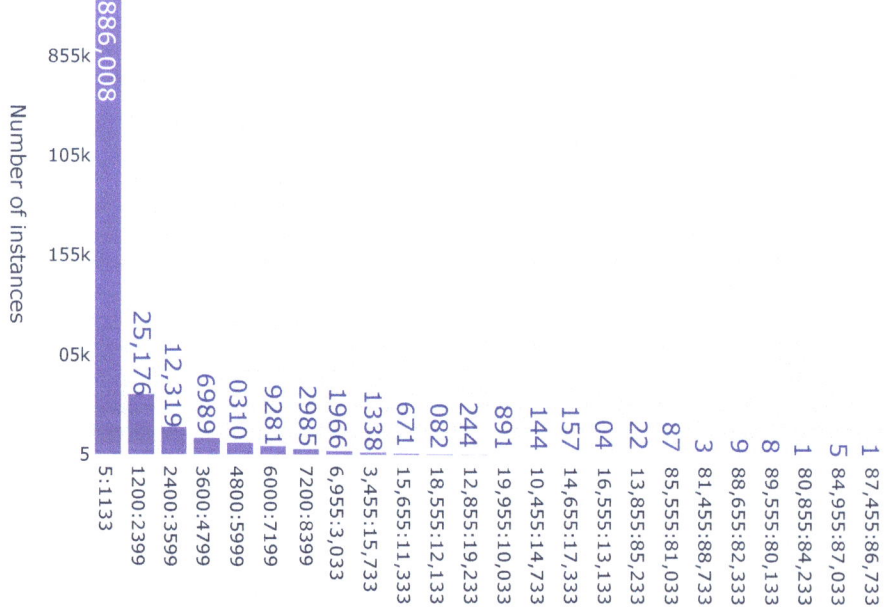

Figure 5. Number of branching points for the instances with 8 jobs.

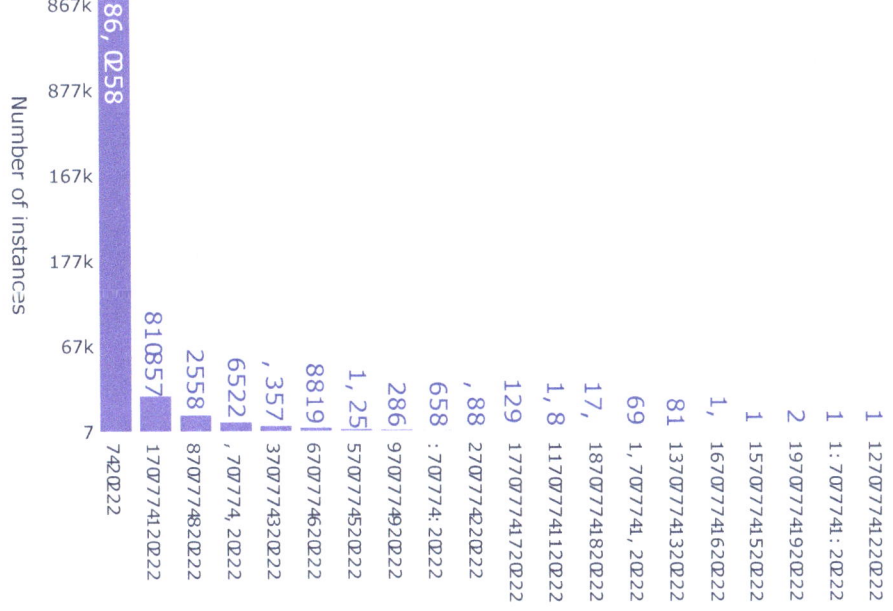

Figure 6. Number of branching points for the instances with 9 jobs.

5. Conclusions

For solving the *NP*-hard problem $1 \mid r_j \mid \varphi_{\max}$ for arbitrary non-decreasing penalty functions, an algorithm has been proposed which implements the branch and bound method. For each sub-instance to be considered, a lower bound on the optimal function value is determined using a solution of the dual problem. The proposed algorithm for solving the dual problem can find a solution in several operations not exceeding $O(n^2)$. The proposed algorithm can find an optimal solution within a time limit of one second for about 98% of the instances for 8 jobs and for about 85% of instances for 9 jobs. Although there are a few instances with a large number of branching points, most instances can be solved very fast by the proposed algorithm. For the hard instances, the execution of the algorithm can be interrupted at any moment, and the current objective function value with the corresponding schedule π can be used as an approximate solution for the instance. However, some generated instances appeared to be very hard. At the moment, we cannot explain this interesting phenomenon. It requires deep additional investigations which are planned in the future.

In addition to the dual problem, the inverse problem has also been solved for the lateness objective function. The algorithm for solving the inverse problem has a complexity of $O(n^2)$ operations. However, in the problem of minimizing maximum lateness, one tries to 'equalize' the lateness while minimizing the maximum, in the inverse problem the lateness values are 'equalized' due to the maximization of the minimum provided that inserted machine idle times are prohibited.

Author Contributions: The results described in this paper were obtained by communications and mutual works of the authors. Conceptualization, A.A.L. and N.P.; data curation, A.A.L. and N.P.; formal analysis, A.A.L., N.P. and F.W.; investigation, A.A.L., N.P. and F.W.; methodology, A.A.L., N.P. and F.W.; project administration, A.A.L.; software, N.P.; supervision, A.A.L. and F.W.; validation, A.A.L., N.P. and F.W.; visualization, N.P.; writing—original draft, A.A.L. and N.P.; writing—review & editing, F.W. All authors have read and agreed to the published version of the manuscript.

Funding: The research was partially supported by RFBR project 18-07-00656 and partially supported by the Basic Research Program of the National Research University Higher School of Economics.

Conflicts of Interest: The authors declare no conflict of interest.

References

1. Lenstra, J.K.; Rinnooy Kan, A.H.G.; Brucker, P. Complexity of Machine Scheduling Problems. *Ann. Discret. Math.* **1977**, *1*, 343–362.
2. Brooks, G.N.; White, C.R. An Algorithm for Finding Optimal or Near-optimal Solutions to the Production Scheduling Problem. *J. Ind. Eng.* **1965**, *16*, 34–40.
3. Schrage, L. Solving Resource-Constrained Network Problems by Implicit Enumeration: Non Preemptive Case. *Oper. Res.* **1970**, *18*, 263–278. [CrossRef]
4. Burduk, V.Y.; Shkurba, V.V. Scheduling Theory. Problems and Solving Methods. *Cybernetics* **1971**, *1*, 89–102. (In Russian)
5. McMahon, G.; Florian, M. On Scheduling with Ready Times and Due Dates to Minimize Maximum Lateness. *Oper. Res.* **1975**, *23*, 475–482. [CrossRef]
6. Korbut, A.A.; Sigal, I.H.; Finkelshtein, Y.Y. Branch and Bound Method (Survey of Theory, Algorithms, Programs and Applications). *Oper. Forsch. Stat. Ser. Optim.* **1977**, *8*, 253–280. (In Russian)
7. Carlier, J. The One-Machine Sequencing Problem. *Eur. J. Oper. Res.* **1982**, *11*, 42–47. [CrossRef]
8. Tanaev, V.S.; Gordon, V.S.; Shafranskiy, Y.M. *Scheduling Theory. Single-Stage Systems Scheduling Theory*; Springer-Science+Business Media, B.V.: Dordrecht, The Netherland, 1994; 374p.
9. Hoogeveen, J. Minimizing Maximum Promptness and Maximum Lateness on a Single Machine. *Math. Oper. Res.* **1996**, *21*, 100–114. [CrossRef]
10. Vakhania, N.; Werner, F. Minimizing Maximum Lateness of Jobs with Naturally Bounded Job Data on a Single Machine. *Theor. Comput. Sci.* **2013**, *501*, 72–81. [CrossRef]
11. Lageweg, B.; Lenstra, J.; Rinnooy Kan, A. Minimizing Maximum Lateness on one machine: Computational experience and applications. *Stat. Neerl.* **1976**, *30*, 25–41. [CrossRef]
12. Liu, Z. Single Machine Scheduling to Minimize Maximum Lateness Subject to Release Dates and Precendence Constraints. *Comp. Oper. Res.* **2010**, *37*, 1537–1543. [CrossRef]
13. Kacem, I.; Kellerer, H. Approximation Schemes for Minimizing Maximum Lateness. *Algorithmica* **2018**, *80*, 3825–3843. [CrossRef]
14. Wang, D.; Xu, D. Single-Machine Scheduling with Workload Dependent Maintenance Duration to Minimize Maximum Lateness. *Math. Probl. Eng.* **2015**, *2015*, 732437. [CrossRef]
15. Sels, V.; Vanhoucke, M. A Hybrid Elektromagnetism-like Mechanism/Tabu Search Procedure for the Single Machine Scheduling Problem with a Maximum Lateness Objective. *Comp. Ind. Eng.* **2014**, *67*, 44–55. [CrossRef]
16. Lazarev, A.A.; Lemtyuznikova, D.; Werner, F. *A General Approximation Approach for Multi-Machine Problems with Minimizing the Maximum penalty*; Preprint 04/19, FMA; OVGU: Magdeburg, Germany, 2019; 25p.
17. Vakhania, N. Scheduling a Single Machine with Primary and Secondary Objectives. *Algorithms* **2018**, *11*, 80. [CrossRef]
18. Lazarev, A.A. *Scheduling Theory. Methods and Algorithms*; ICS RAS: Moscow, Russia, 2019; 408p. (In Russian)
19. Lazarev, A.A. Pareto-optimal set for the NP-hard problem of minimization maximum lateness. Izvestiya of Russian Academy of Sciences. *Theory Syst. Control* **2016**, *6*, 103–110. (In Russian)
20. Lazarev, A.A.; Arkhipov, D.; Werner, F. Scheduling with Equal Processing Times on a Single Machine: Minimizing Maximum Lateness and Makespan. *Optim. Lett.* **2017**, *11*, 165–177. [CrossRef]
21. Aloulou, M.A.; Kovalyov; M.Y.; Portmann, M.-C. Maximization Problems in Single Machine Scheduling. *RAIRO Oper. Res.* **2007**, *41*, 1–18. [CrossRef]
22. Gafarov, E.R.; Lazarev, A.A.; Werner, F. Algorithms for Some Maximization Scheduling Problems on a Single Machine. *Autom. Remote. Control* **2010**, *71*, 2070–2084. [CrossRef]
23. Gafarov, E.R.; Lazarev, A.A.; Werner, F. Single machine Total Tardiness Maximization Problems. *Ann. Oper. Res.* **2013**, *207*, 121–136. [CrossRef]
24. Gafarov, E.R.; Lazarev, A.A.; Werner, F. Transforming a Pseudo-polynomial Algorithm for the Single Machine Total Tardiness Maximization Problem into a Polynomial One. *Ann. Oper. Res.* **2012**, *196*, 247–261. [CrossRef]

25. Lazarev, A.A. Dual of the Maximum Cost Minimization Problem. *J. Sov. Math.* **1989**, *44*, 642–644.
26. Agin, N. Optimum Seeking with Branch and Bounds. *Manag. Sci.* **1966**, *13*, 176–185. [CrossRef]
27. Lawler, E.L.; Wood, D.E. Branch and Bounds Methods: A Survey. *Oper. Res.* **1966**, *14*, 699–719. [CrossRef]
28. Conway, R.W.; Maxwell, W.L.; Miller, L.W. *Theory of Scheduling*; Addison-Wesley: Reading, MA, USA, 1967.
29. Hall, N.G.; Posner, M.E. Generating Experimental Data for Computational Testing with Machine Scheduling Applications. *Oper. Res.* **2001**, *49*, 854–865. [CrossRef]

© 2020 by the authors. Licensee MDPI, Basel, Switzerland. This article is an open access article distributed under the terms and conditions of the Creative Commons Attribution (CC BY) license (http://creativecommons.org/licenses/by/4.0/).

Article

Schedule Execution for Two-Machine Job-Shop to Minimize Makespan with Uncertain Processing Times

Yuri N. Sotskov [1,*], **Natalja M. Matsveichuk** [2,†] **and Vadzim D. Hatsura** [3,†]

1. United Institute of Informatics Problems, National Academy of Sciences of Belarus, Surganova Street 6, 220012 Minsk, Belarus
2. Belorussian State Agrarian Technical University, Nezavisimosti Avenue 99, 220023 Minsk, Belarus; matsveichuk@tut.by
3. Belorussian State University of Informatics and Radioelectronics, P. Brovki Street 6, 220013 Minsk, Belarus; vadimgatsura@gmail.com
* Correspondence: sotskov48@mail.ru; Tel.: +375-17-284-2120

Received: 4 June 2020; Accepted: 29 July 2020; Published: 7 August 2020

Abstract: This study addresses a two-machine job-shop scheduling problem with fixed lower and upper bounds on the job processing times. An exact value of the job duration remains unknown until completing the job. The objective is to minimize a schedule length (makespan). It is investigated how to best execute a schedule, if the job processing time may be equal to any real number from the given (closed) interval. Scheduling decisions consist of the off-line phase and the on-line phase of scheduling. Using the fixed lower and upper bounds on the job processing times available at the off-line phase, a scheduler may determine a minimal dominant set of schedules (minimal DS), which is based on the proven sufficient conditions for a schedule dominance. The DS optimally covers all possible realizations of the uncertain (interval) processing times, i.e., for each feasible scenario, there exists at least one optimal schedule in the minimal DS. The DS enables a scheduler to make the on-line scheduling decision, if a local information on completing some jobs becomes known. The stability approach enables a scheduler to choose optimal schedules for most feasible scenarios. The on-line scheduling algorithms have been developed with the asymptotic complexity $O(n^2)$ for n given jobs. The computational experiment shows the effectiveness of these algorithms.

Keywords: scheduling; job-shop; makespan criterion; uncertain processing times

1. Introduction

Many real-world production planning and scheduling problems have various uncertainties. Different approaches are used for solving the uncertain planning and scheduling problems. In particular, a stability approach [1–4] for solving sequencing and scheduling problems with the interval uncertainty is based on the stability analysis of the optimal job permutations (schedules) to possible variations of the job processing times (durations). In this paper, this approach is applied to the uncertain two-machine job-shop scheduling problem, where a job processing time is only known once the job is completed. Although, the exact value of the job processing time is unknown before scheduling, it is known that the processing time must have a value no less than the lower bound and no greater than the upper bound available before scheduling. It should be noted that uncertainties of the job processing times are due to some external forces in contrast to scheduling problems with controllable processing times [5–7], where the objective is to determine optimal processing times and then to find an optimal schedule for the jobs with the chosen processing times.

1.1. Research Motivation

It is not realistic to assume processing times are exactly known and fixed for many scheduling problems arising in real-world situations. For such an uncertain scheduling problem, job processing times are random variables. Moreover, it is often hard to obtain probability distributions for all random processing times of the jobs to be processed. In such cases, schedules constructed due to assuming certain probability distributions are often not close to the optimal schedule. Although, the probability distribution of the job processing time may not be known before scheduling, the upper and lower bounds on the job processing time are easy to obtain in most practical scheduling environments. The available information on these lower and upper bounds on the job processing times should be utilized in finding optimal schedules for the scheduling problem with an interval uncertainty.

Since there may not exist a unique schedule that remains optimal for all possible realizations of the job processing times (all possible scenarios), it is desirable to construct a minimal dominant set of schedules (permutations of the jobs to be processed), which dominate all other ones. At the off-line phase of scheduling (i.e., before starting an execution of the constructed schedule), a minimal dominant set of schedules may be determined based on the proven dominance relations [8].

If the constructed minimal dominant set of schedules is a singleton, then a single schedule remaining optimal for all possible scenarios exists. Otherwise, one can reduce the size of the determined minimal dominant set of schedules at the on-line phase of scheduling based on the additional information about completing some jobs. This additional on-line information allows a scheduler to find new dominance relations in order to best execute a schedule. It is clear that on-line scheduling decisions must be realized very quickly. In other words, only polynomial algorithms may be applied at the on-line phase of scheduling.

1.2. Contributions of This Research

In this paper, it is shown how to determine a minimal dominant set of schedules that would contain at least one optimal schedule for every scenario that is possible. The necessary and sufficient conditions are proven for the existence of a single pair of job permutations, which is optimal for the two-machine job-shop scheduling problem with any possible scenario. The algorithms have been developed for testing a set of the proven sufficient conditions for a schedule dominance and for the realization of a schedule, which is either optimal or very close to optimal one for the factual scenario. The developed algorithms are polynomial in the number n of the given jobs. Their asymptotic complexities do not exceed $O(n^2)$. The computational experiments on a large number of randomly generated instances of the uncertain (interval) two-machine job-shop scheduling problem show the efficiency and effectiveness of the developed off-line and on-line algorithms and programs. For different distributions of the factual job processing times, the developed on-line algorithms perform with the maximal errors of the achieved makespan less than 1% provided that $n \in \{20, 30, \ldots, 100\}$. For all tested classes of the randomly generated instances, the average makespan errors $\Delta_{ave}\%$ for all tested numbers $n \in \{10, 20, \ldots, 100\}$ of jobs $\mathcal{J}$ are less than 0.02%. Each tested series of 1000 randomly generated instances was solved within no more than one second.

The paper is organized as follows. Settings of the considered scheduling problems with the interval uncertainty and main notation are introduced in Section 2. A literature review is presented in Section 3. The results published for the uncertain (interval) scheduling flow-shop problem are discussed in Section 3.2. These results are used in Section 4 for finding the optimal job permutations at the off-line phase of scheduling. In Section 4.2, the precedence digraphs are described for determining a minimal dominant set of schedules. An illustrative example is considered in Section 4.3. The on-line phase of scheduling is investigated in Section 5, where two theorems for the dominant sets of schedules have been proven. Section 6 contains the algorithms developed for the on-line phase of scheduling, illustrative examples (Section 6.2) and the discussion of the conducted computational experiments (Section 6.3). Appendix B consists of the tables with the detailed computational results. Some concluding remarks are made in Section 7.

2. Settings of Scheduling Problems and Main Notations

A set $\mathcal{J} = \{J_1, J_2, \ldots, J_n\}$ of the given jobs must be processed on different machines from a set $\mathcal{M} = \{M_1, M_2\}$. All jobs are available for processing from the same time $t = 0$. Using the standard notation $\alpha|\beta|\gamma$ [9], this deterministic two-machine job-shop scheduling problem to minimize the makespan is denoted as follows: $J2|n_i \leq 2|C_{max}$, where $\alpha = J2$ means a job-shop processing system with two available different machines and n_i a number of possible stages for processing a job $J_i \in \mathcal{J}$. The criterion $\gamma = C_{max}$ determines the minimization of a schedule length (makespan) as follows:

$$C_{max} := \min_{s \in S} C_{max}(s) = \min_{s \in S} \{\max\{C_i(s) : J_i \in \mathcal{J}\}\}, \tag{1}$$

where $C_i(s)$ denotes the completion time-point of the job $J_i \in \mathcal{J}$ in the schedule s and S denotes a set of all semi-active schedules existing for the deterministic problem $J2|n_i \leq 2|C_{max}$. (A schedule s is called a semi-active one [10–12] if the completion time-point $C_i(s)$ of any job $J_i \in \mathcal{J}$ cannot be reduced without changing an order of the jobs on some machine.)

Let O_{ij} denote an operation of the job $J_i \in \mathcal{J}$ processed on the machine $M_j \in \mathcal{M}$. Each of the available machines can process the job $J_i \in \mathcal{J}$ no more than once, a preemption of the operation O_{ij} being not allowed. The job $J_i \in \mathcal{J}$ has its own processing route through the available machines in set $\mathcal{M}$. The partition $\mathcal{J} = \mathcal{J}_1 \cup \mathcal{J}_2 \cup \mathcal{J}_{1,2} \cup \mathcal{J}_{2,1}$ of the jobs is given and fixed, where each job $J_i \in \mathcal{J}_{1,2}$ must be processed first on machine M_1 and then on machine M_2, i.e., all jobs from the set $\mathcal{J}_{1,2}$ have the same machine route (M_1, M_2). Each job $J_i \in \mathcal{J}_{2,1}$ has an opposite machine route (M_2, M_1). The set $\mathcal{J}_j$, where $j \in \{1, 2\}$, consists of all jobs, which must be processed only on one machine $M_j \in \mathcal{M}$. The following notation $m_h = |\mathcal{J}_h|$ will be used, where $h \in \{1; 2; 1,2; 2,1\}$.

In this research, it is investigated the uncertain (interval) two-machine job-shop scheduling problem denoted as $J2|l_{ij} \leq p_{ij} \leq u_{ij}, n_i \leq 2|C_{max}$, where the duration p_{ij} of each operation O_{ij} is unknown before scheduling. It is only known that the inclusion $p_{ij} \in [l_{ij}, u_{ij}]$ holds for any possible realization of the chosen schedule, where $u_{ij} \geq l_{ij} \geq 0$. It is also assumed that a probability distribution of the random duration of a job from the set $\mathcal{J}$ is also unknown before scheduling. Let a set T of all possible scenarios $p = (p_{1,1}, p_{1,2}, \ldots, p_{n1}, p_{n2})$ of the job processing times be determined as follows:

$$T = \{p : l_{ij} \leq p_{ij} \leq u_{ij}, J_i \in \mathcal{J}, M_j \in \mathcal{M}\}.$$

It should be noted that the problem $J2|l_{ij} \leq p_{ij} \leq u_{ij}, n_i \leq 2|C_{max}$ is mathematically incorrect since one cannot calculate makespan $C_{max}(s)$ in the equality (1) before completing the jobs J_i in the set $\mathcal{J}$ provided that the strict inequality $u_{ij} > l_{ij}$ holds. Moreover, in most cases there does not exist a schedule, which is optimal for all possible scenarios $p \in T$ for the uncertain job-shop problem $J2|l_{ij} \leq p_{ij} \leq u_{ij}, n_i \leq 2|C_{max}$. Therefore, one cannot solve most such uncertain (interval) scheduling problems in the generally accepted sense.

In [13], it is proven that the deterministic job-shop problem $J2|n_i \leq 2|C_{max}$ is solvable in $O(n \log n)$ time. The optimal semi-active schedule for this deterministic problem is determined as the pair (π', π'') of two job permutations (called a Jackson's pair of permutations), where $\pi' = (\pi_{1,2}, \pi_1, \pi_{2,1})$ is an optimal permutation of the jobs $\mathcal{J}_1 \cup \mathcal{J}_{1,2} \cup \mathcal{J}_{2,1}$ processed on machine M_1 and $\pi'' = (\pi_{2,1}, \pi_2, \pi_{1,2})$ is an optimal permutation of the jobs $\mathcal{J}_2 \cup \mathcal{J}_{1,2} \cup \mathcal{J}_{2,1}$ on machine M_2. Such an optimal semi-active schedule is presented in Figure 1. In what follows, it is assumed that job J_i belongs to the permutation π_h, if the following inclusion holds: $J_i \in \mathcal{J}_h$.

In a Jackson's pair of permutations (π', π''), the optimal order for processing jobs from the set $\mathcal{J}_1$ (from the set $\mathcal{J}_2$, respectively) may be arbitrary (due to this, we fix them in the increasing order of their indexes). For the permutation $\pi_{1,2}$ (permutation $\pi_{2,1}$, respectively), the following inequality holds:

$$\min\{p_{i_e1}, p_{i_f2}\} \leq \min\{p_{i_f1}, p_{i_e2}\} \tag{2}$$

for all indexes e and f provided that $1 \leq e < f \leq m_{1,2}$ ($1 \leq f < e \leq m_{2,1}$, respectively). The permutation $\pi_{1,2}$ (permutation $\pi_{2,1}$) is called a Johnson's permutation; see [14].

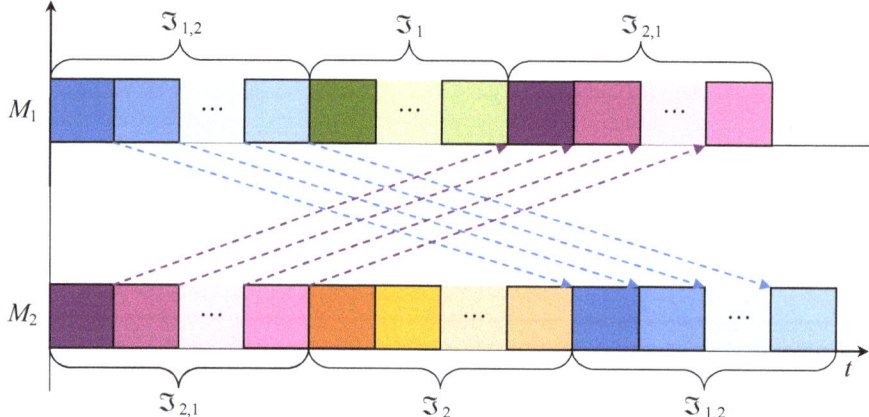

Figure 1. An example of the optimal semi-active schedule without idle times on both machines.

The deterministic scheduling problem $J2|n_i \leq 2|C_{max}$ associated with a fixed scenario p of the job processing times is an individual deterministic problem. In what follows, this problem is denoted as follows: $J2|p, n_i \leq 2|C_{max}$. For any fixed scenario $p \in T$, there exists a Jackson's pair (π', π'') of permutations, which is optimal for the problem $J2|p, n_i \leq 2|C_{max}$, i.e., the equality $C_{max}(\pi', \pi'') = C_{max}^p$ holds, where C_{max}^p denotes the optimal makespan value for the problem $J2|p, n_i \leq 2|C_{max}$.

Let $S_{1,2}$ denote a set of all permutations of $m_{1,2}$ jobs from the set $\mathcal{J}_{1,2}$, where $|S_{1,2}| = m_{1,2}!$. The set $S_{2,1}$ is a set of all permutations of $m_{2,1}$ jobs from the set $\mathcal{J}_{2,1}$, $|S_{2,1}| = m_{2,1}!$.

Let the set $S = <S_{1,2}, S_{2,1}>$ be a subset of the Cartesian product $(S_{1,2}, \pi_1, S_{2,1}) \times (S_{2,1}, \pi_2, S_{1,2})$, each element of the set S being a pair of job permutations $(\pi', \pi'') \in S$, where $\pi' = (\pi_{1,2}^i, \pi_1, \pi_{2,1}^j)$ and $\pi'' = (\pi_{2,1}^j, \pi_2, \pi_{1,2}^i)$ with inequalities $1 \leq i \leq m_{1,2}!$ and $1 \leq j \leq m_{2,1}!$. It is known that the set S determines all semi-active schedules and vice versa; see [12]. Since index i (and index j) is the same in each permutation from the pair $(\pi', \pi'') \in S$ and it is a fixed permutation π_1 (permutation π_2), the equality $|S| = m_{1,2}! \cdot m_{2,1}!$ holds. The following definition of a J-solution is used for the uncertain (interval) job-shop scheduling problem $J2|l_{ij} \leq p_{ij} \leq u_{ij}, n_i \leq 2|C_{max}$.

Definition 1. *An inclusion-minimal set of the pairs of job permutations $S(T) \subseteq S$ is called a J-solution for the uncertain job-shop problem $J2|l_{ij} \leq p_{ij} \leq u_{ij}, n_i \leq 2|C_{max}$ with the set $\mathcal{J}$ of the given jobs, if for each scenario $p \in T$, the set $S(T)$ contains at least one pair $(\pi', \pi'') \in S$ of job permutations that is optimal for the individual deterministic problem $J2|p, n_i \leq 2|C_{max}$ with a fixed scenario p.*

From Definition 1, it follows that for any proper subset S' of the set $S(T)$, $S' \subset S(T)$, there exists a scenario $p' \in T$ such that the set S' does not contain an optimal pair of job permutations for the individual deterministic problem $J2|p', n_i \leq 2|C_{max}$ with a fixed scenario p'.

3. A Literature Review and Closed Results

It should be noted that the uncertain flow-shop scheduling problem denoted as $F2|l_{ij} \leq p_{ij} \leq u_{ij}, n_i \leq 2|C_{max}$ is well studied [15], unlike the uncertain job-shop scheduling problem.

3.1. Approaches to Scheduling Problems with Different Forms of Uncertainties

For the well-known stochastic approach, it is assumed that the job processing times are random variables with certain probability distributions determined before scheduling. There are two types of the stochastic scheduling problems [10], where one is on stochastic jobs and another is on stochastic machines. In the stochastic job scheduling problem, each job processing time is a random variable with a known probability distribution. With the objective of minimizing the expected makespan value, the flow-shop problem was studied in [16–18]. In the stochastic machine scheduling problem, each job processing time is a constant, while each completion time of the given job is a random variable due to the machine breakdown or machine non-availability. In [19–21], the flow-shop scheduling problems to stochastically minimize either makespan or total completion time were investigated.

If it is impossible to determine probability distributions for all random job processing times, other approaches have to be used [11,22–25]. In the approach of seeking a robust schedule [22,26–28], a decision-maker looks for a schedule that hedges against the worst-case possible scenario.

A fuzzy approach [29–35] allows a scheduler to find best schedules with respect to fuzzy processing times of the jobs to be processed. The work of [35] addresses to the job-shop scheduling problem with uncertain processing times modeled as triangle fuzzy numbers, where the criterion is to minimize the expected makespan value. Based on the disjunctive graph model of the job-shop problem, a definition of criticality is proposed for this job-shop problem along with neighborhood structure for a local search. It is shown that the proposed neighborhood structure has two properties: feasibility and connectivity, which allow a scheduler to improve the efficiency of the local search and to ensure asymptotic convergence (in probability) to a globally optimal solution of the uncertain job-shop problem. The conducted computational experiments supported these theoretical results.

The stability approach was developed in [1,4,36,37] for the C_{max} criterion, and in [2,38–40] for the total completion time criterion, $\sum C_i := \sum_{J_i \in \mathcal{J}} C_i(\pi')$. The aim of this approach is to construct a minimal dominant set $S(T)$ of schedules, which optimally covers all feasible scenarios T. The dominant set $S(T)$ is used in the multi-phase decision framework; see [41]. The set $S(T)$ is constructed at the first off-line phase of scheduling. Based on the set $S(T)$, it is possible to find a schedule remaining optimal for most feasible scenarios. The set $S(T)$ enables a scheduler to execute best a schedule in most cases of the uncertain flow-shop scheduling problem $F2|l_{ij} \leq p_{ij} \leq u_{ij}|C_{max}$ [41].

The stability radius of the optimal semi-active schedule was studied in [4], where a formula for calculating the stability radius and corresponding algorithms were described and tested.

In [36], the sufficient conditions were proven when a transposition of the given jobs minimizes the makespan criterion. The work of [42] addressed the objective criterion $\sum C_i$ in the uncertain two-machine flow-shop scheduling problem. The case of separate setup times with the criterion of minimizing a total completion time or makespan was investigated in [43].

For the uncertain flow-shop problem $F2|l_{ij} \leq p_{ij} \leq u_{ij}|C_{max}$, an additional criterion is often introduced. In particular, a robust schedule minimizing the worst-case deviation from the optimal value was proposed in [44] to hedge against the interval or discrete uncertainties. In [45], a binary NP-hardness was proven for finding a pair $(\pi_q, \pi_q) \in S$ of the identical job permutations that minimizes the worst-case absolute regret for the uncertain two-machine flow-shop problem with the criterion C_{max} and only two possible scenarios. In [46], a branch and bound method was developed for the uncertain job-shop scheduling problem to minimize makespan and optimize robustness based on a mixed graph model and the propositions proposed in [47]. The effectiveness of the developed algorithm was clarified by solving test uncertain job-shop scheduling problems.

The work of [48] addresses robust scheduling for a flexible job-shop scheduling problem with a random machine breakdown. Two objectives makespan and robustness were considered. Robustness was indicated by the expected value of the relative difference between the deterministic and factual makespan values. Two measures for robustness have been developed. The first suggested measure considers the probability of machine breakdowns. The second measure considers the location of float times and machine breakdowns. A multi-objective evolutionary algorithm is presented and experimentally compared with several other existing measures.

A function of predictive scheduling in order to obtain a stable and robust schedule for a shop floor was investigated in [49]. An innovative maintenance planning and production scheduling method has been proposed. The proposed method uses a database to collect information about failure-free times, a prediction module of failure-free times, predictive rescheduling module, a module for evaluating the accuracy of prediction and maintenance performance. The proposed approach is based on probability theory and applied for solving a job-shop scheduling problem. For unpredicted failures, a rescheduling procedure was also developed. The evaluation procedure provides information about the degradation of a performance measure and the stability of a schedule.

The simulation and experimental design methods play a useful role in solving job-shop scheduling problems with uncertain parameters (see survey [50], where many studies about dynamic and static job-shop scheduling problems with material handling are described and systematized).

In [51], a quality robustness and a solution robustness were investigated in order to compare the operational efficiency of the job-shop in the events of machine failures. Two well-known proactive approaches were compared to compute the operational efficiency of the job-shop with unpredicted machine failures. In the computational experiments, the predictive-reactive approach (without a prediction) and the proactive-reactive one (with a prediction) were applied for the job-shop model with possible disruptions. The computational results of computer simulations for the above two approaches were compared in order to select better schedules for reducing costs and waste due to machine failures.

The paper [52] presents a methodological pattern to assess the effectiveness of Order Review and Release (ORR) techniques in a job-shop environment. It is presented a comparison among three ORR approaches, i.e., a time bucketing approach, a probabilistic approach and a temporal approach. Simulation results highlighted that the performances of the ORR techniques tested depend on how perturbed the environment, where they are implemented, is. Based on a computer simulation, it was shown that the ORR techniques greatly differ in their robustness against environment perturbations.

The paper [53] presents an effective heuristic algorithm for the job-shop problem with uncertain arrival times of the jobs, processing times, due dates and part priorities. A separable problem formulation that balances modeling accuracy and solution complexity is described with the goal to minimize expected part tardiness and earliness cost. The optimization is subject to arrival times and operation precedence constraints (for each possible realization), and machine capacity constraints (in the expected value sense). The solution algorithm based on a Lagrangian relaxation and stochastic dynamic programming was developed to obtain dual solutions. The computational complexity of the developed algorithm is only slightly higher than the one without considering uncertainties of the numerical parameters. Numerical testing supported by a simulation demonstrated that near optimal solutions were obtained, and uncertainties are effectively handled for problems of practical sizes.

The published results on the application of the stability approach for the uncertain two-machine flow-shop problem are presented in Section 3.2. These results are described in detail since they are used for the uncertain job-shop problem $J2|l_{ij} \leq p_{ij} \leq u_{ij}, n_i \leq 2|C_{max}$ in Sections 4–6.

3.2. Closed Results for Uncertain (Interval) Flow-Shop Scheduling Problems

The uncertain job-shop problem $J2|l_{ij} \leq p_{ij} \leq u_{ij}, n_i \leq 2|C_{max}$ is a generalization of the uncertain flow-shop problem $F2|l_{ij} \leq p_{ij} \leq u_{ij}|C_{max}$, where all given jobs have the same machine route. Two uncertain flow-shop problems are associated with an uncertain job-shop problem $J2|l_{ij} \leq p_{ij} \leq u_{ij}, n_i \leq 2|C_{max}$. In one of these flow-shop problems, an optimal schedule for processing the jobs $\mathcal{J}_{1,2}$ must be determined, i.e., it is assumed that $\mathcal{J}_{2,1} = \mathcal{J}_1 = \mathcal{J}_2 = \emptyset$. In another associated flow-shop problem, an optimal schedule for processing jobs $\mathcal{J}_{2,1}$ must be determined, i.e., it is assumed that $\mathcal{J}_{1,2} = \mathcal{J}_1 = \mathcal{J}_2 = \emptyset$. Our approach to the solution of the uncertain job-shop scheduling problem $J2|l_{ij} \leq p_{ij} \leq u_{ij}, n_i \leq 2|C_{max}$ is based on the following remark.

Remark 1. *The solution of the uncertain job-shop scheduling problem $J2|l_{ij} \leq p_{ij} \leq u_{ij}, n_i \leq 2|C_{max}$ may be based on the solutions of the associated flow-shop scheduling problem $F2|l_{ij} \leq p_{ij} \leq u_{ij}|C_{max}$ with the job set $\mathcal{J} = \mathcal{J}_{1,2}$, where $\mathcal{J}_{2,1} = \mathcal{J}_1 = \mathcal{J}_2 = \emptyset$, and that with the job set $\mathcal{J} = \mathcal{J}_{2,1}$ (i.e., $\mathcal{J}_{1,2} = \mathcal{J}_1 = \mathcal{J}_2 = \emptyset$).*

The sense of Remark 1 becomes clear from Figure 2, where the semi-active schedule s for the job-shop scheduling problem $J2|l_{ij} \leq p_{ij} \leq u_{ij}, n_i \leq 2|C_{max}$ is presented. Indeed, in Figure 2, the length $C_{max}(s)$ of the schedule s is equal to the length of the corresponding semi-active schedule determined for the associated flow-shop scheduling problem $F2|l_{ij} \leq p_{ij} \leq u_{ij}|C_{max}$ with the job set $\mathcal{J} = \mathcal{J}_{1,2}$. Thus, if one will solve both associated flow-shop problem $F2|l_{ij} \leq p_{ij} \leq u_{ij}|C_{max}$ with the job set $\mathcal{J} = \mathcal{J}_{1,2}$ and associated flow-shop problem $F2|l_{ij} \leq p_{ij} \leq u_{ij}|C_{max}$ with the job set $\mathcal{J} = \mathcal{J}_{2,1}$, then the original job-shop scheduling problem $J2|l_{ij} \leq p_{ij} \leq u_{ij}, n_i \leq 2|C_{max}$ will be also solved.

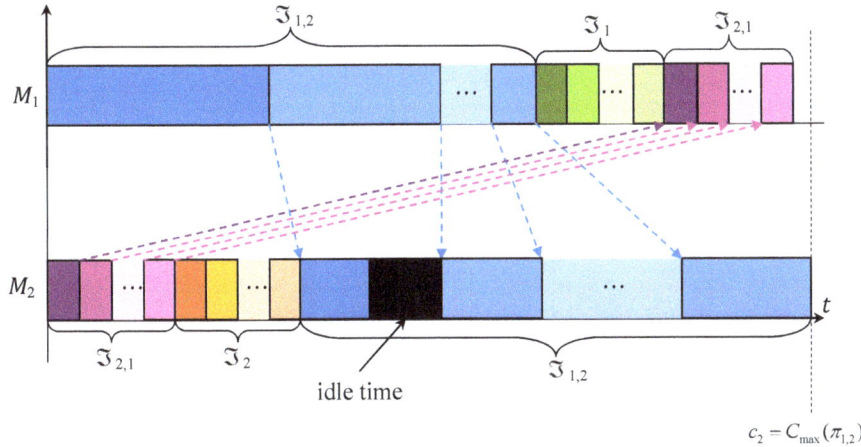

Figure 2. The optimal semi-active schedule for the job-shop scheduling problem.

We next observe in detail the results obtained for the two-machine flowshop problem $F2|l_{ij} \leq p_{ij} \leq u_{ij}|C_{max}$ with the job set $\mathcal{J} = \mathcal{J}_{1,2}$. For using the above notations introduced for the uncertain job-shop problem, we need the following remark for the uncertain flow-shop problem.

Remark 2. *The considered problem $F2|l_{ij} \leq p_{ij} \leq u_{ij}|C_{max}$ has the following two mandatory properties:*

(i) *the set S is a set of $n!$ pairs (π_q, π_q) of the identical permutations of $n = m_{1,2}$ jobs from the set $\mathcal{J} = \mathcal{J}_{1,2}$ since the machine route for processing all jobs $\mathcal{J}_{1,2}$ is the same (M_1, M_2);*

(ii) *the J-solution (see Definition 1) is a set of Johnson's permutations of the jobs $\mathcal{J} = \mathcal{J}_{1,2}$, i.e., for each scenario $p \in T$ the set $S(T)$ contains at least one optimal pair (π_q, π_q) of identical Johnson's permutations π_q such that the inequality (2) holds for all indexes e and f.*

The following Theorems 1 and 2 have been proven in [54].

Theorem 1 ([54]). *There exists a J-solution $S(T)$ for the uncertain flow-shop problem $F2|l_{ij} \leq p_{ij} \leq u_{ij}|C_{max}$ with a fixed order $J_v \to J_w$ of the jobs J_v and J_w in all permutations π_q, $(\pi_q, \pi_q) \in S(T)$, if and only if at least one of the following two conditions hold:*

$$u_{v1} \leq l_{v2} \text{ and } u_{v1} \leq l_{w1}; \quad (3)$$

$$u_{w2} \leq l_{w1} \text{ and } u_{w2} \leq l_{v2}. \quad (4)$$

Theorem 2 provides the necessary and sufficient conditions for existing a single-element J-solution $S(T) = \{(\pi_q, \pi_q)\}$ for the uncertain flow-shop scheduling problem $F2|l_{ij} \leq p_{ij} \leq u_{ij}|C_{max}$. The partition $\mathcal{J} = \mathcal{J}^0 \cup \mathcal{J}^1 \cup \mathcal{J}^2 \cup \mathcal{J}^*$ of the set $\mathcal{J} = \mathcal{J}_{1,2}$ is given, where

$\mathcal{J}^0 = \{J_i \in \mathcal{J} : u_{i1} \leq l_{i2}, u_{i2} \leq l_{i1}\}$,
$\mathcal{J}^1 = \{J_i \in \mathcal{J} : u_{i1} \leq l_{i2}, u_{i2} > l_{i1}\} = \{J_i \in \mathcal{J} \setminus \mathcal{J}^0 : u_{i1} \leq l_{i2}\}$,
$\mathcal{J}^2 = \{J_i \in \mathcal{J} : u_{i1} > l_{i2}, u_{i2} \leq l_{i1}\} = \{J_i \in \mathcal{J} \setminus \mathcal{J}^0 : u_{i2} \leq l_{i1}\}$,
$\mathcal{J}^* = \{J_i \in \mathcal{J} : u_{i1} > l_{i2}, u_{i2} > l_{i1}\}$.

Note that for each job $J_g \in \mathcal{J}^0$, the inequalities $u_{g1} \leq l_{g2}$ and $u_{g2} \leq l_{g1}$ imply the equalities $l_{g1} = u_{g1} = l_{g2} = u_{g2}$. Thus, the equalities $p_{g1} = p_{g2} =: p_g$ hold.

Theorem 2 ([54]). *There exists a single-element J-solution $S(T) \subset S$, $|S(T)| = 1$, for the uncertain flow-shop problem $F2|l_{ij} \leq p_{ij} \leq u_{ij}|C_{max}$, if and only if the following two conditions hold:*

(j) for any pair of jobs J_i and J_j from the set $\mathcal{J}^1$ (from the set $\mathcal{J}^2$, respectively), either $u_{i1} \leq l_{j1}$ or $u_{j1} \leq l_{i1}$ (either $u_{i2} \leq l_{j2}$ or $u_{j2} \leq l_{i2}$, respectively);

(jj) inequality $|\mathcal{J}^*| \leq 1$ holds and for the job $J_{i*} \in \mathcal{J}^*$ both inequalities $l_{i*1} \geq \max\{u_{i1} : J_i \in \mathcal{J}^1\}$, $l_{i*2} \geq \max\{u_{j2} : J_j \in \mathcal{J}^2\}$ hold with inequality $\max\{l_{i*1}, l_{i*2}\} \geq p_g$ valid for each job $J_g \in \mathcal{J}^0$.

Theorem 2 characterizes the simplest case of the uncertain flow-shop problem $F2|l_{ij} \leq p_{ij} \leq u_{ij}|C_{max}$, i.e., there is a job permutation π_q dominating all others.

Let $\mathcal{J} \times \mathcal{J}$ denote a Cartesian product of the set $\mathcal{J}$. If $\mathcal{J}^0 = \emptyset$, then there exists the following binary relation $\mathcal{A}_\prec \subseteq \mathcal{J} \times \mathcal{J}$ over the set $\mathcal{J} = \mathcal{J}_{1,2}$.

Definition 2. *For the jobs $J_x \in \mathcal{J}$ and $J_y \in \mathcal{J}$, the inclusion $(J_x, J_y) \in \mathcal{A}_\prec$ holds if and only if at least one of the conditions (3) and (4) holds with $v = x$ and $w = y$ and neither the condition (3) no the condition (4) holds with $v = y$ and $w = x$ (or at least one of the conditions (3) and (4) holds both with $v = x$ and $w = y$ and with $v = y$, $w = x$ and $x < y$).*

The above relation $(J_x, J_y) \in \mathcal{A}_\prec$ may be represented as follows: $J_x \prec J_y$. The binary relation $\mathcal{A}_\prec$ is a strict order [55] that determines the precedence digraph $\mathcal{G} = (\mathcal{J}, \mathcal{A}_\prec)$ with the vertex set $\mathcal{J}$ and the arc set $\mathcal{A}_\prec$. The permutation $\pi_q = (J_{q_1}, J_{q_2}, \ldots, J_{q_n})$, $(\pi_q, \pi_q) \in S$, is a total strict order over the set $\mathcal{J}$. The total strict order determined by the permutation π_q is a linear extension of the partial strict order $\mathcal{A}_\prec$, if the inclusion $(J_{q_x}, J_{q_y}) \in \mathcal{A}_\prec$ implies the inequality $x < y$. Let $\Pi(\mathcal{G})$ denote a set of all permutations $\pi_q \in S_{1,2}$ determining linear extensions of the partial strict order $\mathcal{A}_\prec$. The equality $\Pi(\mathcal{G}) = \{\pi_q\}$ is characterized in Theorem 2, where the strict order $\mathcal{A}_\prec$ over the set $\mathcal{J}$ is represented as follows: $J_{q_1} \prec \ldots \prec J_{q_i} \prec J_{q_{i+1}} \prec \ldots \prec J_{q_n}$. The following two claims have been proven in [55].

Theorem 3 ([55]). *For any scenario $p \in T$, the set $\Pi(\mathcal{G})$ contains a Johnson's permutation for the deterministic flow-shop problem $F2|p|C_{max}$ with the job set $\mathcal{J} = \mathcal{J}_{1,2} = \mathcal{J}^* \cup \mathcal{J}^1 \cup \mathcal{J}^2$.*

Corollary 1 ([55]). *There exists a J-solution $S(T)$ for the uncertain flow-shop problem $F2|l_{ij} \leq p_{ij} \leq u_{ij}|C_{max}$ with the job set $\mathcal{J} = \mathcal{J}_{1,2} = \mathcal{J}^* \cup \mathcal{J}^1 \cup \mathcal{J}^2$, such that the inclusion $\pi_q \in \Pi(\mathcal{G})$ holds for all pairs of job permutations, where $(\pi_q, \pi_q) \in S(T)$.*

In [55], it is shown how to determine a minimal dominant set $S(T) = \{(\pi_q, \pi_q)\}$ with $\pi_q \in \Pi(\mathcal{G})$. The digraph $\mathcal{G} = (\mathcal{J}, \mathcal{A}_\prec)$ is considered as a condense form of a J-solution for the uncertain flow-shop problem $F2|l_{ij} \leq p_{ij} \leq u_{ij}|C_{max}$. The above results are used in Sections 4–6 for reducing a size of the dominant set $S(T)$ for the uncertain job-shop problem $J2|l_{ij} \leq p_{ij} \leq u_{ij}, n_i \leq 2|C_{max}$.

4. The Off-Line Phase of Scheduling

The above setting of the uncertain job-shop scheduling problem $J2|l_{ij} \leq p_{ij} \leq u_{ij}, n_i \leq n|C_{max}$ implies the following remark.

Remark 3. *The factual value p_{ij}^* of the job processing time p_{ij} becomes known at the time-point $c_j(i)$ when the operation O_{ij} is completed on the machine $M_j \in \mathcal{M}$.*

Due to Remark 3, if all jobs $\mathcal{J}$ are completed on the corresponding machines from the set $\mathcal{M}$, the durations of all operations O_{ij} take on exact values p_{ij}^*, where $l_{ij} \leq p_{ij}^* \leq u_{ij}$, and a unique factual scenario $p^* \in T$ is realized. A pair of job permutations selected for this realization should be optimal for scenario p^*. For constructing such an optimal pair of job permutations, we propose to implement two phases, namely: the off-line phase of scheduling and the on-line phase of scheduling.

The off-line phase is completed before starting a realization of the selected semi-active schedule. At the off-line phase, a scheduler knows the exact lower and upper bounds on the job processing times and the aim is to determine a minimal dominant set of the pairs of job permutations (π', π'').

The on-line phase is started when the corresponding machine starts the processing of the first job in the selected schedule. At this phase, a scheduler can use an additional information on the job processing time, since for each operation O_{ij}, the exact value p_{ij}^* of the processing time $p_{ij} \in T$ becomes known at the completion time $c_j(i)$ of this operation; see Remark 3.

We next consider the off-line phase of scheduling for the uncertain job-shop problem $J2|l_{ij} \leq p_{ij} \leq u_{ij}, n_i \leq 2|C_{max}$ and describe the sufficient conditions for existing a small dominant set of the semi-active schedules. Along with Definition 1, the following one is also used.

Definition 3. *A set of the pairs of job permutations $DS(T) \subseteq S$ is a dominant set for the uncertain job-shop problem $J2|l_{ij} \leq p_{ij} \leq u_{ij}, n_i \leq 2|C_{max}$, if for each scenario $p \in T$ the set $DS(T)$ contains at least one optimal pair of job permutations for the individual deterministic job-shop problem $J2|p, n_i \leq 2|C_{max}$ with scenario p.*

Obviously, the J-solution is a dominant set for the uncertain job-shop problem $J2|l_{ij} \leq p_{ij} \leq u_{ij}, n_i \leq 2|C_{max}$. Before processing the set $\mathcal{J}$ of given jobs, a scheduler does not know the exact values of the job processing times. Nevertheless, it is needed to determine an optimal pair of permutations of the jobs $\mathcal{J}$ for their processing on the machines $\mathcal{M} = \{M_1, M_2\}$.

In Section 4.1, the sufficient conditions are presented for existing a pair of job permutations (π', π'') such that the equality $DS(T) = \{(\pi', \pi'')\}$ holds. Section 4.2 contains the sufficient conditions allowing a scheduler to construct a semi-active schedule (if any), which dominates all other schedules in the set S. If a singleton $DS(T) = \{(\pi', \pi'')\}$ does not exist, a scheduler should construct partial strict orders $A_{\preceq}^{1,2}$ and $A_{\preceq}^{2,1}$ over set $\mathcal{J}_{1,2}$ and set $\mathcal{J}_{2,1}$; see Section 3.

4.1. Conditions for Existing a Single Optimal Pair of Job Permutations

The following conditions for existing an optimal pair of job permutations are proven in [8].

Theorem 4 ([8]). *If one of the following conditions either (5) or (6) holds:*

$$\sum_{J_i \in \mathcal{J}_{1,2}} u_{i1} \leq \sum_{J_i \in \mathcal{J}_{2,1} \cup \mathcal{J}_2} l_{i2} \text{ and } \sum_{J_i \in \mathcal{J}_{1,2}} l_{i2} \geq \sum_{J_i \in \mathcal{J}_{2,1} \cup \mathcal{J}_1} u_{i1}, \tag{5}$$

$$\sum_{J_i \in \mathcal{J}_{2,1}} u_{i2} \leq \sum_{J_i \in \mathcal{J}_{1,2} \cup \mathcal{J}_1} l_{i1} \text{ and } \sum_{J_i \in \mathcal{J}_{2,1}} l_{i1} \geq \sum_{J_i \in \mathcal{J}_{1,2} \cup \mathcal{J}_2} u_{i2}, \tag{6}$$

then any pair of permutations $(\pi', \pi'') \in S$ is a singleton $DS(T) = \{(\pi', \pi'')\}$ for the uncertain job-shop problem $J2|l_{ij} \leq p_{ij} \leq u_{ij}, n_i \leq 2|C_{max}$ with the job set $\mathcal{J} = \mathcal{J}_1 \cup \mathcal{J}_2 \cup \mathcal{J}_{1,2} \cup \mathcal{J}_{2,1}$.

Corollary 2 ([8]). *If the following inequality holds:*

$$\sum_{J_j \in \mathcal{J}_{1,2}} u_{i1} \leq \sum_{J_j \in \mathcal{J}_{2,1} \cup \mathcal{J}_2} l_{i2}, \tag{7}$$

then the set $<\{\pi_{1,2}\}, S_{2,1}>\subseteq S$, where $\pi_{1,2}$ is an arbitrary permutation in the set $S_{1,2}$, is a dominant set for the uncertain job-shop problem $J2|l_{ij} \leq p_{ij} \leq u_{ij}, n_i \leq 2|C_{max}$ with the job set $\mathcal{J} = \mathcal{J}_1 \cup \mathcal{J}_2 \cup \mathcal{J}_{1,2} \cup \mathcal{J}_{2,1}$.

Corollary 3 ([8]). *If the following inequality holds:* $\sum_{J_j \in \mathcal{J}_{2,1}} u_{i2} \leq \sum_{J_j \in \mathcal{J}_{1,2} \cup \mathcal{J}_1} l_{i1}$, *then the set* $<S_{1,2}, \{\pi_{2,1}\}>$, *where* $\pi_{2,1}$ *is an arbitrary permutation in the set* $S_{2,1}$, *is a dominant set for the uncertain job-shop problem* $J2|l_{ij} \leq p_{ij} \leq u_{ij}, n_i \leq 2|C_{max}$ *with the job set* $\mathcal{J} = \mathcal{J}_1 \cup \mathcal{J}_2 \cup \mathcal{J}_{1,2} \cup \mathcal{J}_{2,1}$.

In order to determine an optimal permutation for processing jobs from the set $\mathcal{J}_{2,1}$ (set $\mathcal{J}_{1,2}$, respectively), we consider the uncertain flow-shop problem $F2|l_{ij} \leq p_{ij} \leq u_{ij}|C_{max}$ with the job set $\mathcal{J}_{1,2} \subseteq \mathcal{J}$ and the machine route (M_1, M_2), and that with the job set $\mathcal{J}_{2,1} \subseteq \mathcal{J}$ and the machine route (M_2, M_1). The following theorem has been proven in [8].

Theorem 5 ([8]). *Let the set* $S'_{1,2} \subseteq S_{1,2}$ *be a set of permutations from the dominant set for the flow-shop problem* $F2|l_{ij} \leq p_{ij} \leq u_{ij}|C_{max}$ *with the job set* $\mathcal{J}_{1,2}$, *and the set* $S'_{2,1} \subseteq S_{2,1}$ *be a set of permutations from the dominant set for the flow-shop problem* $F2|l_{ij} \leq p_{ij} \leq u_{ij}|C_{max}$ *with the job set* $\mathcal{J}_{2,1}$. *Then the set* $<S'_{1,2}, S'_{2,1}>\subseteq S$ *is a dominant set for the job-shop problem* $J2|l_{ij} \leq p_{ij} \leq u_{ij}, n_i \leq 2|C_{max}$ *with the job set* $\mathcal{J} = \mathcal{J}_1 \cup \mathcal{J}_2 \cup \mathcal{J}_{1,2} \cup \mathcal{J}_{2,1}$.

4.2. Precedence Digraphs Determining a Minimal Dominant Set of Schedules

Based on Remark 1, the off-line phase of scheduling for the uncertain job-shop problem $J2|l_{ij} \leq p_{ij} \leq u_{ij}, n_i \leq 2|C_{max}$ may be based on solving the uncertain flow-shop problem $F2|l_{ij} \leq p_{ij} \leq u_{ij}|C_{max}$ with the job set $\mathcal{J}_{1,2}$ and that with the job set $\mathcal{J}_{2,1}$. A criterion for the existence of a single-element J-solution for the uncertain flow-shop problem $F2|l_{ij} \leq p_{ij} \leq u_{ij}|C_{max}$ is determined in Theorem 2.

In what follows, it is assumed that the equality $\mathcal{J}_{1,2} = \mathcal{J}^1_{1,2} \cup \mathcal{J}^2_{1,2} \cup \mathcal{J}^*_{1,2}$ holds, i.e., $\mathcal{J}^0_{1,2} = \emptyset$. Using the results presented in Section 3, one can determine a binary relation $A^{1,2}_\prec$ for the uncertain flow-shop problem $F2|l_{ij} \leq p_{ij} \leq u_{ij}|C_{max}$ with the job set $\mathcal{J}_{1,2}$. For the job set $\mathcal{J}_{1,2}$, the binary relation $A^{1,2}_\prec$ determines the digraph $G_{1,2} = (\mathcal{J}_{1,2}, A^{1,2}_\prec)$ with the vertex set $\mathcal{J}_{1,2}$ and the arc set $A^{1,2}_\prec$.

Definition 4. *Two jobs* $J_x \in \mathcal{J}_{1,2}$ *and* $J_y \in \mathcal{J}_{1,2}$, $x \neq y$, *are conflict if they are not in the relation* $A^{1,2}_\prec$, *i.e.,* $(J_x, J_y) \notin A^{1,2}_\prec$ *and* $(J_y, J_x) \notin A^{1,2}_\prec$.

Due to Definition 2, for the conflict jobs $J_x \in \mathcal{J}_{1,2}$ and $J_y \in \mathcal{J}_{1,2}$, $x \neq y$, relations (3) and (4) do not hold for the case $v = x$ with $w = y$, nor for the case $v = y$ with $w = x$.

Definition 5. *The inclusion-minimal set* $\mathcal{J}_x \subseteq \mathcal{J}_{1,2}$ *of the jobs is called a conflict set of the jobs, if for any job* $J_y \in \mathcal{J}_{1,2} \setminus \mathcal{J}_x$ *either relation* $(J_x, J_y) \in A^{1,2}_\prec$ *or relation* $(J_y, J_x) \in A^{1,2}_\prec$ *holds for each job* $J_x \in \mathcal{J}_x$.

There may exist several conflict sets in the set $\mathcal{J}_{1,2}$. Let the strict order $A^{1,2}_\prec$ for the flow-shop problem $F2|l_{ij} \leq p_{ij} \leq u_{ij}|C_{max}$ with the job set $\mathcal{J}_{1,2}$ be represented as follows:

$$J_1 \prec J_2 \prec \ldots \prec J_k \prec \{J_{k+1}, J_{k+2}, \ldots, J_{k+r}\} \prec J_{k+r+1} \prec J_{k+r+2} \prec \ldots \prec J_{m_{1,2}}. \tag{8}$$

Here, an optimal permutation for processing jobs from the set $\{J_1, J_2, \ldots, J_k\}$ (for jobs from the set $\{J_{k+r+1}, J_{k+r+2}, \ldots, J_{m_{1,2}}\}$) is as follows: $(J_1, J_2, \ldots, J_k)$ $((J_{k+r+1}, J_{k+r+2}, \ldots, J_{m_{1,2}})$, respectively). All jobs between braces in the presentation (8) constitute the conflict set of the jobs and they are in relation $A^{1,2}_\prec$ with any job located outside the braces. Due to Theorem 3, the set $\Pi(G_{1,2})$ of the permutations generated by the digraph $G_{1,2}$ includes an optimal permutation for each vector $p_{1,2}$ of the processing times of the jobs $\mathcal{J}_{1,2}$. Due to Corollary 1, the set $S_{1,2}(T) = \{(\pi_{1,2}, \pi_{1,2})\}$ with $\pi_{1,2} \in \Pi(G_{1,2})$ is a J-solution for the flow-shop problem $F2|l_{ij} \leq p_{ij} \leq u_{ij}|C_{max}$ with the job set $\mathcal{J}_{1,2}$. Analogously, the set $S_{2,1}(T) = \{(\pi_{2,1}, \pi_{2,1})\}$ with $\pi_{2,1} \in \Pi(G_{2,1})$ is a J-solution for the problem $F2|l_{ij} \leq p_{ij} \leq u_{ij}|C_{max}$

with the job set $\mathcal{J}_{2,1}$. Due to Theorem 5, one can determine a dominant set for the job-shop problem $J2|l_{ij} \leq p_{ij} \leq u_{ij}, n_i \leq 2|C_{max}$ with the job set $\mathcal{J}$ as follows: $<\Pi(G_{1,2}), \Pi(G_{2,1})> \subseteq S$; see Remark 1.

The following three theorems are proven in [8], where the notation $L_2 := \sum_{J_j \in \mathcal{J}_{2,1} \cup \mathcal{J}_2} l_{j2}$ is used. These theorems allow a scheduler to reduce the cardinality of a dominant set for the uncertain job-shop scheduling problem $J2|l_{ij} \leq p_{ij} \leq u_{ij}, n_i \leq 2|C_{max}$.

Theorem 6 ([8]). *Let the strict order $A_\prec^{1,2}$ over the set $\mathcal{J}_{1,2} = \mathcal{J}_{1,2}^* \cup \mathcal{J}_{1,2}^1 \cup \mathcal{J}_{1,2}^2$ be determined as follows: $J_1 \prec \ldots \prec J_k \prec \{J_{k+1}, J_{k+2}, \ldots, J_{k+r}\} \prec J_{k+r+1} \prec \ldots \prec J_{m_{1,2}}$. If the following inequality holds:*

$$\sum_{i=1}^{k+r} u_{i1} \leq L_2 + \sum_{i=1}^{k} l_{i2}, \tag{9}$$

then the set $S' = <\{\pi_{1,2}\}, \Pi(G_{2,1})> \subset S$, where $\pi_{1,2} \in \Pi(G_{1,2})$, is a dominant set for the job-shop problem $J2|l_{ij} \leq p_{ij} \leq u_{ij}, n_i \leq 2|C_{max}$ with the job set $\mathcal{J} = \mathcal{J}_1 \cup \mathcal{J}_2 \cup \mathcal{J}_{1,2} \cup \mathcal{J}_{2,1}$.

Theorem 7 ([8]). *Let the partial strict order $A_\prec^{1,2}$ over the set $\mathcal{J}_{1,2} = \mathcal{J}_{1,2}^* \cup \mathcal{J}_{1,2}^1 \cup \mathcal{J}_{1,2}^2$ be determined as follows: $J_1 \prec \ldots \prec J_k \prec \{J_{k+1}, J_{k+2}, \ldots, J_{k+r}\} \prec J_{k+r+1} \prec \ldots \prec J_{m_{1,2}}$. If the inequality*

$$u_{k+s,1} \leq L_2 + \sum_{i=1}^{k+s-1}(l_{i2} - u_{i1}) \tag{10}$$

holds for each $s \in \{1, 2, \ldots, r\}$, then the set $S' = <\{\pi_{1,2}\}, S_{2,1}>$, where $\pi_{1,2} = (J_1, \ldots, J_{k-1}, J_k, J_{k+1}, J_{k+2}, \ldots, J_{k+r}, J_{k+r+1}, \ldots, J_{m_{1,2}}) \in \Pi(G_{1,2})$, is a dominant set for the job-shop problem $J2|l_{ij} \leq p_{ij} \leq u_{ij}, n_i \leq 2|C_{max}$ with the job set $\mathcal{J} = \mathcal{J}_1 \cup \mathcal{J}_2 \cup \mathcal{J}_{1,2} \cup \mathcal{J}_{2,1}$.

Theorem 8 ([8]). *Let the partial strict order $A_\prec^{1,2}$ over the set $\mathcal{J}_{1,2} = \mathcal{J}_{1,2}^* \cup \mathcal{J}_{1,2}^1 \cup \mathcal{J}_{1,2}^2$ have the form $J_1 \prec \ldots \prec J_k \prec \{J_{k+1}, J_{k+2}, \ldots, J_{k+r}\} \prec J_{k+r+1} \prec \ldots \prec J_{m_{1,2}}$. If the inequality*

$$\sum_{i=r-s+2}^{r-s+1} l_{k+i,1} \geq \sum_{j=r-s+1}^{r} u_{k+j,2} \tag{11}$$

holds for each $s \in \{1, 2, \ldots, r\}$, then the set $S' = <\{\pi_{1,2}\}, S_{2,1}>$, where $\pi_{1,2} = (J_1, \ldots, J_{k-1}, J_k, J_{k+1}, J_{k+2}, \ldots, J_{k+r}, J_{k+r+1}, \ldots, J_{m_{1,2}}) \in \Pi(G_{1,2})$, is a dominant set for the job-shop problem $J2|l_{ij} \leq p_{ij} \leq u_{ij}, n_i \leq 2|C_{max}$ with the job set $\mathcal{J} = \mathcal{J}_1 \cup \mathcal{J}_2 \cup \mathcal{J}_{1,2} \cup \mathcal{J}_{2,1}$.

One can describe the analogs of Theorems 6–8 for reducing the cardinality of a dominant set for the job-shop problem $J2|l_{ij} \leq p_{ij} \leq u_{ij}, n_i \leq 2|C_{max}$ provided that for the flow-shop problem $F2|l_{ij} \leq p_{ij} \leq u_{ij}|C_{max}$ with the job set $\mathcal{J}_{2,1}$, there exists a partial strict order $A_\prec^{2,1}$ over the set $\mathcal{J}_{2,1} = \mathcal{J}_{2,1}^* \cup \mathcal{J}_{2,1}^1 \cup \mathcal{J}_{2,1}^2$ with the following form: $J_1 \prec \ldots \prec J_k \prec \{J_{k+1}, J_{k+2}, \ldots, J_{k+r}\} \prec J_{k+r+1} \prec \ldots \prec J_{m_{2,1}}$.

If the set $\{J_1, \ldots, J_k\}$ is empty in the constructed job permutation, then it is needed to check the conditions of Theorem 8. If the set $\{J_{k+r+1}, \ldots, J_{m_{1,2}}\}$ is empty, then one needs to check the conditions of Theorem 7. Note that it is enough to test only one permutation for checking the conditions of Theorem 7 and only one permutation for checking the conditions of Theorem 8; see [8].

4.3. An Illustrative Example

To illustrate the above results, we consider Example 1 of the uncertain job-shop scheduling problem $J2|l_{ij} \leq p_{ij} \leq u_{ij}, n_i \leq 2|C_{max}$ with eight jobs $\{J_1, J_2, \ldots, J_8\} = \mathcal{J}$. Let three jobs J_1, J_2 and J_3 have the machine route (M_1, M_2), jobs J_6, J_7 and J_8 have the opposite machine route (M_2, M_1), job J_4 and job J_5 have to be processed only on machine M_1 and machine M_2, respectively. The partition $\mathcal{J} =$

$\mathcal{J}_{1,2} \cup \mathcal{J}_{2,1} \cup \mathcal{J}_1 \cup \mathcal{J}_2$ is given, where $\mathcal{J}_{1,2} = \{J_1, J_2, J_3\}$, $\mathcal{J}_{2,1} = \{J_6, J_7, J_8\}$, $\mathcal{J}_1 = \{J_4\}$ and $\mathcal{J}_2 = \{J_5\}$. The lower and upper bounds on the job processing times are determined in Table 1.

Table 1. The numerical input data for Example 1.

J_j	J_1	J_2	J_3	J_4	J_5	J_6	J_7	J_8
l_{i1}	6	8	7	2	-	1	1	1
u_{i1}	7	9	9	3	-	3	3	3
l_{i2}	6	5	4	-	2	2	3	4
u_{i2}	7	6	5	-	3	4	4	4

To solve this uncertain job-shop scheduling problem, one need to determine an optimal pair (π', π'') of permutations of the eight jobs for their processing on machine M_1 and machine M_2. These permutations π' and π'' have the following forms: $\pi' = (\pi_{1,2}, \pi_1, \pi_{2,1})$, $\pi'' = (\pi_{2,1}, \pi_2, \pi_{1,2})$.

It is necessary to find four permutations $\pi_{1,2}, \pi_{2,1}, \pi_1$ and π_2 of the jobs from the sets $\mathcal{J}_{1,2}, \mathcal{J}_{2,1}, \mathcal{J}_1$ and $\mathcal{J}_2$, respectively. The permutations π_1 and π_2 are determined as follows: $\pi_1 = (J_4)$ and $\pi_2 = (J_5)$.

We test the sufficient conditions given in Section 4.1. The conditions (5) of Theorem 4 do not hold. For testing the conditions (6) of Theorem 4, one can obtain the following relations:

$$\sum_{J_i \in \mathcal{J}_{2,1}} u_{i2} = u_{6,2} + u_{7,2} + u_{8,2} = 4 + 4 + 4 = 12 \leq \sum_{J_i \in \mathcal{J}_{1,2} \cup \mathcal{J}_1} l_{i1} = l_{1,1} + l_{2,1} + l_{3,1} + l_{4,1} = 6 + 8 + 7 + 2 = 23,$$

$$\sum_{J_i \in \mathcal{J}_{2,1}} l_{i1} = l_{6,1} + l_{7,1} + l_{8,1} = 1 + 1 + 1 = 3 \not\geq \sum_{J_i \in \mathcal{J}_{1,2} \cup \mathcal{J}_2} u_{i2} = u_{1,2} + u_{2,2} + u_{3,2} + u_{5,2} = 7 + 6 + 5 + 3 = 21.$$

It should be noted that the case when conditions of Theorem 4 hold was considered in [8].

As the first condition in (6) holds, due to Corollary 3, one can construct permutation $\pi_{2,1} = (J_6, J_7, J_8)$ by arranging the jobs from the set $\mathcal{J}_{2,1}$ in the increasing of their indexes.

For the jobs from the set $\mathcal{J}_{1,2}$, the partition $\mathcal{J}_{1,2} = \mathcal{J}_{1,2}^1 \cup \mathcal{J}_{1,2}^2 \cup \mathcal{J}_{1,2}^*$ holds, where $\mathcal{J}_{1,2}^* = \{J_1\}$ and $\mathcal{J}_{1,2}^2 = \{J_2, J_3\}$. The condition of Theorem 2 holds for these jobs. Therefore, the following optimal permutation: $\pi_{1,2} = (J_1, J_2, J_3)$ is determined.

Thus, there exists a pair of job permutations (π', π''), where $\pi' = (J_1, J_2, J_3, J_4, J_6, J_7, J_8)$ and $\pi'' = (J_6, J_7, J_8, J_5, J_1, J_2, J_3)$, which is optimal for all possible scenarios $p \in T$. Hence, there exists a single-element dominant set $DS(T) = \{(\pi', \pi'')\}$ for Example 1 of the uncertain job-shop problem $J2|l_{ij} \leq p_{ij} \leq u_{ij}, n_i \leq 2|C_{max}$ with the bounds on the job processing times given in Table 1.

The optimal semi-active schedule is constructed for Example 1 at the off-line phase of scheduling, despite of the uncertainty of the job processing times. Such an issue is called as STOP 1 in the scheduling algorithms developed in [8] and used in Section 6 of this paper.

5. The On-line Phase of Scheduling

Due to Remark 3, if the job J_i is completed on the corresponding machine $M_j \in \mathcal{M}$, the duration of the operation O_{ij} takes on exact value p_{ij}^*, where $l_{ij} \leq p_{ij}^* \leq u_{ij}$. A scheduler can use this information on the duration of the operation O_{ij} for a selection of the next job for processing on machine M_j. Since it is on-line phase of scheduling, such a selection should be very quick.

It is first assumed that the set $S' = <\Pi(G_{1,2}), \{\pi_{2,1}^*\}> \subset S$, is a dominant set for the problem $J2|l_{ij} \leq p_{ij} \leq u_{ij}, n_i \leq 2|C_{max}$ with the job set $\mathcal{J}$. In other words, the optimal permutations for processing all jobs from the set $\mathcal{J}_{2,1}$ are already determined at the off-line phase of scheduling.

Let the strict order $A_\prec^{1,2}$ over the set $\mathcal{J}_{1,2} = \mathcal{J}_{1,2}^* \cup \mathcal{J}_{1,2}^1 \cup \mathcal{J}_{1,2}^2$ be determined as follows: $J_1 \prec \ldots \prec J_k \prec \{J_{k+1}, J_{k+2}, \ldots, J_{k+r}\} \prec J_{k+r+1} \prec \ldots \prec J_{m_{1,2}}$. At the initial time $t = 0$, machine M_1 has to start processing jobs from the set $\{J_1, \ldots, J_k\}$ in the following optimal order: $(J_1, \ldots, J_k)$. At the same time $t = 0$, machine M_2 has to start processing jobs from the set $\mathcal{J}_{2,1}$ in the order determined by the permutation $\pi_{2,1}^*$, then jobs from the set $\mathcal{J}_2$ in the arbitrary order, and then jobs from the set $\{J_1,\ldots,J_k\}$ in the following optimal order: $(J_1,\ldots,J_k)$; see Figure 3.

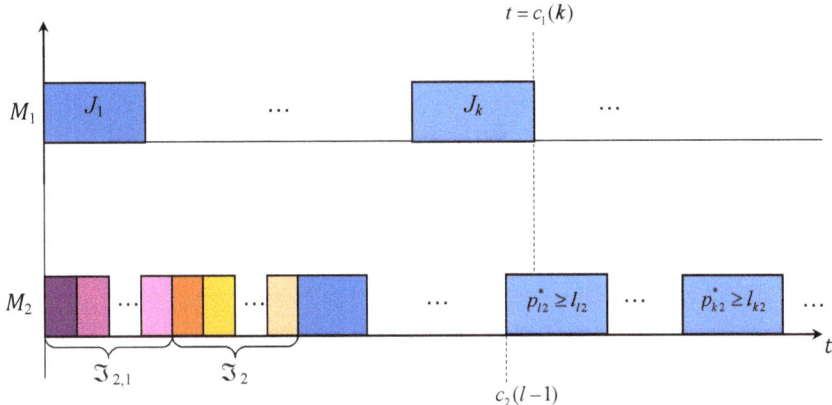

Figure 3. The initial part of the schedule execution.

At the time-point $t = c_1(k)$, machine M_1 completes the operation O_{k1}. Let $\mathcal{J}(i,j)$ denote a set of all jobs processed on machine M_j from the initial part of the schedule till the job J_i, e.g., the set of jobs $\{J_1, J_2, \ldots, J_k\}$ is denoted as $\mathcal{J}(k,1)$; see Figure 3. Due to Remark 3, at the time-point $t = c_1(k)$, the factual values p_{i1}^* of the processing times p_{i1} of all jobs J_i in the set $\mathcal{J}(k,1)$ are already known.

Let machine M_2 process the job $J_l \in \mathcal{J}_{2,1} \cup \mathcal{J}_2 \cup \{J_1, J_2, \ldots, J_k\}$ at the time-point $t = c_1(k)$, i.e., $t = c_1(k) < c_2(l)$. Let $\mathcal{J}(l-1, 2)$ denote a set of all jobs whose processing is completed on machine M_2 before time-point $t = c_1(k)$. Figure 3 depicts this situation for the job $J_{l-1} \in \{J_1, J_2, \ldots, J_k\} \subset \mathcal{J}_{1,2}$.

The factual values p_{i2}^* of the processing times p_{i2} of all jobs J_i in the set $\mathcal{J}(l-1, 2)$ are known at the time-point $t = c_1(k) > c_2(l-1)$, i.e., $p_{i2} = p_{i2}^*$, while the factual values of the processing times p_{j2} of other jobs in the set $\mathcal{J}$ remain unknown at the time-point $t = c_1(k) < c_2(l)$. Thus, at the time-point $t = c_1(k)$, the following subset of possible scenarios:

$$T(k, l-1) = \{p \in T : p_{i1} = p_{i1}^*, p_{j2} = p_{j2}^*, J_i \in \mathcal{J}(k,1), J_j \in \mathcal{J}(l-1,2)\}$$

may be realized instead of the initial set T of all possible scenarios; $T(k, l-1) \subseteq T$.

At the time-point $t = c_1(k)$ (it is called a decision-point), a scheduler has to make a decision about the order for processing jobs from the conflict set $\{J_{k+1}, J_{k+2}, \ldots, J_{k+r}\}$. The sufficient conditions given in Theorems 6 and 7 can be reformulated in the following two theorems. (Note that Theorem 8 cannot be reformulated for the use at the on-line phase of scheduling.)

Theorem 9. *Let the set $S' = <\Pi(G_{1,2}), \{\pi_{2,1}^*\}> \subset S$ be a dominant set for the uncertain problem $J2|l_{ij} \leq p_{ij} \leq u_{ij}, n_i \leq 2|C_{max}$ with the job set $\mathcal{J}$. Let the strict order $A_{\prec}^{1,2}$ over the set $\mathcal{J}_{1,2} = \mathcal{J}_{1,2}^* \cup \mathcal{J}_{1,2}^1 \cup \mathcal{J}_{1,2}^2$ be determined as follows: $J_1 \prec \ldots \prec J_k \prec \{J_{k+1}, J_{k+2}, \ldots, J_{k+r}\} \prec J_{k+r+1} \prec \ldots \prec J_{m_{1,2}}$. If at the time-point $t = c_1(k)$, the following inequality holds:*

$$c_1(k) + \sum_{i=k+1}^{k+r} u_{i1} \leq c_2(l-1) + \sum_{J_i \in (\mathcal{J}_{2,1} \cup \mathcal{J}_2 \cup \mathcal{J}(k,1)) \setminus \mathcal{J}(l-1,2)} l_{i2}, \tag{12}$$

then at the time-point $t = c_1(k)$, the set $S' = <\{\pi_{1,2}\}, \{\pi_{2,1}^*\}> \subset S$, where $\pi_{1,2} \in \Pi(G_{1,2})$, is a dominant set for the problem $J2|l_{ij} \leq p_{ij} \leq u_{ij}, n_i \leq 2|C_{max}$ with the job set $\mathcal{J}$ and the set $T(k, l-1)$ of possible scenarios.

Proof. Let p be an arbitrary vector from the set $T(k, l-1)$ of possible scenarios at the time-point $t = c_1(k)$. Let $C_{\max}^p$ denote the optimal makespan value for the deterministic job-shop problem $J2|p, n_i \leq 2|C_{\max}$ with the set $\mathcal{J}$ of the given jobs and the vector p of the job processing times.

We consider an arbitrary permutation $\pi_{1,2} \in \Pi(G_{1,2})$ and show that the pair of job permutations $(\pi', \pi'') = ((\pi_{1,2}, \pi_1, \pi^*_{2,1}), (\pi^*_{2,1}, \pi_2, \pi_{1,2})) \in S'$ is an optimal one for the deterministic job-shop problem $J2|p, n_i \leq 2|C_{\max}$ with the set $\mathcal{J}$ of the jobs and with any vector $p \in T(k, l-1)$ of the job processing times, i.e., the equality $C_{\max}(\pi', \pi'') = C_{\max}^p$ holds. Since the equality $C_{\max}(\pi', \pi'') = \max\{c_1(\pi'), c_2(\pi'')\}$ holds, one has to consider two possible cases (a) and (b).

Case (a): It is assumed that $c_1(\pi') \geq c_2(\pi'')$. Then, one can obtain the following equalities:

$$C_{\max}(\pi', \pi'') = c_1(\pi') = \max\left\{\sum_{J_i \in \mathcal{J}_{1,2} \cup \mathcal{J}_{2,1} \cup \mathcal{J}_1} p_{i1}, C_{\max}(\pi^*_{2,1})\right\}, \tag{13}$$

where $C_{\max}(\pi^*_{2,1})$ is the value of makespan for the deterministic flow-shop problem $F2|p_{2,1}|C_{\max}$ with the job set $\mathcal{J}_{2,1}$ and the vector $p_{2,1}$ whose components are equal to the corresponding components of the vector p. Due to the conditions of Theorem 9, the permutation $\pi^*_{2,1}$ is optimal for the deterministic flow-shop problem $F2|p_{2,1}|C_{\max}$ with the set $\mathcal{J}_{2,1}$ of the given jobs and with vector $p_{2,1}$ of the job processing times. Therefore, $C_{\max}(\pi^*_{2,1})$ is an optimal makespan value for the deterministic flow-shop problem $F2|p_{2,1}|C_{\max}$ and $C_{\max}(\pi^*_{2,1})$ is a minimal completion time for processing all jobs from the set $\mathcal{J}_{2,1}$ on both machines. From the equalities (13), one can obtain the equality $C_{\max}(\pi', \pi'') = C_{\max}^p$.

Case (b): It is assumed that $c_1(\pi') < c_2(\pi'')$. Then, one can obtain the following equalities:

$$C_{\max}(\pi', \pi'') = c_2(\pi'') = \max\left\{\sum_{J_i \in \mathcal{J}_{2,1} \cup \mathcal{J}_2 \cup \mathcal{J}_{1,2}} p_{i2}, C_{\max}(\pi_{1,2})\right\}, \tag{14}$$

where $C_{\max}(\pi_{1,2})$ is an optimal value of the makespan criterion for the deterministic flow-shop problem $F2|p_{1,2}|C_{\max}$ with the job set $\mathcal{J}_{1,2}$ and with the vector $p_{1,2}$ of the job processing times (the components of this vector are equal to the corresponding components of the vector p). Since $\pi_{1,2} \in \Pi(G_{1,2})$, the initial part of the permutation $\pi_{1,2}$ has the following form: $(J_1, J_2, \ldots, J_k)$. For every pair of jobs from the set $\{J_1, J_2, \ldots, J_k\}$, at least one of the conditions, either (3) or (4), holds, see Theorem 1.

Therefore, for the job processing times determined by the vector p for the jobs $\{J_1, J_2, \ldots, J_k\}$, the inequalities (2) hold. Thus, in the permutation $\pi_{1,2}^{beg} := (J_1, J_2, \ldots, J_k)$, all the jobs are arranged in the Johnson's order. One can conclude that the following value

$$C_{\max}(\pi_{1,2}^{beg}) = \max_{1 \leq m \leq k}\left\{\sum_{i=1}^m p_{i1} + \sum_{i=m}^k p_{i2}\right\} \tag{15}$$

determines an optimal makespan value for the deterministic flow-shop problem $F2|p_{1,2}^{beg}|C_{\max}$ with the job set $\{J_1, J_2, \ldots, J_k\}$ and the corresponding vector $p_{1,2}^{beg}$ of the job processing times (the components of the vector $p_{1,2}^{beg}$ are equal to the corresponding components of the vector p). Therefore, $C_{\max}(\pi_{1,2}^{beg})$ is a minimal makespan value for processing jobs of the set $\{J_1, J_2, \ldots, J_k\}$ on both machines. Then, for the time-point $c_2(k)$ when machine M_2 completes the operation O_{k2}, one can obtain the following equality:

$$c_2(k) = \max\left\{\sum_{J_i \in \mathcal{J}_{2,1} \cup \mathcal{J}_2 \cup \mathcal{J}_{1,2}(k,1)} p_{i2}, C_{\max}(\pi_{1,2}^{beg})\right\}. \tag{16}$$

Due to the inequality (12) and the equality (16), one can obtain the following inequalities for the jobs from the conflict set $\{J_{k+1}, J_{k+2}, \ldots, J_{k+r}\}$:

$$c_1(k) + \sum_{i=k+1}^{k+r} p_{i1} \leq c_1(k) + \sum_{i=k+1}^{k+r} u_{i1} \leq c_2(l-1) + \sum_{J_i \in (\mathcal{J}_{2,1} \cup \mathcal{J}_2 \cup \mathcal{J}(k,1)) \setminus \mathcal{J}(l-1,2)} l_{i2} \leq \tag{17}$$

$$\leq c_2(l-1) + \sum_{J_i \in (\mathcal{J}_{2,1} \cup \mathcal{J}_2 \cup \mathcal{J}(k,1)) \setminus \mathcal{J}(l-1,2)} p_{i2} \leq \max\left\{\sum_{J_i \in \mathcal{J}_{2,1} \cup \mathcal{J}_2 \cup \mathcal{J}_{1,2}(k,1)} p_{i2}, \ C_{\max}(\pi_{1,2}^{beg})\right\} = c_2(k).$$

From the inequalities (17), one can obtain the following inequality:

$$c_1(k) + \sum_{i=k+1}^{k+r} p_{i1} \leq c_2(k). \tag{18}$$

Thus, machine M_2 processes all jobs from the conflict set $\{J_{k+1}, J_{k+2}, \ldots, J_{k+r}\}$ without idle times and without an idle before processing the first job from this conflict set for any order of these conflict jobs. Using the inequality (18), one can conclude that the time-point when machine M_2 completes the processing of the last job from the conflict set $\{J_{k+1}, J_{k+2}, \ldots, J_{k+r}\}$ in the permutation $\pi_{1,2}$ is determined as follows:

$$c_2 = c_2(k) + \sum_{i=k+1}^{k+r} p_{i2}, \tag{19}$$

where c_2 is an optimal makespan value for processing jobs from the set $\{J_1, J_2, \ldots, J_k, J_{k+1}, J_{k+2}, \ldots, J_{k+r}\}$. Next, we consider jobs from the set $\{J_{k+r+1}, \ldots, J_{m_{1,2}}\}$.

Let $\pi_{1,2}^{end} := (J_{k+r+1}, \ldots, J_{m_{1,2}})$ denote the permutation of the jobs $\{J_{k+r+1}, \ldots, J_{m_{1,2}}\}$ in the permutation $\pi_{1,2}$. Analogously as for the job set $\{J_1, J_2, \ldots, J_k\}$, one can obtain that the value of

$$C_{\max}(\pi_{1,2}^{end}) := \max_{k+r+1 \leq m \leq m_{1,2}} \left\{\sum_{i=k+r+1}^{m} p_{i1} + \sum_{i=m}^{m_{1,2}} p_{i2}\right\} \tag{20}$$

is an optimal makespan value for the deterministic flow-shop problem $F2|p_{1,2}^{end}|C_{\max}$ with the job set $\{J_{k+r+1}, \ldots, J_{m_{1,2}}\}$ and with the vector $p_{1,2}^{end}$ whose components are equal to the components of the vector p. Thus, $C_{\max}(\pi_{1,2}^{end})$ is a minimal makespan value for processing all jobs from the set $\{J_{k+r+1}, \ldots, J_{m_{1,2}}\}$ on both machines. The time-point when machine M_2 completes the processing of the last job from the permutation π'' can be calculated as follows:

$$c_2(\pi'') = \max\left\{\sum_{i=1}^{k+r} p_{i1} + C_{\max}(\pi_{1,2}^{end}), \ c_2 + \sum_{i=k+r+1}^{m_{1,2}} p_{i2}\right\} =$$

$$= \max\left\{\sum_{i=1}^{k+r} p_{i1} + C_{\max}(\pi_{1,2}^{end}), \ c_2(k) + \sum_{i=k+1}^{k+r} p_{i2} + \sum_{i=k+r+1}^{m_{1,2}} p_{i2}\right\} =$$

$$= \max\left\{\sum_{i=1}^{k+r} p_{i1} + C_{\max}(\pi_{1,2}^{end}), \ \max\left\{\sum_{J_i \in \mathcal{J}_{2,1} \cup \mathcal{J}_2 \cup \mathcal{J}_{1,2}(k,1)} p_{i2}, \ C_{\max}(\pi_{1,2}^{beg})\right\} + \sum_{i=k+1}^{m_{1,2}} p_{i2}\right\} =$$

$$= \max\left\{\sum_{i=1}^{k+r} p_{i1} + C_{\max}(\pi_{1,2}^{end}), \ \sum_{J_i \in \mathcal{J}_{2,1} \cup \mathcal{J}_2 \cup \mathcal{J}_{1,2}(k,1)} p_{i2} + \sum_{i=k+1}^{m_{1,2}} p_{i2}, \ C_{\max}(\pi_{1,2}^{beg}) + \sum_{i=k+1}^{m_{1,2}} p_{i2}\right\} =$$

$$= \max\left\{C_{\max}(\pi_{1,2}^{beg}) + \sum_{i=k+1}^{m_{1,2}} p_{i2}, \ \sum_{i=1}^{k+r} p_{i1} + C_{\max}(\pi_{1,2}^{end}), \ \sum_{J_i \in \mathcal{J}_{2,1} \cup \mathcal{J}_{1,2}} p_{i2}\right\}, \tag{21}$$

where relations (16) and (19) are used.

Due to Theorem 3, the set $\Pi(G_{1,2})$ contains a Johnson's permutation for the deterministic flow-shop problem $F2|p_{1,2}|C_{\max}$ with the job set $\mathcal{J}_{1,2}$ and with the vector $p_{1,2}$ of the job durations. We denote this Johnson's permutation as $\pi_{1,2}^*$. Since $\pi_{1,2}^* \in \Pi(G_{1,2})$, the permutation $\pi_{1,2}^*$ has the following form: $\pi_{1,2}^* = (J_1, \ldots, J_k, J_{[k+1]}, J_{[k+2]}, \ldots, J_{[k+r]}, J_{k+r+1}, \ldots, J_{m_{1,2}})$, where the set of indexes is determined as follows: $\{[k+1], [k+2], \ldots, [k+r]\} = \{k+1, k+2, \ldots, k+r\}$.

The optimal makespan value $C_{\max}(\pi_{1,2}^*)$ can be calculated as follows:

$$C_{\max}(\pi_{1,2}^*) = \max_{1 \le m \le m_{1,2}} \left\{ \sum_{i=1}^{m} p_{i1} + \sum_{i=m}^{m_{1,2}} p_{i2} \right\} = \max \left\{ \max_{1 \le m \le k} \left\{ \sum_{i=1}^{m} p_{i1} + \sum_{i=m}^{k} p_{i2} + \sum_{i=k+1}^{m_{1,2}} p_{i2} \right\}, \right.$$

$$\max_{[k+1] \le m \le [k+r]} \left\{ \sum_{i=1}^{m} p_{i1} + \sum_{i=m}^{m_{1,2}} p_{i2} \right\}, \quad \max_{k+r+1 \le m \le m_{1,2}} \left\{ \sum_{i=1}^{k+r} p_{i1} + \sum_{i=k+r+1}^{m} p_{i1} + \sum_{i=m}^{m_{1,2}} p_{i2} \right\} \right\} =$$

$$= \max \left\{ C_{\max}(\pi_{1,2}^{beg}) + \sum_{i=k+1}^{m_{1,2}} p_{i2}, \max_{[k+1] \le m \le [k+r]} \left\{ \sum_{i=1}^{m} p_{i1} + \sum_{i=m}^{m_{1,2}} p_{i2} \right\}, \sum_{i=1}^{k+r} p_{i1} + C_{\max}(\pi_{1,2}^{end}) \right\}, \quad (22)$$

where relations (15) and (20) are used. From relations (21) and (22), one can obtain the relations

$$c_2(\pi'') = \max \left\{ C_{\max}(\pi_{1,2}^{beg}) + \sum_{i=k+1}^{m_{1,2}} p_{i2}, \sum_{i=1}^{k+r} p_{i1} + C_{\max}(\pi_{1,2}^{end}), \sum_{J_i \in \mathcal{J}_{2,1} \cup \mathcal{J}_2 \cup \mathcal{J}_{1,2}} p_{i2} \right\} \le$$

$$\le \max \left\{ C_{\max}(\pi_{1,2}^*), \sum_{J_i \in \mathcal{J}_{2,1} \cup \mathcal{J}_2 \cup \mathcal{J}_{1,2}} p_{i2} \right\}. \quad (23)$$

Therefore, relations (14) and (23) imply the equality $C_{\max}(\pi', \pi'') = C_{\max}^p$.

Thus, in both cases (a) and (b), the equality $C_{\max}(\pi', \pi'') = C_{\max}^p$ holds and the pair of permutations $(\pi', \pi'') = ((\pi_{1,2}, \pi_1, \pi_{2,1}^*), (\pi_{2,1}^*, \pi_2, \pi_{1,2}))$ is optimal for the deterministic job-shop problem $J2|p, n_i \le 2|C_{\max}$ with the scenario $p \in T(k, l-1)$. Therefore, the set $S' = <\{\pi_{1,2}\}, \{\pi_{2,1}^*\}>$ contains an optimal pair of job permutations for the job-shop problem $J2|p, n_i \le 2|C_{\max}$ with vector $p \in T(k, l-1)$ of the job processing times. Since the vector p is arbitrarily chosen in the set $T(k, l-1)$, the set S' contains an optimal pair of job permutations for each scenario in the set $T(k, l-1)$.

Due to Definition 3, the set S' is a dominant set for the uncertain job-shop problem $J2|l_{ij} \le p_{ij} \le u_{ij}, n_i \le 2|C_{\max}$ with the job set $\mathcal{J}$ and with the set $T(k, l-1)$ of possible scenarios. □

Theorem 10. Let the set $S' = <\Pi(G_{1,2}), \{\pi_{2,1}^*\}> \subset S$ be a dominant set for the uncertain job-shop problem $J2|l_{ij} \le p_{ij} \le u_{ij}, n_i \le 2|C_{max}$ with the job set $\mathcal{J}$. Let the partial strict order $A_{\prec}^{1,2}$ over the set $\mathcal{J}_{1,2} = \mathcal{J}_{1,2}^* \cup \mathcal{J}_{1,2}^1 \cup \mathcal{J}_{1,2}^2$ be determined as follows: $J_1 \prec \ldots \prec J_k \prec \{J_{k+1}, J_{k+2}, \ldots, J_{k+r}\} \prec J_{k+r+1} \prec \ldots \prec J_{m_{1,2}}$. If at the time-point $t = c_1(k)$, the following inequalities hold:

$$c_1(k) + \sum_{i=k+1}^{k+s} u_{i1} \le c_2(l-1) + \sum_{J_i \in (\mathcal{J}_{2,1} \cup \mathcal{J}_2 \cup \mathcal{J}(k-1,1)) \setminus \mathcal{J}(l-1,2)} l_{i2} + \sum_{i=k}^{k+s-1} l_{i2} \quad (24)$$

for all indexes $s \in \{1, 2, \ldots, r\}$, then at the time-point $t = c_1(k)$, the set $S' = <\{\pi_{1,2}\}, \{\pi_{2,1}^*\}>$, where $\pi_{1,2} = (J_1, \ldots, J_{k-1}, J_k, J_{k+1}, J_{k+2}, \ldots, J_{k+r}, J_{k+r+1}, \ldots, J_{m_{1,2}}) \in \Pi(G_{1,2})$, is a dominant set for the uncertain problem $J2|l_{ij} \le p_{ij} \le u_{ij}, n_i \le 2|C_{max}$ with the job set $\mathcal{J}$ and the set $T(k, l-1)$ of possible scenarios.

Proof. The proof of this theorem is similar to the above proof of Theorem 9 with the exception of the inequalities (17) and (18). From the condition (24) with $s = 1$, one can obtain the following inequality:

$$c_1(k) + u_{k+1,1} \le c_2(l-1) + \sum_{J_i \in (\mathcal{J}_{2,1} \cup \mathcal{J}_2 \cup \mathcal{J}(k-1,1)) \setminus \mathcal{J}(l-1,2)} l_{i2} + l_{k2}. \quad (25)$$

Based on the inequality (25), one can obtain the following relations:

$$c_1(k) + p_{k+1,1} \le c_1(k) + u_{k+1,1} \le c_2(l-1) + \sum_{J_i \in (\mathcal{J}_{2,1} \cup \mathcal{J}_2 \cup \mathcal{J}(k-1,1)) \setminus \mathcal{J}(l-1,2)} l_{i2} + l_{k2} \le$$

$$\leq c_2(l-1) + \sum_{J_i \in (\mathcal{J}_{2,1} \cup \mathcal{J}_2 \cup \mathcal{J}(k,1)) \setminus \mathcal{J}(l-1,2)} p_{i2} = c_2(k). \qquad (26)$$

Due to relations (26), the following inequality holds:

$$c_1(k) + p_{k+1,1} \leq c_2(k). \qquad (27)$$

Thus, machine M_2 processes the job J_{k+1} in permutation $\pi_{1,2}$ without an idle time between the jobs J_k and J_{k+1}. Analogously, using $s \in \{2, 3, \ldots, r\}$, one can show that the following inequalities hold:

$$c_1(k) + p_{k+1,1} + p_{k+2,1} \leq c_2(k+1);$$

$$c_1(k) + p_{k+1,1} + p_{k+2,1} + p_{k+3,1} \leq c_2(k+2);$$

$$\ldots;$$

$$c_1(k) + \sum_{i=k+1}^{k+r} p_{i1} \leq c_2(k+r-1).$$

Therefore, machine M_2 processes jobs from the conflict set $\{J_{k+1}, J_{k+2}, \ldots, J_{k+r}\}$ in permutation $\pi_{1,2}$ without idle times between the jobs J_{k+1} and J_{k+2}, between the jobs J_{k+2} and J_{k+3} and so on, between the jobs J_{k+r-1} and J_{k+r}. Then, the following relations hold:

$$c_2 = c_2(k+r) = c_2(k+r-1) + p_{k+r,2} = c_2(k+r-2) + p_{k+r-1,2} + p_{k+r,2} = \ldots = c_2(k) + \sum_{i=k+1}^{k+r} p_{i2}$$

leading to the equality (19). The rest of the proof is the same as the rest of the proof of Theorem 9.

It is shown that the pair of job permutations $(\pi', \pi'') = ((\pi_{1,2}, \pi_1, \pi_{2,1}^*), (\pi_{2,1}^*, \pi_2, \pi_{1,2})) \in S'$ is optimal for the deterministic job-shop problem $J2|p, n_i \leq 2|C_{max}$ with any vector $p \in T(k, l-1)$ of the job processing times. Due to Definition 3, the set S' is a dominant set for the uncertain job-shop problem $J2|l_{ij} \leq p_{ij} \leq u_{ij}, n_i \leq 2|C_{max}$ with the job set $\mathcal{J}$ and the set $T(k, l-1)$ of possible scenarios. □

It is easy to be convinced that the sufficient conditions given in Theorems 9 and 10 may be tested in polynomial time $O(r^2)$ of the number r of the conflict jobs.

Similarly, one can prove analogs of Theorems 9 and 10 if the set $S' = <\{\pi_{1,2}^*\}, \Pi(G_{2,1})> \subset S$ provided that a dominant set for the uncertain job-shop problem $J2|l_{ij} \leq p_{ij} \leq u_{ij}, n_i \leq 2|C_{max}$ with the job set $\mathcal{J}$ and the partial strict order $A_\prec^{2,1}$ over the set $\mathcal{J}_{2,1} = \mathcal{J}_{2,1}^* \cup \mathcal{J}_{2,1}^1 \cup \mathcal{J}_{2,1}^2$ has the following form: $J_1 \prec \ldots \prec J_k \prec \{J_{k+1}, J_{k+2}, \ldots, J_{k+r}\} \prec J_{k+r+1} \prec \ldots \prec J_{m_{2,1}}$.

6. Scheduling Algorithms and Computational Results

The experimental study was performed on a large number of randomly generated instances of the uncertain job-shop scheduling problem $J2|l_{ij} \leq p_{ij} \leq u_{ij}, n_i \leq 2|C_{max}$. The off-line phase of scheduling was based on Algorithms 1 and 2 developed in [8]. Algorithms 1 and 2 are presented in Appendix A.

Algorithms 3–5 are developed for the on-line phase of scheduling. The input for each of these three algorithms includes the output of Algorithms 1 and 2 [8] applied at the off-line phase of scheduling.

Let outputs of Algorithms 1 and 2 [8] applied at the off-line phase of scheduling consist of the optimal permutation $\pi_{1,2}$ of the jobs $\mathcal{J}_{1,2}$ and the optimal permutation $\pi_{2,1}$ of the jobs $\mathcal{J}_{2,1}$. In such a case, the single-element dominant set $DS(T) = \{(\pi_{1,2}, \pi_1, \pi_{2,1}), (\pi_{2,1}, \pi_2, \pi_{1,2})\}$ is already constructed for the considered instance of the uncertain problem $J2|l_{ij} \leq p_{ij} \leq u_{ij}, n_i \leq 2|C_{max}$. Therefore, the pair $\{(\pi_{1,2}, \pi_1, \pi_{2,1}), (\pi_{2,1}, \pi_2, \pi_{1,2})\}$ of the job permutations is optimal for the deterministic instance $J2|p, n_i \leq 2|C_{max}$ with any scenario $p \in T$. Thus, such an instance of the uncertain problem $J2|l_{ij} \leq p_{ij} \leq u_{ij}, n_i \leq 2|C_{max}$ is optimally solved by Algorithms 1 and 2 at the off-line phase of scheduling.

Hence, there is no need to use the on-line phase of scheduling for such an instance of the uncertain job-shop scheduling problem $J2|l_{ij} \leq p_{ij} \leq u_{ij}, n_i \leq 2|C_{max}$.

In Section 6.1, it shown how to solve instances of the uncertain job-shop problem $J2|l_{ij} \leq p_{ij} \leq u_{ij}, n_i \leq 2|C_{max}$, which cannot be optimally solved at the off-line phase of scheduling.

6.1. Algorithms 3–5 for the On-Line Phase of Scheduling

Let the considered instance of the uncertain job-shop problem $J2|l_{ij} \leq p_{ij} \leq u_{ij}, n_i \leq 2|C_{max}$ cannot be optimally solved by Algorithms 1 and 2 [8] applied at the off-line phase of scheduling. Thus, due to an application of Algorithm 1 or Algorithm 2, one can obtain one of the following three possible outputs:

(a) the permutation $\pi_{2,1}$ of the jobs from set $\mathcal{J}_{2,1}$ and the partial strict order $A_{\prec}^{1,2}$ of the jobs $\mathcal{J}_{1,2}$;
(b) the permutation $\pi_{1,2}$ of the jobs from set $\mathcal{J}_{1,2}$ and the partial strict order $A_{\prec}^{2,1}$ of the jobs $\mathcal{J}_{2,1}$;
(c) the partial strict order $A_{\prec}^{1,2}$ of the jobs $\mathcal{J}_{1,2}$ and the partial strict order $A_{\prec}^{2,1}$ of the jobs $\mathcal{J}_{2,1}$.

Let B denote a number of the conflict sets in a partial strict order (in both partial strict orders) for the obtained output (a), (b) or (c). In other words, B denotes a maximal number of time-points in the decision-making at the on-line phase of scheduling. Let integer b, where $b \leq B$, denote a number of time-points in the decision-making, where optimal orders of the conflict jobs were found using Theorem 9 or Theorem 10. Using these notations, we next describe Algorithm 3 provided that there is no factual processing times of the jobs $\mathcal{J}$ in the input of Algorithm 3; see Remark 3.

Let Algorithm 3 terminate at Step 16, i.e., it has not been constructed an optimal pair of job permutations for the factual scenario $p^* \in T$ randomly determined after completing the on-line phase of scheduling. Therefore, there is a strictly positive error $\Delta(s)$ of the objective function $C_{max}(s)$ calculated for the constructed and realized schedule s. In such a case, the proven sufficient conditions for the optimality of the schedule s do not hold in some decision-points (or in a single decision-point) at the on-line phase of scheduling. If Algorithm 3 terminates at Step 17, then an optimal pair of job permutations has been constructed for the factual scenario $p^* \in T$ randomly generated after completing the on-line phase of scheduling. The optimality of this pair of the job permutations was established only after the schedule execution, since the tested sufficient conditions for the optimality of the schedule s do not hold in some decision-points (or in a single decision-point).

If Algorithm 3 terminates at Step 18, then the tested sufficient conditions hold for all decision-points considered at the on-line phase of scheduling. Therefore, the constructed pair of job permutations is optimal for all factual scenarios $p^* \in T$ which were possible during the on-line phase of scheduling. In this case, the optimal pair of job permutations was established before the end of the schedule execution (after the last decision-point). The described Algorithm 3 must be used if the input (a) is obtained due to the application of Algorithms 1 and 2 [8] at the off-line phase of scheduling. Similarly, one can describe Algorithm 4 with the sufficient conditions from the analogs of Theorems 9 and 10 for their use in the case, when the input (b) is obtained due to the application of Algorithms 1 and 2 at the off-line phase of scheduling.

Similar Algorithm 5 must be used in the case, when the input (c) is obtained due to the application of Algorithms 1 and 2 at the off-line phase of scheduling. In Algorithm 3, a decision-point may occur on machine M_1 and on machine M_2 simultaneously. Therefore, one has to check the conditions of Theorems 9 and 10 or their analogs alternately for the corresponding conflict sets of the jobs from the set $\mathcal{J}_{1,2}$ and those from the set $\mathcal{J}_{2,1}$.

Algorithm 3 for the on-line phase of scheduling

Input: Lower bounds l_{ij} and upper bounds u_{ij} on the durations p_{ij}
of all operations $O_{ij} \in \mathcal{J}$ processed on machines $M_j \in \mathcal{M}$;
a permutation π_1 of the jobs $\mathcal{J}_1$ and a permutation π_2 of the jobs $\mathcal{J}_2$;
an optimal permutation $\pi_{2,1}$ of the jobs from the set $\mathcal{J}_{2,1}$;
a partial strict order $A_\prec^{1,2}$ of the jobs from the set $\mathcal{J}_{1,2}$;
a number B of the conflict sets in the partial strict order $A_\prec^{1,2}$.

Output: Permutation $\pi_{1,2}$ of the jobs from the set $\mathcal{J}_{1,2}$.

Step 1: Set $b = 0$.

Step 2: UNTIL the completion time-point of the last job in the set $\mathcal{J}$,
process the whole linear part of the jobs in the partial strict order $A_\prec^{1,2}$
on the machine M_1 till a conflict set of the jobs is met;
let t denote a time-point of the completion of the linearly ordered set of jobs.

Step 3: Process jobs of the permutation $(\pi_{2,1}, \pi_2)$ and then process the linear part
in the partial strict order $A_\prec^{1,2}$ on the machine M_2 up to time-point t.

Step 4: Check the conditions of Theorem 9 for the conflict set of the jobs.

Step 5: IF the sufficient conditions of Theorem 9 hold
THEN set $b := b + 1$ and choose an arbitrary order π_q of the conflict jobs
GOTO step 11.

Step 6: ELSE set $d_z = l_{z2} - u_{z1}$ for all conflict jobs J_z
and partition the conflict jobs J_z into two subsets X_1 and X_2,
where $J_z \in X_1$ if $d_z \geq 0$, and $J_z \in X_2$ otherwise.

Step 7: Construct the following order π_q of the conflict jobs:
First, arrange the jobs from the set X_1 in the non-decreasing order
of the values of u_{i1}, then arrange the jobs from the set X_2
in the non-increasing order of the values of l_{i2}.

Step 8: Check the conditions of Theorem 10 for the constructed
permutation of the conflict jobs.

Step 9: IF the sufficient conditions of Theorem 10 hold **THEN**
set $b := b + 1$ **GOTO** step 11.

Step 10: Construct a Johnson's permutation π_q of the conflict jobs
based on the inequalities (2) provided that $p_{ij} = (u_{ij} + l_{ij})/2$.

Step 11: Include the permutation π_q of the conflict jobs in the strict order
$A_\prec^{1,2}$ instead of the conflict set of these jobs.

Step 12: RETURN

Step 13: IF $b = B$ **THEN GOTO** step 18.

Step 14: Calculate makespan $C_{\max}(s)$ for the schedule s constructed at steps 1 – 12;
calculate makespan $C_{\max}(s^*)$ for the optimal schedule s^* polynomially
calculated for the corresponding deterministic problem $J2|p^*, n_i \leq 2|C_{max}$,
where the factual processing times p^* are randomly generated for all jobs $\mathcal{J}$.

Step 15: IF $C_{\max}(s) = C_{\max}(s^*)$ **THEN GOTO** step 17.

Step 16: STOP 4: The constructed schedule s is not optimal for the factual
processing times p^* of the jobs $\mathcal{J}$.

Step 17: STOP 3: The optimality of the constructed schedule s for the factual
processing times p^* of the jobs $\mathcal{J}$ was established only after
the execution of the schedule s.

Step 18: STOP 2: The optimality of the constructed schedule s for the factual
processing times p^* of the jobs $\mathcal{J}$ was proven before the end
of the execution of this schedule.

6.2. The Modified Example with Different Factual Scenarios

To demonstrate the on-line phase of scheduling based on Algorithm 3, it is considered Example 2 of the problem $J2|l_{ij} \leq p_{ij} \leq u_{ij}, n_i \leq 2|C_{max}$ with the numerical input data given in Table 1 similarly as for Example 1 with the only one exception. It is assumed that $u_{3,2} = 6$.

The first part of the off-line phase of scheduling for solving Example 2 is similar to that for Example 1 till checking the conditions of Theorem 2. Indeed, the conditions of Theorem 2 do not hold for the jobs from the set $\mathcal{J}_{1,2}$ since the following strict inequalities hold: $u_{2,2} > l_{3,2}$ and $u_{3,2} > l_{2,2}$.

Due to checking the inequalities (3) and (4), one can determine the binary relation $A_{\prec}^{1,2}$ over the set $\mathcal{J}_{1,2}$ in the following form: $J_1 \prec \{J_2, J_3\}$. Thus, the set $\{J_2, J_3\}$ is a conflict set with two jobs; see Definition 5. Then, one can consecutively check the conditions of Theorems 6–8 for the jobs from the set $\mathcal{J}_{1,2}$. After letting $k = 1$, $r = 2$, one can calculate $L_2 = \sum_{J_i \in \mathcal{J}_{2,1} \cup \mathcal{J}_2} l_{i2} = l_{6,2} + l_{7,2} + l_{8,2} + l_{5,2} = 2 + 3 + 4 + 2 = 11$ and then obtain the following relations:

$$\sum_{i=1}^{k+r} u_{i1} = u_{1,1} + u_{2,1} + u_{3,1} = 7 + 9 + 9 = 25 \not\leq L_2 + \sum_{i=1}^{k} l_{i2} = L_2 + l_{1,2} = 11 + 6 = 17.$$

Thus, the condition of Theorem 6 does not hold for Example 2. Next, one can check the conditions of Theorem 7. Similarly as in the previous case, one can obtain that $L_2 = 11$, $k = 1$, and $r = 2$. Due to the condition (10), one can obtain two inequalities as follows: $s = 1$ and $s = 2$. Then, one can check both permutations of the jobs from the set $\mathcal{J}_{1,2}$, which satisfy the partial strict order $A_{\prec}^{1,2}$, as follows: $\Pi(\mathcal{G}_{1,2}) = \{\pi_{1,2}^1, \pi_{1,2}^2\}$, where $\pi_{1,2}^1 = \{J_1, J_2, J_3\}$ and $\pi_{1,2}^2 = \{J_1, J_3, J_2\}$.

Thus, the permutation $\pi_{1,2}^1$ must be tested. One can obtain the following relations:

$$u_{2,1} = 9 \leq L_2 + (l_{1,2} - u_{1,1}) = 11 + (6 - 7) = 10;$$

$$u_{3,1} = 9 \not\leq L_2 + \sum_{i=1}^{2}(l_{i2} - u_{i1}) = L_2 + (l_{1,2} - u_{1,1}) + (l_{2,2} - u_{2,1}) = 11 + (6 - 7) + (5 - 9) = 6.$$

Hence, the condition of Theorem 7 does not hold for the permutation $\pi_{1,2}^1$.

Analogously, for the permutation $\pi_{1,2}^2$, the following relations hold:

$$u_{3,1} = 9 \leq L_2 + (l_{1,2} - u_{1,1}) = 11 + (6 - 7) = 10;$$

$$u_{2,1} = 9 \not\leq L_2 + \sum_{i=1}^{2}(l_{i2} - u_{i1}) = L_2 + (l_{1,2} - u_{1,1}) + (l_{3,2} - u_{3,1}) = 11 + (6 - 7) + (4 - 9) = 5.$$

Hence, the condition of Theorem 7 does not hold for the permutation $\pi_{1,2}^2$ as well.

It is impossible to check the condition of Theorem 8, since the conflict set of the jobs $\{J_2, J_3\}$ is located at the end of the partial strict order $A_{\prec}^{1,2}$. Thus, the off-line phase of scheduling is completed, and the constructed partial strict order $A_{\prec}^{1,2}$ is not a linear order. Therefore, there does not exist a pair of permutations of the jobs, which is optimal for any scenario $p \in T$. In this case, Algorithms 1 and 2 [8] do not terminate with STOP 1. A scheduler needs to use the on-line phase of scheduling for solving Example 2 further.

The output of the off-line phase of scheduling for Example 2 contains the permutation $\pi_{2,1} = (J_6, J_7, J_8)$ of the jobs $\mathcal{J}_{2,1}$ processed on both machines M_1 and M_2. The partial strict order $A_{\prec}^{1,2} = (J_1 \prec \{J_2, J_3\})$ of the jobs $\mathcal{J}_{1,2}$ is constructed. The obtained output (a) of the off-line phase of scheduling shows that Algorithm 3 must be used at the on-line phase of scheduling for solving Example 2.

We next show that Algorithm 3 can be stopped either with STOP 2 (Step 18) or with STOP 3 (Step 17) or with STOP 4 (Step 16) depending on the factual values of the job processing times. Note that $B = 1$; see Algorithm 3.

Case (j): Algorithm 3 is stopped at step 18 (STOP 2).

Consider Step 2 and Step 3 of Algorithm 3. The schedule execution begins as follows: at the initial time-point $t = 0$, machine M_1 starts to process operation $O_{1,1}$, while machine M_2 starts to process operation $O_{6,2}$. This process is continued until the time-point $t = 4$ when machine M_2 completes operation $O_{6,2}$. At this time-point, an exact value of the processing time $p^*_{6,2}$ becomes known, namely: $p^*_{6,2} = 4$. Then, machine M_2 starts to process operation $O_{7,2}$ and machine M_1 continues the processing of operation $O_{1,1}$. At the time-point $t = 6$, machine M_1 completes operation $O_{1,1}$. Therefore, an exact value of the duration of operation $O_{1,1}$ becomes known as follows: $p^*_{1,1} = 6$. At this time-point, a scheduler needs to choose either job J_2 or job J_3 to be processed next on machine M_1. Note that machine M_2 continues to process the operation $O_{7,2}$ for two time units, wherein $l_{7,2} = 3$.

Consider Step 4 of Algorithm 3, where the condition (12) of Theorem 9 is checked for the conflict set of jobs $\{J_2, J_3\}$. Due to equalities $k = 1$, $r = 2$, $c_1(1) = 6$, $c_2(6) = 4$, one can obtain the following relations: $c_1(1) + u_{2,1} + u_{3,1} = 6 + 9 + 9 = 23 \not\leq c_2(6) + l_{7,2} + l_{8,2} + l_{5,2} + l_{1,2} = 4 + 3 + 4 + 2 + 6 = 19$.

At Steps 6 and 7 of Algorithm 3, one can obtain $d_2 = -4$, $d_3 = -5$ and permutation π_q having the following form: $\pi_q = (J_2, J_3)$. At Steps 8 and 9 of Algorithm 3, the conditions of Theorem 10 are checked as follows: $c_1(1) + u_{2,1} = 6 + 9 = 15 \leq c_2(6) + l_{7,2} + l_{8,2} + l_{5,2} + l_{1,2} = 4 + 3 + 4 + 2 + 6 = 19$;

$$c_1(1) + u_{2,1} + u_{3,1} = 6 + 9 + 9 = 24 \leq c_2(6) + l_{7,2} + l_{8,2} + l_{5,2} + l_{1,2} + l_{2,2} = 4 + 3 + 4 + 2 + 6 + 5 = 24.$$

At Step 11 of Algorithm 3, one can obtain the following strict order $A^{1,2}_\prec = (J_1 \prec J_2 \prec J_3)$ along with the permutation $\pi_{1,2} = (J_1, J_2, J_3)$. Since $b = 1 = B$ (see Step 13), Algorithm 3 is stopped at Step 18; see STOP 2. The optimal order of the conflict jobs J_2 and J_3 is found at the time-point $t = 6$ and the pair of job permutations $\pi' = (J_1, J_2, J_3, J_4, J_6, J_7, J_8)$ and $\pi'' = (J_6, J_7, J_8, J_5, J_1, J_2, J_3)$ is optimal for any scenario from the remaining set of possible scenarios $T(1,6) = \{p \in T : p^*_{1,1} = 6, p^*_{6,2} = 4\}$.

Thus, an additional information on the exact values of the processing times $p^*_{6,2}$ and $p^*_{1,1}$ allows a scheduler to find an optimal order of all conflict jobs. It schould be noted that the optimality of the constructed schedule is proven at the time-point $t = 6$, i.e., before the end of the schedule execution.

At the time-point $t = 6$, machine M_1 begins to process operation $O_{2,1}$. Note that all the above checks are performed at the time-point $t = 6$.

Case (jj): Algorithm 3 is stopped at Step 17 (STOP 3).

It is considered another possible realization of the semi-active schedule since another factual processing times are randomly generated at the on-line phase of scheduling for Example 2.

At the time-point $t = 0$, machine M_1 begins to process operation $O_{1,1}$, while machine M_2 begins to process operation $O_{6,2}$. Let machine M_2 complete operation $O_{6,2}$ at the time-point $t = 2.8$. Thus, the exact processing time $p^*_{6,2} = 2.8$ becomes known. Then, machine M_2 begins to process operation $O_{7,2}$ and completes this process at the time-point $t = 6$ (i.e., $p^*_{7,2} = 3.2$), while machine M_1 continues processing operation $O_{1,1}$. Let at the time-point $t = 6.9$, machine M_1 completes operation $O_{1,1}$ (i.e., $p^*_{1,1} = 6.9$). One needs to choose either job J_2 or job J_3 to be processed next on machine M_1. At this time, machine M_2 continues to process the operation $O_{8,2}$ since $t = 6$ and $(6.9 - 6) = 0.9 < 4 = l_{8,2}$.

Based on the checking of the condition (12) of Theorem 9 for the conflict set of the jobs, one can obtain the following relations: $k = 1$, $r = 2$, $c_1(1) = 6.9$, $c_2(7) = 6$;

$$c_1(1) + u_{2,1} + u_{3,1} = 6.9 + 9 + 9 = 23.9 \not\leq c_2(7) + l_{8,2} + l_{5,2} + l_{1,2} = 6 + 4 + 2 + 6 = 18.$$

Similarly as in the previous case (j), one can obtain $d_2 = -4$, $d_3 = -5$, and the permutation π_q having the following form: $\pi_q = (J_2, J_3)$. The conditions of Theorem 10 are checked as follows:

$$c_1(1) + u_{2,1} = 6.9 + 9 = 15.9 \leq c_2(7) + l_{8,2} + l_{5,2} + l_{1,2} = 6 + 4 + 2 + 6 = 18;$$

$$c_1(1) + u_{2,1} + u_{3,1} = 6.9 + 9 + 9 = 24.9 \not\leq c_2(7) + l_{8,2} + l_{5,2} + l_{1,2} + l_{2,2} = 6 + 4 + 2 + 6 + 5 = 23.$$

Thus, the conditions of Theorem 10 do not hold. At Step 10 of Algorithm 3, one can construct a Johnson's permutation π_q of the conflict jobs based on the inequalities (2) for the processing times of all conflict jobs determined as follows: $p_{ij} = (u_{ij} + l_{ij})/2$. For the jobs J_2 and J_3, one can calculate $p_{2,1} = 8.5$, $p_{2,2} = 5.5$, $p_{3,1} = 8$, $p_{3,2} = 5$ and the Johnson's permutation π_q of the conflict jobs in the following form: $\pi_q = (J_2, J_3)$.

At the time-point $t = 6.9$, one can obtain the pair of permutations $\pi' = (J_1, J_2, J_3, J_4, J_6, J_7, J_8)$ and $\pi'' = (J_6, J_7, J_8, J_5, J_1, J_2, J_3)$ of the jobs for their processing on machines $\mathcal{M}$. Therefore, at the time-point $t = 6$, machine M_1 begins to process operation $O_{2,1}$. Then, at the time-point $t = 10$, machine M_2 completes operation $O_{8,2}$ (the exact processing time $p^*_{8,2} = 4$ becomes known), and then begins to process operation $O_{5,2}$ till the time-point $t = 12.4$ (thus, $p^*_{5,2} = 2.4$), and then begins to process operation $O_{1,2}$. At the time-point $t = 15.5$, machine M_1 completes operation $O_{2,1}$ (i.e., the exact processing time $p^*_{2,1} = 8.6$ becomes known), and then begins to process operation $O_{3,1}$.

Then, at the time-point $t = 18.7$, machine M_2 completes operation $O_{1,2}$ (the exact processing time $p^*_{1,2} = 6.3$ becomes known), and then begins to process operation $O_{2,2}$ till the time-point $t = 23.7$ (thus, $p^*_{2,2} = 5$). At this time-point, machine M_1 still processes operation $O_{3,1}$. As a result, machine M_2 has an idle time in the realized schedule.

At the time-point $t = 24.5$, machine M_1 completes operation $O_{3,1}$ (i.e., $p^*_{3,1} = 9$), and then begins to process operation $O_{4,1}$. Machine M_2 begins to process operation $O_{3,2}$ immediately.

At the time-point $t = 26.5$, machine M_1 completes operation $O_{4,1}$ (i.e., $p^*_{4,1} = 2$), and then begins to process operation $O_{6,1}$ till the time-point $t = 27.5$ (i.e., $p^*_{6,1} = 1$). Then, machine M_1 processes operation $O_{7,1}$ till the time-point $t = 28.5$ (i.e., $p^*_{7,1} = 1$), and then begins to process operation $O_{8,1}$.

At the time-point $t = 30.5$, machine M_2 completes operation $O_{3,2}$ (i.e., the exact processing time $p^*_{3,2} = 6$ becomes known). Thus, machine M_2 completes to process all jobs in the realized permutation π'' at the time-point $c_2(3) = 30.5$. At the time-point $t = 31.5$, machine M_1 completes operation $O_{8,1}$ (and the exact processing time $p^*_{8,1} = 3$ becomes known). Thus, machine M_1 completes to process all jobs in the realized permutation π' at the time-point $c_1(8) = 31.5$.

All uncertain processing times $p \in T$ took their factual values p^*_{ij} as follows:

$$p^* = (p^*_{1,1}, p^*_{1,2}, p^*_{2,1}, \ldots, p^*_{7,2}, p^*_{8,1}, p^*_{8,2}) = (6.9, 6.3, 8.6, 5, 9, 6, 2, 0, 0, 2.4, 1, 2.8, 1, 3.2, 3, 4).$$

It should be remind that these factual processing times p^* were randomly generated at the time-points of the completions of the corresponding operations; see Remark 3.

For the constructed and realized schedule (π', π''), the equalities $C_{max}(\pi', \pi'') = \max\{c_1(8), c_2(3)\} = \max\{31.5, 30.5\} = 31.5$ hold; see Step 14 of Algorithm 3.

Now, one can check whether the constructed and realized schedule (π', π'') is optimal for the factual vector p^* of the job processing times. To this end, one can construct the pair of Jackson's permutations (π'_*, π''_*) for the deterministic problem $J2|p^*, n_i \leq 2|C_{max}$ with the factual vector p^* of the job processing times. Then, one can find the optimal makespan value for the deterministic problem $J2|p^*, n_i \leq 2|C_{max}$ as follows: $C_{max}(\pi'_*, \pi''_*) = 31.5$; see Step 15 of Algorithm 3.

The obtained equalities $C_{max}(\pi'_*, \pi''_*) = 31.5 = C_{max}(\pi', \pi'')$ mean that Algorithm 3 has constructed the optimal schedule for the deterministic problem $J2|p^*, n_i \leq 2|C_{max}$ with the factual vector p^* of the job processing times. However, the optimality of this constructed and realized schedule (π', π'') was established after the execution of the whole schedule (π', π''). Indeed, Algorithm 3 is stopped at Step 17; see STOP 3. The constructed and realized schedule (π', π'') is presented in Figure 4 for case (jj) of the randomly generated factual processing times p^* of the jobs $\mathcal{J}$.

Case (jjj): Algorithm 3 is stopped at Step 16 (STOP 4).

It is considered the same process as in the previous case (jj) up to the time-point $t = 28.5$ when machine M_1 begins to process operation $O_{8,1}$ (machine M_2 processes operation $O_{3,2}$ at this time-point).

Let the equality $p^{**}_{8,1} = 1$ hold for the factual processing time $p^{**}_{8,1}$ of the operation $O_{8,1}$ and machine M_1 complete operation $O_{8,1}$. Thus, machine M_1 completes all operations of the jobs $\mathcal{J}$ in the permutation π' at the time-point 29.5. Therefore, the equality $c_1(8) = 29.5$ holds. Similarly as in the previous case, machine M_2 completes operation $O_{3,2}$ at the time-point $t = 30.5$. Thus, $p^*_{3,2} = 6$ and $c_2(3) = 30.5$. The factual vector of the job processing times is randomly generated as follows:

$$p^{**} = (p^*_{1,1}, p^*_{1,2}, p^*_{2,1}, \ldots, p^*_{7,2}, p^{**}_{8,1}, p^*_{8,2}) = (6.9, 6.3, 8.6, 5, 9, 6, 2, 0, 0, 2.4, 1, 2.8, 1, 3.2, 1, 4).$$

The makespan value for the constructed and realized schedule (π', π'') is determined as follows: $C_{max}(\pi', \pi'') = \max\{c_1(8), c_2(3)\} = \max\{29.5, 30.5\} = 30.5$. However, the optimal makespan value for the deterministic problem $J2|p^{**}, n_i \leq 2|C_{max}$ with the factual vector p^{**} of the job processing times is equal to $29.7 < 30.5 = C_{max}(\pi', \pi'')$, since the optimal order of the jobs J_2 and J_3 is determined as follows: (J_3, J_2). Hence, the constructed and realized schedule (π', π'') is not optimal for the factual vector $p^{**} \in T$ of the job processing times. In this case, Algorithm 3 is stopped at Step 16; see STOP 4.

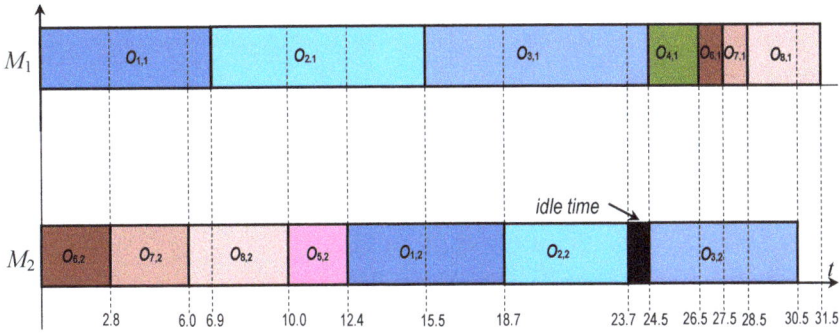

Figure 4. The optimal semi-active schedule for the Example 2 in case (jj).

6.3. Computational Experiments

We describe the computational experiments and computational results obtained for the tested randomly generated instances of the uncertain problem $J2|l_{ij} \leq p_{ij} \leq u_{ij}, n_i \leq 2|C_{max}$. Each tested series consisted of 1000 randomly generated instances with fixed numbers $n \in \{10, 20, \ldots, 100\}$ of the jobs $\mathcal{J}$ and the maximum possible errors $\delta \in \{5\%, 10\%, 20\%, 30\%, 40\%, 50\%, 60\%, 70\%, 80\%, 90\%, 100\%\}$ of the random durations of the operations O_{ij}. The lower bounds l_{ij} and upper bounds u_{ij} on the possible values of the durations p_{ij} of operations O_{ij}, $p_{ij} \in [l_{ij}, u_{ij}]$, were randomly generated as follows. The lower bound l_{ij} was randomly chosen from the segment $[10, 100000]$ using a uniform distribution. The upper bound u_{ij} was determined using the equality $u_{ij} = l_{ij}\left(1 + \frac{\delta}{100}\right)$. The bounds l_{ij} and u_{ij} are decimal fractions with the maximum numbers of digits after the decimal points. The inequality $l_{ij} < u_{ij}$ holds for each job $J_i \in \mathcal{J}$ and each machine $M_j \in \mathcal{M}$.

Algorithms 1 and 2 developed in [8] were used at the off-line phase of scheduling. If the tested instance was not optimally solved using Algorithms 1 and 2, then corresponding Algorithms 3, 4 or 5 was used at the on-line phase of scheduling for solving further the instance of the uncertain problem $J2|l_{ij} \leq p_{ij} \leq u_{ij}, n_i \leq 2|C_{max}$. All developed algorithms were coded in C# and tested on a PC with Intel Core i7-7700 (TM) 4 Quad, 3.6 GHz, 32.00 GB RAM.

In the computational experiments, two procedures were used to generate factual durations of the operations O_{ij} (a factual duration of the job J_i remained unknown until completing this job). In the first part of the computational experiments, the factual duration p^*_{ij} of the operation O_{ij} was randomly generated using a uniform distribution in the range $[l_{ij}, u_{ij}]$. In the second part of the computational experiments, two distribution laws were used in the experiments to determine the factual scenarios. Namely, we used the gamma distribution with parameters $(0.5; 1)$ (we call it as the distribution law with number 1) and the gamma distribution with parameters $(7.5; 1)$ (we call it as the distribution law with number 2). For generating factual processing times for each tested instance, the number of the used distribution was randomly chosen from the possible set $\{1,2\}$.

The sufficient conditions proven in Section 5 are verified in polynomial time $O(n^2)$ of the number n of the jobs $\mathcal{J}$. Therefore, all series of the tested instances in our computational experiments were solved very quickly (less than one second per a series with 1000 instances).

The experiments include testing of 14 classes of the instances of the uncertain problem $J2|l_{ij} \leq p_{ij} \leq u_{ij}, n_i \leq 2|C_{max}$ with different ratios of the numbers m_1, m_2, $m_{1,2}$ and $m_{2,1}$ (where $n = m_1 + m_2 + m_{1,2} + m_{2,1}$) of the jobs in the subsets $\mathcal{J}_1$, $\mathcal{J}_2$, $\mathcal{J}_{1,2}$ and $\mathcal{J}_{2,1}$ of the set $\mathcal{J}$, respectively. Every class of the tested instances of the problem $J2|l_{ij} \leq p_{ij} \leq u_{ij}, n_i \leq 2|C_{max}$ is characterized by the following ratio:

$$\frac{m_1}{n} \cdot 100\% : \frac{m_2}{n} \cdot 100\% : \frac{m_{1,2}}{n} \cdot 100\% : \frac{m_{2,1}}{n} \cdot 100\% \qquad (28)$$

of the percentages of the numbers of jobs in the subsets $\mathcal{J}_1$, $\mathcal{J}_2$, $\mathcal{J}_{1,2}$ and $\mathcal{J}_{2,1}$ of the set $\mathcal{J}$, respectively.

Tables A1–A14 present the computational results obtained for the tested classes of instances with the following ratios (28):

0% : 0% : 10% : 90% (class 1, Table A1); 0% : 0% : 20% : 80% (class 2, Table A2);
0% : 0% : 30% : 70% (class 3, Table A3); 0% : 0% : 40% : 60% (class 4, Table A4);
0% : 0% : 50% : 50% (class 5, Table A5); 5% : 5% : 5% : 85% (class 6, Table A6);
5% : 15% : 5% : 75% (class 7, Table A7); 5% : 20% : 5% : 70% (class 8, Table A8);
10% : 10% : 10% : 70% (class 9, Table A9); 10% : 10% : 40% : 40% (class 10, Table A10);
10% : 20% : 10% : 60% (class 11, Table A11); 10% : 30% : 10% : 50% (class 12, Table A12);
10% : 40% : 10% : 40% (class 13, Table A13); 10% : 60% : 10% : 20% (class 14, Table A14).

All Tables A1–A14 are organized as follows. The procedure for generating factual processing times (the uniform distribution or the gamma distribution) is indicated in the first row of each table. Numbers n of the given jobs $\mathcal{J}$ in the tested instances of the problem $J2|l_{ij} \leq p_{ij} \leq u_{ij}, n_i \leq 2|C_{max}$ are presented in the second row. The maximum possible errors δ of the randomly generated processing times (in percentages) are presented in the first column. For the fixed maximum possible error δ, the obtained computational results are presented in four rows called Stop1, Stop2, Stop3 and Stop4.

The row Stop1 determines the percentage of instances from the tested series, which were optimally solved at the off-line phase of scheduling using either Algorithms 1 or 2 developed in [8]. For such an instance, an optimal pair (π', π'') of the job permutations was constructed before the time-point of starting the first job of the realized schedule, i.e., the equality $C_{max}(\pi', \pi'') = C_{max}(\pi^*, \pi^{**})$ holds, where $(\pi^*, \pi^{**}) \in S$ is an optimal pair of job permutations for the deterministic problem $J2|p^*, n_i \leq 2|C_{max}$ with the factual scenario $p^* \in T$ that is unknown before completing the whole jobs $\mathcal{J}$.

The row Stop2 determines the percentage of instances, which were optimally solved at the on-line phase of scheduling using corresponding Algorithms 3, 4 or 5. For each such an instance, an optimal pair (π', π'') of job permutations for the deterministic problem $J2|p^*, n_i \leq 2|C_{max}$ associated with the factual scenario $p^* \in T$ was constructed by checking sufficient conditions in Theorem 9 or Theorem 10. Remind that the factual scenario $p^* \in T$ for the uncertain problem $J2|l_{ij} \leq p_{ij} \leq u_{ij}, n_i \leq 2|C_{max}$ remains unknown until completing the jobs $\mathcal{J}$.

The row Stop3 determines the percentage of instances, which were optimally solved at the on-line phase of scheduling using Algorithms 3, 4 or 5. In such a case, an optimal pair of job permutations has been constructed for the factual scenario $p^* \in T$. However, the optimality of this pair of job permutations was established only after the execution of the constructed schedule.

The row Stop4 determines the percentage of instances, for which the constructed and realized schedule is not optimal for the deterministic instance $J2|p^*, n_i \leq 2|C_{max}$ with the factual scenario p^*.

6.4. Computational Results

First of all, it is important to determine a total number of the tested instances, for which 3 (or Algorithms 4 and 5) were completed at Step 18 (STOP 2) or at Step 17 (STOP 3). This number shows how many tested instances of the uncertain job-shop scheduling problem have been optimally solved either with the proofs of their optimality before the completion of processing all jobs $\mathcal{J}$ (STOP 2) or the optimality of the obtained schedule was established after the realization of the constructed schedule (STOP 3). For the numbers of jobs from $n = 10$ to $n = 100$ and for each value of the tested errors δ of the processing times, average percentages of the instances optimally solved by Algorithms 1, 2, 3, 4 or 5 (these average percentages summarize the values given in rows Stop1 and Stop2 in all Tables A1–A14) are presented in Table 2 and Figure 5.

Table 2. Average percentages of the instances whose optimality of the constructed permutations was proven at the off-line and on-line phases of scheduling.

δ%	5%	10%	20%	30%	40%	50%	60%	70%	80%	90%	100%
0% : 0% : 10% : 90%	98.38	95.66	74.32	40.09	18.56	8.65	4.4	2.48	1.58	0.95	0.57
0% : 0% : 20% : 80%	99.48	98.81	94.37	74.52	45.33	24.20	12.32	6.55	3.87	2.59	1.65
0% : 0% : 30% : 70%	99.87	99.71	98.99	95.09	81.33	59.33	36.65	21.67	11.93	7.28	4.39
0% : 0% : 40% : 60%	99.97	99.94	99.73	99.02	97.09	90.86	78.5	59.76	40.85	25.93	15.2
0% : 0% : 50% : 50%	100	99.99	99.93	99.73	99.13	97.55	94.21	86.65	72.35	51.41	29.45
5% : 5% : 5% : 85%	99.67	98.45	78.97	43.19	19.26	9.11	4.31	2.02	1.1	0.58	0.27
5% : 15% : 5% : 75%	99.64	98.86	84.44	51.06	24.65	11.88	6.43	3.34	1.95	1.02	0.68
5% : 20% : 5% : 70%	99.57	98.97	86.92	55.79	29.59	14.98	7.97	4.42	2.59	1.45	1.01
10% : 10% : 10% : 70%	99.84	99.48	97.46	83.55	57.09	34.11	18.95	11.14	6.81	4.29	2.8
10% : 10% : 40% : 40%	99.99	100	99.96	99.89	99.69	99.35	98.41	96.55	92.84	85.66	73.22
10% : 20% : 10% : 60%	99.87	99.68	98.37	90.56	71.05	48.26	30.28	18.22	11.36	7.21	4.79
10% : 30% : 10% : 50%	99.9	99.72	99.11	95.33	83.85	66.15	49.34	34.35	24.13	16.64	11.44
10% : 40% : 10% : 40%	99.92	99.75	99.33	97.97	92.52	82.69	70.01	58.6	48.52	40.36	32.76
10% : 60% : 10% : 20%	99.98	99.98	99.93	99.83	99.59	99.01	97.96	96.16	93.01	89.46	85.75

Table 2 shows the total percentages of the optimally solved instances for all classes of the tested instances, for which the optimal schedules were constructed either at the off-line phase of scheduling (STOP 1) or at the on-line phase of scheduling (STOP 2). One can see that for three small values of the maximal errors $\delta \in \{5\%, 10\%, 20\%\}$ for most classes, more than 90% (up to 100%) of the tested instances were optimally solved. For all tested classes with a maximal error $\delta \leq 20\%$, more than 70% tested instances were optimally solved at the off-line or on-line phases of scheduling.

With a further increasing of the maximal error δ, the percentage of solved instances drops rapidly. For most tested classes with the maximal error δ greater than 70%, the percentage of solved instances is less than 10%. However, these indicators differ for different tested classes. For classes 4, 5, 10, 13 and 14 with maximal errors $\delta \leq 70\%$, more than 60% of the tested instances were optimally solved with the proof of the optimality before completing all the jobs. The best computational results are obtained for classes 5, 10 and 14 of the tested instances. More than 80% of the instances from these three classes were optimally solved at the off-line phase of scheduling or at the on-line phases of scheduling provided that the maximal error δ of the given job processing times was no greater than 70%, i.e., for $\delta \in \{5\%, 10\%, 15\%, 20\%, 30\%, 40\%, 50\%, 60\%, 70\%\}$. For both classes 10 and 14 of the tested instances even with an error $\delta = 100\%$, more than 70% of the instances were optimally solved.

On the other hand, for both classes 1 and 6 with a maximal error $\delta = 40\%$, only less than 20% of the tested instances were optimally solved at both off-line phase and on-line phase of scheduling. For classes 1 and 6 with $\delta = 50\%$, less than 10% of the tested instances were optimally solved. Furthermore, these two classes of instances are most difficult ones to find an optimal schedule with the proof of its optimality before completing all the jobs using the on-line phase and off-line phase of scheduling. It should be noted that all tested classes of instances demonstrate a monotonic decrease in the percentages of the optimally solved problems with an increase of the values of the maximal error δ of the job processing times; see Figure 5.

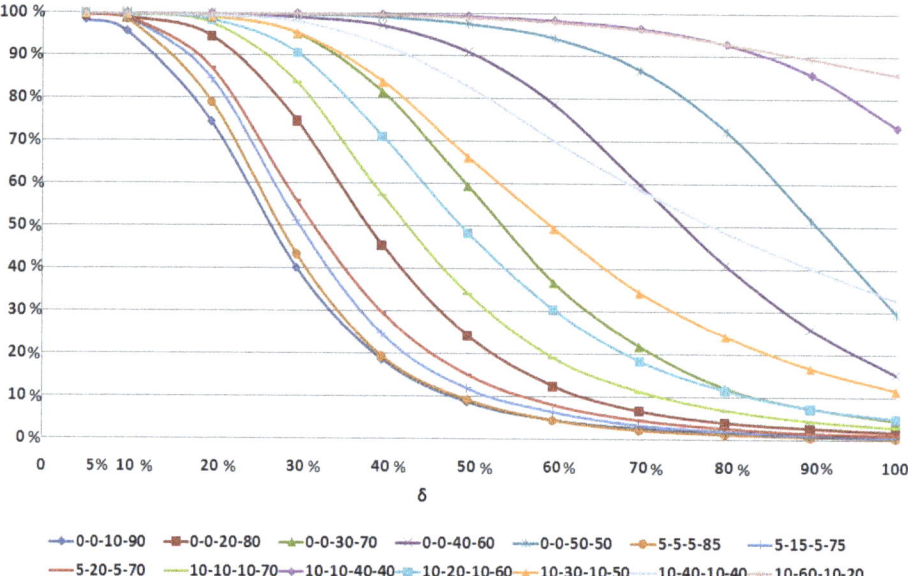

Figure 5. Average percentages of the instances whose optimality of the constructed pair of job permutations was proven at the off-line phase and on-line phase of scheduling.

Let us consider the percentages of the tested instances, for which the optimality of the constructed schedules was proven at the on-line phase of scheduling and the proofs of their optimality being obtained before completing all the jobs. Note that it is novelty of this paper; see rows Stop2 in Tables A1–A14. For all tested numbers of the jobs, $n \in \{10, 20, \ldots, 100\}$, and for all maximal values of the errors $\delta \in \{5\%, 10\%, 20\%, \ldots, 100\%\}$ of the job processing times, the average percentages of the instances, which were optimally solved by Algorithms 3, 4 or 5 at the on-line phase of scheduling are presented in Table 3, where only Stop2 is indicated.

It should be noted that the monotonous increase of the percentages of the optimally solved instances takes place only for classes 10 and 14 of the tested instances. For other tested classes of instances, there is a maximum, and for the different classes of the tested instances, these maximal vales being achieved for different maximal values of the errors δ. Then the percentages of the optimally solved instances decrease again with the increasing of the maximal values δ. The values of the maximal numbers of instances, which optimal solutions have been proven at the on-line phase of scheduling (STOP 2), vary from 0.59% to 8.69% for different classes of instances.

Classes 1–5 are distinguished from the above classes since their maximal numbers of the instances optimally solved at the on-line phase of scheduling vary from 6% to 9%. Average percentages of the instances from these five classes, which were optimally solved by Algorithms 3, 4 or 5 at the on-line phase of scheduling (only Stop2) are shown in Figure 6.

Note that for the difficult classes 1 and 6, the percentages of instances, which were optimally solved at the on-line phase of scheduling with the proofs of their optimality, behave identically with the reaching of the maximum for the maximal error $\delta = 20\%$. However their maximal values differ, namely: from 2.96% for class 6 up to 8.69% for class 1.

Table 3. Average percentages of the instances whose optimality of the constructed permutations was proven at the on-line phase of scheduling.

$\delta\%$	5%	10%	20%	30%	40%	50%	60%	70%	80%	90%	100%
0% : 0% : 10% : 90%	0.97	3.68	8.69	7.26	3.86	1.92	1.06	0.5	0.34	0.2	0.11
0% : 0% : 20% : 80%	0.24	1.24	5.23	8.59	7.85	5.33	2.63	1.49	0.78	0.59	0.32
0% : 0% : 30% : 70%	0.03	0.22	1.33	4.27	6.82	8.13	5.78	4	2.49	1.66	1.07
0% : 0% : 40% : 60%	0.02	0.06	0.31	0.93	2.05	4.15	5.57	6.53	5.6	4.44	3.23
0% : 0% : 50% : 50%	0	0.01	0.06	0.25	0.53	1.24	2.62	4.33	6.06	7.4	6.23
5% : 5% : 5% : 85%	0.33	1.01	2.96	2.31	1.63	0.88	0.49	0.25	0.15	0.07	0.03
5% : 15% : 5% : 75%	0.27	0.84	2.64	2.39	1.78	0.93	0.47	0.3	0.19	0.09	0.08
5% : 20% : 5% : 70%	0.28	0.69	2.38	2.86	1.76	1.07	0.6	0.46	0.26	0.12	0.04
10% : 10% : 10% : 70%	0.06	0.165	0.83	1.87	1.95	1.56	1.08	0.69	0.43	0.32	0.24
10% : 10% : 40% : 40%	0	0	0.02	0.04	0.05	0.14	0.21	0.3	0.45	0.68	0.77
10% : 20% : 10% : 60%	0.05	0.13	0.6	1.29	1.77	1.6	1.22	0.82	0.57	0.41	0.25
10% : 30% : 10% : 50%	0.01	0.07	0.36	0.91	1.49	1.32	1.48	1.03	0.7	0.48	0.36
10% : 40% : 10% : 40%	0	0.03	0.16	0.41	0.78	1.17	1.29	1.01	0.9	0.62	0.46
10% : 60% : 10% : 20%	0	0	0.01	0.05	0.08	0.17	0.28	0.38	0.45	0.51	0.59

For the instances, for which the optimality of the constructed schedules was not proven before completing all the jobs $\mathcal{J}$, the relative errors $\Delta\%$ of the achieved objective function vales for the realized schedules were calculated. Note that the positive errors $\Delta\%$ may occur only if Algorithm 3 (or Algorithms 4 and 5) have been stopped at Step 16; see STOP 4. For all tested numbers of jobs $n \in \{10, 20, \ldots, 100\}$ and for all maximal values of the errors $\delta \in \{5\%, 10\%, 20\%, \ldots, 100\%\}$ of the job processing times, the maximal values of $\Delta_{max}\%$ and the average values of $\Delta_{ave}\%$ were calculated separately for instances with uniform distributions (see Table 4) and gamma distributions (see Table 5).

It can be seen that the values of maximal errors $\Delta_{ave}\%$ significantly differ when applying different distribution laws. With using a uniform distribution, the maximal error Δ_{max} does not exceed 9%, while when using a gamma distribution, the maximal error Δ_{max} could reach a value more than 17%.

It can be seen that for using various distribution laws, Algorithm 3 (Algorithms 4 and 5 as well) terminates at STOP 4 with various combinations of the tested classes and maximal errors $\delta\%$. If a uniform distribution is used, then for classes 1–2, strictly positive errors $\Delta_{ave}\%$ arise for all values of the tested maximal errors $\delta\%$. For classes 9–10 and 11–13, such errors appear more often with increasing the maximal error $\delta\%$.

For a gamma distribution, for all values of $\delta\%$, the error $\Delta_{ave}\%$ arises only for class 1. For classes 2–4, 6, 8, 10, the error $\Delta_{ave}\%$ arises with the growth of maximal errors $\delta\%$. For classes 7, 9, 11–13, on the contrary, the error $\Delta_{ave}\%$ is more common for small values of the maximal errors $\delta\%$.

Table 4. Maximal errors Δ_{max} and average errors Δ_{ave} for all tested instances with factual processing times randomly generated based on a uniform distribution.

Class	$\delta\%$	5%	10%	20%	30%	40%	50%	60%	70%	80%	90%	100%
1	Δ_{max}	0.031511	0.300924	0.691062	0.333292	1.110492	0.881902	2.246299	3.263145	2.729286	3.85936	5.024917
	Δ_{ave}	0.000003	0.000058	0.000225	0.000057	0.000333	0.000613	0.000733	0.00135	0.001868	0.001409	0.002253
2	Δ_{max}	0.0475	0.189872	0	0.125467	0	0.441669	0.243659	1.076096	3.127794	1.286158	1.353086
	Δ_{ave}	0.000005	0.000023	0	0.000013	0	0.000064	0.000045	0.000189	0.000976	0.000221	0.000135
3	Δ_{max}	0	0	0	0	0	0	0	0	0.8199	0	0
	Δ_{ave}	0	0	0	0	0	0	0	0	0.000082	0	0
4	Δ_{max}	0	0	0	0	0	0	0	0	0	0	0
	Δ_{ave}	0	0	0	0	0	0	0	0	0	0	0
5	Δ_{max}	0	0	0	0	0	0	0	0	0	0	0
	Δ_{ave}	0	0	0	0	0	0	0	0	0	0	0
6	Δ_{max}	0	0	0	0	0	0.411415	0.081623	0	0	0	0
	Δ_{ave}	0	0	0	0	0	0.000082	0.000016	0	0	0	0
7	Δ_{max}	0	0	0	0	0	0	0	0	0	0	0
	Δ_{ave}	0	0	0	0	0	0	0	0	0	0	0
8	Δ_{max}	0	0	0	0	0	0	0	0	0	0	0
	Δ_{ave}	0	0	0	0	0	0	0	0	0	0	0
9	Δ_{max}	0	0.068237	0.055299	0	1.31253	0	0.91223	0.893705	1.697913	2.166717	8.617851
	Δ_{ave}	0	0.000007	0.000006	0	0.000244	0	0.000144	0.000167	0.000338	0.000558	0.001348
10	Δ_{max}	0	0	0	0	0	0	0	0	0	0	0
	Δ_{ave}	0	0	0	0	0	0	0	0	0	0	0
11	Δ_{max}	0	0	0	1.47875	3.660297	5.724288	0.810014	3.316178	0.39653	4.42828	4.666154
	Δ_{ave}	0	0	0	0.000148	0.000694	0.000572	0.000081	0.000332	0.000040	0.000799	0.000924
12	Δ_{max}	0	0	0	0	0	0	0	0	0	0	7.243838
	Δ_{ave}	0	0	0	0	0	0	0	0	0	0	0.000724
13	Δ_{max}	0	0	0	0	0	0	0	0	0	0	5.036085
	Δ_{ave}	0	0	0	0	0	0	0	0	0	0	0.000504
14	Δ_{max}	0	0	0	0	0	0	0	0	0	0	0
	Δ_{ave}	0	0	0	0	0	0	0	0	0	0	0

As one can see, using the uniform distribution for the generation of the factual job processing times for classes 4, 5, 7, 10, 14, all tested instances were solved optimally using the developed algorithms and two phases of scheduling. In other words, there are no instances, for which corresponding Algorithms 3, 4 or 5 was stopped at Step 16 (STOP 4). However, for the gamma distribution, there are only two such classes 5 and 14. Thus, classes 5 and 14 can be considered as easy ones, while class 1 is the most difficult one. As for class 1, Algorithms 3, 4 and 5 are stopped at Step 16 (STOP 4) for all values of the tested maximal errors $\delta\%$. Moreover, the maximum makespan error $\Delta_{max}\%$ of more than 5% for the uniform distribution and more than 10% for the gamma distribution is found for classes 1, 9, 11 and 12 of the tested instances (these classes are difficult for the used stability approach).

Table 5. Maximal errors Δ_{max} and average errors Δ_{ave} for all tested instances with factual processing times randomly generated based on a gamma distribution.

Class	$\delta\%$	5%	10%	20%	30%	40%	50%	60%	70%	80%	90%	100%
1	Δ_{max}	0.211802	0.509319	1.995284	2.439105	4.8928	2.670648	6.613935	8.782202	10.59834	9.153295	9.50327
	Δ_{ave}	0.000055	0.000131	0.00031	0.000746	0.001754	0.001746	0.00369	0.005033	0.006755	0.011687	0.01144
2	Δ_{max}	0	0	0	0	0	4.182544	1.361694	1.070905	5.634459	7.845669	7.974282
	Δ_{ave}	0	0	0	0	0	0.000595	0.000317	0.000107	0.001382	0.001737	0.001725
3	Δ_{max}	0	0	0	0	1.32533	0	0	5.566808	0	4.026352	5.385314
	Δ_{ave}	0	0	0	0	0.000133	0	0	0.000557	0	0.000511	0.001354
4	Δ_{max}	0	0	0	0	0	0	0	0	0	6.044646	0
	Δ_{ave}	0	0	0	0	0	0	0	0	0	0.000604	0
5	Δ_{max}	0	0	0	0	0	0	0	0	0	0	0
	Δ_{ave}	0	0	0	0	0	0	0	0	0	0	0
6	Δ_{max}	0	0	0	0.061048	0	0.387884	1.081353	1.125343	0.710307	0.643768	0.67762
	Δ_{ave}	0	0	0	0.000012	0	0.000078	0.000216	0.000401	0.000206	0.000343	0.000136
7	Δ_{max}	0	0	0.143177	0	0	0	0	0	0	0	0
	Δ_{ave}	0	0	0.000029	0	0	0	0	0	0	0	0
8	Δ_{max}	0	0	0.388797	0	0.426346	0	0.289505	0.059146	2.478004	1.167724	4.748751
	Δ_{ave}	0	0	0.000078	0	0.000085	0	0.000029	0.000006	0.000442	0.000234	0.000948
9	Δ_{max}	2.64165	6.620637	5.266738	4.163808	10.56515	0	0	0	0	0	0
	Δ_{ave}	0.000852	0.001946	0.001696	0.001615	0.003941	0	0	0	0	0	0
10	Δ_{max}	0	0	0	0	0	0	0	0.714515	0	0.334513	3.232162
	Δ_{ave}	0	0	0	0	0	0	0	0.000071	0	0.000033	0.000584
11	Δ_{max}	2.988951	10.50113	2.526341	5.632594	7.258956	0	0	0	0	0	0
	Δ_{ave}	0.0003	0.001289	0.000431	0.000887	0.001595	0	0	0	0	0	0
12	Δ_{max}	0	0.095929	1.639148	17.64929	6.3913	0	0	0	0	0	0
	Δ_{ave}	0	0.000010	0.000164	0.002948	0.001737	0	0	0	0	0	0
13	Δ_{max}	0	0.967642	0.7847	0	0	0	0	0	0	0	0
	Δ_{ave}	0	0.000097	0.000078	0	0	0	0	0	0	0	0
14	Δ_{max}	0	0	0	0	0	0	0	0	0	0	0
	Δ_{ave}	0	0	0	0	0	0	0	0	0	0	0

Class 13 of the tested instances is a rather strange one. For using the uniform distribution, a maximum makespan error $\Delta_{max}\%$ of more than 5% was obtained, while when for using the gamma distribution, the maximum makespan error $\Delta_{max}\%$ did not reach even 1%. Note that for all tested classes of the instances, the average makespan errors $\Delta_{ave}\%$ for all tested numbers $n \in \{10, 20, \ldots, 100\}$ of jobs $\mathcal{J}$ are less than 0.02%.

Maximal relative makespan errors $\Delta_{max}\%$ for each tested class and for all values of the tested maximal errors δ are shown in Figure 7 for the instances with uniform distributions and in Figure 8 for the instances with gamma distributions of the factual durations of the given operations.

Figures 7 and 8 also show that the maximal value of the makespan errors $\Delta_{max}\%$ for the constructed and realized schedule for the factual scenarios are achieved for different values of the maximal errors $\delta\%$ for different classes of the tested instances.

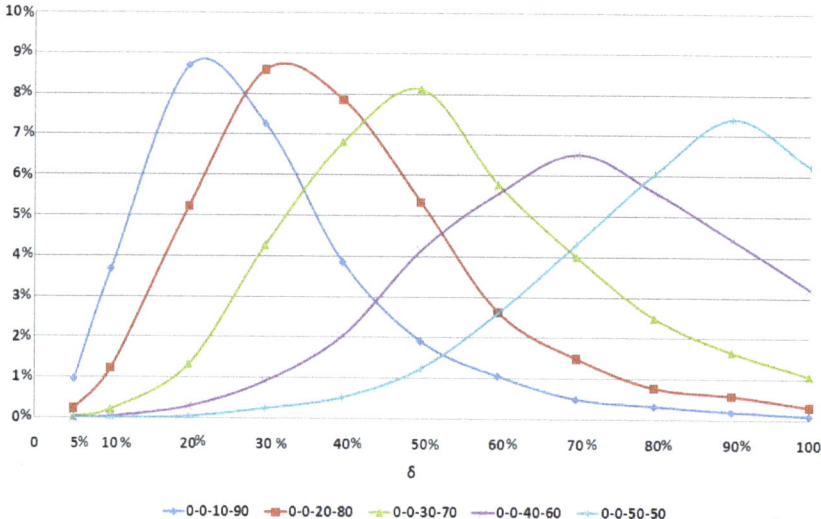

Figure 6. Average percentages of the instances whose optimality of the constructed pair of job permutations was proven at the on-line phase of scheduling.

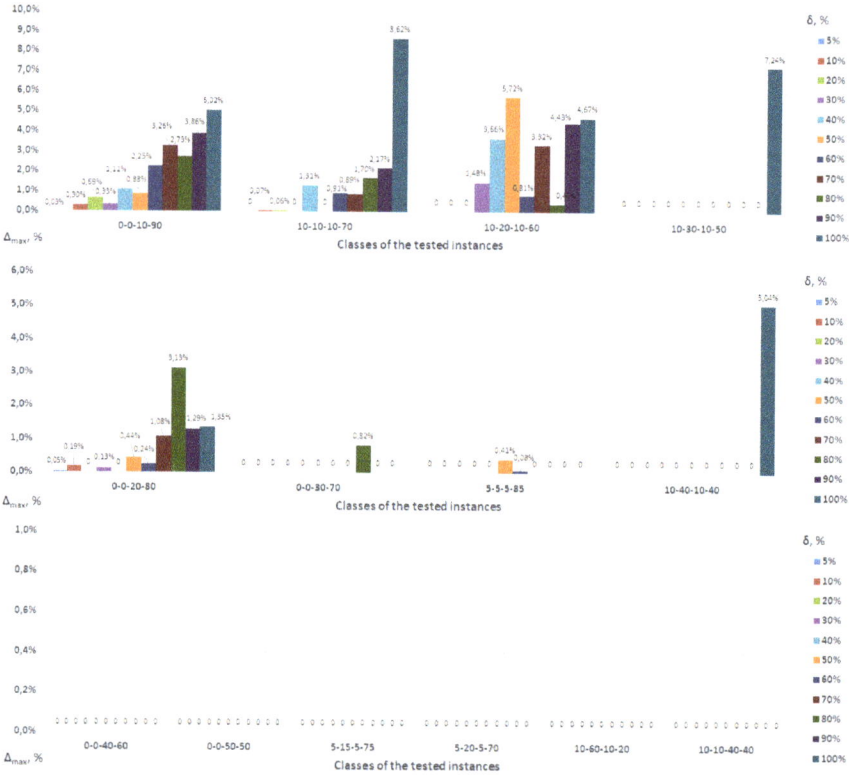

Figure 7. Maximal errors Δ_{max} for the tested instances with a uniform distribution.

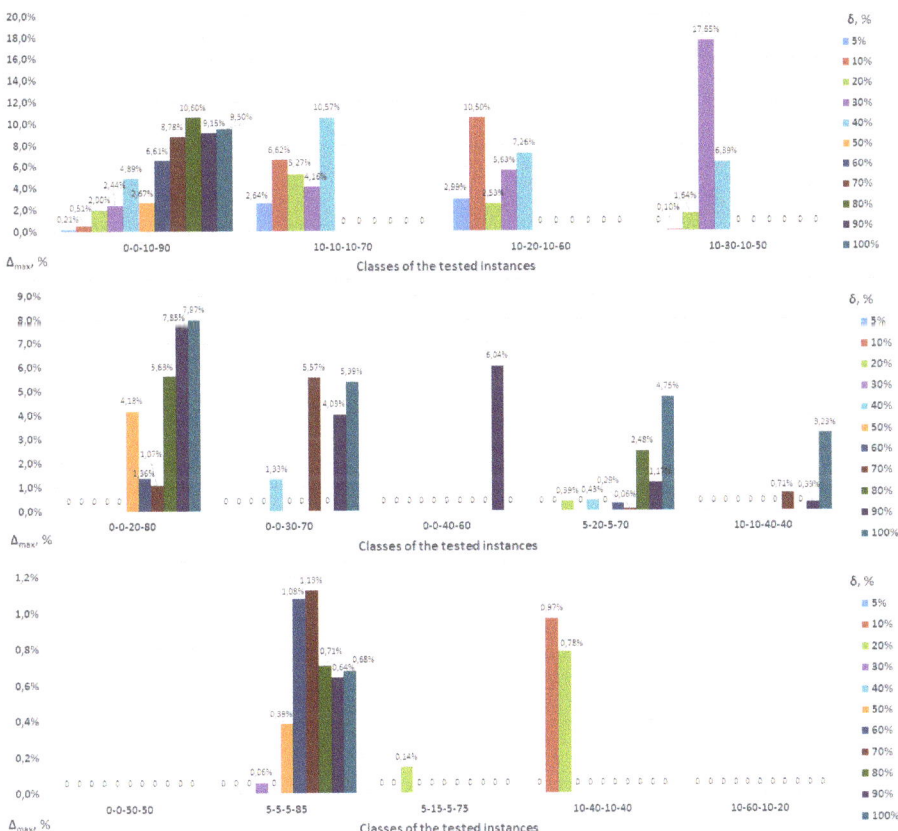

Figure 8. Maximal errors Δ_{max} for the tested instances with a gamma distribution.

7. Concluding Remarks

The uncertain job-shop scheduling problem $J2|l_{ij} \leq p_{ij} \leq u_{ij}, n_i \leq 2|C_{max}$ attract the attention of practitioners and researchers since this problem is applicable in real-life processing systems for some reduction of production costs due to a better utilization of the available machines and resources.

This paper is a continuation of our previous one [8], where only off-line phase of scheduling was investigated and tested for the uncertain problem $J2|l_{ij} \leq p_{ij} \leq u_{ij}, n_i \leq 2|C_{max}$ based on the stability approach. In [8], we tested 15 classes of the randomly generated instances $J2|l_{ij} \leq p_{ij} \leq u_{ij}, n_i \leq 2|C_{max}$. A lot of instances from nine easy classes were optimally solved at the off-line phase of scheduling. If the maximal errors were no greater than 20%, i.e., $\delta \in \{5\%, 10\%, 15\%, 20\%\}$, then more than 80% of the tested instances were optimally solved at the off-line phase of scheduling. If the maximal error was equal to 50%, i.e., $\delta = 50\%$, then 45% of the tested instances were optimally solved.

However, less than 5% of the tested instances with maximal possible error $\delta \geq 20\%$ from six hard tested classes were optimally solved at the off-line phase of scheduling. There were no tested hard instances with the maximal error 50% optimally solved in [8]. All these difficulties were succeeded in Sections 4–6 of this paper, where it is shown that the on-line phase of scheduling allows a scheduler to find either optimal schedule or very close to optimal ones. Additional information on the factual value of the job processing times becomes available once the processing of the job on the machine is completed. Using this information, a scheduler can determine a smaller dominant set of semi-active schedules, which is based on sufficient conditions for schedule dominance. The smaller dominant set enables a scheduler to quickly make an on-line scheduling decision whenever additional information on processing the job becomes available.

In Section 5, it is investigated the optimal pair (π', π'') of job permutations (Theorems 9 and 10). Using the proven analytical results, we derived Algorithms 3–5 for constructing optimal pairs (π', π'') of job permutations for all scenarios $p \in T$ or a small dominant set $S(T)$ of schedules for the uncertain problem $J2|l_{ij} \leq p_{ij} \leq u_{ij}, n_i \leq 2|C_{max}$. At the off-line scheduling phase, Algorithms 1 and 2 [8] are used to determine the partial strict order $A_\prec^{1,2}$ over the job set $\mathcal{J}_{1,2}$ and the partial strict order $A_\prec^{2,1}$ over the job set $\mathcal{J}_{2,1}$. The constructed precedence digraphs $(\mathcal{J}_{1,2}, A_\prec^{1,2})$ and $(\mathcal{J}_{2,1}, A_\prec^{2,1})$ determine a minimal dominant set $S(T)$ of schedules.

In Sections 6, it is shown how to use Algorithms 3–5 for constructing a small dominant set of semi-active schedules that enables a scheduler to make a fast decision whenever information on completing some jobs become available. Based on these algorithms, the problem $J2|l_{ij} \leq p_{ij} \leq u_{ij}, n_i \leq 2|C_{max}$ was solved with very small errors of the obtained objective values. The computational experiments (Section 6.3) show that pairs of job permutations constructed by Algorithms 3–5 are very close to the optimal pairs of job permutations. We tested 14 classes of randomly generated instances. For the tested instances, the percentage of the optimally solved instances slowly decreases with increasing maximal errors δ of the processing times. The developed on-line algorithms perform with the maximal errors of the achieved makespan less than 1% if $n \in \{20, 30, \ldots, 100\}$. For all tested classes of the instances, the average makespan errors for all numbers $n \in \{10, 20, \ldots, 100\}$ of the jobs $\mathcal{J}$ were less than 0.02%.

In a possible further research, one can continue the study of the uncertain job-shop scheduling problem based on the stability approach. It is useful to improve the developed algorithms and to extend them for other machine environments, such as a single machine or processing systems with parallel machines. It is promising to investigate an optimality region of the semi-active schedule and to develop algorithms for constructing a semi-active schedule with the largest optimality region.

It is also useful to apply the stability approach for solving the uncertain flow-shop and job-shop scheduling problems with $|\mathcal{M}| \geq 3$ different machines.

Author Contributions: Y.N.S. and N.M.M. jointly proved theoretical results; Y.N.S. and N.M.M. jointly conceived and designed the algorithm; V.H. performed the experiments; Y.N.S., N.M.M. and V.D.H. jointly analyzed the data; Y.N.S. and N.M.M. jointly wrote the paper. All authors have read and agreed to the published version of the manuscript.

Acknowledgments: We are thankful for useful remarks and suggestions provided by four anonymous reviewers on the earlier drafts of our paper.

Conflicts of Interest: The authors declare no conflict of interest.

Appendix A

Algorithms 1 and 2 Developed in [8].

Algorithm 1

Input: Segments $[l_{ij}, u_{ij}]$ for all jobs $J_i \in \mathcal{J}$ and machines $M_j \in \mathcal{M}$,
a partial strict order $A_{\prec}^{1,2}$ on the set $\mathcal{J}_{1,2} = \mathcal{J}_{1,2}^* \cup \mathcal{J}_{1,2}^1 \cup \mathcal{J}_{1,2}^2$ in the form
$J_1 \prec \ldots \prec J_k \prec \{J_{k+1}, J_{k+2}, \ldots, J_{k+r}\} \prec J_{k+r+1} \prec \ldots \prec J_{m_{1,2}}$.

Output: **EITHER** an optimal job permutation for the problem
$F2|l_{ij} \leq p_{ij} \leq u_{ij}|C_{max}$ with job set $\mathcal{J}_{1,2}$ and any scenario $p \in T$, (see STOP 0).
OR there no permutation $\pi_{1,2}$ of jobs from set $\mathcal{J}_{1,2}$, which is optimal
for all scenarios $p \in T$, (see STOP 1).

Step 1: Set $\delta_s = l_{k+s,2} - u_{k+s,1}$ for all $s \in \{1, 2, \ldots, r\}$.
construct a partition of the set of conflicting jobs into two subsets X_1 and X_2,
where $J_{k+s} \in X_1$ if $\delta_s \geq 0$, and $J_{k+s} \in X_2$, otherwise.

Step 2: Construct a permutation $\pi^1 = (J_1, J_2, \ldots, J_k, \pi_1, \pi_2, J_{k+r+1}, \ldots, J_{m_{1,2}})$, where the permutation
π_1 contains jobs from the set X_1 in the non-decreasing order of the values $u_{k+i,1}$ and the
permutation π_2 contains jobs from the set X_2 in the non-increasing order of the values
$l_{k+i,2}$, renumber jobs in the permutations π_1 and π_2 based on their orders.

Step 3: **IF** for the permutation π^1 conditions of Theorem 7 hold **THEN GOTO** step 8.

Step 4: Set $\delta_s = l_{k+s,1} - u_{k+s,2}$ for all $s \in \{1, 2, \ldots, r\}$.
construct a partition of the set of conflicting jobs into two subsets
Y_1 and Y_2, where $J_{k+s} \in Y_1$ if $\delta_s \geq 0$, and $J_{k+s} \in Y_2$, otherwise.

Step 5: Construct a permutation $\pi^2 = (J_1, J_2, \ldots, J_k, \pi_2, \pi_1, J_{k+r+1}, \ldots, J_{m_{1,2}})$, where the permutation
π_1 contains jobs from the set Y_1 in the non-increasing order of the values $u_{k+i,2}$, and the
permutation π_2 contains jobs from the set Y_2 in the non-decreasing order of the
values $l_{k+i,1}$, renumber jobs in the permutations π_1 and π_2 based on their orders.

Step 6: **IF** for the permutation π^2 conditions of Theorem 8 hold **THEN GOTO** step 9.

Step 7: **ELSE** there is no a single dominant permutation for problem
$F2|l_{ij} \leq p_{ij} \leq u_{ij}|C_{max}$ with job set $\mathcal{J}_{1,2}$ and any scenario $p \in T$ **STOP 1**.

Step 8: **RETURN** permutation π^1, which is a single dominant permutation
for the problem $F2|l_{ij} \leq p_{ij} \leq u_{ij}|C_{max}$ with job set $\mathcal{J}_{1,2}$ **STOP 0**.

Step 9: **RETURN** permutation π^2, which is a single dominant permutation
for the problem $F2|l_{ij} \leq p_{ij} \leq u_{ij}|C_{max}$ with job set $\mathcal{J}_{1,2}$ **STOP 0**.

Algorithm 2

Input:	Lower bounds l_{ij} and upper bounds u_{ij}, $0 < l_{ij} \leq u_{ij}$, of the durations of all operations O_{ij} of jobs $J_i \in \mathcal{J}$ processed on machines $M_j \in \mathcal{M} = \{M_1, M_2\}$.
Output:	**EITHER** pair of permutations $(\pi', \pi'') = ((\pi_{1,2}, \pi_1, \pi_{2,1}), (\pi_{2,1}, \pi_2, \pi_{1,2}))$, where π' is a permutation of jobs from set $\mathcal{J}_{1,2} \cup \mathcal{J}_1 \cup \mathcal{J}_{2,1}$ on machine M_1, π'' is a permutation of jobs from set $\mathcal{J}_{1,2} \cup \mathcal{J}_2 \cup \mathcal{J}_{2,1}$ on machine M_2, such that $\{(\pi', \pi'')\} = DS(T)$, (see STOP 0), **OR** permutation $\pi_{2,1}$ of jobs from set $\mathcal{J}_{2,1}$ on machines M_1 and M_2 and a partial strict order $A_\prec^{1,2}$ of jobs from set $\mathcal{J}_{1,2}$, **OR** permutation $\pi_{1,2}$ of jobs from set $\mathcal{J}_{1,2}$ on machines M_1 and M_2 and a partial strict order $A_\prec^{2,1}$ of jobs from set $\mathcal{J}_{2,1}$, **OR** a partial strict order $A_\prec^{1,2}$ of jobs from set $\mathcal{J}_{1,2}$ and a partial strict order $A_\prec^{2,1}$ of jobs from set $\mathcal{J}_{2,1}$, (see STOP 1).
Step 1:	Determine a partition $\mathcal{J} = \mathcal{J}_1 \cup \mathcal{J}_2 \cup \mathcal{J}_{1,2} \cup \mathcal{J}_{2,1}$ of the job set $\mathcal{J}$, permutation π_1 of jobs from set $\mathcal{J}_1$ and permutation π_2 of jobs from set $\mathcal{J}_2$, arrange the jobs in the increasing order of their indexes.
Step 2:	**IF** the first inequality in condition (5) of Theorem 4 holds **THEN BEGIN** Construct a permutation $\pi_{1,2}$ of jobs from set $\mathcal{J}_{1,2}$, arrange them in the increasing order of their indexes; **IF** the second inequality in condition (5) of Theorem 4 holds **THEN** construct a permutation $\pi_{2,1}$ of jobs from set $\mathcal{J}_{2,1}$, arrange them in the increasing order of their indexes **GOTO** Step 10 **END**
Step 3:	**IF** the first inequality in condition (6) of Theorem 4 holds **THEN BEGIN** Construct a permutation $\pi_{2,1}$ of jobs from set $\mathcal{J}_{2,1}$, arrange them in the increasing order of their indexes; **IF** the second inequality in condition (6) of Theorem 4 holds **THEN** construct a permutation $\pi_{1,2}$ of jobs from set $\mathcal{J}_{1,2}$, arrange the jobs in the increasing order of their indexes **END**
Step 4:	**IF** both permutations $\pi_{1,2}$ and $\pi_{2,1}$ are constructed **THEN GOTO** Step 10.
Step 5:	**IF** permutation $\pi_{1,2}$ is not constructed **THEN** fulfill Procedure 1.
Step 6:	**IF** permutation $\pi_{2,1}$ is not constructed **THEN** fulfill Procedure 2.
Step 7:	**IF** both permutations $\pi_{1,2}$ and $\pi_{2,1}$ are constructed **THEN GOTO** Step 10.
Step 8:	**IF** permutation $\pi_{2,1}$ is constructed **THEN GOTO** Step 11.
Step 9:	**IF** permutation $\pi_{1,2}$ is constructed **THEN GOTO** Step 12 **ELSE GOTO** Step 13.
Step 10:	**RETURN** pair of permutations (π', π''), where π' is the permutation of jobs from set $\mathcal{J}_{1,2} \cup \mathcal{J}_1 \cup \mathcal{J}_{2,1}$ processed on machine M_1 and π'' is the permutation of jobs from set $\mathcal{J}_{1,2} \cup \mathcal{J}_2 \cup \mathcal{J}_{2,1}$ processed on machine M_2 such that $\{(\pi', \pi'')\} = DS(T)$ **STOP 0**.
Step 11:	**RETURN** the permutation $\pi_{2,1}$ of jobs from set $\mathcal{J}_{2,1}$ processed on machines M_1 and M_2, the partial strict order $A_\prec^{1,2}$ of jobs from set $\mathcal{J}_{1,2}$ **GOTO** Step 14.
Step 12:	**RETURN** the permutation $\pi_{1,2}$ of jobs from set $\mathcal{J}_{1,2}$ processed on machines M_1 and M_2, the partial strict order $A_\prec^{2,1}$ of jobs from set $\mathcal{J}_{2,1}$ **GOTO** Step 14.
Step 13:	**RETURN** the partial strict order $A_\prec^{1,2}$ of jobs from set $\mathcal{J}_{1,2}$ and the partial strict order $A_\prec^{2,1}$ of jobs from set $\mathcal{J}_{2,1}$
Step 14:	**STOP 1**.

Appendix B. Tables with Computational Results

Table A1. Computational results for the randomly generated instances with the ratio 0%:0%:10%:90% of the numbers of jobs in the subsets $\mathcal{J}_1$, $\mathcal{J}_2$, $\mathcal{J}_{1,2}$ and $\mathcal{J}_{2,1}$ of the job set $\mathcal{J}$.

		Uniform Distributions									Gamma Distributions								
δ%	n	10	20	30	40	50	60	70	80	90	100	10	20	30	40	50	60	70	80	90	100
5	Stop1	93.9	97.3	98.4	97.6	96.6	97.6	97.9	98.3	98	97.5	93.4	97	97.7	97.1	98.5	98.3	97.6	98.6	98.3	98.7
	Stop2	0.9	1	0.6	1.1	1.6	1.3	1.2	1.1	0.9	1.4	0.8	0.8	0.9	1.1	0.5	0.5	1.3	0.8	0.9	0.6
	Stop3	5.1	1.7	1	1.3	1.8	1.1	0.9	0.6	1.1	1.1	5.3	2.2	1.4	1.8	1	1.2	1.1	0.6	0.8	0.7
	Stop4	0.1	0	0	0	0	0	0	0	0	0	0.5	0	0	0	0	0	0	0	0	0
10	Stop1	85.8	91.8	91.6	93	92.2	92.2	91.9	92.6	93.6	93.8	83.1	93.5	92.8	92.7	93.1	92.3	92.4	93.3	94.2	93.6
	Stop2	2.1	2.5	1.5	3.5	1.1	1	4.2	4.6	4.2	4.2	3.2	2.8	3.1	3.5	3.6	4.3	4.2	3.4	3.5	4.1
	Stop3	11.7	5.7	3.9	3.5	3.7	3.8	3.9	2.8	2.2	2	12.8	3.7	4.1	3.8	3.3	3.4	3.4	3.3	2.3	2.3
	Stop4	0.4	0	0	0	0	0	0	0	0	0	0.9	0	0	0	0	0	0	0	0	0
20	Stop1	70.6	73.2	73.3	71.9	69.6	63.4	63.6	57.7	56.3	53.4	70.2	73.7	74.1	71.5	67.2	64.2	63.4	61.5	56.1	57.5
	Stop2	4.3	6.8	9.4	8.6	8.8	11.1	9.4	10.1	9.2	9.2	4.3	8.1	8.7	10	10	11.3	9.1	9.1	8.9	7.5
	Stop3	24.2	20	17.3	19.5	21.6	25.5	27	32.2	34.5	37.4	24.9	18.2	17.2	18.5	22.8	24.5	27.5	29.4	35	35
	Stop4	0.9	0	0	0	0	0	0	0	0	0	0.6	0	0	0	0	0	0	0	0	0
30	Stop1	53	55.1	47.5	41.1	31	30.7	23.4	20.3	18.5	13.4	52	52.2	44.9	34.9	33.6	26.3	27.3	18.3	18.9	14.2
	Stop2	4.8	8.8	9.1	9.7	9.8	7	6.7	5.2	4.2	4.4	6.2	8.9	10.1	10.4	11	7.9	6	4.6	5.9	4.5
	Stop3	41.9	36.1	43.4	49.2	59.2	62.3	69.9	74.5	77.3	82.2	41	38.9	45	54.7	55.4	65.8	66.7	77.1	75.2	81.3
	Stop4	0.3	0	0	0	0	0	0	0	0	0	0.8	0	0	0	0	0	0	0	0	0
40	Stop1	41.6	32.5	24.7	16.4	10.9	8.9	5.9	3.9	2.1	2.3	37.8	31.8	22.2	18.6	10.6	8.5	6.1	4.3	3	1.9
	Stop2	4.9	7.6	7.7	6.6	4.2	2.9	2	1.1	0.7	0.3	5.2	8.5	9.1	4.8	2.9	3.5	2	1.6	1	0.6
	Stop3	52.7	59.7	67.6	77	84.9	88.2	92.1	95	97.2	97.4	55.3	59.7	68.7	76.6	86.5	88	91.9	94.1	96	97.5
	Stop4	0.8	0.2	0	0	0	0	0	0	0	0	1.7	0	0	0	0	0	0	0	0	0
50	Stop1	29	16.1	9.3	5	2.4	1.4	1.6	0.8	0.5	0	26.2	19.4	10.4	5.6	3.9	1.6	0.6	0.5	0.3	0.1
	Stop2	4.7	5.4	5	2.6	1.2	0.8	0.2	0.3	0.3	0	3.7	5.6	3.8	2.2	0.7	0.7	0.1	0.2	0.1	0.1
	Stop3	64.9	78.5	85.7	92.4	96.4	97.8	98.2	98.9	99.2	100	67.8	75	85.8	92.2	95.4	97.7	98.7	99.4	99.5	99.8
	Stop4	1.4	0	0	0	0	0	0	0	0	0	2.3	0	0	0	0	0	0	0	0	0
60	Stop1	18.1	8.4	4.6	1.4	1	0.3	0.2	0.1	0	0	16.6	9.6	3.6	1.4	1	0.3	0	0.2	0	0
	Stop2	3.8	4	1.6	0.7	0.3	0.1	0	0	0	0	3.2	3.5	2.3	0.9	0.4	0	0.3	0	0.1	0
	Stop3	76.4	87.5	93.8	97.9	98.7	99.6	99.8	99.9	100	100	77.7	86.9	94.1	97.7	98.6	99.7	99.7	99.8	99.9	100
	Stop4	1.7	0.1	0	0	0	0	0	0	0	0	2.5	0	0	0	0	0	0	0	0	0
70	Stop1	12.3	5.1	1.4	0.7	0.1	0	0	0	0	0	12.1	4.7	2.1	0.8	0.2	0	0.1	0	0	0
	Stop2	2.8	1.5	0.7	0.2	0.1	0	0	0	0	0	2	2.1	0.4	0.2	0	0	0	0	0	0
	Stop3	83.2	93.4	97.9	99.1	99.8	100	100	100	100	100	82.8	93.2	97.5	99	99.8	100	99.9	100	100	100
	Stop4	1.7	0	0	0	0	0	0	0	0	0	3.1	0	0	0	0	0	0	0	0	0
80	Stop1	8.9	1.9	0.4	0.3	0.2	0	0	0	0	0	10	2.3	0.6	0.2	0	0	0	0	0	0
	Stop2	1.8	0.8	0.5	0	0	0	0	0	0	0	2.2	1.1	0.2	0	0	0.1	0	0	0	0
	Stop3	87.1	97.1	99.1	99.7	99.8	100	100	100	100	100	84.2	96.5	99.2	99.8	100	99.9	100	100	100	100
	Stop4	2.2	0.2	0	0	0	0	0	0	0	0	3.6	0.1	0	0	0	0	0	0	0	0
90	Stop1	6.9	0.8	0.2	0.1	0	0	0	0	0	0	5.7	0.9	0.3	0	0	0	0	0	0	0
	Stop2	1.2	0.7	0.1	0	0	0	0	0	0	0	1.5	0.5	0	0	0	0	0	0	0	0
	Stop3	90.1	98.4	99.7	99.9	100	100	100	100	100	100	88	98.5	99.7	100	100	100	100	100	100	100
	Stop4	1.8	0.1	0	0	0	0	0	0	0	0	4.8	0.1	0	0	0	0	0	0	0	0
100	Stop1	4.4	0.3	0.1	0.1	0	0	0	0	0	0	4.1	0.3	0	0	0	0	0	0	0	0
	Stop2	1.1	0.2	0	0	0	0	0	0	0	0	0.8	0.2	0	0	0	0	0	0	0	0
	Stop3	92.1	99.4	99.9	99.9	100	100	100	100	100	100	90.2	99.4	100	100	100	100	100	100	100	100
	Stop4	2.4	0.1	0	0	0	0	0	0	0	0	4.9	0.3	0	0	0	0	0	0	0	0

Table A2. Computational results for randomly generated instances with the ratio 0%:0%:20%:80% of the numbers of jobs in the subsets $\mathcal{J}_1$, $\mathcal{J}_2$, $\mathcal{J}_{1,2}$ and $\mathcal{J}_{2,1}$ of the job set $\mathcal{J}$.

		Uniform Distributions										Gamma Distributions									
δ%	n	10	20	30	40	50	60	70	80	90	100	10	20	30	40	50	60	70	80	90	100
5	Stop1	96.7	98.9	98.6	99.5	99.7	99.7	99.9	99.9	99.9	99.9	96.7	98.9	99	99.3	99.6	99.4	99.7	99.8	99.9	99.8
	Stop2	0.8	0.3	0.7	0.1	0	0.2	0.1	0.1	0.1	0.1	0.3	0.4	0.1	0.3	0.1	0.4	0.2	0.1	0.1	0.2
	Stop3	2.4	0.8	0.7	0.4	0.3	0.1	0	0	0	0	3	0.7	0.9	0.4	0.3	0.2	0.1	0.1	0	0
	Stop4	0.1	0	0	0	0	0	0	0	0	0	0	0	0	0	0	0	0	0	0	0
10	Stop1	93.4	96.2	97.6	97.7	98	98.4	98.4	98.7	98.5	99.6	93.5	96.1	97.5	97.6	97.8	98.1	98.4	98.2	99	98.7
	Stop2	1	1.4	1.4	1.4	1.4	1.3	1.1	1.2	1	0.4	1	1.8	1.1	1.3	1.5	1.6	1.3	1.5	0.9	1.1
	Stop3	5.4	2.4	1	0.9	0.6	0.3	0.5	0.1	0.5	0	5.5	2.1	1.4	1.1	0.7	0.3	0.3	0.3	0.1	0.2
	Stop4	0.2	0	0	0	0	0	0	0	0	0	0	0	0	0	0	0	0	0	0	0
20	Stop1	85.4	86.9	89.5	90.2	90.8	90.5	89.1	90	90.3	90.1	83.5	87.5	88.8	88.4	90.6	89.9	91	89.4	90.9	90.1
	Stop2	2.6	4.8	5.3	5	5.8	5.8	5.7	5	5	5.4	3.1	5.7	5.8	7.4	4.7	5.8	4.9	6.5	6.2	4
	Stop3	12	8.3	5.2	4.8	3.4	3.7	5.2	5	4.7	4.5	13.4	6.8	5.4	4.2	4.7	4.3	4.1	4.1	2.9	5.9
	Stop4	0	0	0	0	0	0	0	0	0	0	0	0	0	0	0	0	0	0	0	0
30	Stop1	70.2	76.4	70.1	69.6	69.2	65.7	64.1	60.4	58	56	71.6	73.3	72.9	69	66	63.1	60.4	62.3	62.2	58.1
	Stop2	5.1	6.8	9.7	9.5	8.4	10.5	9.7	9.1	10.8	8.1	3.6	8.7	8.4	9.6	9.8	11.3	8.4	8.3	7.1	8.9
	Stop3	24.6	16.8	20.2	20.9	22.4	23.8	26.2	30.5	31.2	35.9	24.8	18	18.7	21.4	24.2	25.6	31.2	29.4	30.7	33
	Stop4	0.1	0	0	0	0	0	0	0	0	0	0	0	0	0	0	0	0	0	0	0
40	Stop1	56.7	55.3	48.8	41.5	39.6	31.2	31.1	26.5	23.5	21.5	55.1	53.9	46.5	41.5	37.5	31.6	33.2	27.4	26.2	21
	Stop2	5.1	8.2	10.5	10.6	9.6	8.2	7.6	6.9	7.5	5	5.6	9.2	11.5	7.7	8.9	8.7	7.3	7.2	6	5.7
	Stop3	38.2	36.5	40.7	47.9	50.8	60.6	61.3	66.6	69	73.5	39.3	36.9	42	50.8	53.6	59.7	59.5	65.4	67.8	73.3
	Stop4	0	0	0	0	0	0	0	0	0	0	0	0	0	0	0	0	0	0	0	0
50	Stop1	42.6	35.2	27.2	22.9	14.3	11.4	9.9	6.9	6.6	5.9	43.9	37.3	29.2	22	16.3	13.2	10.8	10.5	5.5	5.7
	Stop2	4.4	9.5	9.8	6.5	6	5.2	3.8	3.4	2.2	2.4	6	10	7.3	6.9	6.5	4.8	4.7	2.8	1.9	2.5
	Stop3	52.8	55.3	63	70.6	79.7	83.4	86.3	89.7	91.2	91.7	49.9	52.7	63.5	71.1	77.2	82	84.5	86.7	92.6	91.8
	Stop4	0.2	0	0	0	0	0	0	0	0	0	0.2	0	0	0	0	0	0	0	0	0
60	Stop1	32.9	20.7	14.4	10.7	6.3	3.3	3.3	2.2	1.4	0.8	31.7	23.1	15.7	10.4	7.2	3.9	2.9	1.4	0.9	0.6
	Stop2	5.4	5.5	4.2	2.6	2.3	1.7	0.8	0.6	0.3	0.4	5.6	6	6.4	3.6	2.4	1.5	1.5	0.8	0.3	0.7
	Stop3	61.7	73.8	81.4	86.7	91.4	95	95.9	97.2	98.3	98.8	62.3	70.9	77.9	86	90.4	94.6	95.6	97.8	98.8	98.7
	Stop4	0	0	0	0	0	0	0	0	0	0	0.4	0	0	0	0	0	0	0	0	0
70	Stop1	22.1	13.7	7.6	2.9	1.6	1.1	0.6	0.2	0.3	0.1	23.9	13.1	6.7	3.4	1.6	0.7	0.9	0.3	0.2	0.1
	Stop2	4.9	3.7	2.9	1.7	0.9	0.6	0.4	0.1	0	0.1	4.4	4.5	2	1.5	1.2	0.4	0	0.4	0.1	0.1
	Stop3	72.8	82.6	89.5	95.4	97.5	98.3	99	99.7	99.7	99.8	71.6	82.4	91.3	95.1	97.2	98.9	99.1	99.3	99.7	99.8
	Stop4	0.2	0	0	0	0	0	0	0	0	0	0.1	0	0	0	0	0	0	0	0	0
80	Stop1	17.3	8.3	3.5	1.3	0.5	0.3	0.2	0	0	0	16.4	7.9	3.1	1.9	0.9	0.2	0	0	0	0
	Stop2	2.8	1.7	1.8	0.8	0.6	0	0	0	0	0	3.4	2.4	0.8	0.9	0.1	0.2	0.1	0	0	0
	Stop3	79.4	90	94.7	97.9	98.9	99.7	99.8	100	100	100	79.6	89.7	96.1	97.2	99	99.6	99.9	100	100	100
	Stop4	0.5	0	0	0	0	0	0	0	0	0	0.6	0	0	0	0	0	0	0	0	0
90	Stop1	12.1	4.5	1.4	0.7	0	0.1	0	0	0	0	12.6	5	2.9	0.4	0.1	0.1	0	0	0	0
	Stop2	2.5	2.3	1.3	0.3	0	0.1	0	0	0	0	3.2	1.8	0.2	0.2	0	0	0	0	0	0
	Stop3	85.2	93.2	97.3	99	100	99.8	100	100	100	100	83.7	93.2	96.9	99.4	99.9	99.9	100	100	100	100
	Stop4	0.2	0	0	0	0	0	0	0	0	0	0.5	0	0	0	0	0	0	0	0	0
100	Stop1	10.6	2.7	0.5	0.3	0	0	0	0	0	0	9.1	2.8	0.5	0.1	0	0	0	0	0	0
	Stop2	1.9	1.2	0.3	0	0.2	0	0	0	0	0	2	0.6	0.2	0	0	0	0	0	0	0
	Stop3	87.4	96.1	99.2	99.7	99.8	100	100	100	100	100	88.2	96.6	99.3	99.9	100	100	100	100	100	100
	Stop4	0.1	0	0	0	0	0	0	0	0	0	0.7	0	0	0	0	0	0	0	0	0

Table A3. Computational results for randomly generated instances with the ratio 0%:0%:30%:70% of the numbers of jobs in the subsets $\mathcal{J}_1$, $\mathcal{J}_2$, $\mathcal{J}_{1,2}$ and $\mathcal{J}_{2,1}$ of the job set $\mathcal{J}$.

		Uniform Distributions										Gamma Distributions									
δ%	n	10	20	30	40	50	60	70	80	90	100	10	20	30	40	50	60	70	80	90	100
5	Stop1	99.2	99.6	99.9	99.9	100	100	100	100	100	100	98.6	99.9	99.9	99.9	99.9	100	100	100	100	100
	Stop2	0.2	0.1	0.1	0.1	0	0	0	0	0	0	0	0	0	0	0.1	0	0	0	0	0
	Stop3	0.6	0.3	0	0	0	0	0	0	0	0	1.4	0.1	0.1	0.1	0	0	0	0	0	0
	Stop4	0	0	0	0	0	0	0	0	0	0	0	0	0	0	0	0	0	0	0	0
10	Stop1	97.3	99	99.4	99.9	99.9	100	99.9	99.9	100	100	97.4	98.6	99.1	99.9	99.9	100	99.9	100	99.8	100
	Stop2	0.8	0.4	0.3	0.1	0.1	0	0.1	0.1	0	0	0.5	0.8	0.6	0.1	0.1	0	0.1	0	0.2	0
	Stop3	1.9	0.6	0.3	0	0	0	0	0	0	0	2.1	0.6	0.3	0	0	0	0	0	0	0
	Stop4	0	0	0	0	0	0	0	0	0	0	0	0	0	0	0	0	0	0	0	0
20	Stop1	91.7	95.7	97.1	97.5	98.2	98.6	98.7	98.7	99.7	98.9	91.5	97	97.8	97.2	98.5	98.8	99.1	99.6	99.5	99.4
	Stop2	1.9	2.6	2	2.1	1.4	1.2	1.2	1.1	0.3	0.8	2.3	1.7	1.6	2.1	1.2	1	0.7	0.4	0.5	0.4
	Stop3	6.4	1.7	0.9	0.4	0.4	0.2	0.1	0.2	0	0.3	6.2	1.3	0.6	0.7	0.3	0.2	0.2	0	0	0.2
	Stop4	0	0	0	0	0	0	0	0	0	0	0	0	0	0	0	0	0	0	0	0
30	Stop1	81.9	87.1	89.1	91.9	91.7	92.2	92.2	93.8	93.7	93.9	83	87.2	91.7	91.7	92.4	91.6	92.2	92.8	92.9	93.4
	Stop2	4.6	6.2	5.9	4.5	5.5	4.3	4.3	3.9	2.8	2.5	3.1	5.1	4.1	4.4	3.7	4.4	4.8	4.2	3.5	3.5
	Stop3	13.5	6.7	5	3.6	2.8	3.5	3.5	2.3	3.5	3.6	13.9	7.7	4.2	3.9	3.9	4	3	3	3.6	3.1
	Stop4	0	0	0	0	0	0	0	0	0	0	0	0	0	0	0	0	0	0	0	0
40	Stop1	74.1	74.7	75.9	76	75.1	75.9	73.4	75.8	71.6	72.5	69.1	74.2	76.4	75.6	76.5	76.4	75	74	75.3	72.7
	Stop2	4.4	7.9	7.5	8.2	6.5	6.3	6.9	5.5	6.9	7	6	9.2	7.5	8	7.7	6.4	6.4	7.1	4.6	6.4
	Stop3	21.5	17.4	16.6	15.8	18.4	17.8	19.7	18.7	21.5	20.5	24.8	16.6	16.1	16.4	15.8	17.2	18.6	18.9	20.1	20.9
	Stop4	0	0	0	0	0	0	0	0	0	0	0.1	0	0	0	0	0	0	0	0	0
50	Stop1	61.4	60.7	56.9	53.7	50.4	49.7	47.3	47.3	44.7	39.1	58.2	60.6	57.7	51.7	49.1	50.9	48.3	48.2	43.9	44.3
	Stop2	5.3	9.2	9.7	10.4	10.2	7.7	8.4	7.1	6.7	6.3	5.8	8.7	8.5	12.4	10.2	9.1	7.8	7.1	6.6	5.3
	Stop3	33.3	30.1	33.4	35.9	39.4	42.6	44.3	45.6	48.6	54.6	36	30.7	33.8	35.9	40.7	40	43.9	44.7	49.5	50.4
	Stop4	0	0	0	0	0	0	0	0	0	0	0	0	0	0	0	0	0	0	0	0
60	Stop1	46.7	46.3	39.6	32.9	30.5	29.5	23.9	22.1	18.8	17.8	50.9	42.8	41.7	35.9	30.6	28	24	22.9	19	13.5
	Stop2	6.2	8.1	7.7	7.9	6.2	5.3	4.6	4.1	4.2	2.8	5	9.1	7.5	7.5	7.1	5.1	5.2	4.6	3.7	3.6
	Stop3	47.1	45.6	52.7	59.2	63.3	65.2	71.5	73.8	77	79.4	44.1	48.1	50.8	56.6	62.3	66.9	70.8	72.5	77.3	82.9
	Stop4	0	0	0	0	0	0	0	0	0	0	0	0	0	0	0	0	0	0	0	0
70	Stop1	39.9	34.5	24.5	20.2	17.6	12.2	8.7	8.6	7.6	5.5	38	34.8	23.5	19.1	15.7	13.7	9.5	9	6	4.7
	Stop2	5.7	7.4	6.7	5.5	4.1	2.8	2.7	1.9	1.4	0.9	6	6.8	6.3	5.6	3.9	4.8	3.4	1.4	1.4	1.3
	Stop3	54.4	58.1	68.8	74.3	78.3	85	88.6	89.5	91	93.6	55.9	58.4	70.2	75.3	80.4	81.5	87.1	89.6	92.6	94
	Stop4	0	0	0	0	0	0	0	0	0	0	0.1	0	0	0	0	0	0	0	0	0
80	Stop1	27.9	21.6	14.3	8.3	7.2	4.9	4.2	2.7	0.9	0.8	28.5	21.6	14	9.5	9.3	4.2	3.9	2.2	1.2	1.5
	Stop2	4.8	5	3.6	3	2.6	1.6	0.7	1	0.2	0.3	5.6	5.6	5.2	3.1	1.9	2.3	1.7	0.6	0.7	0.3
	Stop3	67.2	73.4	82.1	88.7	90.2	93.5	95.1	96.3	98.9	98.9	65.9	72.8	80.8	87.4	88.8	93.5	94.4	97.2	98.1	98.2
	Stop4	0.1	0	0	0	0	0	0	0	0	0	0	0	0	0	0	0	0	0	0	0
90	Stop1	22.7	13.2	9.8	4.1	2.1	1.5	1.4	0.7	0.2	0	23	14.4	7.4	3.6	3.4	1.9	1.2	0.7	0.3	0.9
	Stop2	3.4	4.7	3.1	1.9	1.4	0.9	0.5	0.3	0.3	0.2	3.4	5.1	2.9	2	1	0.6	0.7	0.2	0.2	0.3
	Stop3	73.9	82.1	87.1	94	96.5	97.6	98.1	99	99.5	99.8	73.4	80.5	89.7	94.4	95.6	97.5	98.1	99.1	99.5	98.8
	Stop4	0	0	0	0	0	0	0	0	0	0	0.2	0	0	0	0	0	0	0	0	0
100	Stop1	17.1	8.9	3.8	2.1	1.2	0.6	0.2	0.2	0	0	14.7	8	4.7	2.5	1.4	0.3	0.3	0.2	0.1	0.1
	Stop2	2.6	4.3	1.6	1.5	0.4	0.2	0.2	0	0	0	4.6	2.4	1.7	0.6	0.5	0.3	0.3	0.1	0.1	0
	Stop3	80.3	86.8	94.6	96.4	98.4	99.2	99.6	99.8	100	100	80.4	89.6	93.6	96.9	98.1	99.4	99.4	99.7	99.8	99.9
	Stop4	0	0	0	0	0	0	0	0	0	0	0.3	0	0	0	0	0	0	0	0	0

Table A4. Computational results for randomly generated instances with the ratio 0%:0%:40%:60% of the numbers of jobs in the subsets $\mathcal{J}_1$, $\mathcal{J}_2$, $\mathcal{J}_{1,2}$ and $\mathcal{J}_{2,1}$ of the job set $\mathcal{J}$.

		Uniform Distributions										Gamma Distributions									
δ%	n	10	20	30	40	50	60	70	80	90	100	10	20	30	40	50	60	70	80	90	100
5	Stop1	99.8	100	100	100	100	100	100	100	100	100	99.5	99.8	99.9	100	99.9	100	100	100	100	100
	Stop2	0.1	0	0	0	0	0	0	0	0	0	0.1	0	0.1	0	0.1	0	0	0	0	0
	Stop3	0.1	0	0	0	0	0	0	0	0	0	0.4	0.2	0	0	0	0	0	0	0	0
	Stop4	0	0	0	0	0	0	0	0	0	0	0	0	0	0	0	0	0	0	0	0
10	Stop1	99.2	99.7	99.9	100	100	100	100	100	100	100	98.9	100	99.9	100	100	100	100	100	100	100
	Stop2	0.4	0.3	0.1	0	0	0	0	0	0	0	0.2	0	0.1	0	0	0	0	0	0	0
	Stop3	0.4	0	0	0	0	0	0	0	0	0	0.9	0	0	0	0	0	0	0	0	0
	Stop4	0	0	0	0	0	0	0	0	0	0	0	0	0	0	0	0	0	0	0	0
20	Stop1	96.8	98.9	99.4	99.5	99.9	99.8	100	100	100	99.9	96.4	98.6	99.6	99.7	99.9	100	100	100	100	100
	Stop2	1.1	0.6	0.4	0.5	0.1	0.2	0	0	0	0.1	1.6	0.8	0.4	0.3	0	0	0	0	0	0
	Stop3	2.1	0.5	0.2	0	0	0	0	0	0	0	2	0.6	0	0	0.1	0	0	0	0	0
	Stop4	0	0	0	0	0	0	0	0	0	0	0	0	0	0	0	0	0	0	0	0
30	Stop1	91	95.7	97.8	98.7	99.2	99.5	99.4	99.8	99.9	100	91.8	94.9	97.3	98.6	99.1	99.6	99.9	99.9	99.7	100
	Stop2	3.5	1.7	1.2	0.9	0.7	0.2	0.4	0.1	0.1	0	2.3	3	1.8	1.1	0.7	0.3	0.1	0.1	0.3	0
	Stop3	5.5	2.6	1	0.4	0.1	0.3	0.2	0.1	0	0	5.9	2.1	0.9	0.3	0.2	0.1	0	0	0	0
	Stop4	0	0	0	0	0	0	0	0	0	0	0	0	0	0	0	0	0	0	0	0
40	Stop1	85.8	90.4	93.7	95.1	96.4	97.6	97.2	98.2	98.3	98.5	85.4	91.2	93	94.8	96.5	97.4	97.3	98	98.2	98
	Stop2	2.7	4.5	3.5	1.6	1.9	1.3	2.1	0.6	1	1	3.7	3.7	3.2	2.9	1.8	1.6	0.8	1.2	0.7	1.1
	Stop3	11.5	5.1	2.8	3.3	1.7	1.1	0.7	1.2	0.7	0.5	10.9	5.1	3.8	2.3	1.7	1	1.9	0.8	1.1	0.9
	Stop4	0	0	0	0	0	0	0	0	0	0	0	0	0	0	0	0	0	0	0	0
50	Stop1	75.3	82.5	84.8	87.4	87	88.2	89.7	90.7	90.5	92.3	76.3	82.5	82.4	85.4	87.5	89.7	89.6	90.4	90.7	91.3
	Stop2	5.1	5.5	6	4.4	5.1	3.8	3.7	3.3	3	3	4.1	5.9	6	5.9	3.9	3.5	3.3	2.4	2.7	2.4
	Stop3	19.6	12	9.2	8.2	7.9	8	6.6	6	6.5	4.7	19.6	11.6	11.6	8.7	8.6	6.8	7.1	7.2	6.6	6.3
	Stop4	0	0	0	0	0	0	0	0	0	0	0	0	0	0	0	0	0	0	0	0
60	Stop1	63.4	70.1	72.3	74.7	75.8	73.9	72.5	75.4	75.4	76.6	66.3	71	73.7	75.2	72.9	73.2	73.7	71.9	75.5	75.2
	Stop2	6.6	7	7.8	6.6	5.8	4.3	6.6	4.2	3.9	4.5	4.7	7.4	7	5.9	5.8	5.4	4.2	5.2	5.3	3.1
	Stop3	30	22.9	19.9	18.7	18.4	21.8	20.9	20.4	20.7	18.9	29	21.6	19.3	18.9	21.3	21.4	22.1	22.9	19.2	21.7
	Stop4	0	0	0	0	0	0	0	0	0	0	0	0	0	0	0	0	0	0	0	0
70	Stop1	51.6	60	57.2	55.7	56.5	52.9	50.5	50.9	50.8	50	49.5	57	54.6	55.1	54.8	53.2	52.8	49.9	50	51.5
	Stop2	6.3	7.5	8.3	7.3	6.1	5.8	7.2	6.2	5.7	4.2	6.3	8.6	8	7.2	6.5	5.6	7	5.9	5.2	5.7
	Stop3	42.1	32.5	34.5	37	37.4	41.3	42.3	42.9	43.5	45.8	44.2	34.4	37.4	37.7	38.7	41.2	40.2	44.2	44.8	42.8
	Stop4	0	0	0	0	0	0	0	0	0	0	0	0	0	0	0	0	0	0	0	0
80	Stop1	39.8	43.9	39.4	39	37.8	34.8	31.7	31.1	30	28.1	39	43.7	40.1	40.3	41.1	31.3	31.5	28.1	26	28.2
	Stop2	6.3	7.3	8.5	7.4	6.1	4.9	3.7	4.4	4.2	3.3	5.9	8.1	7.8	5.9	6.6	5.7	4.7	3.6	4.6	3
	Stop3	53.9	48.8	52.1	53.6	56.1	60.3	64.6	64.5	65.8	68.6	55.1	48.2	52.1	53.8	52.3	63	63.8	68.3	69.4	68.8
	Stop4	0	0	0	0	0	0	0	0	0	0	0	0	0	0	0	0	0	0	0	0
90	Stop1	33.7	28.9	27.6	24.1	20.7	18.2	16.8	13.5	14.1	12.4	31.7	32.1	28.8	23.8	23.8	20.1	17.1	15.9	14.2	12.3
	Stop2	6	7.7	6.9	5.3	4.5	3.3	3.6	2.8	2.6	1.7	5.7	7.1	5.9	6	4.7	3.5	3.8	3	3.1	1.6
	Stop3	60.3	63.4	65.5	70.6	74.8	78.5	79.6	83.7	83.3	85.9	62.5	60.8	65.3	70.2	71.5	76.4	79.1	81.1	82.7	86.1
	Stop4	0	0	0	0	0	0	0	0	0.1	0	0	0	0	0	0	0	0	0	0	0
100	Stop1	26.1	21.4	15.6	13.1	11.1	7.9	7.3	6	4.4	4.6	24.1	21.2	16.7	16.2	11	8.2	8.3	6.7	5.7	3.8
	Stop2	3.9	6.4	5.4	4.6	3.2	2.9	1.8	1.6	0.7	1.3	5	6.2	5	2.9	3.4	3.4	2.5	1.5	1.9	1
	Stop3	70	72.2	79	82.3	85.7	89.2	90.9	92.4	94.9	94.1	70.9	72.6	78.3	80.9	85.6	88.4	89.2	91.8	92.4	95.2
	Stop4	0	0	0	0	0	0	0	0	0	0	0	0	0	0	0	0	0	0	0	0

Table A5. Computational results for randomly generated instances with the ratio 0%:0%:50%:50% of the numbers of jobs in the subsets $\mathcal{J}_1$, $\mathcal{J}_2$, $\mathcal{J}_{1,2}$ and $\mathcal{J}_{2,1}$ of the job set $\mathcal{J}$.

		Uniform Distributions										Gamma Distributions									
$\delta\%$	n	10	20	30	40	50	60	70	80	90	100	10	20	30	40	50	60	70	80	90	100
5	Stop1	100	100	100	100	100	100	100	100	100	100	100	100	100	100	100	100	100	100	100	100
	Stop2	0	0	0	0	0	0	0	0	0	0	0	0	0	0	0	0	0	0	0	0
	Stop3	0	0	0	0	0	0	0	0	0	0	0	0	0	0	0	0	0	0	0	0
	Stop4	0	0	0	0	0	0	0	0	0	0	0	0	0	0	0	0	0	0	0	0
10	Stop1	99.8	100	100	100	100	100	100	100	100	100	99.9	100	100	100	100	100	100	100	100	100
	Stop2	0.1	0	0	0	0	0	0	0	0	0	0.1	0	0	0	0	0	0	0	0	0
	Stop3	0.1	0	0	0	0	0	0	0	0	0	0	0	0	0	0	0	0	0	0	0
	Stop4	0	0	0	0	0	0	0	0	0	0	0	0	0	0	0	0	0	0	0	0
20	Stop1	99.1	99.6	100	100	100	100	100	100	100	100	98.9	99.9	100	100	100	100	100	100	100	100
	Stop2	0.3	0.2	0	0	0	0	0	0	0	0	0.5	0.1	0	0	0	0	0	0	0	0
	Stop3	0.6	0.2	0	0	0	0	0	0	0	0	0.6	0	0	0	0	0	0	0	0	0
	Stop4	0	0	0	0	0	0	0	0	0	0	0	0	0	0	0	0	0	0	0	0
30	Stop1	96.2	99.5	99.6	99.8	99.9	100	100	100	100	100	96.1	98.7	99.8	100	100	100	100	100	100	100
	Stop2	1.5	0.5	0.4	0.2	0.1	0	0	0	0	0	1.1	1.1	0.1	0	0	0	0	0	0	0
	Stop3	2.3	0	0	0	0	0	0	0	0	0	2.8	0.2	0.1	0	0	0	0	0	0	0
	Stop4	0	0	0	0	0	0	0	0	0	0	0	0	0	0	0	0	0	0	0	0
40	Stop1	91.3	97.4	99.2	99.2	99.6	99.9	100	100	100	100	91.4	96.7	98.9	99.1	99.7	99.6	99.9	100	100	100
	Stop2	3	1	0.2	0.5	0.1	0.1	0	0	0	0	2.1	1.7	0.6	0.7	0.3	0.2	0.1	0	0	0
	Stop3	5.7	1.6	0.6	0.3	0.3	0	0	0	0	0	6.5	1.6	0.5	0.2	0	0.2	0	0	0	0
	Stop4	0	0	0	0	0	0	0	0	0	0	0	0	0	0	0	0	0	0	0	0
50	Stop1	83.4	92.3	95.3	97.9	98.2	98.8	99.3	99.9	99.8	100	81.7	90	95.5	97.5	98.7	98.9	99.4	99.8	99.9	99.9
	Stop2	3.5	3.3	1.9	1.1	0.9	0.6	0.1	0.1	0.1	0	3.8	4.6	2.1	1.3	0.5	0.5	0.4	0	0	0
	Stop3	13.1	4.4	2.8	1	0.9	0.6	0.6	0	0.1	0	14.5	5.4	2.4	1.2	0.8	0.6	0.2	0.2	0.1	0.1
	Stop4	0	0	0	0	0	0	0	0	0	0	0	0	0	0	0	0	0	0	0	0
60	Stop1	69.4	83.1	87.1	93.8	94.5	96.1	96.4	97.8	98.6	98.3	71.2	81.5	89.2	92.6	93.5	95.1	98	98.2	98.5	98.9
	Stop2	5.7	6.7	5	2.5	2.2	1.3	0.8	0.7	0.4	0.6	6.5	6.8	3.3	3.3	2.3	2.1	0.5	0.8	0.7	0.1
	Stop3	24.9	10.2	7.9	3.7	3.3	2.6	2.8	1.5	1	1.1	22.3	11.7	7.5	4.1	4.2	2.8	1.5	1	0.8	1
	Stop4	0	0	0	0	0	0	0	0	0	0	0	0	0	0	0	0	0	0	0	0
70	Stop1	59.2	70.2	75.5	81.1	81.6	88.1	87.9	92.4	92.8	93.6	58.4	72.7	74.8	81.5	85.8	87.8	88	91	90.3	93.7
	Stop2	5.8	7.4	6.6	6	5.2	4.1	3.7	1.7	1.1	2.4	4.3	8.3	7.6	4.4	3.6	3.2	4.1	2.4	2.6	2
	Stop3	35	22.4	17.9	12.9	13.2	7.8	8.4	5.9	6.1	4	37.3	19	17.6	14.1	10.6	9	7.9	6.6	7.1	4.3
	Stop4	0	0	0	0	0	0	0	0	0	0	0	0	0	0	0	0	0	0	0	0
80	Stop1	45	57.2	61.8	62.7	67.8	68.7	71	74.9	75.5	77	46.8	54.8	61.1	64.1	69.4	71	71.4	73.8	75.8	76
	Stop2	7	8.5	7.2	7.4	7.5	6	6	5.2	4.2	3.9	5.5	7.5	7	7.7	6.4	5.3	6.4	4.5	4.4	3.5
	Stop3	48	34.3	31	29.9	24.7	25.3	23	19.9	20.3	19.1	47.7	37.7	31.9	28.2	24.2	23.7	22.2	21.7	19.8	20.5
	Stop4	0	0	0	0	0	0	0	0	0	0	0	0	0	0	0	0	0	0	0	0
90	Stop1	35.2	41.2	38.7	44.2	42.5	45.6	46.2	45.2	50.3	50	37.2	37.3	41.3	43.8	43.3	46.5	48	46.7	47.5	49.4
	Stop2	6	8.3	9.8	7.7	7.8	7.4	6.4	6.2	6.4	5.5	5.4	8.7	9.8	8.9	9.3	9.2	6.4	6.9	6.2	5.7
	Stop3	58.8	50.5	51.5	48.1	49.7	47	47.4	48.6	43.3	44.5	57.4	54	48.9	47.3	47.4	44.3	45.6	46.4	46.3	44.9
	Stop4	0	0	0	0	0	0	0	0	0	0	0	0	0	0	0	0	0	0	0	0
100	Stop1	25.7	28.8	25.2	22.6	21.4	21.4	20.2	20.5	19	22.1	26.5	26	25.2	25.5	25.1	22.1	22.8	19.9	23.8	20.7
	Stop2	4.1	8.6	8.1	6.9	7.9	5.9	6.6	5.4	5.5	4.3	3.7	9.2	7.8	6	7.2	7.1	5.4	5.1	4.8	4.9
	Stop3	70.2	62.6	66.7	70.5	70.7	72.7	73.2	74.1	75.5	73.6	69.8	64.8	67	68.5	67.7	70.8	71.8	75	71.4	74.4
	Stop4	0	0	0	0	0	0	0	0	0	0	0	0	0	0	0	0	0	0	0	0

Table A6. Computational results for randomly generated instances with the ratio 5%:5%:5%:85% of the numbers of jobs in the subsets $\mathcal{J}_1$, $\mathcal{J}_2$, $\mathcal{J}_{1,2}$ and $\mathcal{J}_{2,1}$ of the job set $\mathcal{J}$.

		Uniform Distributions					Gamma Distributions				
$\delta\%$	n	20	40	60	80	100	20	40	60	80	100
5	Stop1	99	99.5	99.7	99.3	99.4	98.1	99.7	99.8	99.2	99.7
	Stop2	0.3	0.3	0.2	0.4	0.6	0.4	0.2	0	0.6	0.3
	Stop3	0.7	0.2	0.1	0.3	0	1.5	0.1	0.2	0.2	0
	Stop4	0	0	0	0	0	0	0	0	0	0
10	Stop1	96.4	97.4	98.4	97.8	97.1	95.4	97.6	97.7	98	98.6
	Stop2	0.9	1.1	0.7	1.2	0.7	1.5	1.4	1	1.3	0.3
	Stop3	2.7	1.5	0.9	1	2.2	3.1	1	1.3	0.7	1.1
	Stop4	0	0	0	0	0	0	0	0	0	0
20	Stop1	84.3	81.4	76.5	71.7	67.9	86.5	80.5	73.8	71	66.5
	Stop2	3.1	3.5	3.7	2.5	2.1	2.9	3.5	2.9	3.7	1.7
	Stop3	12.6	15.1	19.8	25.8	30	10.6	16	23.3	25.3	31.8
	Stop4	0	0	0	0	0	0	0	0	0	0
30	Stop1	61.8	52.9	38.4	27.1	20.9	65.1	53	40	29.3	20.3
	Stop2	5.7	2.9	2.1	2.1	0.3	3.4	2.6	1.4	1.4	1.2
	Stop3	32.5	44.2	59.5	70.8	78.8	31.4	44.4	58.6	69.3	78.5
	Stop4	0	0	0	0	0	0.1	0	0	0	0
40	Stop1	38.6	25.5	14.1	7.3	3.1	40	24.4	13.6	6.5	3.2
	Stop2	5.4	1.2	1.2	0.4	0.3	4.5	2.4	0.4	0.4	0.1
	Stop3	56	73.3	84.7	92.3	96.6	55.5	73.2	86	93.1	96.7
	Stop4	0	0	0	0	0	0	0	0	0	0
50	Stop1	27.4	8.4	3.2	1.7	0.3	25.9	10.9	3.3	0.9	0.3
	Stop2	2.3	1.4	0.3	0	0	3.7	1.1	0	0	0
	Stop3	70.2	90.2	96.5	98.3	99.7	70.4	88	96.7	99.1	99.7
	Stop4	0.1	0	0	0	0	0	0	0	0	0
60	Stop1	14.3	3.4	0.5	0.2	0	15.7	3.2	0.6	0.3	0
	Stop2	2.3	0.4	0	0	0	1.8	0.1	0.3	0	0
	Stop3	83.4	96.2	99.5	99.8	100	82.4	96.7	99.1	99.7	100
	Stop4	0	0	0	0	0	0.1	0	0	0	0
70	Stop1	8	1.1	0.3	0	0	6.5	1.7	0.1	0	0
	Stop2	0.9	0	0	0	0	1.5	0	0.1	0	0
	Stop3	91.1	98.9	99.7	100	100	91.8	98.3	99.8	100	100
	Stop4	0	0	0	0	0	0.2	0	0	0	0
80	Stop1	4	0.1	0	0.1	0	5.1	0.2	0	0	0
	Stop2	0.5	0.1	0	0	0	0.9	0	0	0	0
	Stop3	95.5	99.8	100	99.9	100	93.8	99.8	100	100	100
	Stop4	0	0	0	0	0	0.2	0	0	0	0
90	Stop1	2.6	0.3	0	0	0	2.2	0	0	0	0
	Stop2	0.3	0	0	0	0	0.4	0	0	0	0
	Stop3	97.1	99.7	100	100	100	97	100	100	100	100
	Stop4	0	0	0	0	0	0.4	0	0	0	0
100	Stop1	0.9	0	0	0	0	1.5	0	0	0	0
	Stop2	0.3	0	0	0	0	0	0	0	0	0
	Stop3	98.8	100	100	100	100	98.4	100	100	100	100
	Stop4	0	0	0	0	0	0.1	0	0	0	0

Table A7. Computational results for randomly generated instances with the ratio 5%:15%:5%:75% of the numbers of jobs in the subsets $\mathcal{J}_1$, $\mathcal{J}_2$, $\mathcal{J}_{1,2}$ and $\mathcal{J}_{2,1}$ of the job set $\mathcal{J}$.

		Uniform Distributions					Gamma Distributions				
$\delta\%$	n	20	40	60	80	100	20	40	60	80	100
5	Stop1	98.9	99.4	99.3	99.7	99.5	98.4	99.7	99.4	99.7	99.7
	Stop2	0.2	0.3	0.4	0.3	0.3	0.4	0.2	0.2	0.2	0.2
	Stop3	0.9	0.3	0.3	0	0.2	1.2	0.1	0.4	0.1	0.1
	Stop4	0	0	0	0	0	0	0	0	0	0
10	Stop1	97	98.2	97.7	98.6	99.3	95.5	98.6	98.6	98.4	98.3
	Stop2	1	0.8	1.1	0.7	0.3	1.3	0.8	0.8	0.7	0.9
	Stop3	2	1	1.2	0.7	0.4	3.2	0.6	0.6	0.9	0.8
	Stop4	0	0	0	0	0	0	0	0	0	0
20	Stop1	86	86.7	83.3	76.9	75.4	88.3	86.1	83.3	77.4	74.6
	Stop2	2.6	2.8	2.5	3.1	2.5	2.5	2.3	3.1	2.7	2.3
	Stop3	11.4	10.5	14.2	20	22.1	9.2	11.6	13.6	19.9	23.1
	Stop4	0	0	0	0	0	0	0	0	0	0
30	Stop1	67.1	58.1	47.1	36.1	29.3	69	64.5	48.2	38.6	28.7
	Stop2	3.1	3.2	2.8	2.6	1.5	3.2	2.1	2.2	2	1.2
	Stop3	29.8	38.7	50.1	61.3	69.2	27.8	33.4	49.6	59.4	70.1
	Stop4	0	0	0	0	0	0	0	0	0	0
40	Stop1	48.2	30.1	18.9	11.1	8.3	45.6	31.8	17.6	10.5	6.6
	Stop2	2.7	3.7	1.2	0.3	0.1	4.3	2.6	1.9	0.7	0.3
	Stop3	49.1	66.2	79.9	88.6	91.6	50.1	65.6	80.5	88.8	93.1
	Stop4	0	0	0	0	0	0	0	0	0	0
50	Stop1	30.1	15.1	7.4	3	1.2	29.1	13.2	6	2.7	1.7
	Stop2	2.8	1.3	0.7	0.3	0	2.6	1.1	0.3	0.2	0
	Stop3	67.1	83.6	91.9	96.7	98.8	68.3	85.7	93.7	97.1	98.3
	Stop4	0	0	0	0	0	0	0	0	0	0
60	Stop1	19.2	7.6	1.9	0.5	0	21.1	5.9	2.6	0.5	0.3
	Stop2	1.5	0.4	0	0	0.1	2	0.5	0.2	0	0
	Stop3	79.3	92	98.1	99.5	99.9	76.9	93.6	97.2	99.5	99.7
	Stop4	0	0	0	0	0	0	0	0	0	0
70	Stop1	11.4	2.6	0.9	0	0	12	3.1	0.3	0	0.1
	Stop2	1.9	0.1	0	0	0	0.8	0.2	0	0	0
	Stop3	86.7	97.3	99.1	100	100	87.2	96.7	99.7	100	99.9
	Stop4	0	0	0	0	0	0	0	0	0	0
80	Stop1	7.6	1.2	0.1	0	0	7.9	0.6	0.2	0	0
	Stop2	0.7	0.1	0	0	0	1	0.1	0	0	0
	Stop3	91.7	98.7	99.9	100	100	91.1	99.3	99.8	100	100
	Stop4	0	0	0	0	0	0	0	0	0	0
90	Stop1	3.4	0.6	0.1	0	0	4.6	0.5	0.1	0	0
	Stop2	0.4	0	0	0	0	0.5	0	0	0	0
	Stop3	96.2	99.4	99.9	100	100	94.9	99.5	99.9	100	100
	Stop4	0	0	0	0	0	0	0	0	0	0
100	Stop1	2.7	0.2	0	0	0	2.9	0.2	0	0	0
	Stop2	0.5	0	0	0	0	0.3	0	0	0	0
	Stop3	96.8	99.8	100	100	100	96.8	99.8	100	100	100
	Stop4	0	0	0	0	0	0	0	0	0	0

Table A8. Computational results for randomly generated instances with the ratio 5%:20%:5%:70% of the numbers of jobs in the subsets $\mathcal{J}_1$, $\mathcal{J}_2$, $\mathcal{J}_{1,2}$ and $\mathcal{J}_{2,1}$ of the job set $\mathcal{J}$.

		Uniform Distributions					Gamma Distributions				
$\delta\%$	n	20	40	60	80	100	20	40	60	80	100
5	Stop1	98.6	99.2	99.6	99.4	99.5	98.3	99.2	99.7	99.8	99.6
	Stop2	0.1	0.4	0.3	0.4	0.5	0.5	0.1	0.1	0.1	0.3
	Stop3	1.3	0.4	0.1	0.2	0	1.2	0.7	0.2	0.1	0.1
	Stop4	0	0	0	0	0	0	0	0	0	0
10	Stop1	96.1	98.5	99.7	99.1	99.1	96.6	98.2	98.2	98.9	98.4
	Stop2	1.1	0.6	0.2	0.7	0.4	0.3	0.9	1.2	0.6	0.9
	Stop3	2.8	0.9	0.1	0.2	0.5	3.1	0.9	0.6	0.5	0.7
	Stop4	0	0	0	0	0	0	0	0	0	0
20	Stop1	88.8	87.6	84.6	80.1	77.6	89.5	89.8	84.6	83.5	79.3
	Stop2	1.9	2.4	2.6	2.2	2.8	3	1.9	2.7	1.9	2.4
	Stop3	9.3	10	12.8	17.7	19.6	7.5	8.3	12.7	14.6	18.3
	Stop4	0	0	0	0	0	0	0	0	0	0
30	Stop1	72	62.3	52.3	43.2	33.5	70.6	63.1	53	42.4	36.9
	Stop2	3.4	2.2	3.9	3	1.8	3.9	3.4	3.9	2.5	0.6
	Stop3	24.6	35.5	43.8	53.8	64.7	25.5	33.5	43.1	55.1	62.5
	Stop4	0	0	0	0	0	0	0	0	0	0
40	Stop1	50.6	37.1	25.9	15	9.9	50.7	36.4	25.9	16.5	10.3
	Stop2	3.7	2.1	1.8	0.7	0.4	3.5	2.7	1.3	1	0.4
	Stop3	45.7	60.8	72.3	84.3	89.7	45.8	60.9	72.8	82.5	89.3
	Stop4	0	0	0	0	0	0	0	0	0	0
50	Stop1	36.2	19.5	9.4	4.5	1.7	33.2	17.9	9.6	4.3	2.8
	Stop2	3.1	1.6	0	0.1	0.2	3.9	0.9	0.6	0.3	0
	Stop3	60.7	78.9	90.6	95.4	98.1	62.9	81.2	89.8	95.4	97.2
	Stop4	0	0	0	0	0	0	0	0	0	0
60	Stop1	25.2	7.7	3.1	1.2	0.3	24	7.6	3.5	0.6	0.5
	Stop2	2	0.7	0.3	0	0	1.6	1.2	0.2	0	0
	Stop3	72.8	91.6	96.6	98.8	99.7	74.4	91.2	96.3	99.4	99.5
	Stop4	0	0	0	0	0	0	0	0	0	0
70	Stop1	12	3.3	0.8	0.2	0.4	16.6	4.1	1.6	0.5	0.1
	Stop2	2.3	0.4	0.1	0	0	1.1	0.7	0	0	0
	Stop3	85.7	96.3	99.1	99.8	99.6	82.3	95.2	98.4	99.5	99.9
	Stop4	0	0	0	0	0	0	0	0	0	0
80	Stop1	9.4	1.8	0.3	0	0	9.9	1.8	0.1	0	0
	Stop2	1.4	0.1	0	0	0	1	0.1	0	0	0
	Stop3	89.2	98.1	99.7	100	100	89	98.1	99.9	100	100
	Stop4	0	0	0	0	0	0.1	0	0	0	0
90	Stop1	5.6	0.3	0.1	0.1	0	6.5	0.6	0.1	0	0
	Stop2	0.7	0.1	0	0	0	0.3	0.1	0	0	0
	Stop3	93.7	99.6	99.9	99.9	100	93.2	99.3	99.9	100	100
	Stop4	0	0	0	0	0	0	0	0	0	0
100	Stop1	4.5	0.5	0	0	0	4.5	0.2	0	0	0
	Stop2	0.2	0	0	0	0	0.2	0	0	0	0
	Stop3	95.3	99.5	100	100	100	95.3	99.8	100	100	100
	Stop4	0	0	0	0	0	0	0	0	0	0

Table A9. Computational results for randomly generated instances with the ratio 10%:10%:10%:70% of the numbers of jobs in the subsets $\mathcal{J}_1$, $\mathcal{J}_2$, $\mathcal{J}_{1,2}$ and $\mathcal{J}_{2,1}$ of the job set $\mathcal{J}$.

		Uniform Distributions										Gamma Distributions									
$\delta\%$	n	10	20	30	40	50	60	70	80	90	100	10	20	30	40	50	60	70	80	90	100
5	Stop1	98.9	99.5	100	99.8	100	99.9	100	100	100	99.9	98.4	99.4	99.9	100	99.9	99.9	100	100	100	100
	Stop2	0.4	0	0	0.2	0	0	0	0	0	0	0.1	0	0.3	0.1	0	0	0.1	0	0	0
	Stop3	0.7	0.5	0	0	0	0.1	0	0	0	0	1.6	0.3	0	0	0.1	0	0	0	0	0
	Stop4	0	0	0	0	0	0	0	0	0	0	0	0	0	0	0	0	0	0	0	0
10	Stop1	95.9	98.9	98.9	100	99.8	100	100	100	99.9	99.9	96.2	98.4	99.5	99.7	99.8	99.9	99.9	99.8	99.8	99.9
	Stop2	0.2	0.5	0.6	0	0.2	0	0	0	0.1	0	0.3	0.3	0.2	0.1	0.1	0.1	0.1	0.2	0.2	0.1
	Stop3	3.8	0.6	0.5	0	0	0	0	0	0	0.1	3.4	1.3	0.3	0.2	0.1	0	0	0	0	0
	Stop4	0.1	0	0	0	0	0	0	0	0	0	0.1	0	0	0	0	0	0	0	0	0
20	Stop1	88.8	97	97.1	97.8	97.8	97.5	98.2	98.7	97.5	97.9	88.5	95.3	96.5	97.2	97.6	97.3	97.9	98.3	98.2	97.5
	Stop2	1.6	1.2	1.1	0.5	0.6	1	0.6	0.4	0.2	0.2	1.6	1.6	1	1.3	0.5	0.7	0.8	0.7	0.6	0.4
	Stop3	9.5	1.8	1.8	1.7	1.6	1.5	1.2	0.9	2.3	1.9	9.8	3.1	2.5	1.5	1.9	2	1.3	1	1.2	2.1
	Stop4	0.1	0	0	0	0	0	0	0	0	0	0.1	0	0	0	0	0	0	0	0	0
30	Stop1	79.3	85.8	86.6	85.1	81.4	83.3	82.6	79.2	77.4	75.9	80.4	86.3	86.5	85.1	81.2	81.6	80.8	80	79	76.1
	Stop2	3.7	2.2	2.5	1.7	2.9	1.8	1.5	1.2	1.5	1	2.4	2.8	2.8	1.7	1.9	1.4	1.4	1.3	1	0.6
	Stop3	17	12	10.9	13.2	15.7	14.9	15.9	19.6	21.1	23.1	16.9	10.9	10.7	13.2	16.9	17	17.8	18.7	20	23.3
	Stop4	0	0	0	0	0	0	0	0	0	0	0.3	0	0	0	0	0	0	0	0	0
40	Stop1	66.6	71.7	65.8	62.2	58.2	53.1	48.4	44	42	38.3	65.8	70.1	65.8	60.9	57.4	52.7	53.9	46.8	40.6	38.5
	Stop2	3	2.6	2.4	2.8	2.2	2.4	1.7	2	1.1	0.9	2.6	2.7	2.6	1.8	1.6	1.5	2.1	1	0.9	1.1
	Stop3	30.2	25.7	31.8	35	39.6	44.5	49.9	54	56.9	60.8	31.1	27.2	31.6	37.3	41	45.8	44	52.2	58.5	60.4
	Stop4	0.2	0	0	0	0	0	0	0	0	0	0.5	0	0	0	0	0	0	0	0	0
50	Stop1	55.8	55.3	46.9	38.9	30.9	25.5	22.2	19.9	15.1	13.2	56	52.9	44.3	39.4	34.1	27.4	22.6	18.2	17.5	14.9
	Stop2	2.6	2.9	2.4	2.5	1.7	1	0.8	0.3	0.7	0.6	3.3	3.5	2.4	1.4	1.7	1.1	0.7	0.5	0.4	0.7
	Stop3	41.6	41.8	50.7	58.6	67.4	73.5	77	79.8	84.2	86.2	40.5	43.6	53.3	59.2	64.2	71.5	76.7	81.3	82.1	84.4
	Stop4	0	0	0	0	0	0	0	0	0	0	0.2	0	0	0	0	0	0	0	0	0
60	Stop1	44.4	35.9	27.8	20.1	15.6	12.7	8.6	7.1	7.8	3.3	43.6	36.5	26.2	20.7	14.6	9.9	8.8	6.4	4.2	3.2
	Stop2	3.2	2.7	1.6	0.9	0.8	0.7	0.2	0.1	0.2	0.1	3.3	2	1.4	1.8	1.7	0.2	0.2	0.3	0.1	0
	Stop3	52.2	61.4	70.6	79	83.6	86.6	91.2	92.8	92	96.6	52.5	61.5	72.4	77.5	83.7	89.9	91	93.3	95.7	96.8
	Stop4	0.2	0	0	0	0	0	0	0	0	0	0.6	0	0	0	0	0	0	0	0	0
70	Stop1	33.5	25.4	17	11.2	6.6	2.8	2	1.2	1.1	0.7	36.7	24	17.3	11.6	5.7	5.2	3.5	1.5	0.7	1.1
	Stop2	2.5	2	1.5	0.4	0.5	0.1	0.3	0.2	0.1	0	2.4	1.9	1	0.5	0.2	0.2	0	0.1	0	0
	Stop3	63.8	72.6	81.5	88.4	92.9	97.1	97.7	98.6	98.8	99.3	60.5	74.1	81.7	87.9	94.1	94.6	96.5	98.4	99.3	98.9
	Stop4	0.2	0	0	0	0	0	0	0	0	0	0.4	0	0	0	0	0	0	0	0	0
80	Stop1	29.8	13.5	10	5.1	2.9	1.2	0.3	0.4	0	0.1	27.9	18.6	8.4	4	2.5	1.4	0.9	0.4	0.2	0.1
	Stop2	1.9	1.4	0.3	0.4	0.1	0	0	0.1	0	0	2.5	0.8	0.5	0.2	0.1	0.2	0	0	0	0
	Stop3	68.1	85.1	89.7	94.5	97	98.8	99.7	99.5	100	99.9	68.8	80.6	91.1	95.8	97.4	98.4	99.1	99.6	99.8	99.9
	Stop4	0.2	0	0	0	0	0	0	0	0	0	0.8	0	0	0	0	0	0	0	0	0
90	Stop1	22.4	10.5	3.9	1.7	0.7	0.1	0.3	0.2	0	0	20.8	10.2	4.5	3.1	0.7	0.2	0.3	0	0	0
	Stop2	1.9	0.6	0.6	0.1	0	0	0	0	0	0	1.6	0.9	0.5	0	0	0.1	0	0	0	0
	Stop3	75.3	88.9	95.5	98.2	99.3	99.9	99.7	99.8	100	100	76.8	88.9	95	96.9	99.3	99.7	99.7	100	100	100
	Stop4	0.4	0	0	0	0	0	0	0	0	0	0.8	0	0	0	0	0	0	0	0	0
100	Stop1	15.6	5.9	1.5	0.9	0.6	0.1	0.1	0	0	0	15.9	6.7	2.6	0.9	0.2	0.1	0.2	0	0	0
	Stop2	1.6	0.7	0.2	0	0	0	0	0	0	0	1.7	0.2	0.2	0.1	0	0	0	0	0	0
	Stop3	82.3	93.4	98.3	99.1	99.4	99.9	99.9	100	100	100	81.4	93.1	97.2	99	99.8	99.9	99.8	100	100	100
	Stop4	0.5	0	0	0	0	0	0	0	0	0	1	0	0	0	0	0	0	0	0	0

143

Table A10. Computational results for randomly generated instances with the ratio 10%:10%:40%:40% of the numbers of jobs in the subsets $\mathcal{J}_1$, $\mathcal{J}_2$, $\mathcal{J}_{1,2}$ and $\mathcal{J}_{2,1}$ of the job set $\mathcal{J}$.

		Uniform Distributions										Gamma Distributions									
δ%	n	10	20	30	40	50	60	70	80	90	100	10	20	30	40	50	60	70	80	90	100
5	Stop1	99.9	100	100	100	100	100	100	100	100	100	100	100	100	100	100	100	100	100	100	100
	Stop2	0	0	0	0	0	0	0	0	0	0	0	0	0	0	0	0	0	0	0	0
	Stop3	0.1	0	0	0	0	0	0	0	0	0	0	0	0	0	0	0	0	0	0	0
	Stop4	0	0	0	0	0	0	0	0	0	0	0	0	0	0	0	0	0	0	0	0
10	Stop1	100	100	100	100	100	100	100	100	100	100	100	100	100	100	100	100	100	100	100	100
	Stop2	0	0	0	0	0	0	0	0	0	0	0	0	0	0	0	0	0	0	0	0
	Stop3	0	0	0	0	0	0	0	0	0	0	0	0	0	0	0	0	0	0	0	0
	Stop4	0	0	0	0	0	0	0	0	0	0	0	0	0	0	0	0	0	0	0	0
20	Stop1	99.3	100	100	100	100	100	100	100	100	100	99.5	100	100	100	100	100	100	100	100	100
	Stop2	0.2	0	0	0	0	0	0	0	0	0	0.1	0	0	0	0	0	0	0	0	0
	Stop3	0.5	0	0	0	0	0	0	0	0	0	0.4	0	0	0	0	0	0	0	0	0
	Stop4	0	0	0	0	0	0	0	0	0	0	0	0	0	0	0	0	0	0	0	0
30	Stop1	98.4	99.9	100	100	100	100	100	100	100	100	99	99.9	100	100	100	100	100	100	100	100
	Stop2	0.5	0.1	0	0	0	0	0	0	0	0	0.1	0	0	0	0	0	0	0	0	0
	Stop3	1.1	0	0	0	0	0	0	0	0	0	0.9	0.1	0	0	0	0	0	0	0	0
	Stop4	0	0	0	0	0	0	0	0	0	0	0	0	0	0	0	0	0	0	0	0
40	Stop1	97	99.8	99.9	100	100	100	100	100	100	100	96.4	99.8	100	100	100	100	100	100	100	100
	Stop2	0.1	0	0.1	0	0	0	0	0	0	0	0.5	0.2	0	0	0	0	0	0	0	0
	Stop3	2.9	0.2	0	0	0	0	0	0	0	0	3.1	0	0	0	0	0	0	0	0	0
	Stop4	0	0	0	0	0	0	0	0	0	0	0	0	0	0	0	0	0	0	0	0
50	Stop1	93	99	99.9	99.9	100	100	100	100	100	100	92.8	99.6	99.9	100	100	100	100	100	100	100
	Stop2	1.2	0.5	0.1	0	0	0	0	0	0	0	0.9	0.1	0	0	0	0	0	0	0	0
	Stop3	5.8	0.5	0	0.1	0	0	0	0	0	0	6.3	0.3	0.1	0	0	0	0	0	0	0
	Stop4	0	0	0	0	0	0	0	0	0	0	0	0	0	0	0	0	0	0	0	0
60	Stop1	86.6	97.9	99.6	99.5	99.8	99.8	99.8	99.9	100	100	85.8	97.1	99.3	99.6	99.8	99.7	99.9	100	100	100
	Stop2	1.4	0.5	0	0.2	0	0.1	0	0	0	0	1.5	0.4	0	0	0	0	0	0	0	0
	Stop3	12	1.6	0.4	0.3	0.2	0.1	0.2	0.1	0	0	12.7	2.5	0.7	0.4	0.2	0.3	0.1	0	0	0
	Stop4	0	0	0	0	0	0	0	0	0	0	0	0	0	0	0	0	0	0	0	0
70	Stop1	82.3	92.5	96.7	97.2	98.6	98.1	99	99.2	99.5	99.7	80	92.7	96.5	97.5	97.9	99.2	99.5	99.5	99.8	99.5
	Stop2	1.8	1.1	0.2	0.2	0	0	0.1	0	0	0.1	1.7	0.5	0	0.2	0.1	0	0	0	0	0
	Stop3	15.9	6.4	3.1	2.6	1.4	1.9	0.9	0.8	0.5	0.2	18.3	6.8	3.5	2.3	2	0.8	0.5	0.5	0.2	0.5
	Stop4	0	0	0	0	0	0	0	0	0	0	0	0	0	0	0	0	0	0	0	0
80	Stop1	71.8	85.9	91.5	93.8	94.4	95.3	95.6	97.7	97.3	98.1	73.1	87.2	91	93.1	95.1	95.9	96.7	97.3	98.1	98.8
	Stop2	1.9	1.2	0.6	0.2	0.5	0	0	0.1	0	0.1	2	0.9	0.2	0.7	0.1	0.2	0	0.3	0	0
	Stop3	26.3	12.9	7.9	6	5.1	4.7	4.4	2.2	2.7	1.8	24.9	11.9	8.8	6.2	4.8	3.9	3.3	2.4	1.9	1.2
	Stop4	0	0	0	0	0	0	0	0	0	0	0	0	0	0	0	0	0	0	0	0
90	Stop1	63.2	77.8	81.6	84.2	85	88.7	89.9	91.5	93.6	93.6	61.7	77.2	83.3	85	86.4	88	90.4	91.4	92.7	94.4
	Stop2	2.6	1	0.8	1.1	0.2	0.2	0.6	0.2	0.1	0.2	2.6	1.9	0.9	0.2	0.5	0.2	0.1	0.1	0	0.1
	Stop3	34.2	21.2	17.6	14.7	14.8	11.1	9.5	8.3	6.3	6.2	35.7	20.9	15.8	14.8	13.1	11.8	9.5	8.5	7.3	5.5
	Stop4	0	0	0	0	0	0	0	0	0	0	0	0	0	0	0	0	0	0	0	0
100	Stop1	53.1	66.7	68.3	70.3	74.1	74.1	75.6	78.3	81	84.1	51.8	67	69.8	71	72.5	75.9	75.7	79.6	79.8	80.3
	Stop2	2.1	1.7	0.6	0.8	0.3	0.7	0.3	0.2	0.1	0.3	2.4	0.9	1	0.7	1.1	0.5	0.3	0.6	0.3	0.5
	Stop3	44.8	31.6	31.1	28.9	25.6	25.2	24.1	21.5	18.9	15.6	45.8	32.1	29.2	28.3	26.4	23.6	24	19.8	19.9	19.2
	Stop4	0	0	0	0	0	0	0	0	0	0	0	0	0	0	0	0	0	0	0	0

Table A11. Computational results for randomly generated instances with the ratio 10%:20%:10%:60% of the numbers of jobs in the subsets $\mathcal{J}_1$, $\mathcal{J}_2$, $\mathcal{J}_{1,2}$ and $\mathcal{J}_{2,1}$ of the job set $\mathcal{J}$.

		Uniform Distributions										Gamma Distributions									
$\delta\%$	n	10	20	30	40	50	60	70	80	90	100	10	20	30	40	50	60	70	80	90	100
5	Stop1	99.2	99.6	99.8	100	99.9	100	100	100	100	100	98.5	99.8	99.9	99.9	100	100	100	100	100	100
	Stop2	0	0.1	0.2	0	0.1	0	0	0	0	0	0.2	0.1	0.1	0.1	0	0	0	0	0	0
	Stop3	0.8	0.3	0	0	0	0	0	0	0	0	1.3	0.1	0	0	0	0	0	0	0	0
	Stop4	0	0	0	0	0	0	0	0	0	0	0	0	0	0	0	0	0	0	0	0
10	Stop1	98.1	99.2	99.4	99.9	99.9	99.8	100	100	100	100	96.4	99.2	99.8	99.7	99.8	99.9	99.9	100	100	100
	Stop2	0.1	0.5	0.4	0.1	0.1	0	0	0	0	0	0.6	0.3	0.2	0.1	0	0.1	0.1	0	0	0
	Stop3	1.8	0.3	0.2	0	0	0.2	0	0	0	0	3	0.5	0	0.2	0.2	0	0	0	0	0
	Stop4	0	0	0	0	0	0	0	0	0	0	0	0	0	0	0	0	0	0	0	0
20	Stop1	91.3	97.7	97.7	97.8	99.1	98.7	98.8	98.7	99.1	99.6	90.3	96.5	97.6	98	98.4	99.1	99.4	99.4	99.2	99
	Stop2	1.4	0.8	0.8	0.7	0.4	0.7	0.4	0.4	0.1	0.1	0.3	1.3	1.4	0.9	0.7	0.6	0.1	0	0.6	0.3
	Stop3	7.3	1.5	1.5	1.5	0.5	0.6	0.8	0.9	0.8	0.3	9.4	2.2	1	1.1	0.9	0.3	0.5	0.6	0.2	0.7
	Stop4	0	0	0	0	0	0	0	0	0	0	0	0	0	0	0	0	0	0	0	0
30	Stop1	83.4	90.5	92	91.4	89.7	90.8	90.4	88.7	88.8	87	82.8	92.2	91.8	92.1	90.5	90.5	88.7	89.8	87.6	86.9
	Stop2	1.7	1.7	1.7	1.1	1.2	1.6	0.6	1.4	1.1	1	1.6	1.2	1.5	0.8	1.5	1.8	0.6	1.2	1.3	1.1
	Stop3	14.8	7.8	6.3	7.5	9.1	7.6	9	9.9	10.1	12	15.6	6.6	6.7	7.1	8	7.7	10.7	9	11.1	12
	Stop4	0.1	0	0	0	0	0	0	0	0	0	0	0	0	0	0	0	0	0	0	0
40	Stop1	75.8	78.3	76.6	74	72	69.2	65.5	62.7	57.6	59.2	74.3	78.2	78.1	76.3	70.5	68.3	64	65.5	61.2	58.2
	Stop2	1.9	3	3	1.9	1.6	1.4	1.8	1	0.9	1.2	1.6	2.7	2.3	1.3	2.5	1.8	1.7	0.9	1.4	1.5
	Stop3	22	18.7	20.4	24.1	26.4	29.4	32.7	36.3	41.5	39.6	24	19.1	19.6	22.4	27	29.9	34.3	33.6	37.4	40.3
	Stop4	0.3	0	0	0	0	0	0	0	0	0	0.1	0	0	0	0	0	0	0	0	0
50	Stop1	64.4	65	59.6	53.1	46.9	40.6	40.1	35	31.4	30.4	64.6	63.3	57.6	53.9	49.4	42.6	38.3	36.6	32.4	28
	Stop2	1.9	2.8	1.9	2.1	1.7	1.5	0.8	1.1	1.5	0.7	1.6	3.8	2.4	1.5	1.3	1.1	0.8	1.1	1.3	1.1
	Stop3	33.7	32.2	38.5	44.8	51.4	57.9	59.1	63.9	67.1	68.9	33.6	32.9	40	44.6	49.3	56.3	60.9	62.3	66.3	70.9
	Stop4	0	0	0	0	0	0	0	0	0	0	0.2	0	0	0	0	0	0	0	0	0
60	Stop1	54.4	52.2	42.3	33	28.1	21.7	19.9	15.8	13.7	11.3	54.4	49.2	43.4	33.6	26.7	22.2	20.2	14.9	13	11.1
	Stop2	1.4	2.9	1.8	1.6	1.4	0.7	0.4	0.3	0.2	0.3	2.4	2.4	2.3	2.8	0.9	0.9	0.7	0.4	0.3	0.3
	Stop3	44.1	44.9	55.9	65.4	70.5	77.6	79.7	83.9	86.1	88.4	43	48.4	54.3	63.6	72.4	76.9	79.1	84.7	86.7	88.6
	Stop4	0.1	0	0	0	0	0	0	0	0	0	0.2	0	0	0	0	0	0	0	0	0
70	Stop1	47	37.9	25.3	21	14.1	11	7.5	6	4.4	3.1	44.2	34.1	28.5	19.4	14	11.3	7.5	4.4	4.1	3.1
	Stop2	1.5	1.4	1.3	1.5	0.7	0.5	0.3	0.2	0.1	0	2.1	2.9	1.4	0.7	0.9	0.4	0.1	0.2	0.1	0.1
	Stop3	51.5	60.7	73.4	77.5	85.2	88.5	92.2	93.8	95.5	96.9	53.4	63	70.1	79.9	85.1	88.3	92.4	95.4	95.8	96.8
	Stop4	0	0	0	0	0	0	0	0	0	0	0.3	0	0	0	0	0	0	0	0	0
80	Stop1	38	26.7	16	10.5	6.4	4.9	2.5	1.9	0.8	0.7	35	26.1	17.6	10.4	8.1	4.2	2.2	2.1	1.3	0.5
	Stop2	1.3	1.7	1.3	1	0.9	0.1	0	0.1	0	0	1.6	1.3	0.7	0.7	0.3	0.4	0	0	0	0
	Stop3	60.6	71.6	82.7	88.5	92.7	95	97.5	98	99.2	99.3	63.1	72.6	81.7	88.9	91.6	95.4	97.8	97.9	98.7	99.5
	Stop4	0.1	0	0	0	0	0	0	0	0	0	0.3	0	0	0	0	0	0	0	0	0
90	Stop1	29	15.7	9.2	4.7	3.8	1.7	1.6	0	0.2	0.3	29.5	16.7	10.2	5.9	3.2	2.1	1.1	0.7	0.3	0.2
	Stop2	1.5	1.4	0.7	0.4	0.1	0.2	0.1	0	0	0.1	1.7	1.2	0.4	0.3	0	0	0	0	0	0
	Stop3	69.3	82.9	90.1	94.9	96.1	98.1	98.3	100	99.8	99.6	68.6	82.1	89.4	93.8	96.8	97.9	98.9	99.3	99.7	99.8
	Stop4	0.2	0	0	0	0	0	0	0	0	0	0.2	0	0	0	0	0	0	0	0	0
100	Stop1	21.6	12.2	6.7	2.4	1.6	0.6	0.2	0.1	0.1	0	23	11	5.8	3	1.7	0.5	0.2	0.3	0	0
	Stop2	1.2	0.8	0.1	0.1	0.1	0	0	0.1	0	0	1.1	0.6	0.5	0.2	0.1	0	0	0	0	0
	Stop3	77	87	93.2	97.5	98.3	99.4	99.8	99.9	99.8	100	75.3	88.4	93.7	96.8	98.2	99.5	99.8	99.7	100	100
	Stop4	0.2	0	0	0	0	0	0	0	0	0	0.6	0	0	0	0	0	0	0	0	0

Table A12. Computational results for randomly generated instances with the ratio 10%:30%:10%:50% of the numbers of jobs in the subsets $\mathcal{J}_1$, $\mathcal{J}_2$, $\mathcal{J}_{1,2}$ and $\mathcal{J}_{2,1}$ of the job set $\mathcal{J}$.

		Uniform Distributions										Gamma Distributions									
$\delta\%$	n	10	20	30	40	50	60	70	80	90	100	10	20	30	40	50	60	70	80	90	100
5	Stop1	98.6	99.9	100	100	100	100	100	100	100	100	99.3	100	100	100	100	100	100	100	100	100
	Stop2	0.2	0	0	0	0	0	0	0	0	0	0	0	0	0	0	0	0	0	0	0
	Stop3	1.2	0.1	0	0	0	0	0	0	0	0	0.7	0	0	0	0	0	0	0	0	0
	Stop4	0	0	0	0	0	0	0	0	0	0	0	0	0	0	0	0	0	0	0	0
10	Stop1	97.8	99.4	99.8	99.8	100	99.8	100	100	100	100	97.3	99.7	99.7	100	100	99.9	100	100	99.9	100
	Stop2	0.2	0.1	0.1	0.2	0	0.2	0	0	0	0	0.2	0.1	0.1	0	0	0	0	0	0.1	0
	Stop3	2	0.5	0.1	0	0	0	0	0	0	0	2.5	0.2	0.2	0	0	0.1	0	0	0	0
	Stop4	0	0	0	0	0	0	0	0	0	0	0	0	0	0	0	0	0	0	0	0
20	Stop1	93.5	98.1	98.1	99.4	99.1	99.7	99.6	99.9	99.7	99.4	95	97.5	98.8	99	99.5	99.5	99.8	99.7	99.6	100
	Stop2	0.4	0.8	0.6	0.2	0.4	0.2	0.2	0.1	0.3	0.2	0.6	0.7	0.6	0.5	0.5	0.3	0.1	0.2	0.3	0
	Stop3	6.1	1.1	1.3	0.4	0.5	0.1	0.2	0	0	0.4	4.4	1.8	0.6	0.5	0	0.2	0.1	0.1	0.1	0
	Stop4	0	0	0	0	0	0	0	0	0	0	0	0	0	0	0	0	0	0	0	0
30	Stop1	85.9	94.2	96.1	95.2	94.3	95.3	96.1	95.8	95.8	95.2	89.9	94.2	94.2	95.3	95.4	95.9	95.2	95.2	92.8	96.3
	Stop2	0.6	1.4	1.1	1.1	1.2	0.5	0.7	1.1	0.3	0.4	0.9	0.8	1.4	1	1.5	1	1.2	0.6	1.1	0.3
	Stop3	13.5	4.4	2.8	3.7	4.5	4.2	3.2	3.1	3.9	4.4	9.2	5	4.4	3.7	3.1	3.1	3.6	4.2	6.1	3.4
	Stop4	0	0	0	0	0	0	0	0	0	0	0	0	0	0	0	0	0	0	0	0
40	Stop1	80.5	85.7	85.5	83.4	85.4	81.9	82.3	77.7	77.8	77.5	81.7	85.9	88.3	85.6	82.1	84.3	81.6	79.1	80.9	80.1
	Stop2	1.3	2.2	1.9	1.9	1.3	1.4	1.6	1.7	0.7	1.2	1.4	2.2	1.5	2.3	2	1.3	0.4	1.6	1	0.9
	Stop3	18.2	12.1	12.6	14.7	13.3	16.7	16.1	20.6	21.5	21.3	16.9	11.9	10.2	12.1	15.9	14.4	18	19.3	18.1	19
	Stop4	0	0	0	0	0	0	0	0	0	0	0	0	0	0	0	0	0	0	0	0
50	Stop1	74.1	74.8	72.6	70	67	62	59.2	58.5	54.9	52.7	73	75.9	73.6	69.7	67.6	60.9	62.6	59.1	56.6	52
	Stop2	1.2	2.6	2.5	1.8	1.3	1.5	1.5	0.9	0.9	0.6	0.9	2.1	2.1	1.3	1.2	1.1	0.8	1	0.6	0.4
	Stop3	24.7	22.6	24.9	28.2	31.7	36.5	39.3	40.6	44.2	46.7	26.1	22	24.3	29	31.2	38	36.6	39.9	42.8	47.6
	Stop4	0	0	0	0	0	0	0	0	0	0	0	0	0	0	0	0	0	0	0	0
60	Stop1	66.9	64.3	60.9	52.3	45	46.7	40	37.4	34.6	32.2	63.8	64.2	58.4	55.9	48.2	42.8	40.1	35.9	36.3	31.2
	Stop2	1.3	3.2	1.6	2.5	1.5	0.9	1.8	0.9	0.8	0.9	0.8	2	1.9	1.6	1.8	1.4	1.2	1.3	0.9	1.3
	Stop3	31.8	32.5	37.5	45.2	53.5	52.4	58.2	61.7	64.6	66.9	35.4	33.8	39.7	42.5	50	55.8	58.7	62.8	62.8	67.5
	Stop4	0	0	0	0	0	0	0	0	0	0	0	0	0	0	0	0	0	0	0	0
70	Stop1	58.4	52.6	42.7	38.7	34.6	29.4	21.1	21.5	17.6	16.3	57.7	50.8	45	37.9	31	27.6	26.2	21.6	20.5	15.1
	Stop2	1.4	2.2	1	2.3	0.8	0.7	0.5	0.3	0.2	0.1	0.9	3.5	2.4	1.1	1.3	0.6	0.6	0.4	0.3	0
	Stop3	40.2	45.2	56.3	59	64.6	69.9	78.4	78.2	82.2	83.6	41.3	45.7	52.6	61	67.7	71.8	73.2	78	79.2	84.9
	Stop4	0	0	0	0	0	0	0	0	0	0.1	0	0	0	0	0	0	0	0	0	0
80	Stop1	51.9	40.7	34.1	28.3	21.4	14.2	13.9	9.9	9.8	8	49.7	41.7	33.3	27.1	24.1	17.4	14.1	12.6	8.8	7.5
	Stop2	0.9	1.7	1.4	1.4	0.2	0.1	0.2	0.3	0.1	0	0.8	2.5	1.6	0.9	0.5	0.3	0.6	0.5	0	0
	Stop3	47.2	57.6	64.5	70.3	78.4	85.7	85.9	89.8	90.1	92	49.4	55.8	65.1	72	75.4	82.3	85.3	86.9	91.2	92.5
	Stop4	0	0	0	0	0	0	0	0	0	0	0.1	0	0	0	0	0	0	0	0	0
90	Stop1	44.2	31.3	24.6	17.3	12.3	11	7.1	4.7	3.6	4.2	44	32.7	24.8	16.3	12.7	10	7.9	6.5	4.2	3.8
	Stop2	1.3	1.3	0.4	0.8	0.2	0.1	0.1	0.1	0.1	0	1	1.4	1.2	1	0.3	0	0.1	0	0.1	0.1
	Stop3	54.5	67.4	75	81.9	87.5	88.9	92.8	95.2	96.3	95.8	54.6	65.9	74	82.7	87	90	92	93.5	95.7	96.1
	Stop4	0	0	0	0	0	0	0	0	0	0.4	0	0	0	0	0	0	0	0	0	0
100	Stop1	36.1	24.3	17.8	12.4	7.5	6.2	3	2.9	2.3	1.1	35.5	25.9	15.3	10.6	6.9	5.1	3.1	2	1.8	1.7
	Stop2	0.8	1.6	0.8	0.6	0.2	0.1	0	0	0.1	0	1	1.1	0.3	0.3	0.2	0	0.1	0	0	0
	Stop3	63.1	74.1	81.4	87	92.3	93.7	97	97.1	97.6	98.9	62.9	73	84.4	89.1	92.9	94.9	96.8	98	98.2	98.3
	Stop4	0	0	0	0	0	0	0	0	0	0.6	0	0	0	0	0	0	0	0	0	0

Table A13. Computational results for randomly generated instances with the ratio 10%:40%:10%:40% of the numbers of jobs in the subsets $\mathcal{J}_1$, $\mathcal{J}_2$, $\mathcal{J}_{1,2}$ and $\mathcal{J}_{2,1}$ of the job set $\mathcal{J}$.

		Uniform Distributions										Gamma Distributions									
$\delta\%$	n	10	20	30	40	50	60	70	80	90	100	10	20	30	40	50	60	70	80	90	100
5	Stop1	98.9	100	100	100	100	100	100	100	100	100	99.6	99.9	100	100	100	100	100	100	100	100
	Stop2	0	0	0	0	0	0	0	0	0	0	0	0	0	0	0	0	0	0	0	0
	Stop3	1.1	0	0	0	0	0	0	0	0	0	0.4	0.1	0	0	0	0	0	0	0	0
	Stop4	0	0	0	0	0	0	0	0	0	0	0	0	0	0	0	0	0	0	0	0
10	Stop1	97.6	99.6	99.9	99.9	100	100	100	100	100	100	98.2	99.5	99.9	99.9	100	100	100	99.9	100	100
	Stop2	0	0	0.1	0.1	0	0	0	0	0	0	0	0	0.1	0.1	0	0	0	0.1	0	0
	Stop3	2.4	0.4	0	0	0	0	0	0	0	0	1.8	0.5	0	0	0	0	0	0	0	0
	Stop4	0	0	0	0	0	0	0	0	0	0	0	0	0	0	0	0	0	0	0	0
20	Stop1	95.8	98.6	99	99.7	99.7	99.9	99.7	100	100	99.9	95.5	97.6	99.7	99.3	99.5	99.8	100	99.8	99.9	100
	Stop2	0	0.4	0.4	0.1	0.3	0.1	0.2	0	0	0.1	0.1	0.4	0	0.5	0.2	0.2	0	0.1	0.1	0
	Stop3	4.2	1	0.6	0.2	0	0	0.1	0	0	0	4.4	2	0.3	0.2	0.3	0	0	0.1	0	0
	Stop4	0	0	0	0	0	0	0	0	0	0	0	0	0	0	0	0	0	0	0	0
30	Stop1	93.2	94.8	97.1	97.8	98.1	98.9	99	99	98.5	99.3	93.4	94.9	96.8	98.3	98.5	97.4	98.8	99.4	99.1	99
	Stop2	0	0.8	0.5	0.9	0.2	0.1	0	0.4	0.5	0.1	0.5	1	0.7	0.3	0.4	0.9	0.2	0.1	0.5	0
	Stop3	6.8	4.4	2.4	1.3	1.7	1	1	0.6	1	0.6	6.1	4.1	2.5	1.4	1.1	1.7	1	0.5	0.4	1
	Stop4	0	0	0	0	0	0	0	0	0	0	0	0	0	0	0	0	0	0	0	0
40	Stop1	88.6	91.1	91	91.9	94.5	91.2	91.7	92	93.3	93	87.7	90.5	91.6	92	92.5	93	93.1	92.6	92	91.5
	Stop2	0.3	1.4	1.1	0.7	0.2	0.8	0.9	0.7	0.9	0.5	0.2	1.7	1	1.3	0.8	0.4	0.9	0.6	0.4	0.7
	Stop3	11.1	7.5	7.9	7.4	5.3	8	7.4	7.3	5.8	6.5	12.1	7.8	7.4	6.7	6.7	6.6	6	6.8	7.6	7.8
	Stop4	0	0	0	0	0	0	0	0	0	0	0	0	0	0	0	0	0	0	0	0
50	Stop1	84.6	86.9	84.6	84.8	81	82	79.3	79.7	77.8	77.8	85	84.9	82	81.6	82	83.1	77.2	79.7	78.5	78.1
	Stop2	0.6	0.9	1.2	1.6	1.6	1.1	1	0.8	1.1	0.9	0	2.2	1.8	2	1	1.2	1.3	0.9	1.1	1
	Stop3	14.8	12.2	14.2	13.6	17.4	16.9	19.7	19.5	21.1	21.3	15	12.9	16.2	16.4	17	15.7	21.5	19.4	20.4	20.9
	Stop4	0	0	0	0	0	0	0	0	0	0	0	0	0	0	0	0	0	0	0	0
60	Stop1	79.7	77.8	75.4	71.5	71.1	69.2	63.5	61.3	59	58.8	77.5	75.1	75.4	71.8	70.7	66.4	66.3	63.5	60.1	60.3
	Stop2	0.2	1.9	2	1.9	1.9	1.2	0.8	1.1	0.9	0.6	0.8	1.7	2.1	1.5	1.4	1.8	0.5	0.8	1.7	1
	Stop3	20.1	20.3	22.6	26.6	27	29.6	35.7	37.6	40.1	40.6	21.7	23.2	22.5	26.7	27.9	31.8	33.2	35.7	38.2	38.7
	Stop4	0	0	0	0	0	0	0	0	0	0	0	0	0	0	0	0	0	0	0	0
70	Stop1	73.5	68.5	65.8	65.9	57.7	54.2	49.6	47.9	44.8	45.3	75.4	71.4	65.7	62.8	59.4	55.9	50.7	49.5	44.6	43.2
	Stop2	0.6	1.7	1.6	1.4	1.5	0.7	0.9	0.7	0.7	0.8	0.2	1.5	1.2	1.2	1.4	0.7	1	0.7	1	0.7
	Stop3	25.9	29.8	32.6	32.7	40.8	45.1	49.5	51.4	54.5	53.9	24.3	27.1	33.1	36	39.2	43.4	48.3	49.8	54.4	56.1
	Stop4	0	0	0	0	0	0	0	0	0.1	0	0	0	0	0	0	0	0	0	0	0
80	Stop1	67.6	60.5	58.7	52.8	46.1	43.5	40.1	36.7	35.4	34.2	66.6	62.4	53.4	50.8	49	43	41.4	37.8	38.7	33.7
	Stop2	0.3	2.7	1.2	1.1	0.9	0.6	0.8	0.4	0.2	0.6	0.4	2.1	2	1.1	1.4	1	0.4	0.3	0.3	0.2
	Stop3	32.1	36.8	40.1	46.1	53	55.9	59.1	62.9	64.4	65.2	32.9	35.5	44.6	48.1	49.6	56	58.2	61.9	61	66.1
	Stop4	0	0	0	0	0	0	0	0	0	0.1	0	0	0	0	0	0	0	0	0	0
90	Stop1	63.3	50.9	49.5	42.8	36.3	36.2	33.7	28.7	28.2	28.9	58.2	54.1	51.9	44.1	37.3	33.8	31.6	30.8	27	27.5
	Stop2	0.8	1.5	1.7	0.9	0.5	0.1	0.3	0.2	0.2	0.3	0.3	1.4	1.2	1	0.7	0.5	0.4	0.1	0.3	0
	Stop3	35.9	47.6	48.8	56.3	63.2	63.7	66	71.1	71.6	70.8	41.5	44.5	46.9	54.9	62	65.7	68	69.1	72.7	72.5
	Stop4	0	0	0	0	0	0	0	0	0	0	0	0	0	0	0	0	0	0	0	0
100	Stop1	58.1	48.4	41.1	32.5	30.9	27.2	24.2	21.2	23.1	20.4	55.5	44.4	38.2	32.2	32.3	27.7	24.1	21.8	22.4	20.3
	Stop2	0.2	1	0.6	0.8	0.3	0.4	0.2	0.4	0	0	0.7	1.5	0.7	0.8	0.6	0.5	0.2	0.1	0.2	0
	Stop3	41.6	50.6	58.3	66.7	68.8	72.4	75.6	78.4	76.9	79.6	43.8	54.1	61.1	67	67.1	71.8	75.7	78.1	77.4	79.7
	Stop4	0.1	0	0	0	0	0	0	0	0	0	0	0	0	0	0	0	0	0	0	0

Table A14. Computational results for randomly generated instances with the ratio 10%:60%:10%:20% of the numbers of jobs in the subsets $\mathcal{J}_1$, $\mathcal{J}_2$, $\mathcal{J}_{1,2}$ and $\mathcal{J}_{2,1}$ of the job set $\mathcal{J}$.

		Uniform Distributions										Gamma Distributions									
$\delta\%$	n	10	20	30	40	50	60	70	80	90	100	10	20	30	40	50	60	70	80	90	100
5	Stop1	100	99.8	100	100	100	100	100	100	100	100	100	99.9	100	100	100	100	100	100	100	100
	Stop2	0	0	0	0	0	0	0	0	0	0	0	0	0	0	0	0	0	0	0	0
	Stop3	0	0.2	0	0	0	0	0	0	0	0	0	0.1	0	0	0	0	0	0	0	0
	Stop4	0	0	0	0	0	0	0	0	0	0	0	0	0	0	0	0	0	0	0	0
10	Stop1	100	99.9	100	100	100	100	100	100	100	100	100	99.8	99.9	100	100	100	100	100	100	100
	Stop2	0	0	0	0	0	0	0	0	0	0	0	0	0	0	0	0	0	0	0	0
	Stop3	0	0.1	0	0	0	0	0	0	0	0	0	0.2	0.1	0	0	0	0	0	0	0
	Stop4	0	0	0	0	0	0	0	0	0	0	0	0	0	0	0	0	0	0	0	0
20	Stop1	100	99.4	99.9	100	100	100	100	100	100	100	100	99.3	99.9	100	100	100	100	100	100	100
	Stop2	0	0	0	0	0	0	0	0	0	0	0	0	0.1	0	0	0	0	0	0	0
	Stop3	0	0.6	0.1	0	0	0	0	0	0	0	0	0.7	0	0	0	0	0	0	0	0
	Stop4	0	0	0	0	0	0	0	0	0	0	0	0	0	0	0	0	0	0	0	0
30	Stop1	99.9	98.7	99.7	99.9	100	100	100	100	100	100	100	98.6	99.5	99.5	99.9	100	100	100	100	100
	Stop2	0	0.3	0	0	0	0	0	0	0	0	0	0.2	0.1	0.3	0	0	0	0	0	0
	Stop3	0.1	1	0.3	0.1	0	0	0	0	0	0	0	1.2	0.4	0.2	0.1	0	0	0	0	0
	Stop4	0	0	0	0	0	0	0	0	0	0	0	0	0	0	0	0	0	0	0	0
40	Stop1	100	97.3	99.5	99.7	99.7	100	99.8	100	99.9	99.9	100	97.2	98.6	99.1	99.8	100	100	100	99.9	100
	Stop2	0	0.3	0.3	0	0	0	0.1	0	0	0	0	0.1	0.5	0.2	0	0	0	0	0	0
	Stop3	0	2.4	0.2	0.3	0.3	0	0.1	0	0.1	0.1	0	2.7	0.9	0.7	0.2	0	0	0	0.1	0
	Stop4	0	0	0	0	0	0	0	0	0	0	0	0	0	0	0	0	0	0	0	0
50	Stop1	99.9	96.1	97	98.3	99.2	99	99.3	99.5	99.8	100	100	95.8	96.9	98.4	99.2	99.2	99.7	99.9	99.6	100
	Stop2	0	0.2	0.5	0.2	0.3	0.5	0.2	0	0.2	0	0	0.2	0.5	0.4	0	0	0.1	0	0	0
	Stop3	0.1	3.7	2.5	1.5	0.5	0.5	0.5	0.5	0	0	0	4	2.6	1.2	0.8	0.8	0.2	0.1	0.4	0
	Stop4	0	0	0	0	0	0	0	0	0	0	0	0	0	0	0	0	0	0	0	0
60	Stop1	99.7	94.6	96.1	96.8	97.8	98.2	97.9	98.3	98.7	99	99.8	94.9	95.3	97.6	97.9	98.3	96.9	98.8	98.4	98.6
	Stop2	0	0.4	1.1	0.1	0.3	0.2	0.7	0	0.1	0.1	0	0.1	0.2	0.3	0.2	0.3	0.6	0.3	0.4	0.2
	Stop3	0.3	5	2.8	3.1	1.9	1.6	1.4	1.7	1.2	0.9	0.2	5	4.5	2.1	1.9	1.4	2.5	0.9	1.2	1.2
	Stop4	0	0	0	0	0	0	0	0	0	0	0	0	0	0	0	0	0	0	0	0
70	Stop1	99.6	94.4	94.2	94.8	95.9	94.6	95.6	95.8	96.5	96.7	99.6	94	94.9	93.4	94.5	96	97	96.2	96	96
	Stop2	0	0.4	0.6	0.5	0.6	0.6	0.3	0.4	0.4	0	0	0.2	0.7	0.5	0.4	0.8	0.2	0.2	0.4	0.3
	Stop3	0.4	5.2	5.2	4.7	3.5	4.8	4.1	3.8	3.1	3.3	0.4	5.8	4.4	6.1	5.1	3.2	2.8	3.6	3.6	3.7
	Stop4	0	0	0	0	0	0	0	0	0	0	0	0	0	0	0	0	0	0	0	0
80	Stop1	99.8	90.9	92	91.2	91.2	92	92	92.3	91.9	92	99.1	89.5	92	91.6	90.8	92.6	92.4	93.4	92.6	92.1
	Stop2	0	0.2	0.6	0.8	0.7	0.2	0.5	0.6	0.4	0.3	0	0.4	0.6	0.7	1.1	0.6	0.5	0.2	0.4	0.1
	Stop3	0.2	8.9	7.4	8	8.1	7.8	7.5	7.1	7.7	7.7	0.9	10.1	7.4	7.7	8.1	6.8	7.1	6.4	7	7.8
	Stop4	0	0	0	0	0	0	0	0	0	0	0	0	0	0	0	0	0	0	0	0
90	Stop1	98.9	88.3	89.5	88.3	87.6	89	86.1	87.8	86.9	86.2	98.6	89.9	87.3	88.8	88.5	88.7	87.8	85.1	88.5	87.2
	Stop2	0	0.2	1	0.9	1.2	0.5	0.8	0.6	0.2	0.3	0	0.1	0.5	0.3	0.4	0.5	0.7	0.4	0.6	0.9
	Stop3	1.1	11.5	9.5	10.8	11.2	10.5	13.1	11.6	12.9	13.5	1.4	10	12.2	10.9	11.1	10.8	11.5	14.5	10.9	11.9
	Stop4	0	0	0	0	0	0	0	0	0	0	0	0	0	0	0	0	0	0	0	0
100	Stop1	97.8	88.4	86.6	83.1	85.7	83.4	82.8	81.3	82.1	82.4	97.6	90.2	84.4	85	84.2	81.7	80.2	82.2	80.7	83.2
	Stop2	0	0.4	1.1	1	1	0.6	0.7	0.5	0.8	0.2	0	0.2	0.7	0.9	1.2	0.6	0.6	0.5	0.4	0.5
	Stop3	2.2	11.2	12.3	15.9	13.3	16	16.5	18.2	17.1	17.4	2.4	9.6	14.9	14.1	14.6	17.7	19.2	17.3	18.9	16.3
	Stop4	0	0	0	0	0	0	0	0	0	0	0	0	0	0	0	0	0	0	0	0

References

1. Lai, T.-C.; Sotskov, Yu.N. Sequencing with uncertain numerical data for makespan minimization. *J. Oper. Res. Soc.* **1999**, *50*, 230–243. [CrossRef]
2. Lai, T.-C.; Sotskov, Yu.N.; Sotskova, N.Y.; Werner, F. Mean flow time minimization with given bounds of processing times. *Eur. J. Oper. Res.* **2004**, *159*, 558–573. [CrossRef]
3. Sotskov, Yu.N.; Egorova, N.M.; Lai, T.-C. Minimizing total weighted flow time of a set of jobs with interval processing times. *Math. Comput. Model.* **2009**, *50*, 556–573. [CrossRef]
4. Lai, T.-C.; Sotskov, Yu.N.; Sotskova, N.Y.; Werner, F. Optimal makespan scheduling with given bounds of processing times. *Math. Comput. Model.* **1997**, *26*, 67–86.
5. Cheng, T.C.E.; Shakhlevich, N.V. Proportionate flow shop with controllable processing times. *J. Sched.* **1999**, *27*, 253–265. [CrossRef]
6. Cheng, T.C.E.; Kovalyov, M.Y.; Shakhlevich, N.V. Scheduling with controllable release dates and processing times: Makespan minimization. *Eur. J. Oper. Res.* **2006**, *175*, 751–768. [CrossRef]
7. Jansen, K.; Mastrolilli, M.; Solis-Oba, R. Approximation schemes for job shop scheduling problems with controllable processing times. *Eur. J. Oper. Res.* **2005**, *167*, 297–319. [CrossRef]
8. Sotskov, Y.N.; Matsveichuk, N.M.; Hatsura V.D. Two-machine job-shop scheduling problem to minimize the makespan with uncertain job durations. *Algorithms* **2020**, *13*, 4. [CrossRef]
9. Graham, R.L.; Lawler, E.L.; Lenstra, J.K.; Rinnooy Kan, A.H.G. Optimization and approximation in deterministic sequencing and scheduling. *Ann. Discret. Appl. Math.* **1979**, *5*, 287–326.
10. Pinedo, M. *Scheduling: Theory, Algorithms, and Systems*; Prentice-Hall: Englewood Cliffs, NJ, USA, 2002.
11. Sotskov, Yu.N.; Werner, F. *Sequencing and Scheduling with Inaccurate Data*; Nova Science Publishers: Hauppauge, NY, USA, 2014.
12. Tanaev, V.S.; Sotskov, Yu.N.; Strusevich, V.A. *Scheduling Theory: Multi-Stage Systems*; Kluwer Academic Publishers: Dordrecht, The Netherlands, 1994.
13. Jackson, J.R. An extension of Johnson's results on job lot scheduling. *Nav. Res. Logist. Q.* **1956**, *3*, 201–203. [CrossRef]
14. Johnson, S.M. Optimal two and three stage production schedules with set up times included. *Nav. Res. Logist. Q.* **1954**, *1*, 61–68. [CrossRef]
15. Gonzalez-Neira, E.M.; Montoya-Torres, J.R.; Barrera, D. Flow shop scheduling problem under uncertainties: Review and trends. *Int. J. Ind. Engin. Comput.* **2017**, *8*, 399–426. [CrossRef]
16. Elmaghraby, S; Thoney, K.A. Two-machine flowshop problem with arbitrary processing time distributions. *IIE Trans.* **2000**, *31*, 467–477.
17. Kamburowski, J. Stochastically minimizing the makespan in two-machine flow shops without blocking. *Eur. J. Oper. Res.* **1999**, *112*, 304–309. [CrossRef]
18. Ku, P.S.; Niu, S.C. On Johnson's two-machine flow-shop with random processing times. *Oper. Res.* **1986**, *34*, 130–136. [CrossRef]
19. Allahverdi, A. Stochastically minimizing total flowtime in flowshops with no waiting space. *Eur. J. Oper. Res.* **1999**, *113*, 101–112. [CrossRef]
20. Allahverdi, A.; Mittenthal, J. Two-machine ordered flowshop scheduling under random breakdowns. *Math. Comput. Model.* **1999**, *20*, 9–17. [CrossRef]
21. Portougal, V.; Trietsch, D. Johnson's problem with stochastic processing times and optimal service level. *Eur. J. Oper. Res.* **2006**, *169*, 751–760. [CrossRef]
22. Daniels, R.L.; P. Kouvelis. Robust scheduling to hedge against processing time uncertainty in single stage production. *Manag. Sci.* **1995**, *41*, 363–376. [CrossRef]
23. Sabuncuoglu, I.; Goren, S. Hedging production schedules against uncertainty in manufacturing environment with a review of robustness and stability research. *Int. J. Comput. Integr. Manuf.* **2009**, *22*, 138–157. [CrossRef]
24. Subramaniam, V.; Raheja, A.S.; Rama Bhupal Reddy, K. Reactive repair tool for job shop schedules. *Int. J. Prod. Res.* **2005**, *1*, 1–23. [CrossRef]
25. Gur, S.; Eren, T. Scheduling and planning in service systems with goal programming: Literature review. *Mathematics* **2018**, *6*, 265. [CrossRef]
26. Pereira, J. The robust (minmax regret) single machine scheduling with interval processing times and total weighted completion time objective. *Comput. Oper. Res.* **2016**, *66*, 141–152. [CrossRef]

27. Kasperski, A.; P. Zielinski. A 2-approximation algorithm for interval data minmax regret sequencing problems with total flow time criterion. *Oper. Res. Lett.* **2008**, *36*, 343–344. [CrossRef]
28. Wu, Z.; Yu, S; Li, T. A meta-model-based multi-objective evolutionary approach to robust job shop scheduling. *Mathematics* **2019**, *7*, 529. [CrossRef]
29. Kuroda, M.; Wang, Z. Fuzzy job shop scheduling. *Int. J. Prod. Econ.* **1996**, *44*, 45–51. [CrossRef]
30. Grabot, B.; Geneste, L. Dispatching rules in scheduling: A fuzzy approach. *Int. J. Prod. Res.* **1994**, *32*, 903–915. [CrossRef]
31. Özelkan, E.C.; Duckstein, L. Optimal fuzzy counterparts of scheduling rules. *Eur. J. Oper. Res.* **1999**, *113*, 593–609. [CrossRef]
32. Navakkoli-Moghaddam, R.; Safaei, N.; Kah, MMO. Accessing feasible space in a generalized job shop scheduling problem with the fuzzy processing times: A fuzzy-neural approach. *J. Oper. Res. Soc.* **2008**, *59*, 431–442. [CrossRef]
33. Al-Atroshi, A.M.; Azez, S.T.; Bhnam, B.S. An effective genetic algorithm for job shop scheduling with fuzzy degree of satisfaction. *Int. J. Comput. Sci. Issues* **2013**, *10*, 180–185.
34. Kasperski, A.; Zielinski, P. Possibilistic minmax regret sequencing problems with fuzzy parameteres. *IEEE Trans. Fuzzy Syst.* **2011**, *19*, 1072–1082. [CrossRef]
35. Gonzalez-Rodriguez, I.; Vela, C.R.; Puente, J.; Varela, R. A new local search for the job shop problem with uncertain durations. In Proceedings of the Eighteenth International Conference on Automated Planning and Scheduling (ICAPS 2008), Sydney, Australia, 14–18 September 2008; Association for the Advancement of Artificial Intelligence: Menlo Park, CA, USA, 2008; pp. 124–131.
36. Allahverdi, A.; Sotskov, Yu.N. Two-machine flowshop minimum-lenght scheduling problem with random and bounded processing times. *Int. Trans. Oper. Res.* **2003**, *10*, 65–76. [CrossRef]
37. Matsveichuk, N.M.; Sotskov, Yu.N. A stability approach to two-stage scheduling problems with uncertain processing times. In *Sequencing and Scheduling with Inaccurate Data*; Yu, N., Sotskov, Werner, F., Eds.; Nova Science Publishers: Hauppauge, NY, USA, 2014; pp. 377–407.
38. Lai, T.-C.; Sotskov, Yu.N.; Egorova, N.G.; Werner, F. The optimality box in uncertain data for minimising the sum of the weighted job completion times. *Int. J. Prod. Res.* **2018**, *56*, 6336–6362. [CrossRef]
39. Sotskov, Yu.N.; Egorova, N.M. Single machine scheduling problem with interval processing times and total completion time objective. *Algorithms* **2018**, *75*, 66. [CrossRef]
40. Sotskov, Yu.N.; Lai, T.-C. Minimizing total weighted flow time under uncertainty using dominance and a stability box. *Comput. Oper. Res.* **2012**, *39*, 1271–1289. [CrossRef]
41. Matsveichuk, N.M.; Sotskov, Y.N.; Egorova, N.G.; Lai T.-C. Schedule execution for two-machine flow-shop with interval processing times. *Math. Comput. Model.* **2009**, *49*, 991–1011. [CrossRef]
42. Sotskov, Yu.N.; Allahverdi, A.; Lai, T.-C. Flowshop scheduling problem to minimize total completion time with random and bounded processing times. *J. Oper. Res. Soc.* **2004**, *55*, 277–286. [CrossRef]
43. Allahverdi, A.; Aldowaisan, T.; Sotskov, Yu.N. Two-machine flowshop scheduling problem to minimize makespan or total completion time with random and bounded setup times. *Int. J. Math. Math. Sci.* **2003**, *39*, 2475–2486. [CrossRef]
44. Kouvelis, P.; Yu, G. *Robust Discrete Optimization and Its Application*; Kluwer Academic Publishers: Boston, MA, USA, 1997.
45. Kouvelis, P.; Daniels, R.L.; Vairaktarakis, G. Robust scheduling of a two-machine flow shop with uncertain processing times. *IEEE Trans.* **2000**, *32*, 421–432. [CrossRef]
46. Kuwata, Y.; Morikawa, K.; Takahashi, K.; Nakamura, N. Robustness optimisation of the minimum makespan schedules in a job shop. *Int. J. Manuf. Technol. Manag.* **2003**, *5*, 1–9. [CrossRef]
47. Carlier, J; Pinson, E. An algorithm for solving the job-shop problem. *Manag. Sci.* **1989**, *35*, 164–176. [CrossRef]
48. Xiong, J.; Xing, L.; Chen, Y.-W. Robust scheduling for multi-objective flexible job-shop problems with random machine breakdowns. *Int. J. Prod. Econ.* **2013**, *141*, 112–126. [CrossRef]
49. Paprocka, I. The model of maintenance planning and production scheduling for maximising robustness. *Int. J. Prod. Res.* **2019**, *57*, 1–22. [CrossRef]
50. Xie, C.; Allen, T.T. Simulation and experimental design methods for job shop scheduling with material handling: A survey. *Int. J. Adv. Manuf. Technol.* **2015**, *80*, 233–243. [CrossRef]
51. Paprocka, I. Evaluation of the effects of a machine failure on the robustness of a job shop system—Proactive approaches. *Sustainability* **2019**, *11*, 65. [CrossRef]

52. Cigolini, R.; Perona, M.; Portioli, A. Comparison of order and release techniques in a dynamic and uncertain job shop environment. *Int. J. Prod. Res.* **1998**, *36*, 2931–2951. [CrossRef]
53. Luh, P.B.; Chen, D.; Thakur, L.S. An effective approach for job-shop scheduling with uncertain processing requirements. *IEEE Trans. Robot. Autom.* **1999**, *15*, 328–339. [CrossRef]
54. Ng, C.T.; Matsveichuk, N.M.; Sotskov, Y.N.; Cheng, T.C.E. Two-machine flow-shop minimum-length scheduling with interval processing times. *Asia-Pac. J. Oper. Res.* **2009**, *26*, 1–20. [CrossRef]
55. Matsveichuk, N.M.; Sotskov, Y.N.; Werner, F. The dominance digraph as a solution to the two-machine flow-shop problem with interval processing times. *Optimization* **2011**, *60*, 1493–1517. [CrossRef]

© 2020 by the authors. Licensee MDPI, Basel, Switzerland. This article is an open access article distributed under the terms and conditions of the Creative Commons Attribution (CC BY) license (http://creativecommons.org/licenses/by/4.0/).

Article

On the Degree-Based Topological Indices of Some Derived Networks

Haidar Ali [1], Muhammad Ahsan Binyamin [1], Muhammad Kashif Shafiq [1] and Wei Gao [2,*]

[1] Department of Mathematics, Government College University, Faisalabad 38023, Pakistan
[2] School of Information Science and Technology, Yunnan Normal University, Kunming 650500, China
* Correspondence: gaowei@ynnu.edu.cn

Received: 23 May 2019; Accepted: 22 June 2019; Published: 10 July 2019

Abstract: There are numeric numbers that define chemical descriptors that represent the entire structure of a graph, which contain a basic chemical structure. Of these, the main factors of topological indices are such that they are related to different physical chemical properties of primary chemical compounds. The biological activity of chemical compounds can be constructed by the help of topological indices. In theoretical chemistry, numerous chemical indices have been invented, such as the Zagreb index, the Randić index, the Wiener index, and many more. Hex-derived networks have an assortment of valuable applications in drug store, hardware, and systems administration. In this analysis, we compute the Forgotten index and Balaban index, and reclassified the Zagreb indices, ABC_4 index, and GA_5 index for the third type of hex-derived networks theoretically.

Keywords: forgotten index; balaban index; reclassified the zagreb indices; ABC_4 index; GA_5 index; $HDN_3(m)$; $THDN_3(m)$; $RHDN_3(m)$

1. Introduction

Topological indices are very useful tools for chemists which are provided by Graph Theory. In a molecular graph, vertices denotes the atoms and edges are represented as chemical bonds in the terms of graph theory. To predict bioactivity of the chemical compounds, the topological indices such as ABC index, Wiener index, Randić index, Szeged index and Zagreb indices are very useful.

A graph ζ is a tuple, which consists of the n-connected vertex set $|V(\zeta)|$ and the edge set $|E(\zeta)|$. $\tau(m)$ denotes the degree of a vertex 'm' in a graph ζ. A graph can be represented by the polynomials, numeric numbers, a sequence of numbers, or a matrix. Throughout this article, all graphs examined are simple, finite, and connected.

As a chemical descriptor, the topological index has an integer attached to the graph which features the graph, and there is no change under graph automorphism. Previously, interest in the computing chemistry domain has grown in terms of topological descriptors and is mainly associated with the use of unusual quantities, the relationship between the structure property, and the relationship of the structure quantity. The topological indices that are based on distance, degree, and polynomials are some of the main classes of these indices. In a number of these segments, degree-based displayers are widely important and chemical graphs play an integral part in theory and theoretical chemistry.

In this article, we consider some important topological indices and some important derived graphs. We examine their chemical behavior by the help of topological indices. These topological indices are of use to chemists.

Chen et al. [1] gleaned a hexagonal mesh which consists of triangles. Triangle graphs are called oxide graphs in terms of chemistry. We can construct a *hexagonal mesh* by joining these triangles, as shown in Figure 1. There does not exist any hexagonal mesh whose dimension equals 1. By the joining of six triangles, we make a hexagonal mesh of dimension 2, $HX(2)$ (see Figure 1 (1)). By putting the triangles around the all sides of $HX(2)$, we obtain hexagonal mesh of dimension 3, $HX(3)$ (see Figure 1 (2)). Furthermore, we assemble the nth hexagonal mesh by putting n triangles around the boundary of each hexagon.

Drawing Algorithm of Third Type of Hex-Derived Networks HDN_3

Step-1: For HDN_3, we should draw a hexagonal mesh of dimension m.

Step-2: Draw a K_3 graph in each subgraph of K_3 and join all the vertices to the outer vertices of each K_3. The new graph is called an $HDN3$ (see Figure 2) network.

Step-3: By HDN_3 network, we can simply design $THDN_3$ (see Figure 3) and $RHDN_3$ (see Figure 4).

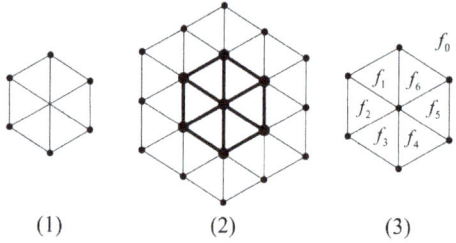

Figure 1. Hexagonal meshes: (1) HX$_2$, (2) HX$_3$, and (3), all facing HX$_2$.

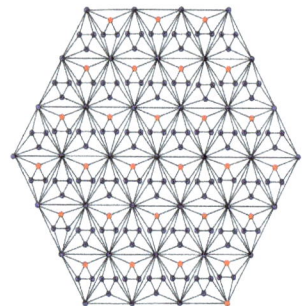

Figure 2. Third type of hex-derived network ($HDN_3(4)$).

In this paper, 'ζ' is taken as a simple connected graph and the degree of any vertex $\acute{m} \in V(\zeta)$ is stands for $\tau(\acute{m})$.

The oldest, most desired and supremely studied degree-based topological index was introduced by Milan Randić and is known as *Randić index* [2] denoted by $R_{-\frac{1}{2}}(\zeta)$ and described as

$$R_{-\frac{1}{2}}(\zeta) = \sum_{\acute{m}\acute{n}\in E(\zeta)} \frac{1}{\sqrt{\tau(\acute{m})\tau(\acute{n})}}. \tag{1}$$

The *Forgotten index*, also called F-index, was discovered by Furtula and Ivan Gutman [3] and described as

$$F(\zeta) = \sum_{\acute{m}\acute{n}\in E(\zeta)} ((\tau(\acute{m}))^2 + (\tau(\acute{n}))^2). \tag{2}$$

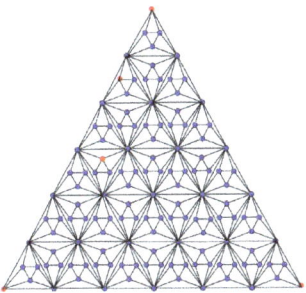

Figure 3. Third type of triangular hex-derived network ($THDN_3(7)$).

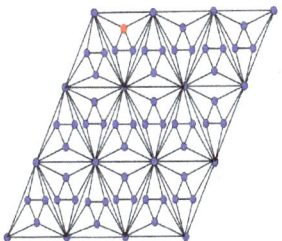

Figure 4. Third type of rectangular hex-derived network ($RHDN_3(4,4)$).

In 1982, Balaban [4,5] found another important index known as *Balaban index*. For a graph ξ of 'n' vertices and 'm' edges, and is described as

$$J(\xi) = \left(\frac{m}{m-n+2}\right) \sum_{\acute{m}\acute{n} \in E(\xi)} \frac{1}{\sqrt{\tau(\acute{m}) \times \tau(\acute{n})}}. \tag{3}$$

The reclassified the Zagreb indices which are proposed by Ranjini et al. [6], is of three types. For a graph ξ, it is described as

$$ReZG_1(\xi) = \sum_{\acute{m}\acute{n} \in E(\xi)} \left(\frac{\tau(\acute{m}) \times \tau(\acute{n})}{\tau(\acute{m}) + \tau(\acute{n})}\right), \tag{4}$$

$$ReZG_2(\xi) = \sum_{\acute{m}\acute{n} \in E(\xi)} \left(\frac{\tau(\acute{m}) + \tau(\acute{n})}{\tau(\acute{m}) \times \tau(\acute{n})}\right), \tag{5}$$

$$ReZG_3(\xi) = \sum_{\acute{m}\acute{n} \in E(\xi)} (\tau(\acute{m}) \times \tau(\acute{n}))(\tau(\acute{m}) + \tau(\acute{n})). \tag{6}$$

The atom-bond connectivity (ABC) index is a useful predictive index in the study of the heat of formation in alkanes [7] and is introduced by Estrada et al. [8].

Ghorbani et al. [9] introduced the ABC_4 index and is described as

$$ABC_4(\xi) = \sum_{\acute{m}\acute{n} \in E(\xi)} \sqrt{\frac{S_{\acute{m}} + S_{\acute{n}} - 2}{S_{\acute{m}} S_{\acute{n}}}}. \tag{7}$$

Graovac et al. [10] introduced the GA_5 index and is described as

$$GA_5(\xi) = \sum_{\acute{m}\acute{n} \in E(\xi)} \frac{2\sqrt{S_{\acute{m}} S_{\acute{n}}}}{(S_{\acute{m}} + S_{\acute{n}})}. \tag{8}$$

2. Main Results

Simonraj et al. [11] created the new network which is named as third type of hex-derived networks. Chang-Cheng Wei et al. [12] found some topological indices of certain new derived networks. In this paper, we compute the exact results for all the above descriptors. For these results on different degree-based topological descriptors for a variety of graphs, we recommend [13–20]. For the basic notations and definitions, see [21,22].

2.1. Results for $HDN_3(m)$

In this part, the Forgotten index, Balaban index, reclassified the Zagreb indices, ABC_4 index, and GA_5 index are under consideration for the third type of hex-derived network.

Theorem 1. *Consider the third type of hex-derived network $HDN_3(m)$; its Forgotten index is equal to*

$$F(HDN_3(m)) = 6(5339 - 8132n + 3108n^2).$$

Proof. Let ξ_1 be the hex-derived network of Type 3, $HDN_3(m)$ shown in Figure 2, where $m \geq 4$. The hex derived network ξ_1 has $21m^2 - 39m + 19$ vertices and the edge set of ξ_1 is divided into nine partitions based on the degrees of end vertices as shown in Table 1.

Forgotten index can be calculated by using Table 1. Thus, from (2), it follows,

$$\begin{aligned}F(\xi_1) &= 32|E_1(\xi_1)| + 65|E_2(\xi_1)| + 116|E_3(\xi_1)| + 340|E_4(\xi_1)| + 149|E_5(\xi_1)| + 373|E_6(\xi_1)| + \\ & \quad 200|E_7(\xi_1)| + 424|E_8(\xi_1)| + 648|E_9(\xi_1)|.\end{aligned}$$

After some calculations, we have the final result

$$\implies F(\xi_1) = 6(5339 - 8132n + 3108n^2).$$

□

Table 1. Edge partition of third type of hex-derived network $HDN_3(m)$, based on degrees of end vertices of each edge.

$(\tau_{\acute{m}}, \tau_{\acute{n}})$ Where $\acute{m}\acute{n} \in E(\xi_1)$	Number of Edges	$(\tau_{\acute{m}}, \tau_{\acute{n}})$ Where $\acute{m}\acute{n} \in E(\xi_1)$	Number of Edges
(4,4)	$18m^2 - 36m + 18$	(7,18)	6
(4,7)	24	(10,10)	$6m - 18$
(4,10)	$36m - 72$	(10,18)	$12m - 24$
(4,18)	$36m^2 - 108m + 84$	(18,18)	$9m^2 - 33m + 30$
(7,10)	12	-	-

In the subsequent theorem, we compute the Balaban index of the third type of hex-derived network, ξ_1.

Theorem 2. *For the third type of hex-derived network ξ_1, the Balaban index is equal to*

$$\begin{aligned}J(\xi_1) &= \left(\frac{1}{70(43 - 84m + 42m^2)}\right)((20 - 41m + 21m^2)(1595.47 + 7(-307 - 270\sqrt{2} + 12\sqrt{5} + \\ & \quad 54\sqrt{10})m) + 210(5 + 3\sqrt{2})m^2)\end{aligned}$$

Proof. Let ξ_1 be the third type of hex-derived network $HDN_3(m)$. The Balaban index can be calculated by using (3) and with the help of Table 1, we have.

$$J(\xi_1) = \left(\frac{63n^2 - 123n + 60}{43 - 84n + 42n^2}\right)\left(\frac{1}{4}|E_1(\xi_1)| + \frac{1}{2\sqrt{7}}|E_2(\xi_1)| + \frac{1}{2\sqrt{10}}|E_3(\xi_1)| + \frac{1}{6\sqrt{2}}|E_4(\xi_1)| + \frac{1}{\sqrt{70}}|E_5(\xi_1)| + \frac{1}{3\sqrt{14}}|E_6(\xi_1)| + \frac{1}{10}|E_7(\xi_1)| + \frac{1}{6\sqrt{5}}|E_8(\xi_1)| + \frac{1}{18}|E_9(\xi_1)|\right).$$

After some calculations, we have the result

$$\Longrightarrow J(\xi_1) = \left(\frac{1}{70(43 - 84m + 42m^2)}\right)((20 - 41m + 21m^2)(1595.47 + 7(-307 - 270\sqrt{2} + 12\sqrt{5} + 54\sqrt{10})m) + 210(5 + 3\sqrt{2})m^2).$$

□

Now, we compute $ReZG_1$, $ReZG_2$ and $ReZG_3$ indices of the third type of hex-derived network ξ_1.

Theorem 3. *Let ξ_1 be the third type of hex-derived network, then*

- $ReZG_1(\xi_1) = 19 - 39m + 21m^2$,
- $ReZG_2(\xi_1) = \frac{115452}{425} - \frac{5637m}{11} + \frac{2583m^2}{11}$,
- $ReZG_3(\xi_1) = 12(27381 - 38996m + 13692m^2)$.

Proof. Reclassified Zagreb index can be calculated by using Table 1, the $ReZG_1(\xi_1)$ by using Equation (4) as follows.

$$ReZG_1(\xi_1) = 2|E_1(\xi_1)| + \frac{28}{11}|E_2(\xi_1)| + \frac{20}{7}|E_3(\xi_1)| + \frac{36}{11}|E_4(\xi_1)| + \frac{70}{17}|E_5(\xi_1)| + \frac{126}{25}|E_6(\xi_1)| + 5|E_7(\xi_1)| + \frac{45}{7}|E_8(\xi_1)| + 9|E_9(\xi_1)|.$$

After some calculations, we have

$$\Longrightarrow ReZG_1(\xi_1) = 19 - 39m + 21m^2.$$

The $ReZG_2(\xi_1)$ can be calculated by using (5) as follows.

$$ReZG_2(\xi_1) = \frac{1}{2}|E_1(\xi_1)| + \frac{11}{28}|E_2(\xi_1)| + \frac{7}{20}|E_3(\xi_1)| + \frac{11}{36}|E_4(\xi_1)| + \frac{17}{70}|E_5(\xi_1)| + \frac{25}{126}|E_6(\xi_1)| + \frac{1}{5}|E_7(\xi_1)| + \frac{7}{45}|E_8(\xi_1)| + \frac{1}{9}|E_9(\xi_1)|.$$

After some calculations, we have

$$\Longrightarrow ReZG_2(\xi_1) = \frac{115452}{425} - \frac{5637m}{11} + \frac{2583m^2}{11}.$$

The $ReZG_3(\xi_1)$ index can be calculated from (6) as follows.

$$ReZG_3(\xi_1) = \sum_{\acute{m}\acute{n} \in E(\xi_1)} (\tau(\acute{m}) \times \tau(\acute{n}))(\tau(\acute{m}) + \tau(\acute{n})) = \sum_{\acute{m}\acute{n} \in E_j(\xi_1)} \sum_{j=1}^{9} (\tau(\acute{m}) \times \tau(\acute{n}))(\tau(\acute{m}) + \tau(\acute{n}))$$

$$ReZG_3(\xi_1) = 128|E_1(\xi_1)| + 308|E_2(\xi_1)| + 560|E_3(\xi_1)| + 1584|E_4(\xi_1)| + 1190|E_5(\xi_1)| + 3150|E_6(\xi_1)| + 2000|E_7(\xi_1)| + 5040|E_8(\xi_1)| + 11664|E_9(\xi_1)|.$$

After some calculations, we have

$$\Longrightarrow ReZG_3(\xi_1) = 12(27381 - 38996m + 13692m^2).$$

□

Now, we find ABC_4 and GA_5 indices of third type of hex-derived network ξ_1.

Theorem 4. *Let ξ_1 be the third type of hex-derived network, then*

- $ABC_4(\xi_1) = 51.706 + \frac{3}{20}\sqrt{\frac{79}{2}}(-5+m) + 3\sqrt{\frac{53}{70}}(-4+m) + \frac{3}{5}\sqrt{\frac{109}{14}}(-4+m) + \sqrt{\frac{114}{5}}(-4+m) + \frac{3}{35}\sqrt{\frac{139}{2}}(-4+m) + 3\sqrt{\frac{14}{65}}(-3+m) + 12\sqrt{\frac{26}{55}}(-3+m) + 2\sqrt{\frac{174}{35}}(-3+m) + \sqrt{\frac{62}{7}}(-3+m) + \sqrt{\frac{78}{11}}(-2+m) + \frac{9}{11}\sqrt{\frac{43}{2}}(-2+m)^2 + \frac{1}{3}\sqrt{\frac{35}{2}}(-5+2m) + \frac{1}{26}\sqrt{\frac{155}{2}}(24-17m+3m^2) + 3\sqrt{\frac{6}{13}}(19-15m+3m^2);$
- $GA_5(\xi_1) = 315.338 + \frac{288}{29}\sqrt{5}(-4+m) + \frac{48}{11}\sqrt{7}(-4+m) + \frac{16}{9}\sqrt{35}(-4+m) + \frac{9}{2}\sqrt{7}(-3+m) + \frac{36}{11}\sqrt{35}(-3+m) + \frac{48}{23}\sqrt{385}(-3+m) + \frac{12}{37}\sqrt{1365}(-3+m) + \frac{18}{5}\sqrt{11}(-2+m) - 99m + 27m^2 + \frac{12}{25}\sqrt{429}(19-15m+3m^2).$

Proof. The $ABC_4(\xi_1)$ index can be calculated by using (7) and by Table 2, as follows.

$$\begin{aligned}
ABC_4(\xi_1) &= \frac{2}{5}\sqrt{\frac{14}{33}}|E_{10}(\xi_1)| + \frac{\sqrt{59}}{30}|E_{11}(\xi_1)| + \frac{1}{15}\sqrt{\frac{77}{6}}|E_{12}(\xi_1)| + \frac{36}{11}\frac{2}{\sqrt{77}}|E_{13}(\xi_1)| + \\
&\quad \frac{1}{6}\sqrt{\frac{31}{14}}|E_{14}(\xi_1)| + \frac{1}{14}\sqrt{\frac{103}{11}}|E_{15}(\xi_1)| + \frac{1}{4}\sqrt{\frac{53}{70}}|E_{16}(\xi_1)| + \frac{1}{6}\sqrt{\frac{67}{33}}|E_{17}(\xi_1)| + \\
&\quad \frac{1}{9}\sqrt{\frac{85}{22}}|E_{18}(\xi_1)| + \frac{4}{3}\sqrt{\frac{10}{473}}|E_{19}(\xi_1)| + \frac{1}{18}\sqrt{\frac{32}{2}}|E_{20}(\xi_1)| + \frac{1}{2}\sqrt{\frac{13}{66}}|E_{21}(\xi_1)| + \\
&\quad \frac{1}{2}\sqrt{\frac{37}{231}}|E_{22}(\xi_1)| + \frac{1}{4}\sqrt{\frac{19}{30}}|E_{23}(\xi_1)| + \frac{1}{6}\sqrt{\frac{163}{129}}|E_{24}(\xi_1)| + \frac{1}{2}\sqrt{\frac{29}{210}}|E_{25}(\xi_1)| + \\
&\quad \frac{1}{22}\sqrt{\frac{43}{2}}|E_{26}(\xi_1)| + \frac{1}{2}\sqrt{\frac{57}{473}}|E_{27}(\xi_1)| + \frac{1}{2}\sqrt{\frac{13}{110}}|E_{28}(\xi_1)| + \frac{1}{2}\sqrt{\frac{3}{26}}|E_{29}(\xi_1)| + \\
&\quad \frac{1}{3}\sqrt{\frac{43}{154}}|E_{30}(\xi_1)| + \frac{1}{9}\sqrt{\frac{181}{86}}|E_{31}(\xi_1)| + \frac{1}{4}\sqrt{\frac{31}{77}}|E_{32}(\xi_1)| + 2\sqrt{\frac{17}{3311}}|E_{33}(\xi_1)| + \\
&\quad \frac{1}{14}\sqrt{\frac{43}{11}}|E_{34}(\xi_1)| + \frac{1}{40}\sqrt{\frac{79}{2}}|E_{35}(\xi_1)| + \frac{1}{20}\sqrt{\frac{109}{14}}|E_{36}(\xi_1)| + \frac{1}{2}\sqrt{\frac{99}{1505}}|E_{37}(\xi_1)| + \\
&\quad \frac{1}{6}\sqrt{\frac{283}{559}}|E_{38}(\xi_1)| + \frac{1}{70}\sqrt{\frac{139}{2}}|E_{39}(\xi_1)| + \frac{1}{2}\sqrt{\frac{7}{130}}|E_{40}(\xi_1)| + \frac{1}{78}\sqrt{\frac{155}{2}}|E_{41}(\xi_1)|.
\end{aligned}$$

After some calculations, we have

$$\begin{aligned}
\Longrightarrow ABC_4(\xi_1) &= 51.706 + \frac{3}{20}\sqrt{\frac{79}{2}}(-5+m) + 3\sqrt{\frac{53}{70}}(-4+m) + \frac{3}{5}\sqrt{\frac{109}{14}}(-4+m) + \\
&\quad \sqrt{\frac{114}{5}}(-4+m) + \frac{3}{35}\sqrt{\frac{139}{2}}(-4+m) + 3\sqrt{\frac{14}{65}}(-3+m) + 12\sqrt{\frac{26}{55}}(-3+m) + \\
&\quad 2\sqrt{\frac{174}{35}}(-3+m) + \sqrt{\frac{62}{7}}(-3+m) + \sqrt{\frac{78}{11}}(-2+m) + \frac{9}{11}\sqrt{\frac{43}{2}}(-2+m)^2 + \\
&\quad \frac{1}{3}\sqrt{\frac{35}{2}}(-5+2m) + \frac{1}{26}\sqrt{\frac{155}{2}}(24-17m+3m^2) + 3\sqrt{\frac{6}{13}}(19-15m+3m^2).
\end{aligned}$$

The $GA_5(\xi_1)$ index can be determined from (8) as follows.

$$\begin{aligned}GA_5(\xi_1) &= \frac{5}{29}\sqrt{33}|E_{10}(\xi_1)| + \frac{60}{11}|E_{11}(\xi_1)| + \frac{30}{79}\sqrt{6}|E_{12}(\xi_1)| + \frac{5}{51}\sqrt{77}|E_{13}(\xi_1)| + \\ &\quad \frac{3}{8}\sqrt{7}|E_{14}(\xi_1)| + \frac{4}{15}\sqrt{11}|E_{15}(\xi_1)| + \frac{4}{27}\sqrt{35}|E_{16}(\xi_1)| + \frac{4}{23}\sqrt{33}|E_{17}(\xi_1)| + \\ &\quad \frac{6}{29}\sqrt{22}|E_{18}(\xi_1)| + \frac{1}{31}\sqrt{957}|E_{19}(\xi_1)| + |E_{20}(\xi_1)| + \frac{3}{10}\sqrt{11}|E_{21}(\xi_1)| + \\ &\quad \frac{12}{113}\sqrt{77}|E_{22}(\xi_1)| + \frac{12}{29}\sqrt{5}|E_{23}(\xi_1)| + \frac{4}{55}\sqrt{129}|E_{24}(\xi_1)| + \frac{3}{22}\sqrt{35}|E_{25}(\xi_1)| + \\ &\quad |E_{26}(\xi_1)| + \frac{4}{173}\sqrt{1419}|E_{27}(\xi_1)| + \frac{1}{23}\sqrt{385}|E_{28}(\xi_1)| + \frac{1}{25}\sqrt{429}|E_{29}(\xi_1)| + \\ &\quad \frac{6}{131}\sqrt{462}|E_{30}(\xi_1)| + \frac{6}{61}\sqrt{86}|E_{31}(\xi_1)| + \frac{8}{157}\sqrt{385}|E_{32}(\xi_1)| + \frac{1}{103}\sqrt{9933}|E_{33}(\xi_1)| + \\ &\quad \frac{4}{31}\sqrt{55}|E_{34}(\xi_1)| + |E_{35}(\xi_1)| + \frac{4}{11}\sqrt{7}|E_{36}(\xi_1)| + \frac{4}{269}\sqrt{4515}|E_{37}(\xi_1)| + \\ &\quad \frac{4}{95}\sqrt{559}|E_{38}(\xi_1)| + |E_{39}(\xi_1)| + \frac{1}{37}\sqrt{1365}|E_{40}(\xi_1)| + |E_{41}(\xi_1)|.\end{aligned}$$

After some calculations, we have

$$\begin{aligned}\Longrightarrow GA_5(\xi_1) &= 315.338 + \frac{288}{29}\sqrt{5}(-4+m) + \frac{48}{11}\sqrt{7}(-4+m) + \frac{16}{9}\sqrt{35}(-4+m) + \frac{9}{2} \\ &\quad \sqrt{7}(-3+m) + \frac{36}{11}\sqrt{35}(-3+m) + \frac{48}{23}\sqrt{385}(-3+m) + \frac{12}{37}\sqrt{1365}(-3+m) + \\ &\quad \frac{18}{5}\sqrt{11}(-2+m) - 99m + 27m^2 + \frac{12}{25}\sqrt{429}(19 - 15m + 3m^2).\end{aligned}$$

□

Table 2. Edge partition of the third type of hex-derived network $HDN_3(m)$ based on sum of degrees of end vertices of each edge.

$(\tau_{\acute{m}}, \tau_{\acute{n}})$ Where $\acute{m}\acute{n} \in E(\xi_1)$	Number of Edges	$(\tau_{\acute{m}}, \tau_{\acute{n}})$ Where $\acute{m}\acute{n} \in E(\xi_1)$	Number of Edges
(25, 33)	12	(44, 44)	$18m^2 - 72m + 72$
(25, 36)	12	(44, 129)	36
(25, 54)	12	(44, 140)	$48m - 144$
(25, 77)	12	(44, 156)	$36m^2 - 180m + 228$
(28, 36)	$12m - 36$	(54, 77)	12
(28, 77)	12	(54, 129)	6
(28, 80)	$12m - 48$	(77, 80)	12
(33, 36)	12	(77, 129)	12
(33, 54)	12	(77, 140)	12
(33, 129)	12	(80, 80)	$6m - 30$
(36, 36)	$12m - 30$	(80, 140)	$12m - 48$
(36, 44)	$12m - 24$	(129, 140)	12
(36, 77)	48	(129, 156)	6
(36, 80)	$24m - 96$	(140, 140)	$6m - 24$
(36, 129)	24	(140, 156)	$12m - 36$
(36, 140)	$24m - 72$	(156, 156)	$9m^2 - 51m + 72$

2.2. Results for Third Type of Triangular Hex-Derived Network $THDN_3(m)$

Now, we discuss the third type of rectangular hex-derived network and compute exact results for Forgotten index and Balaban index, and reclassified the Zagreb indices, ABC_4 index, and GA_5 index for $THDN_3(m)$.

Theorem 5. *Consider the third type of triangular hex-derived network of $THDN_3(n)$; its Forgotten index is equal to*

$$F(THDN_3(m)) = 12(990 - 997m + 259m^2).$$

Proof. Let ζ_2 be the third type of triangular hex-derived network, $THDN_3(m)$ shown in Figure 3, where $m \geq 4$. The third type of triangular hex-derived network ζ_2 has $\frac{7m^2-11m+6}{2}$ vertices and the edge set of ζ_2 is divided into six partitions based on the degree of end vertices as shown in Table 3.

By using edge partition from Table 3, we get. Thus, from (2) it follows that

$$F(\zeta_2) = 32|E_1(\zeta_2)| + 116|E_2(\zeta_2)| + 340|E_3(\zeta_2)| + 200|E_4(\zeta_2)| + 424|E_5(\zeta_2)| + 648|E_6(\zeta_2)|.$$

By doing some calculations, we get

$$\Longrightarrow F(\zeta_2) = 12(990 - 997m + 259m^2).$$

□

Table 3. Edge partition of the third type of triangular hex-derived network $THDN_3(m)$ based on degrees of end vertices of each edge.

(τ_x, τ_y) Where $\acute{m}\acute{n} \in E(\zeta_1)$	Number of Edges	(τ_u, τ_v) Where $\acute{m}\acute{n} \in E(\zeta_1)$	Number of Edges
(4, 4)	$3m^2 - 6m + 9$	(10, 10)	$3m - 6$
(4, 10)	$18m - 30$	(10, 18)	$6m - 18$
(4, 18)	$6m^2 - 30m + 36$	(18, 18)	$\frac{3m^2 - 21m + 36}{2}$

In the following theorem, we compute the Balaban index of the third type of triangular hex-derived network, ζ_2.

Theorem 6. *For the third type of triangular hex-derived network ζ_2, the Balaban index is equal to*

$$J(\zeta_2) = \left(\frac{1}{40(8 - 14m + 7m^2)}\right)\left(6 - 13m + 7m^2\right)(159 + 1802\sqrt{2} - 36\sqrt{5} - 90\sqrt{10} + (-107 - 150\sqrt{2} + 12\sqrt{5} + 54\sqrt{10})m + 10(5 + 3\sqrt{2})m^2\right).$$

Proof. Let ζ_2 be the third type of triangular hex-derived network $THDN_3(m)$. By using edge partition from Table 3, the result follows. The Balaban index can be calculated by using (3) as follows.

$$J(\zeta_2) = \frac{3}{2}\left(\frac{6 - 13m + 7m^2}{8 - 14m + 7m^2}\right)\left(\frac{1}{4}|E_1(\zeta_2)| + \frac{1}{2\sqrt{10}}|E_2(\zeta_2)| + \frac{1}{6\sqrt{2}}|E_3(\zeta_2)| + \frac{1}{10}|E_4(\zeta_2)| + \frac{1}{6\sqrt{5}}|E_5(\zeta_2)| + \frac{1}{18}|E_6(\zeta_2)|\right).$$

After some calculation, we have

$$\Longrightarrow J(\zeta_2) = \left(\frac{1}{40(8 - 14m + 7m^2)}\right)\left(6 - 13m + 7m^2\right)(159 + 1802\sqrt{2} - 36\sqrt{5} - 90\sqrt{10} + (-107 - 150\sqrt{2} + 12\sqrt{5} + 54\sqrt{10})m + 10(5 + 3\sqrt{2})m^2\right).$$

□

Now, we compute $ReZG_1$, $ReZG_2$ and $ReZG_3$ indices of third type of triangular hex-derived network ζ_2.

Theorem 7. Let ξ_2 be the third type of triangular hex-derived network, then

- $ReZG_1(\xi_2) = \frac{3}{154}(3408 - 5117m + 2009m^2)$,
- $ReZG_2(\xi_2) = \frac{1}{2}(6 - 11m + 7m^2)$,
- $ReZG_3(\xi_2) = 24(6192 - 5185m + 1141m^2)$.

Proof. By using edge partition given in Table 3, the $ReZG_1(\xi_2)$ can be calculated by using (4) as follows.

$$ReZG_1(\xi_2) = 2|E_1(\xi_2)| + \frac{20}{7}|E_2(\xi_2)| + \frac{36}{11}|E_3(\xi_2)| + 5|E_4(\xi_2)| + \frac{45}{7}|E_5(\xi_2)| + 9|E_6(\xi_2)|.$$

After some calculation, we have

$$\implies ReZG_1(\xi_2) = \frac{3}{154}(3408 - 5117m + 2009m^2).$$

The $ReZG_2(\xi_2)$ can be calculated by using (5) as follows.

$$ReZG_2(\xi_2) = \frac{1}{2}|E_1(\xi_2)| + \frac{7}{20}|E_2(\xi_2)| + \frac{11}{36}|E_3(\xi_2)| + \frac{1}{5}|E_4(\xi_2)| + \frac{7}{45}|E_5(\xi_2)| + \frac{1}{9}|E_6(\xi_2)|.$$

After some calculation, we have

$$\implies ReZG_2(\xi_2) = \frac{1}{2}(6 - 11m + 7m^2).$$

The $ReZG_3(\xi_2)$ index can be calculated from (6) as follows.

$$ReZG_3(\xi_2) = 128|E_1(\xi_2)| + 560|E_2(\xi_2)| + 1584|E_3(\xi_2)| + 2000|E_4(\xi_2)| + 5040|E_5(\xi_2)| + 11664|E_6(\xi_2)|.$$

After some calculation, we have

$$\implies ReZG_3(\xi_2) = 24(6192 - 5185m + 1141m^2).$$

□

Now, we compute ABC_4 and GA_5 indices of third type of triangular hex-derived network ξ_2.

Theorem 8. Let ξ_2 be the third type of triangular hex-derived network, then

- $ABC_4(\xi_2) = 24.131 + 3\sqrt{\frac{7}{130}}(-6+m) + 6\sqrt{\frac{26}{55}}(-5+m) + \sqrt{\frac{174}{35}}(-5+m) + \frac{3}{10}\sqrt{\frac{109}{14}}(-5+m) + \frac{3}{40}\sqrt{\frac{79}{2}}(-5+m) + \frac{3}{70}\sqrt{\frac{139}{2}}(-5+m) + \frac{3}{2}\sqrt{\frac{53}{70}}(-4+m) + \sqrt{\frac{39}{22}}(-4+m) + \sqrt{\frac{57}{10}}(-4+m) + \frac{3}{22}\sqrt{\frac{43}{2}}(-4+m)^2 + \frac{1}{3}\sqrt{\frac{35}{2}}(-3+m) + 2\sqrt{\frac{7}{11}}(-2+m) + \frac{1}{52}\sqrt{\frac{155}{2}}(42-13m+m^2) + 3\sqrt{\frac{3}{26}}(30-11m+m^2)$;
- $GA_5(\xi_2) = 110.66 + \frac{6}{37}\sqrt{1365}(-6+m) + \frac{24}{11}\sqrt{7}(-5+m) + \frac{18}{11}\sqrt{35}(-5+m) + \frac{24}{23}\sqrt{385}(-5+m) + \frac{144}{29}\sqrt{5}(-4+m) + \frac{9}{5}\sqrt{11}(-4+m) + \frac{8}{9}\sqrt{35}(-4+m) + \frac{36}{29}\sqrt{22}(-2+m) - 12m + 3m^2 + \frac{3}{2}(42-13m+m^2) + \frac{6}{25}\sqrt{429}(30-11m+m^2)$.

Proof. By using the edge partition given in Table 4, the $ABC_4(\xi_2)$ index can be calculated by using (7) as follows.

$$ABC_4(\xi_2) = \frac{1}{11}\sqrt{\frac{21}{2}}|E_7(\xi_2)| + \sqrt{\frac{6}{77}}|E_8(\xi_2)| + \frac{1}{3}\sqrt{\frac{7}{11}}|E_9(\xi_2)| + \frac{1}{11}\sqrt{\frac{43}{6}}|E_{10}(\xi_2)| +$$
$$\sqrt{\frac{23}{462}}|E_{11}(\xi_2)| + \frac{1}{4}\sqrt{\frac{53}{70}}|E_{12}(\xi_2)| + \frac{1}{18}\sqrt{\frac{32}{2}}|E_{13}(\xi_2)| + \frac{1}{2}\sqrt{\frac{13}{66}}|E_{14}(\xi_2)| +$$
$$\frac{5}{3}\frac{1}{\sqrt{6}}|E_{15}(\xi_2)| + \frac{1}{4}\sqrt{\frac{19}{30}}|E_{16}(\xi_2)| + \frac{1}{6}\sqrt{\frac{79}{62}}|E_{17}(\xi_2)| + \frac{1}{2}\sqrt{\frac{29}{210}}|E_{18}(\xi_2)| +$$
$$\frac{1}{22}\sqrt{\frac{43}{2}}|E_{19}(\xi_2)| + \frac{1}{2}\sqrt{\frac{83}{682}}|E_{20}(\xi_2)| + \frac{1}{2}\sqrt{\frac{13}{110}}|E_{21}(\xi_2)| + \frac{1}{2}\sqrt{\frac{3}{26}}|E_{22}(\xi_2)| +$$
$$\frac{1}{33}\sqrt{\frac{65}{2}}|E_{23}(\xi_2)| + \sqrt{\frac{3}{110}}|E_{24}(\xi_2)| + \sqrt{\frac{47}{2046}}|E_{25}(\xi_2)| + \frac{1}{40}\sqrt{\frac{79}{2}}|E_{26}(\xi_2)| +$$
$$\frac{1}{4}\sqrt{\frac{101}{310}}|E_{27}(\xi_2)| + \frac{1}{20}\sqrt{\frac{109}{14}}|E_{28}(\xi_2)| + \frac{1}{2}\sqrt{\frac{131}{2170}}|E_{29}(\xi_2)| + \frac{1}{70}\sqrt{\frac{139}{2}}|E_{30}(\xi_2)| +$$
$$\frac{1}{2}\sqrt{\frac{7}{130}}|E_{31}(\xi_2)| + \frac{1}{78}\sqrt{\frac{155}{2}}|E_{32}(\xi_2)|.$$

After some calculation, we have

$$\implies ABC_4(\xi_2) = 24.131 + 3\sqrt{\frac{7}{130}}(-6+m) + 6\sqrt{\frac{26}{55}}(-5+m) + \sqrt{\frac{174}{35}}(-5+m) +$$
$$\frac{3}{10}\sqrt{\frac{109}{14}}(-5+m) + \frac{3}{40}\sqrt{\frac{79}{2}}(-5+m) + \frac{3}{70}\sqrt{\frac{139}{2}}(-5+m) + \frac{3}{2}\sqrt{\frac{53}{70}}(-4+m) +$$
$$\sqrt{\frac{39}{22}}(-4+m) + \sqrt{\frac{57}{10}}(-4+m) + \frac{3}{22}\sqrt{\frac{43}{2}}(-4+m)^2 + \frac{1}{3}\sqrt{\frac{35}{2}}(-3+m) +$$
$$2\sqrt{\frac{7}{11}}(-2+m) + \frac{1}{52}\sqrt{\frac{155}{2}}(42 - 13m + m^2) + 3\sqrt{\frac{3}{26}}(30 - 11m + m^2).$$

The $GA_5(\xi_2)$ index can be calculated from (8) as follows.

$$GA_5(\xi_2) = 1|E_7(\xi_2)| + \frac{2}{25}\sqrt{154}|E_8(\xi_2)| + \frac{6}{29}\sqrt{22}|E_9(\xi_2)| + \frac{1}{2}\sqrt{3}|E_{10}(\xi_2)| +$$
$$\frac{2}{47}\sqrt{462}|E_{11}(\xi_2)| + \frac{4}{27}\sqrt{35}|E_{12}(\xi_2)| + 1|E_{13}(\xi_2)| + \frac{3}{10}\sqrt{11}|E_{14}(\xi_2)| +$$
$$\frac{2}{17}\sqrt{66}|E_{15}(\xi_2)| + \frac{12}{29}\sqrt{5}|E_{16}(\xi_2)| + \frac{3}{20}\sqrt{31}|E_{17}(\xi_2)| + \frac{3}{22}\sqrt{35}|E_{18}(\xi_2)| +$$
$$1|E_{19}(\xi_2)| + \frac{1}{21}\sqrt{341}|E_{20}(\xi_2)| + \frac{1}{23}\sqrt{385}|E_{21}(\xi_2)| + \frac{1}{25}\sqrt{429}|E_{22}(\xi_2)| +$$
$$1|E_{23}(\xi_2)| + \frac{4}{73}\sqrt{330}|E_{24}(\xi_2)| + \frac{2}{95}\sqrt{2046}|E_{25}(\xi_2)| + 1|E_{26}(\xi_2)| +$$
$$\frac{4}{51}\sqrt{155}|E_{27}(\xi_2)| + \frac{4}{11}\sqrt{7}|E_{28}(\xi_2)| + \frac{1}{33}\sqrt{1085}|E_{29}(\xi_2)| + 1|E_{30}(\xi_2)| +$$
$$\frac{1}{37}\sqrt{1365}|E_{31}(\xi_2)| + 1|E_{32}(\xi_2)|.$$

After some calculation, we have

$$\implies GA_5(\xi_2) = 110.66 + \frac{6}{37}\sqrt{1365}(-6+m) + \frac{24}{11}\sqrt{7}(-5+m) + \frac{18}{11}\sqrt{35}(-5+m) + \frac{24}{23}\sqrt{385}(-5+$$
$$m) + \frac{144}{29}\sqrt{5}(-4+m) + \frac{9}{5}\sqrt{11}(-4+m) + \frac{8}{9}\sqrt{35}(-4+m) + \frac{36}{29}\sqrt{22}(-2+m) - 12m +$$
$$3m^2 + \frac{3}{2}(42 - 13m + m^2) + \frac{6}{25}\sqrt{429}(30 - 11m + m^2).$$

□

Table 4. Edge partition of the third type of triangular hex-derived network $THDN_3(m)$ based on the sum of degrees of end vertices of each edge.

(τ_x, τ_y) Where $\acute{m}\acute{n} \in E(\xi_2)$	Number of Edges	(τ_u, τ_v) Where $\acute{m}\acute{n} \in E(\xi_2)$	Number of Edges
(22, 22)	3	(44, 124)	12
(22, 28)	12	(44, 140)	$24m - 120$
(22, 36)	6	(44, 156)	$6m^2 - 66m + 180$
(22, 66)	$6m - 12$	(66, 66)	3
(28, 66)	24	(66, 80)	6
(28, 80)	$6m - 24$	(66, 124)	6
(36, 36)	$6m - 18$	(80, 80)	$3m - 15$
(36, 44)	$6m - 24$	(80, 124)	6
(36, 66)	12	(80, 140)	$6m - 30$
(36, 80)	$12m - 48$	(124, 140)	6
(36, 124)	24	(140, 140)	$3m - 15$
(36, 140)	$12m - 60$	(140, 156)	$6m - 36$
(44, 44)	$3m^2 - 24m + 48$	(156, 156)	$\frac{3m^2 - 39m + 126}{2}$

2.3. Results for Third Type of Rectangular Hex-Derived Network, $RHDN_3(m, n)$

In this section, we calculate certain degree-based topological indices of the third type of rectangular hex-derived network, $RHDN_3(m, n)$ of dimension $m = n$. We compute Forgotten index and Balaban index, and reclassified the Zagreb indices, forth version of ABC index, and fifth version of GA index in the coming theorems of $RHDN_3(m, n)$.

Theorem 9. *Consider the third type of rectangular hex-derived network $RHDN_3(m)$, its Forgotten index is equal to*

$$F(RHDN_3(m)) = 19726 - 20096m + 6216m^2.$$

Proof. Let ξ_3 be the third type of rectangular hex-derived network, $RHDN_3(m)$ shown in Figure 4, where $m = s \geq 4$. The third type of rectangular hex-derived network ξ_3 has $7m^2 - 12m + 6$ vertices and the edge set of ξ_3 is divided into nine partitions based on the degree of end vertices as shown in Table 5.

Table 5. Edge partition of the third type of rectangular hex-derived network, $RHDN_3(m)$ based on degrees of end vertices of each edge.

$(\tau_{\acute{m}}, \tau_{\acute{n}})$ Where $\acute{m}\acute{n} \in E(\xi_1)$	Number of Edges	$(\tau_{\acute{m}}, \tau_{\acute{n}})$ Where $\acute{m}\acute{n} \in E(\xi_1)$	Number of Edges
(4, 4)	$6m^2 - 12m + 10$	(7, 18)	2
(4, 7)	8	(10, 10)	$4m - 10$
(4, 10)	$24m - 44$	(10, 18)	$8m - 20$
(4, 18)	$12m^2 - 48m + 48$	(18, 18)	$3m^2 - 16m + 21$
(7, 10)	4	-	-

Thus, from (2), it follows that.

$$F(G) = \sum_{\acute{m}\acute{n} \in E(\xi)} ((\tau(\acute{m}))^2 + (\tau(\acute{n}))^2)$$

Let ξ_3 be the third type of rectangular hex-derived network, $THDN_3(m)$. By using edge partition from Table 5, the result follows.

$$F(\xi_3) = \sum_{\acute{m}\acute{n} \in E(\xi_3)} ((\tau(\acute{m}))^2 + (\tau(\acute{n}))^2) = \sum_{\acute{m}\acute{n} \in E_j(\xi_3)} \sum_{j=1}^{9} ((\tau(\acute{m}))^2 + (\tau(\acute{n}))^2)$$

$$F(\zeta_3) = 32|E_1(\zeta_3)| + 65|E_2(\zeta_3)| + 116|E_3(\zeta_3)| + 340|E_4(\zeta_3)| + 149|E_5(\zeta_3)| + 373|E_6(\zeta_3)| + 200|E_7(\zeta_3)| + 424|E_8(\zeta_3)| + 648|E_9(\zeta_3)|.$$

After some calculation, we have

$$\Longrightarrow F(\zeta_3) = 19726 - 20096m + 6216m^2.$$

□

In the following theorem, we compute the Balaban index of the third type of rectangular hex-derived network, ζ_3.

Theorem 10. *For the third type of rectangular hex-derived network ζ_3, the Balaban index is equal to*

$$J(\zeta_3) = \left(\frac{1}{315(15 - 28m + 14m^2)}\right)7(-157 - 180\sqrt{2} + 12\sqrt{5} + 54\sqrt{10})m + 105(5 + 3\sqrt{2})m^2)(19 - 40m + 21m^2)(3(280 + 420\sqrt{2} - 70\sqrt{5} + 60\sqrt{7} - 231\sqrt{10} + 5\sqrt{14} + 6\sqrt{70})).$$

Proof. Let ζ_3 be the rectangular hex-derived network $RHDN_3(m)$. By using edge partition from Table 5, the result follows. The Balaban index can be calculated by using (3) as follows.

$$J(\zeta_3) = \left(\frac{m}{m-n+2}\right) \sum_{\acute{m}\acute{n} \in E(\zeta_3)} \frac{1}{\sqrt{\tau(\acute{m}) \times \tau(\acute{n})}} = \left(\frac{m}{m-n+2}\right) \sum_{\acute{m}\acute{n} \in E_j(\zeta_3)} \sum_{j=1}^{9} \frac{1}{\sqrt{\tau(\acute{m}) \times \tau(\acute{n})}}$$

$$J(\zeta_3) = \left(\frac{19 - 40m + 21m^2}{15 - 28m + 14m^2}\right)\left(\frac{1}{4}|E_1(\zeta_3)| + \frac{1}{2\sqrt{7}}|E_2(\zeta_3)| + \frac{1}{2\sqrt{10}}|E_3(\zeta_3)| + \frac{1}{6\sqrt{2}}|E_4(\zeta_3)| + \frac{1}{\sqrt{70}}|E_5(\zeta_3)| + \frac{1}{3\sqrt{14}}|E_6(\zeta_3)| + \frac{1}{10}|E_7(\zeta_3)| + \frac{1}{6\sqrt{5}}|E_8(\zeta_3)| + \frac{1}{18}|E_9(\zeta_3)|\right).$$

After some calculation, we have

$$\Longrightarrow J(\zeta_3) = \left(\frac{1}{315(15 - 28m + 14m^2)}\right)7(-157 - 180\sqrt{2} + 12\sqrt{5} + 54\sqrt{10})m + 105(5 + 3\sqrt{2})m^2)$$
$$(19 - 40m + 21m^2)(3(280 + 420\sqrt{2} - 70\sqrt{5} + 60\sqrt{7} - 231\sqrt{10} + 5\sqrt{14} + 6\sqrt{70})).$$

□

Now, we compute $ReZG_1$, $ReZG_2$ and $ReZG_3$ indices of the third type of rectangular hex-derived network ζ_3.

Theorem 11. *Let ζ_3 be the third type of rectangular hex-derived network, then*

- $ReZG_1(\zeta_3) = \frac{10102843}{32725} - \frac{2036m}{11} + \frac{861m^2}{11}$,
- $ReZG_2(\zeta_3) = 56 - 12m + 7m^2$,
- $ReZG_3(\zeta_3) = 4(50785 - 50608m + 13692m^2)$.

Proof. By using the edge partition given in Table 5, the $ReZG_1(\zeta_3)$ can be calculated by using (4) as follows.

$$ReZG_1(\zeta) = \sum_{\acute{m}\acute{n} \in E(\zeta_3)} \left(\frac{\tau(\acute{m}) \times \tau(\acute{n})}{\tau(\acute{m}) + \tau(\acute{n})}\right) = \sum_{j=1}^{9} \sum_{\acute{m}\acute{n} \in E_j(\zeta_3)} \left(\frac{\tau(\acute{m}) \times \tau(\acute{n})}{\tau(\acute{m}) + \tau(\acute{n})}\right)$$

$$ReZG_1(\xi_3) = 2|E_1(\xi_3)| + \frac{28}{11}|E_2(\xi_3)| + \frac{20}{7}|E_3(\xi_3)| + \frac{36}{11}|E_4(\xi_3)| + \frac{70}{17}|E_5(\xi_3)| + \frac{126}{25}|E_6(\xi_3)| + $$
$$5|E_7(\xi_3)| + \frac{45}{7}|E_8(\xi_3)| + 9|E_9(\xi_3)|.$$

After some calculation, we have

$$\implies ReZG_1(\xi_3) = \frac{10102843}{32725} - \frac{2036m}{11} + \frac{861m^2}{11}.$$

The $ReZG_2(\xi_3)$ can be calculated by using (5) as follows.

$$ReZG_2(\xi_3) = \sum_{\acute{m}\acute{n}\in E(\xi_3)} \left(\frac{\tau(\acute{m}) + \tau(\acute{n})}{\tau(\acute{m}) \times \tau(\acute{n})} \right) = \sum_{\acute{m}\acute{n}\in E_j(\xi_3)} \sum_{j=1}^{9} \left(\frac{\tau(\acute{m}) + \tau(\acute{n})}{\iota(\acute{m}) \times \iota(\acute{n})} \right)$$

$$ReZG_2(\xi_3) = \frac{1}{2}|E_1(\xi_3)| + \frac{11}{28}|E_2(\xi_3)| + \frac{7}{20}|E_3(\xi_3)| + \frac{11}{36}|E_4(\xi_3)| + \frac{17}{70}|E_5(\xi_3)| + \frac{25}{126}|E_6(\xi_3)| +$$
$$\frac{1}{5}|E_7(\xi_3)| + \frac{7}{45}|E_8(\xi_3)| + \frac{1}{9}|E_9(\xi_3)|.$$

After some calculation, we have

$$\implies ReZG_2(\xi_3) = 56 - 12m + 7m^2.$$

The $ReZG_3(\xi_3)$ index can be calculated from (6) as follows.

$$ReZG_3(\xi_3) = \sum_{\acute{m}\acute{n}\in E(\xi_3)} (\tau(\acute{m}) \times \tau(\acute{n}))(\tau(\acute{m}) + \tau(\acute{n})) = \sum_{\acute{m}\acute{n}\in E_j(\xi_3)} \sum_{j=1}^{9} (\tau(\acute{m}) \times \tau(\acute{n}))(\tau(\acute{m}) + \tau(\acute{n}))$$

$$ReZG_3(\xi_3) = 128|E_1(\xi_3)| + 308|E_2(\xi_3)| + 560|E_3(\xi_3)| + 1584|E_4(\xi_3)| + 1190|E_5(\xi_3)| +$$
$$3150|E_6(\xi_3)| + 2000|E_7(\xi_3)| + 5040|E_8(\xi_3)| + 11664|E_9(\xi_3)|.$$

After some calculation, we have

$$\implies ReZG_3(\xi_3) = 4(50785 - 50608m + 13692m^2).$$

□

Now, we compute ABC_4 and GA_5 indices of the third type of rectangular hex-derived network ξ_3.

Theorem 12. *Let ξ_3 be the third type of rectangular hex-derived network, then*

- $ABC_4(\xi_3) = 22.459 + 8\sqrt{\frac{26}{55}}(-4+m) + 4\sqrt{\frac{58}{105}}(-4+m) + \frac{4}{7}\sqrt{\frac{67}{15}}(-4+m) + 3\sqrt{\frac{6}{13}}(-4+m)^2 + 2\sqrt{\frac{26}{33}}(-3+m) + \frac{3}{11}\sqrt{\frac{43}{2}}(-3+m)^2 + \sqrt{\frac{14}{65}}(-9+2m) + \frac{1}{35}\sqrt{\frac{139}{2}}(-9+2m) + \frac{1}{3}\sqrt{\frac{62}{7}}(-5+2m) + \frac{4}{63}\sqrt{31}(-5+2m) + \frac{4}{9}\sqrt{\frac{97}{7}}(-3+2m) + \frac{2}{21}\sqrt{89}(-3+2m) + \frac{1}{9}\sqrt{\frac{35}{2}}(-11+4m) + \frac{1}{78}\sqrt{\frac{155}{2}}(65 - 28m + 3m^2);$
- $GA_5(\xi_3) = 173.339 + \frac{96}{29}\sqrt{5}(-4+m) + \frac{24}{11}\sqrt{35}(-4+m) + \frac{32}{23}\sqrt{385}(-4+m) + \frac{12}{25}\sqrt{429}(-4+m)^2 + \frac{12}{5}\sqrt{11}(-3+m) - 48m + 9m^2 + \frac{4}{37}\sqrt{1365}(-9+2m) + \frac{3}{2}\sqrt{7}(-5+2m) + \frac{48}{13}(-3+2m) + \frac{32}{11}\sqrt{7}(-3+2m).$

Proof. By using the edge partition given in Table 6, the $ABC_4(\xi_3)$ can be calculated by using (7) as follows.

$$ABC_4(\xi_3) = \sum_{\acute{m}\acute{n}\in E(\xi_3)} \sqrt{\frac{S_{\acute{m}}+S_{\acute{n}}-2}{S_{\acute{m}}S_{\acute{n}}}} = \sum_{\acute{m}\acute{n}\in E_j(\xi_3)}\sum_{j=10}^{44} \sqrt{\frac{S_{\acute{m}}+S_{\acute{n}}-2}{S_{\acute{m}}S_{\acute{n}}}}$$

$$\begin{aligned}ABC_4(\xi_3) &= \frac{1}{11}\sqrt{\frac{21}{2}}|E_{10}(\xi_3)| + \sqrt{\frac{6}{77}}|E_{11}(\xi_3)| + \frac{1}{3}\sqrt{\frac{83}{154}}|E_{12}(\xi_3)| + \frac{1}{5}\sqrt{\frac{46}{33}}|E_{13}(\xi_3)| + \\
&\quad \frac{1}{30}\sqrt{59}|E_{14}(\xi_3)| + \frac{1}{15}\sqrt{\frac{77}{6}}|E_{15}(\xi_3)| + \frac{1}{15}\sqrt{\frac{86}{7}}|E_{16}(\xi_3)| + \frac{1}{6}\sqrt{\frac{31}{14}}|E_{17}(\xi_3)| + \\
&\quad \frac{1}{42}\sqrt{89}|E_{18}(\xi_3)| + \frac{1}{6}\sqrt{\frac{67}{33}}|E_{19}(\xi_3)| + \frac{1}{9}\sqrt{\frac{85}{22}}|E_{20}(\xi_3)| + \frac{4}{3}\sqrt{\frac{10}{473}}|E_{21}(\xi_3)| + \\
&\quad \frac{1}{18}\sqrt{\frac{35}{2}}|E_{22}(\xi_3)| + \frac{1}{2}\sqrt{\frac{13}{66}}|E_{23}(\xi_3)| + \frac{1}{18}\sqrt{\frac{97}{7}}|E_{24}(\xi_3)| + \frac{1}{6}\sqrt{\frac{79}{62}}|E_{25}(\xi_3)| + \\
&\quad \frac{1}{6}\sqrt{\frac{163}{129}}|E_{26}(\xi_3)| + \frac{1}{2}\sqrt{\frac{29}{210}}|E_{27}(\xi_3)| + \frac{1}{22}\sqrt{\frac{43}{2}}|E_{28}(\xi_3)| + \frac{1}{2}\sqrt{\frac{83}{682}}|E_{29}(\xi_3)| + \\
&\quad \frac{1}{2}\sqrt{\frac{57}{473}}|E_{30}(\xi_3)| + \frac{1}{2}\sqrt{\frac{13}{110}}|E_{31}(\xi_3)| + \frac{1}{2}\sqrt{\frac{3}{26}}|E_{32}(\xi_3)| + \frac{1}{9}\sqrt{\frac{115}{42}}|E_{33}(\xi_3)| + \\
&\quad \frac{1}{9}\sqrt{\frac{181}{86}}|E_{34}(\xi_3)| + \frac{1}{63}\sqrt{31}|E_{35}(\xi_3)| + \frac{1}{6}\sqrt{\frac{185}{217}}|E_{36}(\xi_3)| + \frac{1}{3}\sqrt{\frac{190}{903}}|E_{37}(\xi_3)| + \\
&\quad \frac{1}{14}\sqrt{\frac{67}{15}}|E_{38}(\xi_3)| + \frac{1}{2}\sqrt{\frac{131}{2170}}|E_{39}(\xi_3)| + \frac{1}{2}\sqrt{\frac{89}{1505}}|E_{40}(\xi_3)| + \frac{1}{70}\sqrt{\frac{283}{559}}|E_{41}(\xi_3)| + \\
&\quad \frac{1}{70}\sqrt{\frac{139}{2}}|E_{42}(\xi_3)| + \frac{1}{2}\sqrt{\frac{7}{130}}|E_{43}(\xi_3)| + \frac{1}{78}\sqrt{\frac{155}{2}}|E_{44}(\xi_3)|.\end{aligned}$$

After some calculation, we have

$$\begin{aligned}\Longrightarrow ABC_4(\xi_3) &= 22.459 + 8\sqrt{\frac{26}{55}}(-4+m) + 4\sqrt{\frac{58}{105}}(-4+m) + \frac{4}{7}\sqrt{\frac{67}{15}}(-4+m) + 3\sqrt{\frac{6}{13}} \\
&\quad (-4+m)^2 + 2\sqrt{\frac{26}{33}}(-3+m) + \frac{3}{11}\sqrt{\frac{43}{2}}(-3+m)^2 + \sqrt{\frac{14}{65}}(-9+2m) + \frac{1}{35}\sqrt{\frac{139}{2}} \\
&\quad (-9+2m) + \frac{1}{3}\sqrt{\frac{62}{7}}(-5+2m) + \frac{4}{63}\sqrt{31}(-5+2m) + \frac{4}{9}\sqrt{\frac{97}{7}}(-3+2m) + \frac{2}{21}\sqrt{89} \\
&\quad (-3+2m) + \frac{1}{9}\sqrt{\frac{35}{2}}(-11+4m) + \frac{1}{78}\sqrt{\frac{155}{2}}(65-28m+3m^2).\end{aligned}$$

The $GA_5(\xi_3)$ index can be calculated from (8) as follows.

$$GA_5(\xi_3) = \sum_{\acute{m}\acute{n}\in E(\xi_3)} \frac{2\sqrt{S_{\acute{m}}S_{\acute{n}}}}{(S_{\acute{m}}+S_{\acute{n}})} = \sum_{\acute{m}\acute{n}\in E_j(\xi_3)}\sum_{j=10}^{44} \frac{2\sqrt{S_{\acute{m}}S_{\acute{n}}}}{(S_{\acute{m}}+S_{\acute{n}})}$$

$$\begin{aligned}
GA_5(\zeta_3) &= 1|E_{10}(\zeta_3)| + \frac{2}{25}\sqrt{154}|E_{11}(\zeta_3)| + \frac{6}{85}\sqrt{154}|E_{12}(\zeta_3)| + \frac{5}{29}\sqrt{33}|E_{13}(\zeta_3)| + \\
&\quad \frac{60}{61}|E_{14}(\zeta_3)| + \frac{30}{79}\sqrt{6}|E_{15}(\zeta_3)| + \frac{15}{44}\sqrt{7}|E_{16}(\zeta_3)| + \frac{3}{8}\sqrt{7}|E_{17}(\zeta_3)| + \frac{12}{13}|E_{18}(\zeta_3)| + \\
&\quad \frac{4}{23}\sqrt{33}|E_{19}(\zeta_3)| + \frac{6}{29}\sqrt{22}|E_{20}(\zeta_3)| + \frac{1}{27}\sqrt{473}|E_{21}(\zeta_3)| + 1|E_{22}(\zeta_3)| + \\
&\quad \frac{3}{10}\sqrt{11}|E_{23}(\zeta_3)| + \frac{4}{11}\sqrt{7}|E_{24}(\zeta_3)| + \frac{3}{20}\sqrt{31}|E_{25}(\zeta_3)| + \frac{4}{55}\sqrt{129}|E_{26}(\zeta_3)| + \\
&\quad \frac{3}{22}\sqrt{35}|E_{27}(\zeta_3)| + |E_{28}(\zeta_3)| + \frac{1}{21}\sqrt{341}|E_{29}(\zeta_3)| + \frac{4}{173}\sqrt{1419}|E_{30}(\zeta_3)| + \\
&\quad \frac{1}{23}\sqrt{385}|E_{31}(\zeta_3)| + \frac{1}{25}\sqrt{429}|E_{32}(\zeta_3)| + \frac{2}{13}\sqrt{42}|E_{33}(\zeta_3)| + \frac{6}{61}\sqrt{86}|E_{34}(\zeta_3)| + \\
&\quad 1|E_{35}(\zeta_3)| + \frac{12}{187}\sqrt{217}|E_{36}(\zeta_3)| + \frac{1}{32}\sqrt{903}|E_{37}(\zeta_3)| + \frac{12}{29}\sqrt{5}|E_{38}(\zeta_3)| + \\
&\quad \frac{1}{33}\sqrt{1085}|E_{39}(\zeta_3)| + \frac{4}{269}\sqrt{4515}|E_{40}(\zeta_3)| + \frac{4}{95}\sqrt{559}|E_{41}(\zeta_3)| + 1|E_{42}(\zeta_3)| + \\
&\quad \frac{1}{37}\sqrt{1365}|E_{43}(\zeta_3)| + 1|E_{44}(\zeta_3)|.
\end{aligned}$$

After some calculations, we have

$$\begin{aligned}
\implies GA_5(\zeta_3) &= 173.339 + \frac{96}{29}\sqrt{5}(-4+m) + \frac{24}{11}\sqrt{35}(-4+m) + \frac{32}{23}\sqrt{385}(-4+m) + \frac{12}{25}\sqrt{429} \\
&\quad (-4+m)^2 + \frac{12}{5}\sqrt{11}(-3+m) - 48m + 9m^2 + \frac{4}{37}\sqrt{1365}(-9+2m) + \frac{3}{2}\sqrt{7}(-5+2m) + \\
&\quad \frac{48}{13}(-3+2m) + \frac{32}{11}\sqrt{7}(-3+2m).
\end{aligned}$$

□

The graphical representations of topological indices of these networks are depicted in Figures 5 and 6 for certain values of m. By varying the different values of m, the graphs are increasing. These graphs show the correctness of the results.

Table 6. Edge partition of the third type of rectangular hex-derived network $RHDN_3(m)$ based on the sum of degrees of end vertices of each edge.

(τ_x, τ_y) Where $\acute{m}\acute{n} \in E(\zeta_3)$	Number of Edges	(τ_u, τ_v) Where $\acute{m}\acute{n} \in E(\zeta_3)$	Number of Edges
(22, 22)	2	(44, 44)	$6m^2 - 36m + 54$
(22, 28)	8	(44, 124)	8
(22, 63)	4	(44, 129)	12
(25, 33)	4	(44, 140)	$32m - 128$
(25, 36)	4	(44, 156)	$12m^2 - 96m + 192$
(25, 54)	4	(54, 63)	4
(25, 63)	4	(54, 129)	2
(28, 36)	$8m - 20$	(63, 63)	$4m - 10$
(28, 63)	$8m - 12$	(63, 124)	8
(33, 36)	4	(63, 129)	4
(33, 54)	4	(63, 140)	$8m - 32$
(33, 129)	4	(124, 140)	4
(36, 36)	$8m - 22$	(129, 140)	4
(36, 44)	$8m - 24$	(129, 156)	2
(36, 63)	$16m - 40$	(140, 140)	$4m - 18$
(36, 124)	16	(140, 156)	$8m - 36$
(36, 129)	8	(156, 156)	$3m^2 - 28m + 65$
(36, 140)	$16m - 64$	-	-

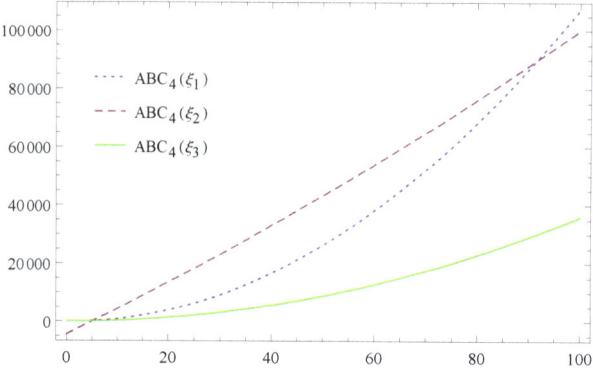

Figure 5. Comparison of ABC_4 index for ξ_1, ξ_2 and ξ_3.

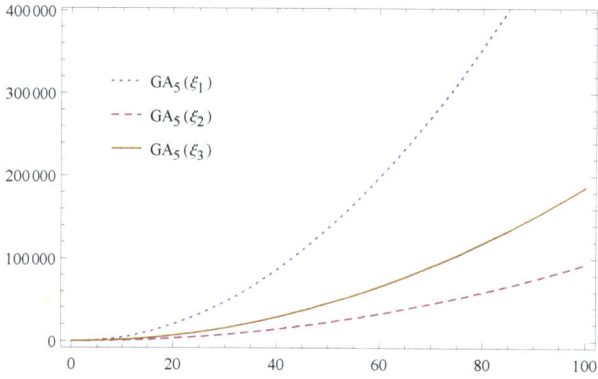

Figure 6. Comparison of GA_5 index for ξ_1, ξ_2 and ξ_3.

3. Conclusions

The study of topological descriptors are very useful to acquire the basic topologies of networks. In this paper, we find the exact results for Forgotten index, Balaban index, reclassified the Zagreb indices, ABC_4 index and GA_5 index of the Hex-derived networks of type 3. Due to their fascinating and challenging features, hex-derived networks have studied literature in relation to different graph-ideological parameters. However, their developmental circulatory features have been read for the foremost in this paper.

We are also very keen in designing some new networks and then study their topological indices which will be quite helpful to understand their primary priorities.

Author Contributions: Software, M.A.B.; validation, M.K.S. writing—original draft preparation, H.A.; writing—review and editing, W.G.; supervision, M.K.S.; funding acquisition, W.G.

Funding: This work has been partially supported by National Science Foundation of China (11761083).

Conflicts of Interest: The authors declare no conflict of interest.

References

1. Chen, M.S.; Shin, K.G.; Kandlur, D.D. Addressing, routing, and broadcasting in hexagonal mesh multiprocessors. *IEEE Trans. Comput.* **1990**, *39*, 10–18. [CrossRef]
2. Randić, M. On Characterization of molecular branching. *J. Am. Chem. Soc.* **1975**, *97*, 6609–6615. [CrossRef]
3. Furtula, B.; Gutman, I. A forgotten topological index. *J. Math. Chem.* **2015**, *53*, 1184–1190. [CrossRef]

4. Balaban, A.T. Highly discriminating distance-based topological index. *Chem. Phys. Lett.* **1982**, *89*, 399–404. [CrossRef]
5. Balaban, A.T.; Quintas, L.V. The smallest graphs, trees, and 4-trees with degenerate topological index. *J. Math. Chem.* **1983**, *14*, 213–233.
6. Ranjini, P.S.; Lokesha, V.; Usha, A. Relation between phenylene and hexagonal squeez using harmonic index. *Int. J. Graph Theory* **2013**, *1*, 116–121.
7. Estrada, E. Atom-bond connectivity and the energetic of branched alkanes. *Chem. Phys. Lett.* **2008**, *463*, 422–425. [CrossRef]
8. Estrada, E.; Torres, L.; Rodrigueza, L.; Gutman, I. An atom-bond connectivity index: modelling the enthalpy of formation of alkanes. *Indian J. Chem.* **1998**, *37*, 849–855.
9. Ghorbani, M.; Hosseinzadeh, M.A. Computing ABC_4 index of nanostar dendrimers. *Optoelectron. Adv. Mater. Rapid Commun.* **2010**, *4*, 1419–1422.
10. Graovac, A.; Ghorbani, M.; Hosseinzadeh, M.A. Computing fifth geometric-arithmetic index for nanostar dendrimers. *J. Math. Nanosci.* **2011**, *1*, 33–42.
11. Simonraj, F.; George, A. On the Metric Dimension of HDN3 and PHDN3. In Proceedings of the IEEE International Conference on Power, Control, Signals and Instrumentation Engineering (ICPCSI), Chennai, India, 21–22 September 2017; pp. 1333–1336.
12. Wei, C.C.; Ali, H.; Binyamin, M.A.; Naeem, M.N.; Liu, J.B. Computing Degree Based Topological Properties of Third Type of Hex-Derived Networks. *Mathematics* **2019**, *7*, 368. [CrossRef]
13. Ali, H.; Sajjad, A.; On further results of hex derived networks, *Open J. Discret. Appl. Math.* **2019**, *2(1)*, 32–40.
14. Bača, M.; Horváthová, J.; Mokrišová, M.; Semaničová-Feňovxcxíkovxax, A.; Suhányiovă, A. On topological indices of carbon nanotube network. *Can. J. Chem.* **2015**, *93*, 1157–1160. [CrossRef]
15. Baig, A.Q.; Imran, M.; Ali, H.; Omega, Sadhana and PI polynomials of benzoid carbon nanotubes, Optoelectron. *Adv. Mater. Rapid Commun.* **2015**, *9*, 248–255.
16. Baig, A.Q.; Imran, M.; Ali, H. On Topological Indices of Poly Oxide, Poly Silicate, DOX and DSL Networks. *Can. J. Chem.* **2015**, *93*, 730–739. [CrossRef]
17. Imran, M.; Baig, A.Q.; Ali, H. On topological properties of dominating David derived networks. *Can. J. Chem.* **2015**, *94*, 137–148. [CrossRef]
18. Imran, M.; Baig, A.Q.; Rehman, S.U.; Ali, H.; Hasni, M. Computing topological polynomials of mesh-derived networks. *Discret. Math. Algorithms Appl.* **2018**, *10*, 1850077. [CrossRef]
19. Imran, M.; Baig, A.Q.; Siddiqui, H.M.A.; Sarwar, M. On molecular topological properties of diamond like networks. *Can. J. Chem.* **2017**, *95*, 758–770. [CrossRef]
20. Simonraj, F.; George, A. Embedding of poly honeycomb networks and the metric dimension of star of david network. *GRAPH-HOC* **2012**, *4*, 11–28. [CrossRef]
21. Diudea, M.V.; Gutman, I.; Lorentz, J. *Molecular Topology*; Nova Science Publishers: Huntington, NY, USA, 2001.
22. Wiener, H. Structural determination of paraffin boiling points. *J. Am. Chem. Soc.* **1947**, *69*, 17–20. [CrossRef]

© 2019 by the authors. Licensee MDPI, Basel, Switzerland. This article is an open access article distributed under the terms and conditions of the Creative Commons Attribution (CC BY) license (http://creativecommons.org/licenses/by/4.0/).

Article

On Extended Adjacency Index with Respect to Acyclic, Unicyclic and Bicyclic Graphs

Bin Yang [1], Vinayak V. Manjalapur [2,*], Sharanu P. Sajjan [3] and Madhura M. Mathai [4] and Jia-Bao Liu [5]

1. Department of Computer Science and Technology, Hefei University, Hefei 230601, China
2. Department of Mathematics, KLE Society's, Basavaprabhu Kore Arts, Science and Commerce College, Chikodi 591201, Karnataka, India
3. Department of Computer Science, Government First Grade College for Women, Jamkhandi 587301, India
4. Department of Mathematics, KLE Society's, Raja Lakhamagouda Science Institute, Belgaum 590001, Karnataka, India
5. School of Mathematics and Physics, Anhui Jianzhu University, Hefei 230601, China
* Correspondence: vinu.m001@gmail.com

Received: 9 June 2019; Accepted: 18 July 2019; Published: 20 July 2019

Abstract: For a (molecular) graph G, the extended adjacency index $EA(G)$ is defined as Equation (1). In this paper we introduce some graph transformations which increase or decrease the extended adjacency (EA) index. Also, we obtain the extremal acyclic, unicyclic and bicyclic graphs with minimum and maximum of the EA index by a unified method, respectively.

Keywords: degree of vertex; extended adjacency index

1. Introduction

Molecular descriptors are playing an important role in Chemistry, Pharmacology, etc. Among them, topological indices have a prominent place. Topological indices (molecular structure descriptor) are numerical quantities of a molecular graphs (or simple graphs), that are invariant under graph isomorphism. And, are used to correlate with various physico chemical properties, chemical reactivity or biological activity. There are hundreds of topological indices that have found some applications in theoretical chemistry, especially in QSPR/QSAR research. Among all topological indices one of the most investigated are the degree based topological indices, among them, the old and widely studied topological index is Randić index [1], see the recent articles [2,3] and references cited there in. Recently researchers are studying various degree based topological indices such as Zagreb group indices [4–9], forgotten index [10–13], etc.

Let $G = (V, E)$ be a simple graph without loops and multiple edges. Let $V(G)$ and $E(G)$ be the vertex set and the edge set of G, respectively. The degree of a vertex u in G is the number of edges incident to it and is denoted by $d_G(u)$. For $v \in V(G)$ and $e \in E(G)$, let $N_G(v)$ be the set of all neighbors of v in G.

Extended adjacency index is one of the degree based topological descriptors which has been proposed by the authors Yang et al. [14] in 1994 and defined as, for any graph G extended adjacency (EA) index is:

$$EA = EA(G) = \sum_{uv \in E(G)} \frac{1}{2}\left(\frac{d_G(u)}{d_G(v)} + \frac{d_G(v)}{d_G(u)}\right). \tag{1}$$

In [14] Yang et al. described that EA index exhibits high discriminating power and correlate well with a number of physico chemical properties and biological activities of organic compounds. There are a couple of topological indices in the literature (see [15]) which are closely related to the extended adjacency index, and they are

$$EA^*(G) = \sum_{i<j}\left(\frac{d_i}{d_j}+\frac{d_j}{d_i}\right) = 2|E|\sum_{j=1}^{n}\frac{1}{d_j} - n$$

and

$$\hat{R}(G) = \sum_{i<j}\left(\frac{d_i}{d_j}+\frac{d_j}{d_i}\right)R_{ij}$$

where R_{ij} is the effective resistance between vertices i and j. Obviously, $EA(G) \leq EA^*(G)$, and all upper bounds for the inverse degree index $\rho(G) = \sum_{j=1}^{n}\frac{1}{d_j}$ can be used to furnish upper bonds for $EA^*(G)$ and $EA(G)$, even though they may not be tight for $EA(G)$.

Since 1994, neither extended adjacency matrix nor the extended adjacency index was taken into the consideration but in recent years only few articles have come out with its algebraic approach [16–18]. Ramane et al. determined the bounds for the EA index and characterizes graphs extremal with respect to them. Also, obtained relation between EA index and other well known topological indices. Moreover, determined the new results on EA index from an algebraic view point [19]. As an application, one can find a unified approach for some degree based topological descriptors in [20–25]. For other undefined notations refer [26,27].

Let S_n, P_n and C_n be the star, path and cycle on n vertices, respectively. Let $G - V$ be a subgraph of graph G by deleting vertex v and $G - e$ be a subgraph of graph G by deleting edge e. Let G_0 be a nontrivial graph and u be its vertex. If G is obtained by G_0 amalgamating a tree T at u. Then we say that T is a subtree of G and u is its root. Let $u \circ v$ denote the amalgamating two vertices u and v of G.

In the present work, we obtain extremal properties of the EA index. In Section 2, we present some graph transformations which increase or decrease EA index. In Section 3, we obtain extremal acyclic, unicyclic and bicyclic graphs with minimum and maximum EA index by a unified method, respectively.

2. Some Graph Transformations

In this section, we present some graph transformations which increase or decrease the EA index and these graph transformations play an important role to determine the extremal graphs of the EA index among acyclic, unicyclic and bicyclic graphs, respectively.

Transformation I. Let G_0 be a non-trivial connected graph and v is a given vertex in G_0. Let G_1 be a graph obtained from G_0 by attaching at v two paths $p : vu_1u_2\ldots u_k$ of length k and $Q : vw_1w_2\ldots w_l$ of length l. Let G_2 be a graph which is obtained from the graph G_1, by Transformation I, $G_2 = G_1 - vw_1 + u_kw_1$.

Lemma 1. *Let G_2 be a graph obtained from G_1 by Transformation I as shown in Figure 1, then*

$$EA(G_1) > EA(G_2).$$

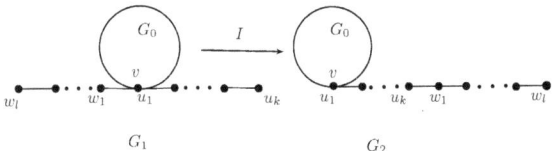

Figure 1. Transformation I.

Proof. In *Transformation I* degree of the vertex v is decreased and the degrees of its neighbor vertices $N_{G_0}(v)$ remains same value. Let us assume that $d_{G_1}(v) > 0$. Then by the definition of EA index, we have

$$EA(G_1) - EA(G_2) > \begin{bmatrix} \frac{1}{2}\left(\frac{d_{G_1}(v)}{d_{G_1}(w_1)} + \frac{d_{G_1}(w_1)}{d_{G_1}(v)}\right) + \frac{1}{2}\left(\frac{d_{G_1}(u_1)}{d_{G_1}(v)} + \frac{d_{G_1}(v)}{d_{G_1}(u_1)}\right) \\ + \frac{1}{2}\left(\frac{d_{G_1}(u_{k-1})}{d_{G_1}(u_k)} + \frac{d_{G_1}(u_k)}{d_{G_1}(u_{k-1})}\right) \end{bmatrix}$$

$$- \begin{bmatrix} \frac{1}{2}\left(\frac{d_{G_2}(u_k)}{d_{G_2}(w_1)} + \frac{d_{G_2}(w_1)}{d_{G_2}(u_k)}\right) + \frac{1}{2}\left(\frac{d_{G_2}(u_1)}{d_{G_2}(v)} + \frac{d_{G_2}(v)}{d_{G_2}(u_1)}\right) \\ + \frac{1}{2}\left(\frac{d_{G_2}(u_{k-1})}{d_{G_2}(u_k)} + \frac{d_{G_2}(u_k)}{d_{G_2}(u_{k-1})}\right) \end{bmatrix}$$

$$= \frac{(2+d_{G_0}(v))^2+4}{2(2+d_{G_0}(v))} + \frac{5}{4} - \left[\frac{(1+d_{G_0}(v))^2+4}{4(1+d_{G_0}(v))} + 2\right]$$

$$= \frac{d_{G_0}(v)}{4(1+d_{G_0}(v))}\left[\frac{d_{G_0}(v)^2}{(2+d_{G_0}(v))} + 3\right] > 0.$$

□

Remark 1. *By continuing the process of Transformation I, any tree T of size t connected to a graph G_1 can be changed into a path P with size $(t + 1)$ (i.e., P_{t+1}). From this process, we infer that EA index is strictly decreases.*

Transformation II. Let G_1 be a connected graph with an edge uv and $d_{G_1}(v) \geq 2$. Suppose that $N_{G_1}(u) = \{v, w_1, w_2, \ldots, w_t\}$ and $w_1, w_2, \ldots, w_t$ are pendent vertices. Let $G_2 = G_1 - \{uw_1, uw_2, \ldots, uw_t\} + \{vw_1, vw_2, \ldots, vw_t\}$.

We now show that *Transformation II* strictly increases the EA index of a graph.

Lemma 2. *Let G_2 be a graph obtained from G_1 by Transformation II as shown in Figure 2. Then*

$$EA(G_2) > EA(G_1).$$

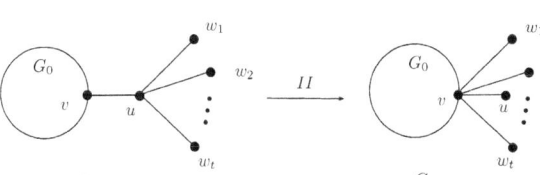

Figure 2. Transformation II.

Proof. Let $d_{G_0}(v) > 0$. In *Transformation II* $d_{G_2}(v) > d_{G_1}(v)$. So similar to the proof of Lemma 1, we have

$$EA(G_2) - E(G_1) > \left[\sum_{i=1}^{t} \frac{1}{2}\left(\frac{d_{G_2}(v)}{d_{G_2}(w_i)} + \frac{d_{G_2}(w_i)}{d_{G_2}(v)}\right) + \frac{1}{2}\left(\frac{d_{G_2}(u)}{d_{G_2}(v)} + \frac{d_{G_2}(v)}{d_{G_2}(u)}\right)\right]$$
$$- \left[\sum_{i=1}^{t} \frac{1}{2}\left(\frac{d_{G_1}(u)}{d_{G_1}(w_i)} + \frac{d_{G_1}(w_i)}{d_{G_1}(u)}\right) + \frac{1}{2}\left(\frac{d_{G_1}(u)}{d_{G_1}(v)} + \frac{d_{G_1}(v)}{d_{G_1}(u)}\right)\right]$$
$$= \frac{1}{2}\sum_{i=1}^{t}\left[\left(\frac{d_{G_2}(v)}{d_{G_2}(w_i)} + \frac{d_{G_2}(w_i)}{d_{G_2}(v)}\right) - \left(\frac{d_{G_1}(u)}{d_{G_1}(w_i)} + \frac{d_{G_1}(w_i)}{d_{G_1}(u)}\right)\right]$$
$$> 0.$$

□

Remark 2. *By continuing the process of Transformation II, any tree T of size t connected to a graph G_1 can be changed into a star S_{t+1}. And from this process EA index increases.*

Transformation III. Let G_1 be a non-trivial connected graph, u and v be two vertices of G_1. Let $P_l = v_1(=u)v_2\ldots v_l(=v)$ is a non-trivial path of length t connected to the vertices u and v in G_1. If $G_2 = G_1 - \{v_1v_2, v_2v_3,\ldots, v_{l-1}v_l\} + \{w(=u\circ v)v_1, wv_2,\ldots, wv_l\}$, see the Figure 3.

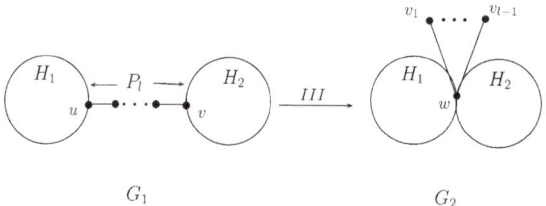

Figure 3. Transformation III.

Lemma 3. *Let G_2 be a connected graph obtained from G_1 by Transformation III as shown in Figure 3. Then*

$$EA(G_2) > EA(G_1).$$

Proof. Let $d_{H_1}(u) = x$ and $d_{H_2}(v) = y$, while w be the new vertex by merging u and v with $d_{G_2}(w) = x + y + l - 1$, with $l \geq 2$. We can easily get that $EA(G_2) - EA(G_1) > 0$, for $l = 2$. We now show that $EA(G_2) - EA(G_1) > 0$, for $l > 2$. From (1), we have

$$EA(G_2) - EA(G_1) > \frac{1}{2}\sum_{i=1}^{l-1}\left(\frac{d_{G_2}(w)}{d_{G_2}(v_i)} + \frac{d_{G_2}(v_i)}{d_{G_2}(w)}\right) - \left[\frac{1}{2}\left(\frac{x}{2} + \frac{2}{x}\right) + \frac{1}{2}\left(\frac{y}{2} + \frac{2}{y}\right) + (l-3)\right]$$
$$= (l-1)\frac{1}{2}\left(\frac{(x+y+l-1)}{1} + \frac{1}{(x+y+l-1)}\right)$$
$$- \left(\frac{x^2+4}{4x}\right) - \left(\frac{y^2+4}{4y}\right) - (l-3)$$
$$> \left[\frac{(x+y+l-1)^2+1}{2(x+y+l-1)} - \frac{x^2+4}{4x}\right] + \left[\frac{(x+y+l-1)^2+1}{2(x+y+l-1)} - \frac{y^2+4}{4y}\right] > 0.$$

□

Transformation IV. Let G_1 be a non-trivial connected graph and $x > 3$, $y > 3$ are two neighbors of vertex v_1. Assume that a pendent path $P = v_1v_2, v_2v_3, \ldots, v_{t-1}v_t$ is attached at v_1 in graph G_1, then $G_2 = G_1 - xv_1 + xv_t$, see Figure 4.

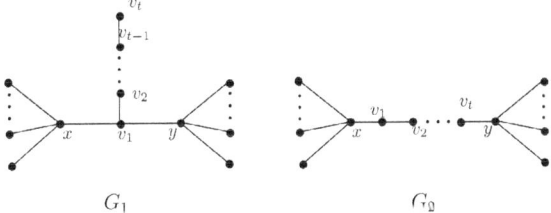

Figure 4. Transformation IV.

Lemma 4. *Let G_2 be a connected graph obtained from G_1 by Transformation IV. Then*

$$EA(G_2) > EA(G_1). \qquad (2)$$

Proof. By the definition of EA index, we have

$$
\begin{aligned}
EA(G_2) - E(G_1) >~ & \left[\tfrac{1}{2}\left(\tfrac{d_{G_2}(x)}{d_{G_2}(v_1)} + \tfrac{d_{G_2}(v_1)}{d_{G_2}(x)} \right) + \tfrac{1}{2}\left(\tfrac{d_{G_2}(v_1)}{d_{G_2}(v_2)} + \tfrac{d_{G_2}(v_2)}{d_{G_2}(v_1)} \right) \right. \\
& \left. + \tfrac{1}{2}\left(\tfrac{d_{G_2}(v_{t-1})}{d_{G_2}(v_t)} + \tfrac{d_{G_2}(v_t)}{d_{G_2}(v_{t-1})} \right) + \tfrac{1}{2}\left(\tfrac{d_{G_2}(v_t)}{d_{G_2}(y)} + \tfrac{d_{G_2}(y)}{d_{G_2}(v_t)} \right) \right] \\
& - \left[\tfrac{1}{2}\left(\tfrac{d_{G_1}(x)}{d_{G_1}(v_1)} + \tfrac{d_{G_1}(v_1)}{d_{G_1}(x)} \right) + \tfrac{1}{2}\left(\tfrac{d_{G_1}(v_1)}{d_{G_1}(v_2)} + \tfrac{d_{G_1}(v_2)}{d_{G_1}(v_1)} \right) \right. \\
& \left. + \tfrac{1}{2}\left(\tfrac{d_{G_1}(v_{t-1})}{d_{G_1}(v_t)} + \tfrac{d_{G_1}(v_t)}{d_{G_1}(v_{t-1})} \right) + \tfrac{1}{2}\left(\tfrac{d_{G_1}(v_1)}{d_{G_1}(y)} + \tfrac{d_{G_1}(y)}{d_{G_1}(v_1)} \right) \right] \\
=~ & \left[\tfrac{1}{2}\left(\tfrac{x}{2} + \tfrac{2}{x} \right) + 2 + \tfrac{1}{2}\left(\tfrac{2}{y} + \tfrac{y}{2} \right) \right] \\
& - \left[\tfrac{1}{2}\left(\tfrac{x}{3} + \tfrac{3}{x} \right) + \tfrac{13}{12} + \tfrac{5}{4} + \tfrac{1}{2}\left(\tfrac{3}{y} + \tfrac{y}{3} \right) \right] \\
=~ & \tfrac{x^2+4}{4x} - \left(\tfrac{x^2+9}{6x} \right) + \tfrac{y^2+4}{4x} - \left(\tfrac{y^2+9}{6y} \right) - \tfrac{1}{3} > 0.
\end{aligned}
$$

□

Transformation V: Let G_0 be a non-trivial connected graph. Let u and v be a pair of equivalent vertices in G_0 with $d_{G_0}(u) = d_{G_0}(v) = x$ and G_1 be a graph obtained by attaching S_{k+1} and S_{l+1} at the vertices u and v of G_0 with $k \geq l$, respectively. If G_2 is the graph obtained by deleting the l pendent vertices at v in G_1 and connecting them to the vertex u of G, respectively, see Figure 5.

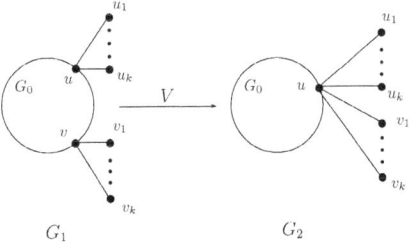

Figure 5. Transformation V.

Lemma 5. *Let G_2 be a connected graph obtained from G_1 by Transformation V. Then*
$$EA(G_2) > EA(G_1).$$

Proof. Let $k \geq l \geq 1$. By (1), we have

$$\begin{aligned}
EA(G_2) - EA(G_1) &> \frac{1}{2}\left(\sum_{i=1}^{k}\frac{d_{G_2}(u)}{d_{G_2}(u_i)} + \frac{d_{G_2}(u_i)}{d_{G_2}(u)}\right) - \frac{1}{2}\left(\sum_{i=1}^{k}\frac{d_{G_1}(u)}{d_{G_1}(u_i)} + \frac{d_{G_1}(u_i)}{d_{G_1}(u)}\right) \\
&+ \frac{1}{2}\left(\sum_{i=1}^{l}\frac{d_{G_2}(u)}{d_{G_2}(v_i)} + \frac{d_{G_2}(v_i)}{d_{G_2}(u)}\right) - \frac{1}{2}\left(\sum_{i=1}^{l}\frac{d_{G_1}(v)}{d_{G_1}(v_i)} + \frac{d_{G_1}(v_i)}{d_{G_1}(v)}\right) \\
&= k\left(\frac{1}{2}\left(\frac{d_{G_2}(u)}{1} + \frac{1}{d_{G_2}(u)}\right)\right) - \frac{1}{2}\left(\frac{d_{G_1}(u)}{1} + \frac{1}{d_{G_1}(u)}\right) \\
&+ l\left(\frac{1}{2}\left(\frac{d_{G_2}(u)}{1} + \frac{1}{d_{G_2}(u)}\right)\right) - \frac{1}{2}\left(\frac{d_{G_1}(v)}{1} + \frac{1}{d_{G_1}(v)}\right) \\
&= k + l > 0.
\end{aligned}$$

□

Remark 3. *From Lemmas 3–5, we can say that Transformation III, Transformation IV and Transformation V increases the EA index of a graph respectively.*

3. Main Results

In this section, we determine the extremal EA index of graphs from $\mathcal{A}_n, \mathcal{U}_n$ and $\mathcal{B}_n$, respectively by a unified method.

Let $\mathcal{A}_n, \mathcal{U}_n$ and $\mathcal{B}_n$ are the set of connected acyclic, unicyclic and bicyclic graphs of order n respectively. Let $C_n(p,q)$ be the graph contains two cycles C_p and C_q having a common vertex with $p+q-1 = n$, $P_n^{k,l,m}$ be the graph obtained by connecting two cycles C_k and C_m with a path P_l with $k+l+m-2 = n$ and $C_n(r,l,t)$ be the graph obtained by joining two triples of pendent vertices of three paths P_l, P_r and P_t to two vertices with $l+r+t-4 = n$. (without loss of generality, we set $2 \leq l \leq r \leq t$). If a bicyclic graph contains one of the three graphs which are depicted in Figure 6 as its subgraph then we have three subsets of $\mathcal{B}_n$ as $\mathcal{B}_n^1 = \{C_n(p,q) : p+q-1 = n\}$, $\mathcal{B}_n^2 = \{C_n(r,l,t) : l+r+t-4 = n\}$ and $\mathcal{B}_n^3 = \{P_n^{k,l,m} : k+l+m-2 = n\}$. So the set $\mathcal{B}_n$ can be partitioned into three subsets $\mathcal{B}_n^1, \mathcal{B}_n^2$ and $\mathcal{B}_n^3$.

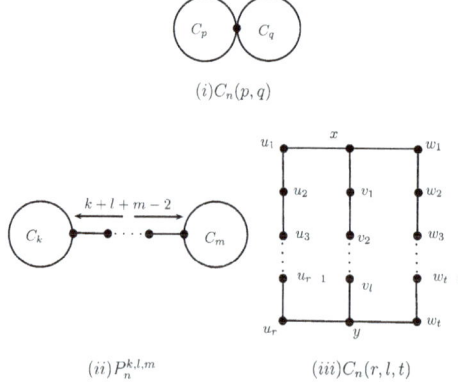

Figure 6. Subgraphs of $\mathcal{B}_n$.

The following theorem gives the minimum and maximum value of the EA index.

Theorem 1. *Let G be a acyclic connected graph with order n. Then*

$$EA(P_n) \leq EA(G) \leq EA(S_n).$$

The lower bound and upper bound is attained iff $G \cong P_n$ and $G \cong S_n$ respectively.

Proof. By using Lemmas 1 and 2 above inequalities holds good.
□

The graphs which are depicted in Figure 7 will be used in the following proof.

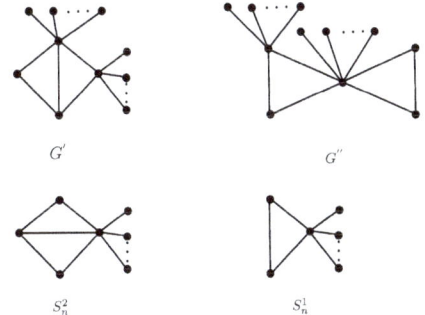

Figure 7. The graphs which are used in the later proof.

Theorem 2. *Let G be a unicyclic graph with order n. Then*

$$EA(C_n) \leq EA(G) \leq EA(S_n^1),$$

where the lower bound and upper bound is attained iff $G \cong C_n$ and $G \cong S_n^1$ respectively.

Proof. Let G contains a uniquely cycle C_l and by Lemma 3 we obtain the graph G_2 in which the size of the cycle is three and its EA index is strictly increased. Moreover, from Lemma 5, we can get the uniquely maximum graph S_n^1 with respect to EA index (see Figure 7). On the other hand, by Lemma 1 we conclude that the minimum graph is C_n.
□

Theorem 3. *Let G be a bicyclic graph with n vertices. Then*

$$n + \frac{3}{2} \leq EA(G) \leq \frac{[(n-1)^2 + 1][3n - 8] + 17}{6(n-1)} + \frac{13}{6}, \qquad (3)$$

where the lower bound and upper bound is attained iff $G \in \{P_n^{k,l,m} : l \geq 3\} \cup \{C_n(r,l,t) : l \geq 2\}$ and $G \cong S_n^2$ respectively.

Proof. Firstly, we have to prove the upper bound for the bicyclic graph with respect to EA index. Suppose G is isomorphic to S_n^2 (or $G \cong S_n^2$), then from (1), we get

$$EA(G) = \frac{[(n-1)^2 + 1][3n - 8] + 17}{6(n-1)} + \frac{13}{6}.$$

Next, we show that $EA(G) < EA(S_n^2)$ for G is not isomorphic to S_n^2.

Case 1: $K_4 - e$ is the subgraph of G.

If $K_4 - e$ is the subgraph of a graph G, then from Lemmas 2 and 5 we obtain G' as a new (bicyclic) graph whose EA index is more than that of G (see Figure 7). One can easily check that $EA(G) = \frac{[(n-1)^2+1][3n-8]+17}{6(n-1)} + \frac{13}{6}$, equality attains iff $G \cong S_n^2$.

Case 2: $K_4 - e$ is not the subgraph of G.

From Lemma 3 we can say that may be there are a bicyclic graph whose EA index is more than that of graph G has the subgraph $K_4 - e$. Hence following two subcases exist.

Subcase 2.1: G contains $C_s(3,2,m)$ as a subgraph.

By Lemma 3 **Subcase 2.1** deduce to **Case 1**.

Subcase 2.2: $C_s(3,2,m)$ is not a subgraph of G.

If $C_s(3,2,m)$ is not a subgraph of G, then from Lemmas 2, 3 and 5, we will have a new graph G'' whose EA index is more than that of G, see Figure 7. It is easy to verify that $EA(G) < \frac{[(n-1)^2+1][3n-8]+17}{6(n-1)} + \frac{13}{6}$.

Furthermore, We have to show the lower bound. By Lemmas 1, 2 and 4, we infer that the extremal graph of the minimum EA index in bicyclic graphs must be the element which belongs to the set $\{\mathcal{B}_n^1, \mathcal{B}_n^2, \mathcal{B}_n^3\}$.

We easily get that $EA(C_n(p,q)) = n+2$; $EA(P_n^{k,l,m}) = n + \frac{17}{12}$ if $l = 2$ and $EA(P_n^{k,l,m}) = n + \frac{3}{2}$, otherwise; $EA(C_n(r,l,t)) = n + \frac{3}{2}$ if $l \geq 2$. Hence the lower bound and the equality attains iff $G \in \{P_n^{k,l,m} : l \geq 3\} \cup \{C_n(r,l,t) : l \geq 2\}$. □

Author Contributions: All the authors contributed equally to preparing this article.

Funding: This work is partially supported by the China Postdoctoral Science Foundation under grant No. 2017M621579 and the Postdoctoral Science Foundation of Jiangsu Province under grant No. 1701081B. Project of Anhui Jianzhu University under Grant no. 2016QD116 and 2017dc03. Supported by Major University Science Research Project of Anhui Province (KJ2016A605), Major Nature Science Project of Hefei University Research and Development Foundation (16ZR13ZDA).

Acknowledgments: The authors are extremely grateful to the anonymous referees for their valuable comments and suggestions, which have led to an improved version of the paper.

Conflicts of Interest: The authors declare no conflict of interest.

References

1. Randić, M. On characterization of molecular branching. *J. Am. Chem. Soc.* **1975**, *97*, 6609–6615. [CrossRef]
2. Gutman, I.; Furtula, B.; Katanić, V. Randic index and information. *AKCE Int. J. Graphs Comb.* **2018**, *15*, 307–312. [CrossRef]
3. Suil, O.; Shi, Y. Sharp bounds for the Randic index of graphs with given minimum and maximum degree. *Discrete Appl. Math.* **2018**, *247*, 111–115.
4. Bedratyuk, L.; Savenko, O. The star sequence and the general first Zagreb index. *MATCH Commun. Math. Comput. Chem.* **2018**, *79*, 407–414.
5. Gutman, I.; Milovanović, E.; Milovanović, I. Beyond the Zagreb indices. *AKCE Int. J. Graphs Comb.* **2018**, in press. [CrossRef]
6. Liu, J.B.; Javed, S.; Javaid, M.; Shabbir, K. Computing first general Zagreb index of operations on graphs. *IEEE Acess* **2019**, *7*, 47494–47502. [CrossRef]
7. Liu, J.B.; QAli, B; Malik, A.M.; Hafiz, S.A.M.; Imran, M. Reformulated Zagreb indices of some derived graphs. *Mathematics* **2019**, *7*, 366. [CrossRef]
8. Liu, J.B.; Wang, C.; Wang, S. Zagreb indices and multiplicative Zagreb indices of Eulerian graphs. *Bull. Malays. Math. Sci. Soc.* **2019**, *42*, 67–78. [CrossRef]
9. Ramane H.S.; Manjalapur, V.V.; Gutman, I. General sum-connectivity index, general product-connectivity index, general Zagreb index and coindices of line graph of subdivision graphs. *AKCE Int. J. Graphs Comb.* **2017**, *14*, 92–100. [CrossRef]

10. Akhter, S.; Imran, M. Computing the forgotten topological index of four operations on graphs. *AKCE Int. J. Graphs Comb.* **2017**, *14*, 70–79. [CrossRef]
11. Che, Z.; Chen, Z. Lower and upper bounds of the forgotten topological index. *MATCH Commun. Math. Comput. Chem.* **2017**, *76*, 635–648.
12. Elumalai, S.; Mansour, T. A short note on tetracyclic graphs with extremal values of Randić index. *Asian-Eur. J. Math.* **2019**, in press. [CrossRef]
13. Elumalai, S.; Mansour, T.; Rostami, A.M. On the bounds of the forgotten topological index. *Turk. J. Math.* **2017**, *41*, 1687–1702. [CrossRef]
14. Yang, Y.Q.; Xu, L.; Hu, C.Y. Extended adjacency matrix indices and their applications. *J. Chem. Inf. Comput. Sci.* **1994**, *34*, 1140–1145. [CrossRef]
15. Bianchi, M.; Cornaro, A.; Palacios, J.L.; Torriero, A. Upper and lower bounds for the mixed degree-Kirchhoff index. *Filomat* **2016**, *30*, 2351–2358. [CrossRef]
16. Adiga, C.; Rakshit, B.R. Upper bounds for extended energy of graphs and some extended equienergetic graphs. *Opusc. Math.* **2018**, *38*, 5–13. [CrossRef]
17. Gutman, I. Relation between energy and extended energy of a graph. *Int. J. Appl. Graph Theory* **2017**, *1*, 42–48.
18. Gutman, I.; Furtula, B.; Das, K.C. Extended energy and its dependence on molecular structure. *Can. J. Chem.* **2017**, *95*, 526–529. [CrossRef]
19. Ramane, H.S.; Manjalapur, V.V.; Gudodagi, G.A. On extended adjacency index and its spectral properties. **2019**, Communicated.
20. Ali, A.; Dimitrov, D. On the extremal graphs with respect to bond incident degree indices. *Discrete Appl. Math.* **2018**, *238*, 32–40. [CrossRef]
21. Deng, H.Y. A unified approach to the extremal Zagreb indices for trees, unicyclic graphs and bicyclic graphs. *MATCH Commun. Math. Comput. Chem.* **2007**, *57*, 597–616.
22. Ji, S.; Li, X.; Huo, B. On reformulated Zagreb indices with respect to acyclic, unicyclic and bicyclic graphs. *MATCH Commun. Math. Comput. Chem.* **2014**, *72*, 723–732.
23. Liu, J.B.; Pan, X.F. A unified approach to the asymptotic topological indices of various lattices. *Appl. Math. Comput.* **2015**, *270*, 62–73. [CrossRef]
24. Pei, L.; Pan, X. Extremal values on Zagreb indices of trees with given distance k-domination number. *J. Inequal. Appl.* **2018**, *2018*, 16. [CrossRef] [PubMed]
25. Xu, K.; Hua, H. A unified approach to extremal multiplicative Zagreb indices for trees, unicyclic and bicyclic graphs. *MATCH Commun. Math. Comput. Chem.* **2012**, *68*, 241–256.
26. Bondy, J.A.; Murty, U.S.R. *Graph Theory and Its Applications*; MacMillan: London, UK, 1976.
27. Harary, F. *Graph Theory*; Addison Wesley Reading: Reading, MA, USA, 1969.

 © 2019 by the authors. Licensee MDPI, Basel, Switzerland. This article is an open access article distributed under the terms and conditions of the Creative Commons Attribution (CC BY) license (http://creativecommons.org/licenses/by/4.0/).

Article
Join Products $K_{2,3} + C_n$

Michal Staš

Faculty of Electrical Engineering and Informatics, Technical University of Košice, 042 00 Košice, Slovakia; michal.stas@tuke.sk

Received: 7 May 2020; Accepted: 1 June 2020; Published: 5 June 2020

Abstract: The crossing number $cr(G)$ of a graph G is the minimum number of edge crossings over all drawings of G in the plane. The main goal of the paper is to state the crossing number of the join product $K_{2,3} + C_n$ for the complete bipartite graph $K_{2,3}$, where C_n is the cycle on n vertices. In the proofs, the idea of a minimum number of crossings between two distinct configurations in the various forms of arithmetic means will be extended. Finally, adding one more edge to the graph $K_{2,3}$, we also offer the crossing number of the join product of one other graph with the cycle C_n.

Keywords: graph; join product; crossing number; cyclic permutation; arithmetic mean

1. Introduction

For the first time, P. Turán described the brick factory problem. He was forced to work in a brickyard and his task was to push the bricks of the wagons along the line from the kiln to the storage location. The factory contained several furnaces and storage places, between which sidewalks passed through the floor. Turán found it difficult to move the wagon through the track passage, and in his mind he began to consider how the factory could be redesigned to minimize these crossings. Since then, the topic has steadily grown and research into the number of crosses has become one of the main areas of graph theory. This problem of reducing the number of crossings on the edges of graphs were studied in many areas.

The crossing number $cr(G)$ of a simple graph G with the vertex set $V(G)$ and the edge set $E(G)$ is the minimum possible number of edge crossings in a drawing of G in the plane (for the definition of a drawing see [1].) It is easy to see that a drawing with minimum number of crossings (an optimal drawing) is always a good drawing, meaning that no edge crosses itself, no two edges cross more than once, and no two edges incident with the same vertex cross. Let D ($D(G)$) be a good drawing of the graph G. We denote the number of crossings in D by $cr_D(G)$. Let G_i and G_j be edge-disjoint subgraphs of G. We denote the number of crossings between edges of G_i and edges of G_j by $cr_D(G_i, G_j)$, and the number of crossings among edges of G_i in D by $cr_D(G_i)$. It is easy to see that for any three mutually edge-disjoint subgraphs G_i, G_j, and G_k of G, the following equations hold:

$$cr_D(G_i \cup G_j) = cr_D(G_i) + cr_D(G_j) + cr_D(G_i, G_j),$$

$$cr_D(G_i \cup G_j, G_k) = cr_D(G_i, G_k) + cr_D(G_j, G_k).$$

By Garey and Johnson [2] we already know that calculating the crossing number of a simple graph is an NP-complete problem. Recently, the exact values of the crossing numbers are known only for some special classes of graphs. In [3], Ho gave the characterization for a few multipartite graphs. So, the main purpose of this work is to extend the results concerning this topic for the complete bipartite graph $K_{2,3}$ on five vertices. In this paper we use definitions and notations of the crossing number theory presented by Klešč in [4]. In the proofs we will also use the Kleitman's result [5] on the crossing numbers of the complete bipartite graphs. He estimated that

$$\mathrm{cr}(K_{m,n}) = \left\lfloor \frac{m}{2} \right\rfloor \left\lfloor \frac{m-1}{2} \right\rfloor \left\lfloor \frac{n}{2} \right\rfloor \left\lfloor \frac{n-1}{2} \right\rfloor, \quad \text{with } \min\{m,n\} \leq 6.$$

Again using Kleitman's result [5], the exact values of the crossing numbers for the join product of two paths, the join product of two cycles, and also for the join product of a path and a cycle were proved in [4]. Further, some values for crossing numbers of $G + D_n$, $G + P_n$, and of $G + C_n$ for arbitrary graph G at most on four vertices are estimated in [6,7]. Let us note that the exact values for the crossing numbers of the join product G with P_n and C_n were also investigated for a few graphs G of order five and six in [1,8–12]. In all mentioned cases, the graph G contains usually at least one cycle and it is connected.

It is important to note that the methods in this paper will mostly use several combinatorial properties on cyclic permutations. If we place the graph $K_{2,3}$ on the surface of the sphere, from the topological point of view, the resulting number of crossings of $K_{2,3} + C_n$ does not matter which of the regions in the subdrawing of $K_{2,3} \cup T^i$ is unbounded, but on how the subgraph T^i crosses or does not cross the edges of $K_{2,3}$ (the description of T^i will be justified in Section 2). This representation of T^i can best be described by the idea of a configuration utilizing some cyclic permutation on the pre-numbered vertices of the graph $K_{2,3}$. We introduce a new idea of various form of arithmetic means on a minimum number of crossings between two corresponding subgraphs T^i and T^j. Certain parts of proofs can be also simplified with the help of software which generates all cyclic permutations of five elements due to Berežný and Buša [13].

2. Possible Drawings of $K_{2,3}$ and Preliminary Results

Let us first consider the join product of the complete bipartite graph $K_{2,3}$ with the discrete graph D_n considered on n vertices. It is not difficult to see that the graph $K_{2,3} + D_n$ contains just one copy of the graph $K_{2,3}$ and n vertices $t_1, \ldots, t_n$, where each vertex t_i, $i = 1, \ldots, n$, is adjacent to every vertex of $K_{2,3}$. For $1 \leq i \leq n$, let T^i denote the subgraph which is uniquely induced by the five edges that are incident with the fixed vertex t_i. This means that the graph $T^1 \cup \cdots \cup T^n$ is isomorphic to the graph $K_{5,n}$ and we obtain

$$K_{2,3} + D_n = K_{2,3} \cup K_{5,n} = K_{2,3} \cup \left(\bigcup_{i=1}^{n} T^i \right). \quad (1)$$

The graph $K_{2,3} + C_n$ contains $K_{2,3} + D_n$ as a subgraph. For all subgraphs of the graph $K_{2,3} + C_n$ which are also subgraphs of the graph $K_{2,3} + nK_1$ we can use the same notations as above. Let C_n^* denote the cycle induced on n vertices of $K_{2,3} + C_n$ but which do not belong to the subgraph $K_{2,3}$. Hence, C_n^* consists of the vertices $t_1, t_2, \ldots, t_n$ and of the edges $\{t_i, t_{i+1}\}$ and $\{t_n, t_1\}$ for $i = 1, \ldots, n-1$. So we get

$$K_{2,3} + C_n = K_{2,3} \cup K_{5,n} \cup C_n^* = K_{2,3} \cup \left(\bigcup_{i=1}^{n} T^i \right) \cup C_n^*. \quad (2)$$

In the paper, the definitions and notation of the cyclic permutations and of the corresponding configurations of subgraphs for a good drawing D of the graph $K_{2,3} + D_n$ presented in [14] are used. By Hernández-Vélez et al. [15], the cyclic permutation that records the (cyclic) counter-clockwise order in which the edges leave a vertex t_i is said to be the rotation $\mathrm{rot}_D(t_i)$ of the vertex t_i. On the basis of this, we use the notation (12345) if the counter-clockwise order the edges incident with the vertex t_i is $t_iv_1, t_iv_2, t_iv_3, t_iv_4$, and t_iv_5. Recall that any such rotation is a cyclic permutation. For our research, we will separate all subgraphs T^i of $K_{2,3} + D_n$, $i = 1, 2, \ldots, n$, into three families of subgraphs depending on how many times are edges of $K_{2,3}$ crossed by the edges of the considered subgraph T^i in D. For $i = 1, 2, \ldots, n$, let $R_D = \{T^i : \mathrm{cr}_D(K_{2,3}, T^i) = 0\}$, and $S_D = \{T^i : \mathrm{cr}_D(K_{2,3}, T^i) = 1\}$. The edges of $K_{2,3}$ are crossed at least twice by each other subgraph T^i in D. For $T^i \in R_D \cup S_D$, let

F^i denote the subgraph $K_{2,3} \cup T^i$, $i \in \{1, 2, \ldots, n\}$, of $K_{2,3} + D_n$. Clearly, the idea of dividing the subgraphs T^i into three mentioned families is also retained in all drawings of the graph $K_{2,3} + C_n$. In [14], there are two possible non isomorphic drawings of the graph $K_{2,3}$, but only with the possibility of obtaining a subgraph $T^i \in R_D$ in D. Due to the arguments in the proof of Theorem 2, if we wanted to get an optimal drawing D of $K_{2,3} + C_n$, then the subdrawing $D(K_{2,3})$ of the graph $K_{2,3}$ induced by D with at least three crossings among the edges of $K_{2,3}$ forces that the set R_D must be nonempty. But, in the cases of $\mathrm{cr}_D(K_{2,3}) \le 2$, just one of the sets R_D or S_D can be empty. With these assumptions, we obtain four non isomorphic drawings of the graph $K_{2,3}$ as shown in Figure 1. The vertex notation of $K_{2,3}$ will be substantiated later in all mentioned drawings, and wherein two disjoint independent sets of vertices of the complete bipartite graph $K_{2,3}$ will be also highlighted by filled and non filled rings.

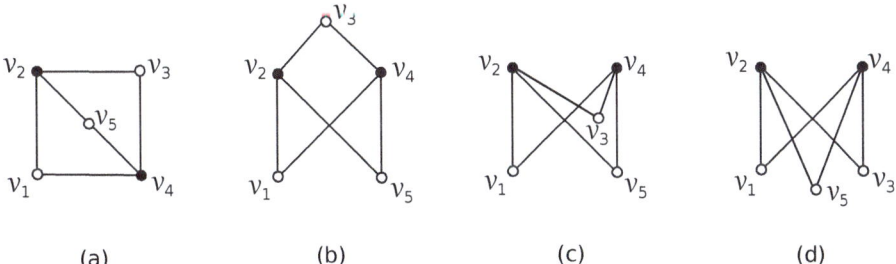

Figure 1. Four possible non isomorphic drawings of the graph $K_{2,3}$. (**a**) the planar drawing of $K_{2,3}$; (**b**) the drawing of $K_{2,3}$ with one crossing on edges of $K_{2,3}$; (**c**) the drawing of $K_{2,3}$ with $\mathrm{cr}_D(K_{2,3}) = 2$; (**d**) the drawing of $K_{2,3}$ with $\mathrm{cr}_D(K_{2,3}) = 3$.

3. The Crossing Number of $K_{2,3} + C_n$

In the proofs of the paper, several parts are based on the Theorem 1 presented in [14].

Theorem 1. $\mathrm{cr}(K_{2,3} + D_n) = 4 \left\lfloor \frac{n}{2} \right\rfloor \left\lfloor \frac{n-1}{2} \right\rfloor + n$ *for any* $n \ge 1$.

Now we are able to prove the main results of the paper. The exact values of the crossing numbers of the small graphs $K_{2,3} + C_3$, $K_{2,3} + C_4$, and $K_{2,3} + C_5$ can be estimated using the algorithm located on the website http://crossings.uos.de/ provided that it uses an ILP formulation based on Kuratowski subgraphs and also generates verifiable formal proofs, for more see [16].

Lemma 1. $\mathrm{cr}(K_{2,3} + C_3) = 10$, $\mathrm{cr}(K_{2,3} + C_4) = 15$, *and* $\mathrm{cr}(K_{2,3} + C_5) = 24$.

Recall that two vertices t_i and t_j of $K_{2,3} + C_n$ are antipodal in a drawing of $K_{2,3} + C_n$ if the subgraphs T^i and T^j do not cross, and a drawing is said to be antipodal-free if it does not have antipodal vertices. The result in the following Theorem 2 has already been claimed by Yuan [17]. The correctness of an article written in Chinese cannot be verified because compilers cannot handle it. Therefore, such results can only be considered as unconfirmed hypotheses.

Theorem 2. $\mathrm{cr}(K_{2,3} + C_n) = 4 \left\lfloor \frac{n}{2} \right\rfloor \left\lfloor \frac{n-1}{2} \right\rfloor + n + 3$ *for* $n \ge 3$.

Proof of Theorem 2. In Figure 2 there is the drawing of $K_{2,3} + C_n$ with $4 \left\lfloor \frac{n}{2} \right\rfloor \left\lfloor \frac{n-1}{2} \right\rfloor + n + 3$ crossings. Thus, $\mathrm{cr}(K_{2,3} + C_n) \le 4 \left\lfloor \frac{n}{2} \right\rfloor \left\lfloor \frac{n-1}{2} \right\rfloor + n + 3$. Theorem 2 is true for $n = 3$, $n = 4$, and $n = 5$ by Lemma 1. Assume $n \ge 6$. We prove the reverse inequality by contradiction. Suppose now that there is a drawing D of $K_{2,3} + C_n$ with

$$\mathrm{cr}_D(K_{2,3} + C_n) < 4 \left\lfloor \frac{n}{2} \right\rfloor \left\lfloor \frac{n-1}{2} \right\rfloor + n + 3 \tag{3}$$

and that
$$\text{cr}(K_{2,3} + C_m) \geq 4\left\lfloor\frac{m}{2}\right\rfloor\left\lfloor\frac{m-1}{2}\right\rfloor + m + 3 \quad \text{for each } 3 \leq m < n. \tag{4}$$

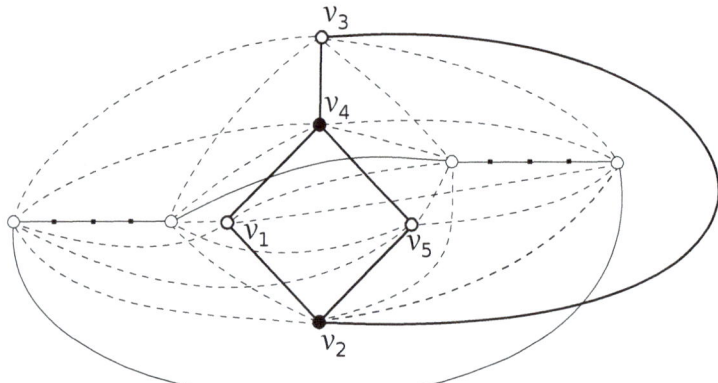

Figure 2. The good drawing of $K_{2,3} + C_n$ with $4\lfloor\frac{n}{2}\rfloor\lfloor\frac{n-1}{2}\rfloor + n + 3$ crossings.

Let us show that the considered drawing D must be antipodal-free. For a contradiction suppose, with the rest of paper, that $\text{cr}_D(T^{n-1}, T^n) = 0$. It is not difficult to verify in Figure 1 that if at least one of T^{n-1} and T^n, say T^n, does not cross the edges of the graph $K_{2,3}$, then the edges of T^{n-1} must cross the edges of $K_{2,3} \cup T^n$ at least twice, that is, $\text{cr}_D(K_{2,3}, T^{n-1} \cup T^n) \geq 2$. By [5], we already know that $\text{cr}(K_{5,3}) = 4$, which yields that each T^k, $k \neq n-1, n$, have to cross the edges of the subgraph $T^{n-1} \cup T^n$ at least four times. On the basis of this, we have

$$\text{cr}_D(K_{2,3} + C_n) = \text{cr}_D\left(K_{2,3} + C_{n-2}\right) + \text{cr}_D(K_{5,n-2}, T^{n-1} \cup T^n) + \text{cr}_D(K_{2,3}, T^{n-1} \cup T^n) + \text{cr}_D(T^{n-1} \cup T^n)$$

$$\geq 4\left\lfloor\frac{n-2}{2}\right\rfloor\left\lfloor\frac{n-3}{2}\right\rfloor + n - 2 + 3 + 4(n-2) + 2 + 0 = 4\left\lfloor\frac{n}{2}\right\rfloor\left\lfloor\frac{n-1}{2}\right\rfloor + n + 3.$$

This contradicts the assumption (3) and consequently confirms that D must be antipodal-free. As the graph $K_{2,3} + D_n$ is a subgraph of the graph $K_{2,3} + C_n$, by Theorem 1, the edges of $K_{2,3} + C_n$ are crossed at least $4\lfloor\frac{n}{2}\rfloor\lfloor\frac{n-1}{2}\rfloor + n$ times, and therefore, at most two edges of the cycle C_n^* can be crossed in D. This also enforces that the vertices t_i of the cycle C_n^* must be placed at most in two different regions in the considered good subdrawing of $K_{2,3}$. If $r = |R_D|$ and $s = |S_D|$, then the assumption (4) together with $\text{cr}(K_{5,n}) = 4\lfloor\frac{n}{2}\rfloor\lfloor\frac{n-1}{2}\rfloor$ enforce that there is at least one subgraph T^i which crosses the edges of $K_{2,3}$ at most once in the drawing D. To be precise,

$$\text{cr}_D(K_{2,3}) + \text{cr}_D(K_{2,3}, K_{5,n}) \leq \text{cr}_D(K_{2,3}) + 0r + s + 2(n - r - s) < n + 3,$$

that is,
$$\text{cr}_D(K_{2,3}) + s + 2(n - r - s) \leq n + 2. \tag{5}$$

This implies that $2r + s \geq n + \text{cr}_D(K_{2,3}) - 2$. Further, if $\text{cr}_D(K_{2,3}) = 0$ and $r = 0$, then $s \geq n - 2$. Now, we will deal with the possibilities of obtaining a subgraph $T^i \in R_D \cup S_D$ in D and we will exhibit that in all mentioned cases a contradiction with the assumption (3) is achieved.

Case 1: $\text{cr}_D(K_{2,3}) = 0$. In this case, without lost of generality, we can assume the drawing with the vertex notation of $K_{2,3}$ as shown in Figure 1a. The unique subdrawing of $K_{2,3}$ induced by D contains three different regions. Hence, let us denote these three regions by $\omega_{1,4,3,2}$, $\omega_{1,4,5,2}$, and $\omega_{2,5,4,3}$ depending on which of vertices are located on the boundary of the corresponding region. Since the vertices of C_n^* do not have to be placed in the same region in the considered subdrawing of $K_{2,3}$, two possible subcases may occur:

(a) All vertices of C_n^* are placed in two regions of subdrawing of $K_{2,3}$ induced by D. In the rest of paper, based on their symmetry, we can suppose that all vertices t_i of C_n^* are placed in $\omega_{1,4,3,2} \cup \omega_{1,4,5,2}$. Of course, there is no possibility to obtain a subdrawing of $K_{2,3} \cup T^i$ for a $T^i \in R_D$, that is, $r = 0$. Clearly, the edges of C_n^* must cross the edges of $K_{2,3}$ exactly twice. This fact, with the property (5) in the form $0 + 1s + 2(n-s) < n+1$, confirms that $s = n$, which yields that each subgraph T^i cross the edges of $K_{2,3}$ just once. If some vertices t_i of C_n^* are placed in $\omega_{1,4,3,2}$, then we deal with the configurations $\mathcal{A}_k$, $k \in \{1,2,3,4\}$ (they have been already introduced in [14]). For $t_i \in \omega_{1,4,5,2}$, there are four other ways for how to obtain the subdrawing of F^i depending on which edge of $K_{2,3}$ is crossed by the edge $t_i n_3$ provided by there is only one subdrawing of $F^i \setminus \{v_3\}$ represented by the rotation (1452). These four possibilities can be denoted by $\mathcal{A}_k$, for $k = 5,6,7,8$ and they are represent by the cyclic permutations (14532), (13452), (14523), and (14352), respectively. Consequently, we denote by $\mathcal{M}_D$ the subset of all configurations that exist in the drawing D belonging to the set $\mathcal{M} = \{\mathcal{A}_k : k = 1, \ldots, 8\}$. Using the same arguments like in [14], the resulting lower bounds for the number of crossings of two configurations from $\mathcal{M}$ can be established in Table 1 (here, $\mathcal{A}_k$ and $\mathcal{A}_l$ are configurations of the subgraphs F^i and F^j, where $k, l \in \{1, \ldots, 8\}$).

Table 1. The necessary number of crossings between two different subgraphs T^i and T^j for the configurations $\mathcal{A}_k$ and $\mathcal{A}_l$.

-	$\mathcal{A}_1$	$\mathcal{A}_2$	$\mathcal{A}_3$	$\mathcal{A}_4$	$\mathcal{A}_5$	$\mathcal{A}_6$	$\mathcal{A}_7$	$\mathcal{A}_8$
$\mathcal{A}_1$	4	2	3	3	1	3	2	2
$\mathcal{A}_2$	2	4	3	3	3	1	2	2
$\mathcal{A}_3$	3	3	4	2	2	2	1	3
$\mathcal{A}_4$	3	3	2	4	2	2	3	1
$\mathcal{A}_5$	1	3	2	2	4	2	3	3
$\mathcal{A}_6$	3	1	2	2	2	4	3	3
$\mathcal{A}_7$	2	2	1	3	3	3	4	2
$\mathcal{A}_8$	2	2	3	1	3	3	2	4

Let us first assume that $\{\mathcal{A}_k, \mathcal{A}_{k+4}\} \subseteq \mathcal{M}_D$ for some $k \in \{1,2,3,4\}$. In the rest of paper, let us assume two different subgraphs $T^{n-1}, T^n \in S_D$ such that F^{n-1} and F^n have different configurations $\mathcal{A}_1$ and $\mathcal{A}_5$, respectively. Then, $\mathrm{cr}_D(K_{2,3} \cup T^{n-1} \cup T^n, T^i) \geq 1 + 5 = 6$ holds for any $T^i \in S_D$ with $i \neq n-1, n$ by summing the values in all columns in two considered rows of Table 1. Hence, by fixing the subgraph $K_{2,3} \cup T^{n-1} \cup T^n$, we have

$$\mathrm{cr}_D(K_{2,3} + C_n) \geq \mathrm{cr}_D(K_{5,n-2}) + \mathrm{cr}_D(K_{5,n-2}, K_{2,3} \cup T^{n-1} \cup T^n) + \mathrm{cr}_D(K_{2,3} \cup T^{n-1} \cup T^n)$$

$$\geq 4 \left\lfloor \frac{n-2}{2} \right\rfloor \left\lfloor \frac{n-3}{2} \right\rfloor + 6(n-2) + 2 + 1 \geq 4 \left\lfloor \frac{n}{2} \right\rfloor \left\lfloor \frac{n-1}{2} \right\rfloor + n + 3.$$

Due to the symmetry of three remaining pairs of configurations, we also obtain a contradiction in D by applying the same process. Now, let us turn to the good drawing D of the graph $K_{2,3} + C_n$ in which $\{\mathcal{A}_k, \mathcal{A}_{k+4}\} \not\subseteq \mathcal{M}_D$ for any $k = 1,2,3,4$. Further, let us also suppose that the number of subgraphs with the configuration $\mathcal{A}_k \in \mathcal{M}_D$ is at least equal to the number of subgraphs with the configuration $\mathcal{A}_l \in \mathcal{M}_D$, for each possible $l \neq k$, and let $T^i \in S_D$ be such a subgraph with the configuration $\mathcal{A}_k$ of F^i. Hence,

$$\mathrm{cr}_D(K_{5,n-1}, T^i) = \sum_{j \neq i} \mathrm{cr}_D(T^j, T^i) \geq 3(n-2) + 2 - 2\left\lfloor \frac{n}{7} \right\rfloor \geq \frac{5}{2}n - \frac{5}{2},$$

where an idea of the arithmetic mean of the values four, three and two of Table 1 could be exploited. Thus, by fixing the subgraph T^i, we have

$$\mathrm{cr}_D(K_{2,3}+C_n) = \mathrm{cr}_D(K_{2,3}+C_{n-1}) + \mathrm{cr}_D(K_{5,n-1}, T^i) + \mathrm{cr}_D(K_{2,3}, T^i)$$

$$\geq 4\left\lfloor\frac{n-1}{2}\right\rfloor\left\lfloor\frac{n-2}{2}\right\rfloor + n - 1 + 3 + \frac{5}{2}n - \frac{5}{2} + 1 \geq 4\left\lfloor\frac{n}{2}\right\rfloor\left\lfloor\frac{n-1}{2}\right\rfloor + n + 3.$$

(b) All vertices t_i of C_n^* are placed in the same region of subdrawing of $K_{2,3}$ induced by D. In the rest of paper, based also on their symmetry, we suppose that $t_i \in \omega_{1,4,3,2}$ for each $i = 1, \ldots, n$. Whereas the set R_D is again empty, there are at least $n - 2$ subgraphs $T^i \in S_D$ provided by the property (5) in the form $s + 2(n - s) < n + 3$. For $T^i \in S_D$, we consider only one from the configurations $\mathcal{A}_k$, for $k = 1, 2, 3, 4$. Again, let us also assume that the number of subgraphs with the configuration $\mathcal{A}_k$ is at least equal to the number of subgraphs with the configuration $\mathcal{A}_l$, for each possible $l \neq k$, and let $T^i \in S_D$ be such a subgraph with the configuration $\mathcal{A}_k$ of F^i. Then, by fixing the subgraph T^i, we have

$$\mathrm{cr}_D(K_{2,3}+C_n) = \mathrm{cr}_D(K_{2,3}+C_{n-1}) + \mathrm{cr}_D(K_{5,n-1}, T^i) + \mathrm{cr}_D(K_{2,3}, T^i)$$

$$\geq 4\left\lfloor\frac{n-1}{2}\right\rfloor\left\lfloor\frac{n-2}{2}\right\rfloor + n - 1 + 3 + 3(s-2) + 2 + 1(n-s) + 1 = 4\left\lfloor\frac{n-1}{2}\right\rfloor\left\lfloor\frac{n-2}{2}\right\rfloor$$

$$+ 2n + 2s - 1 \geq 4\left\lfloor\frac{n-1}{2}\right\rfloor\left\lfloor\frac{n-2}{2}\right\rfloor + 2n + 2(n-2) - 1 \geq 4\left\lfloor\frac{n}{2}\right\rfloor\left\lfloor\frac{n-1}{2}\right\rfloor + n + 3,$$

wherein a simplified form of the idea of the arithmetic mean of the values of Table 1 is applied.

Case 2: $\mathrm{cr}_D(K_{2,3}) = 1$. We can choose the drawing with the vertex notation of $K_{2,3}$ like in Figure 1b. Similarly as in the previous case, we will discuss two subcases:

(a) The cycle C_n^* is crossed by some edge of the graph $K_{2,3}$. As the edges of C_n^* cross the edges of $K_{2,3}$ exactly twice, there is a subgraph T^i which does not cross the edges of $K_{2,3}$ provided by the property (5) in the form $1 + s + 2(n - r - s) \leq n$. For a $T^i \in R_D$, the reader can easily verify that the subgraph $F^i = K_{2,3} \cup T^i$ is uniquely represented by $\mathrm{rot}_D(t_i) = (12345)$, and $\mathrm{cr}_D(T^i, T^j) \geq 4$ holds for any $T^j \in R_D$ with $j \neq i$ provided that $\mathrm{rot}_D(t_i) = \mathrm{rot}_D(t_j)$, for more see [18]. Moreover, it is not difficult to verify in possible regions of $D(K_{2,3} \cup T^i)$ that $\mathrm{cr}_D(K_{2,3} \cup T^i, T^j) \geq 4$ is true for each $T^j \in S_D$. Thus, by fixing the subgraph $K_{2,3} \cup T^i$, we have

$$\mathrm{cr}_D(K_{2,3}+C_n) \geq \mathrm{cr}_D(K_{5,n-1}) + \mathrm{cr}_D(K_{5,n-1}, K_{2,3} \cup T^i) + \mathrm{cr}_D(K_{2,3} \cup T^i)$$

$$\geq 4\left\lfloor\frac{n-1}{2}\right\rfloor\left\lfloor\frac{n-2}{2}\right\rfloor + 4(r-1) + 4s + 3(n-r-s) + 1 = 4\left\lfloor\frac{n-1}{2}\right\rfloor\left\lfloor\frac{n-2}{2}\right\rfloor$$

$$+ 3n + (r+s) - 3 \geq 4\left\lfloor\frac{n-1}{2}\right\rfloor\left\lfloor\frac{n-2}{2}\right\rfloor + 3n + 4 - 3 \geq 4\left\lfloor\frac{n}{2}\right\rfloor\left\lfloor\frac{n-1}{2}\right\rfloor + n + 3,$$

where $r + s \geq 4$ holds also due to the property (5).

(b) None edge of C_n^* is crossed by the edges of $K_{2,3}$. Since all vertices t_i of the cycle C_n^* are placed in the same region of subdrawing of $K_{2,3}$ induced by D, they must be placed in the outer region of $D(K_{2,3})$. If there is a $T^i \in R_D$, then the edges of $K_{2,3} \cup T^i$ are crossed at least four times by any subgraph T^j with the placement of the vertex t_j in the outer region of $D(K_{2,3})$, which yields that the similar idea as in the previous subcase can be used by fixing the subgraph $K_{2,3} \cup T^i$

$$\mathrm{cr}_D(K_{2,3}+C_n) \geq 4\left\lfloor\frac{n-1}{2}\right\rfloor\left\lfloor\frac{n-2}{2}\right\rfloor + 4(r-1) + 4(n-r) + 1$$

$$= 4\left\lfloor\frac{n-1}{2}\right\rfloor\left\lfloor\frac{n-2}{2}\right\rfloor + 4n - 3 \geq 4\left\lfloor\frac{n}{2}\right\rfloor\left\lfloor\frac{n-1}{2}\right\rfloor + n + 3.$$

To finish the proof of this case, let us suppose that the set R_D is empty. Whereas the set R_D is empty, there are at least $n-1$ subgraphs $T^i \in S_D$ according to the property (5). Since the edges v_2v_3, v_3v_4, v_2v_5, and v_1v_4 of $K_{2,3}$ can be crossed by the edges t_iv_4, t_iv_2, t_iv_4, and t_iv_2, respectively, these four ways can be denoted by $\mathcal{B}_k$, for $k = 1, 2, 3, 4$. So, the configurations $\mathcal{B}_1$, $\mathcal{B}_2$, $\mathcal{B}_3$, and $\mathcal{B}_4$ are uniquely described by the cyclic permutations (12435), (13245), (12354), and (13452), respectively., and the aforementioned properties of the cyclic rotations imply all lower-bounds of number of crossings in Table 2.

Table 2. The necessary number of crossings between two different subgraphs T^i and T^j for the configurations $\mathcal{B}_k$ and $\mathcal{B}_l$.

-	$\mathcal{B}_1$	$\mathcal{B}_2$	$\mathcal{B}_3$	$\mathcal{B}_4$
$\mathcal{B}_1$	4	2	2	2
$\mathcal{B}_2$	2	4	2	2
$\mathcal{B}_3$	2	2	4	2
$\mathcal{B}_4$	2	2	2	4

Now, let us also suppose that the number of subgraphs with the configuration $\mathcal{B}_k$ is at least equal to the number of subgraphs with the configuration $\mathcal{B}_l$, for each possible $l \neq k$, and let $T^i \in S_D$ be such a subgraph with the configuration $\mathcal{B}_k$ of F^i. Hence,

$$\sum_{T^j \in S_D,\, j \neq i} \mathrm{cr}_D(T^i, T^j) \geq 3(s-2) + 2 - 2\left\lfloor \frac{s}{4} \right\rfloor \geq \frac{5}{2}s - 4,$$

where again an idea of the arithmetic mean of the values four and two of Table 2 could be exploited. Thus, by fixing the subgraph T^i, we have

$$\mathrm{cr}_D(K_{2,3} + C_n) = \mathrm{cr}_D(K_{2,3} + C_{n-1}) + \mathrm{cr}_D(K_{5,n-1}, T^i) + \mathrm{cr}_D(K_{2,3}, T^i)$$

$$\geq 4\left\lfloor \frac{n-1}{2} \right\rfloor \left\lfloor \frac{n-2}{2} \right\rfloor + n - 1 + 3 + \frac{5}{2}s - 4 + 1(n-s) + 1 = 4\left\lfloor \frac{n-1}{2} \right\rfloor \left\lfloor \frac{n-2}{2} \right\rfloor$$

$$+ 2n + \frac{3}{2}s - 1 \geq 4\left\lfloor \frac{n-1}{2} \right\rfloor \left\lfloor \frac{n-2}{2} \right\rfloor + 2n + \frac{3}{2}(n-1) - 1 \geq 4\left\lfloor \frac{n}{2} \right\rfloor \left\lfloor \frac{n-1}{2} \right\rfloor + n + 3.$$

Case 3: $\mathrm{cr}_D(K_{2,3}) = 2$. In the rest of paper, we choose the drawing with the vertex notation of $K_{2,3}$ like in Figure 1c. Obviously the set R_D must be empty. As $s = n$ by the property (5), all vertices t_i of the subgraphs $T^i \in S_D$ must be placed in the region of $D(K_{2,3})$ with four vertices v_1, v_2, v_4, and v_5 of the graph $K_{2,3}$ on its boundary. For $T^i \in S_D$, there is only one possible subdrawing of $F^i \setminus \{v_3\}$ described by the rotation (1245), which yields that there are exactly three ways of obtaining the subdrawing of $K_{2,3} \cup T^i$ depending on which edge of $K_{2,3}$ may be crossed by t_iv_3. In all cases of $T^i \in S_D$ represented by either (12345) or (12453) or (12435), it is not difficult to verify using cyclic permutations that $\mathrm{cr}_D(T^i, T^j) \geq 2$ is fulfilling for each $T^j \in S_D$, $j \neq i$. Thus, by fixing the subgraph T^i, we have

$$\mathrm{cr}_D(K_{2,3} + C_n) = \mathrm{cr}_D(K_{2,3} + C_{n-1}) + \mathrm{cr}_D(K_{5,n-1}, T^i) + \mathrm{cr}_D(K_{2,3}, T^i)$$

$$\geq 4\left\lfloor \frac{n-1}{2} \right\rfloor \left\lfloor \frac{n-2}{2} \right\rfloor + n - 1 + 3 + 2(n-1) + 1 \geq 4\left\lfloor \frac{n}{2} \right\rfloor \left\lfloor \frac{n-1}{2} \right\rfloor + n + 3.$$

Case 4: $\mathrm{cr}_D(K_{2,3}) = 3$. We assume the drawing with the vertex notation of $K_{2,3}$ like in Figure 1d. As the property (5) enforces $r \geq 1$ and $r + s \geq 4$, the proof can proceed in the similar way as in the Subcase 2a).

We have shown, in all cases, that there is no good drawing D of the graph $K_{2,3} + C_n$ with fewer than $4\left\lfloor \frac{n}{2} \right\rfloor \left\lfloor \frac{n-1}{2} \right\rfloor + n + 3$ crossings. This completes the proof. □

Finally, in Figure 2, we are able to add the edge $v_1 v_5$ to the graph $K_{2,3}$ without additional crossings, and we obtain one new graph H in Figure 3. So, the drawing of the graph $H + C_n$ with $4\lfloor \frac{n}{2} \rfloor \lfloor \frac{n-1}{2} \rfloor + n + 3$ crossings is obtained. On the other hand, $K_{2,3} + C_n$ is a subgraph of $H + C_n$, and therefore, $\mathrm{cr}(H + C_n) \geq \mathrm{cr}(K_{2,3} + C_n)$. Thus, the next result is an immediate consequence of Theorem 2.

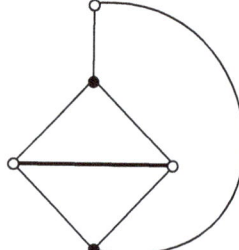

Figure 3. The graph H by adding one edge to the graph $K_{2,3}$.

Corollary 1. $\mathrm{cr}(H + C_n) = 4\lfloor \frac{n}{2} \rfloor \lfloor \frac{n-1}{2} \rfloor + n + 3$ *for any* $n \geq 3$.

4. Conclusions

We suppose that the application of various forms of arithmetic means can be used to estimate the unknown values of the crossing numbers for join products of some graphs on five vertices with the paths, and also with the cycles. The same we expect for larger graphs, namely for a lot of symmetric graphs of order six.

Funding: This research received no external funding.

Conflicts of Interest: The author declares no conflict of interest.

References

1. Klešč, M. The crossing numbers of join of the special graph on six vertices with path and cycle. *Discret. Math.* **2010**, *310*, 1475–1481. [CrossRef]
2. Garey, M.R.; Johnson, D.S. Crossing number is NP-complete. *SIAM J. Algebr. Discret. Methods* **1983**, *4*, 312–316. [CrossRef]
3. Ho, P.T. On the crossing number of some complete multipartite graphs. *Utilitas Math* **2009**, *79*, 125–143.
4. Klešč, M. The join of graphs and crossing numbers. *Electron. Notes Discret. Math.* **2007**, *28*, 349–355. [CrossRef]
5. Kleitman, D.J. The crossing number of $K_{5,n}$. *J. Comb. Theory* **1970**, *9*, 315–323. [CrossRef]
6. Klešč, M.; Schrötter, Š. The crossing numbers of join products of paths with graphs of order four. *Discuss. Math. Graph Theory* **2011**, *31*, 321–331. [CrossRef]
7. Klešč, M. The crossing numbers of join of cycles with graphs of order four. In Proceedings of the Aplimat 2019: 18th Conference on Applied Mathematics, Bratislava, Slovakia, 5–7 February 2019; pp. 634–641.
8. Klešč, M.; Kravecová, D.; Petrillová, J. The crossing numbers of join of special graphs. *Electr. Eng. Inform.* **2011**, *2*, 522–527.
9. Berežný, Š.; Staš, M. Cyclic permutations and crossing numbers of join products of two symmetric graphs of order six. *Carpathian J. Math.* **2019**, *35*, 137–146.
10. Draženská, E. On the crossing number of join of graph of order six with path. In Proceedings of the CJS 2019: 22nd Czech-Japan Seminar on Data Analysis and Decision Making, Nový Světlov, Czechia, 25–28 September 2019; pp. 41–48.
11. Klešč, M.; Schrötter, Š. The crossing numbers of join of paths and cycles with two graphs of order five. In *Lecture Notes in Computer Science: Mathematical Modeling and Computational Science*; Springer: Berlin/Heidelberg, Germany, 2012; Volume 7125, pp. 160–167.

12. Staš, M.; Petrillová, J. On the join products of two special graphs on five vertices with the path and the cycle. *J. Math. Model. Geom.* **2018**, *6*, 1–11.
13. Berežný, Š.; Buša, J., Jr. Algorithm of the Cyclic-Order Graph Program (Implementation and Usage). *J. Math. Model. Geom.* **2019**, *7*, 1–8. [CrossRef]
14. Staš, M. Alternative proof on the crossing number of $K_{2,3,n}$. *J. Math. Model. Geom.* **2019**, *7*, 13–20.
15. Hernández-Vélez, C.; Medina, C.; Salazar G. The optimal drawing of $K_{5,n}$. *Electron. J. Comb.* **2014**, *21*, 29.
16. Chimani, M.; Wiedera, T. An ILP-based proof system for the crossing number problem. In Proceedings of the 24th Annual European Symposium on Algorithms (ESA 2016), Aarhus, Denmark, 22–24 August 2016; Volume 29, pp. 1–13.
17. Yuan, X. On the crossing numbers of $K_{2,3} \vee C_n$. *J. East China Norm. Uni. Nat. Sci.* **2011**, *5*, 21–24.
18. Woodall, D.R. Cyclic-order graphs and Zarankiewicz's crossing number conjecture. *J. Graph Theory* **1993**, *17*, 657–671. [CrossRef]

© 2020 by the author. Licensee MDPI, Basel, Switzerland. This article is an open access article distributed under the terms and conditions of the Creative Commons Attribution (CC BY) license (http://creativecommons.org/licenses/by/4.0/).

Article

Method for Developing Combinatorial Generation Algorithms Based on AND/OR Trees and Its Application

Yuriy Shablya [1,*], Dmitry Kruchinin [1] and Vladimir Kruchinin [2]

1. Department of Complex Information Security of Computer Systems, Tomsk State University of Control Systems and Radioelectronics, Tomsk 634050, Russia; kdv@fb.tusur.ru
2. Institute of Innovation, Tomsk State University of Control Systems and Radioelectronics, Tomsk 634050, Russia; kru@2i.tusur.ru
* Correspondence: shablya-yv@mail.ru

Received: 6 May 2020; Accepted: 9 June 2020; Published: 12 June 2020

Abstract: In this paper, we study the problem of developing new combinatorial generation algorithms. The main purpose of our research is to derive and improve general methods for developing combinatorial generation algorithms. We present basic general methods for solving this task and consider one of these methods, which is based on AND/OR trees. This method is extended by using the mathematical apparatus of the theory of generating functions since it is one of the basic approaches in combinatorics (we propose to use the method of compositae for obtaining explicit expression of the coefficients of generating functions). As a result, we also apply this method and develop new ranking and unranking algorithms for the following combinatorial sets: permutations, permutations with ascents, combinations, Dyck paths with return steps, labeled Dyck paths with ascents on return steps. For each of them, we construct an AND/OR tree structure, find a bijection between the elements of the combinatorial set and the set of variants of the AND/OR tree, and develop algorithms for ranking and unranking the variants of the AND/OR tree.

Keywords: combinatorial generation; method; algorithm; AND/OR tree; Euler–Catalan's triangle; labeled Dyck path; ranking algorithm; unranking algorithm

MSC: 68R05; 05C05; 05A15

---

1. Introduction

Many information objects have a hierarchical or recursive structure. In this case, a tree structure is a convenient form of representing such information objects. This allows us to describe an information object by a combinatorial set and apply combinatorial generation algorithms for it. A combinatorial set is a finite set whose elements have some structure and there is an algorithm for constructing the elements of this set. The elements of combinatorial sets (combinatorial objects) like combinations, permutations, partitions, compositions, paths, graphs, trees, etc. play an important role in mathematics and computer science.

Knuth [1] gives a detailed overview of the formation and development of the direction related to designing combinatorial algorithms. In this overview, special attention is paid to the procedure for traversing all possible elements of a given combinatorial set. This problem can be studied as enumerating, listing, and generating elements of a given combinatorial set. On the other hand, Ruskey [2] introduces the concept of combinatorial generation and distinguishes the following four tasks in this area:

1. Listing: generating elements of a given combinatorial set sequentially;

2. Ranking: ranking (numbering) elements of a given combinatorial set;
3. Unranking: generating elements of a given combinatorial set in accordance with their ranks;
4. Random selection: generating elements of a given combinatorial set in random order.

Before applying combinatorial generation algorithms, it is necessary to develop them. General methods for developing combinatorial generation algorithms were studied by such researches as E.M. Reingold [3], D.L. Kreher [4], E. Barcucci [5,6], S. Bacchelli [7,8], A. Del Lungo [9,10], V. Vajnovszki [11–13], P. Flajolet [14,15], C. Martinez and X. Molinero [16–20], B.Y. Ryabko and Y.S. Medvedeva [21–23], and V.V. Kruchinin [24].

There are several basic general methods for developing combinatorial generation algorithms:

- backtracking [3,4]: this method is used for exhaustive generation and based on constructing a state space tree for a combinatorial set and obtaining feasible solutions on level l of the tree that are built up from partial solutions on level $l - 1$;
- ECO-method [5–13]: this method is used for exhaustive generation and based on producing a set of new objects of size $n + 1$, using a succession rule and starting from an object of size n;
- Flajolet's method [14–20]: this method can be applied for listing, ranking, and unranking a large family of unlabeled and labeled admissible combinatorial classes that can be obtained by using admissible combinatorial operators (disjoint union, Cartesian unlabeled product, labeled product, sequence, set, cycle, powerset, substitution, etc.);
- Ryabko's method [21–23]: this method can be applied for ranking and unranking combinatorial objects presented in the form of a word, this method operates with word prefixes and uses the divide-and-conquer paradigm;
- Kruchinin's method [24]: this method can be applied for listing, ranking, and unranking a combinatorial set represented in the form of an AND/OR tree structure for which the total number of its variants is equal to the value of the cardinality function of the combinatorial set.

Each of these methods claims the universality of its application in the development of new combinatorial generation algorithms. The study of these methods have shown the following results connected with their limitations and requirements [25]:

- the main characteristic, which is necessary for developing combinatorial generation algorithms, is the cardinality function of a combinatorial set;
- some methods (backtracking and ECO-method) are aimed only at the development of listing algorithms;
- there are restrictions on applying some methods (ECO-method and Flajolet's method) for combinatorial sets that are described by more than one parameter;
- most methods require the representation of a combinatorial object in a special form (for example, as a word, a sequence, a specification, or an AND/OR tree), but this is not always a trivial task and requires additional research;
- there are requirements for additional information describing a combinatorial set.

Also, there are many combinatorial generation algorithms that are based on features of the applied combinatorial set or that are based on simple counting techniques (for example, see [26–28]). Therefore, the methods used for developing such algorithms cannot be universal (they cannot be applied to develop new combinatorial generation algorithms for other combinatorial sets).

Thus, there is no universal general method that can be applied for developing new combinatorial generation algorithms. The main purpose of our research is to derive and improve general methods for developing combinatorial generation algorithms. In this paper, we consider and extend Kruchinin's method for developing combinatorial generation algorithms, which is based on the use of AND/OR trees. This method:

- allows us to develop all types of combinatorial generation algorithms (listing, ranking, and unranking algorithms);
- has no restrictions on the number of parameters that describe combinatorial sets (this allows us to consider complex discrete structures);
- requires only an expression of the cardinality function as additional information describing a combinatorial set.

However, to apply this method, it is necessary to know an expression of the cardinality function of a combinatorial set that must satisfy the following conditions:

- the expression of the cardinality function can contain only positive integers (let $\mathbb{N}$ denotes the set of natural numbers);
- the expression of the cardinality function can contain only such algebraic operations as addition (which is denoted by $+$) and multiplication (which is denoted by $\times$);
- the cardinality function may be recursively defined in terms of itself (let R denotes a primitive recursive function).

If an expression of the cardinality function f of a combinatorial set A satisfies the conditions presented above, then we will say that f belongs to the algebra $\{\mathbb{N}, +, \times, R\}$. The requirement of the cardinality function that belongs to the algebra $\{\mathbb{N}, +, \times, R\}$ is the main restriction of Kruchinin's method. If the required form of the cardinality function is unknown for a given combinatorial set, then this method cannot be applied to develop combinatorial generation algorithms.

The organization of this paper is as follows. Section 2 of this paper is devoted to a brief description of the main theoretical points of the used method for developing combinatorial generation algorithms. In Section 3, our modification of the original method is presented. The modification is based on applying the method of compositae from the theory of generating functions. To confirm the effectiveness of using the proposed modification of the original method, we develop new ranking and unranking algorithms for the following combinatorial sets: permutations, permutations with ascents, combinations, Dyck paths with return steps, labeled Dyck paths with ascents on return steps. The obtained results are shown in Section 4.

2. Method for Developing Combinatorial Generation Algorithms Based on AND/OR Trees

Kruchinin [24] introduces a method for developing combinatorial generation algorithms, which is based on the use of AND/OR trees. This method is based on representing a combinatorial set in the form of an AND/OR tree structure for which the total number of its variants is equal to the value of the cardinality function of the combinatorial set. Using an AND/OR tree structure, it is possible to develop listing, ranking, and unranking algorithms for a given combinatorial set. The effectiveness of this method is shown in the development of combinatorial generation algorithms for a large number of combinatorial sets (for example, permutations, combinations, partitions, compositions, the Fibonacci numbers, the Catalan numbers, the Stirling numbers, tree structures, and formal languages).

An AND/OR tree is a tree structure that contains nodes of two types: AND nodes and OR nodes. Figure 1 shows a way for representing nodes in an AND/OR tree.

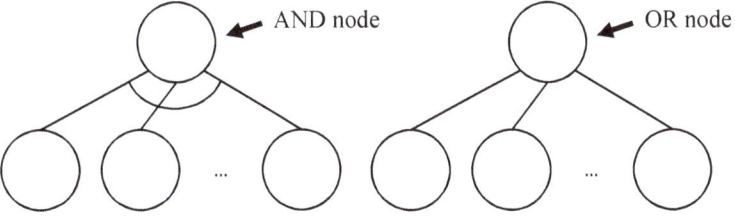

Figure 1. The representation of nodes in an AND/OR tree.

A variant of an AND/OR tree is a tree structure obtained by removing all edges except one for each OR node. Figure 2 shows an example of an AND/OR tree and all its variants.

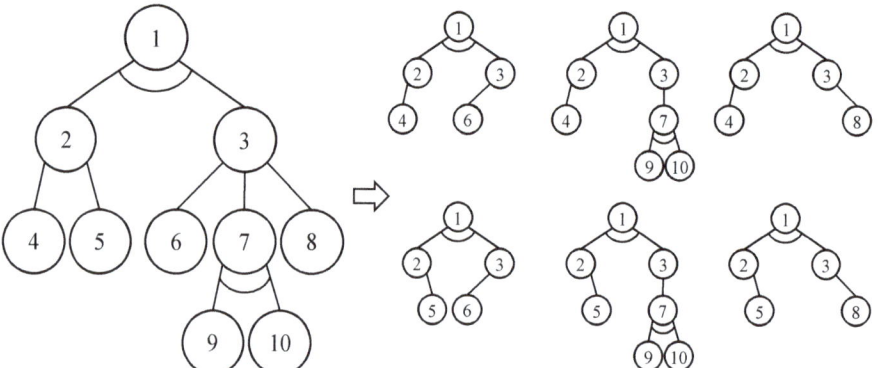

Figure 2. An AND/OR tree and all its variants.

If we know the cardinality function f of a combinatorial set A that belongs to the algebra $\{\mathbb{N}, +, \times, R\}$, then we can construct an AND/OR tree structure for which the total number of its variants is equal to the value of the cardinality function (Theorems 1–3 and Corollaries 1–2 in [24]. To do this, it is necessary to perform the following steps for the cardinality function f:

- each addition $+$ from f must be represented as an OR node of the AND/OR tree where all terms are represented as sons of the OR node;
- each multiplication $\times$ from f must be represented as an AND node of the AND/OR tree where all factors are represented as sons of the AND node;
- each coefficient $k \in \mathbb{N}$ from f must be represented as an OR node of the AND/OR where all sons are leaves and their number is equal to k;
- each recursive operation from f must be represented as recursion in the AND/OR tree (it will be denoted by a node with a triangle).

Thus, the method for developing combinatorial generation algorithms based on AND/OR trees can be written in the following form:

Input: The cardinality function f of a combinatorial set A that belongs to the algebra $\{\mathbb{N}, +, \times, R\}$.
Output: The combinatorial generation algorithms

$$\texttt{RankVariant}(v) : W(D) \to \mathbb{N}_{|W(D)|}$$

and

$$\texttt{UnrankVariant}(r) : \mathbb{N}_{|W(D)|} \to W(D),$$

where each variant v of an AND/OR tree D constructed for the combinatorial set A must correspond to a specific combinatorial object $a \in A$. That is, the bijection $A \leftrightarrow W(D)$ must be defined, where $W(D)$ is the set of all the variants v of the AND/OR tree D. The combination of this bijection with the algorithms $\texttt{RankVariant}(v)$ and $\texttt{UnrankVariant}(r)$ represents the desired combinatorial generation algorithms $\texttt{Rank}(a) : A \to \mathbb{N}$ and $\texttt{Unrank}(r) : \mathbb{N} \to A$ (for ranking and unranking elements of the combinatorial set A).

The ranking algorithm $\texttt{RankVariant}(v)$ allows us to associate each variant $v \in W(D)$ of the AND/OR tree D with a unique number $r \in \mathbb{N}_{|W(D)|} = \{0, 1, \ldots, |W(D)| - 1\}$ called the rank. The rank r of a combinatorial object a represented as a variant v of the corresponding AND/OR tree D is determined by the value of $l(z)$ for the node z of the variant v that is the root of the AND/OR tree D.

The value of $l(z)$ corresponds to a number for the node z that satisfies the condition

$$0 \leq l(z) < w(z),$$

where $w(z)$ shows the number of variants in the subtree of the node z.

The value of $l(z)$ is calculated according to the following rules:

1. If a node z of the variant v is a leaf of the AND/OR tree D, then

$$l(z) = 0;$$

2. If a node z of the variant v is an AND node of the AND/OR tree D, then

$$l(z) = l(s_1^{(z)}) + w(s_1^{(z)}) \left(l(s_2^{(z)}) + w(s_2^{(z)}) \left(\ldots \left(l(s_{n-1}^{(z)}) + w(s_{n-1}^{(z)}) l(s_n^{(z)}) \right) \ldots \right) \right),$$

where n is equal to the number of sons for the node z and $s_i^{(z)}$ is the i-th son of the node z;

3. If a node z of the variant v is an OR node of the AND/OR tree D, then

$$l(z) = l(s_k^{(z)}) + \sum_{i=1}^{k-1} w(s_i^{(z)}),$$

where k is the position of the son $s_k^{(z)}$ of the node z chosen in the variant v among all its sons in the AND/OR tree D.

For the general case, the rules for ranking the variants of an AND/OR tree was formalized and presented as Algorithm 1. To run this algorithm, it is necessary to start from the command RankVariant$(root, v, D)$, where $root$ is the root of an AND/OR tree D.

Algorithm 1: A general algorithm for ranking the variants of an AND/OR tree.

1 RankVariant (z, v, D)
2 **begin**
3   **if** $z =$ a leaf of the AND/OR tree D **then** $l := 0$
4   **if** $z =$ an AND node of the AND/OR tree D **then**
5       $n :=$ the number of sons for the node z
6       $l :=$ RankVariant $(s_n^{(z)}, v, D)$
7   **for** $i := n-1$ **to** 1 **do** $l :=$ RankVariant $(s_i^{(z)}, v, D) + w(s_i^{(z)}) \cdot l$
8   **end**
9   **if** $z =$ an OR node of the AND/OR tree D **then**
10       $k :=$ the position of the son of the node z chosen in the variant v among all its sons
11       $l :=$ RankVariant $(s_k^{(z)}, v, D)$
12   **for** $i := 1$ **to** $k-1$ **do** $l := l + w(s_i^{(z)})$
13   **end**
14   **return** l
15 **end**

The unranking algorithm UnrankVariant(r) performs the inverse operation to the algorithm RankVariant(v). That is, the algorithm UnrankVariant(r) allows us to associate each rank $r \in \mathbb{N}_{|W(D)|}$ with a unique variant v of the AND/OR tree D. The algorithm UnrankVariant(r) is based on the inverse actions to the calculations used in the algorithm RankVariant(v). For the general case, the rules for unranking the variants of an AND/OR tree were formalized and presented as Algorithm 2.

Algorithm 2: A general algorithm for unranking the variants of an AND/OR tree.

1   UnrankVariant (r, D)
2   **begin**
3   Push $(r, root)$ onto the stack
4   Add *root* to the variant v
5   **while** *the stack is not empty* **do**
6   Pop (l, z) from the stack
7   **if** $z =$ *an AND node of the AND/OR tree D* **then**
8           $n :=$ the number of sons for the node z
9   **for** $i := 1$ **to** n **do**
10             $l_i := l \mod w(s_i^{(z)})$
11   Push $(l_i, s_i^{(z)})$ onto the stack
12   Add $s_i^{(z)}$ to the variant v
13             $l := \left\lfloor \dfrac{l}{w(s_i^{(z)})} \right\rfloor$
14   **end**
15   **end**
16   **if** $z =$ *an OR node of the AND/OR tree D* **then**
17           $k := 1$
18           $sum := 0$
19   **while** $sum + w(s_k^{(z)}) \leq l$ **do**
20             $sum := sum + w(s_i^{(z)})$
21             $k := k + 1$
22   **end**
23           $l := l - sum$
24   Push $(l, s_k^{(z)})$ onto the stack
25   Add $s_k^{(z)}$ to the variant v
26   **end**
27   **end**
28   **return** v
29   **end**

3. Modification of the Method for Developing Combinatorial Generation Algorithms

The use of the method for developing combinatorial generation algorithms based on AND/OR trees has the following two restrictions:

Firstly, if we do not know the cardinality function of a combinatorial set that belongs to the algebra $\{\mathbb{N}, +, \times, R\}$, then we cannot construct the corresponding AND/OR tree for the combinatorial set. Therefore, this method for developing combinatorial generation algorithms cannot be applied without an AND/OR tree structure.

If an AND/OR tree structure is constructed for a combinatorial set, then there is a new problem. This problem is associated with finding a bijection between the elements of the combinatorial set and the set of variants of the AND/OR tree (that is, each variant v of an AND/OR tree D constructed for the combinatorial set A must correspond to a specific combinatorial object $a \in A$, and vice versa). A general approach for solving this problem does not exist, since each combinatorial set has its own and completely unique characteristics. We propose to use the following recommendation: it is necessary to consider the changes that occur in the structure of combinatorial objects when moving from one node of an AND/OR tree to another (considering the type of a node, the label of a node and the selected sons) and show these changes in the bijection.

For solving the first problem, we propose to apply the theory of generating functions, since it is one of the basic approaches in modern combinatorics and generating functions are already known for many combinatorial sets. An ordinary generating function of a sequence $(a_n)_{n\geq 0}$ is the following formal power series [29]:

$$A(t) = a_0 + a_1 t + a_2 t^2 + \ldots = \sum_{n\geq 0} a_n t^n.$$

For the coefficients of the powers of generating functions, the notion of the compositae of a generating function was introduced in [30]. The composita of an ordinary generating function

$$C(t) = \sum_{n>0} g_n t^n$$

is the following function with two variables $G^\Delta(n,k)$, which is a coefficients function of the k-th power of the generating function $G(t)$:

$$(G(t))^k = \sum_{n\geq k} G^\Delta(n,k) t^n.$$

This mathematical apparatus provides such operations on compositae as shift, addition, multiplication, composition, reciprocation, and compositional inversion of generating functions. Such operations on compositae allow us to obtain explicit expressions for the coefficients of generating functions. To obtain an explicit expression for the coefficients of a generating function using the method of compositae, it is necessary to decompose the given generating function into functions for that the compositae are known and apply the corresponding operations to them. More detailed information about the compositae can be found in [30–35].

If for a given combinatorial set A we consider its subset $A_n \subset A$, which contains only the combinatorial objects of size n, then the cardinality function $f(n) = |A_n|$ of this combinatorial set A_n can be described by a generating function

$$F(t) = \sum_{n\geq 0} f_n t^n = \sum_{n\geq 0} f(n) t^n = \sum_{n\geq 0} |A_n| t^n.$$

Hence, to obtain an expression for the cardinality function $f(n)$ of a combinatorial set for which a generating function is known, we can apply the method of compositae for obtaining an explicit expression of the coefficients f_n of the generating function [25]. Moreover, the operations on compositae can be extended to the case of multivariate generating functions. This makes it possible to obtain expressions for the cardinality functions of combinatorial sets that are described by more than one parameter.

The obtained method for developing combinatorial generation algorithms with its modification is presented as a sequence of the following steps:

1. If the cardinality function f of a combinatorial set A that belongs to the algebra $\{\mathbb{N}, +, \times, R\}$ is known, then go to Step 4.
2. If the generating function F for the values of the cardinality function f of the combinatorial set A is known, then apply the method of compositae for obtaining an explicit expression of the coefficients of the generating function. Otherwise, the method for developing combinatorial generation algorithms cannot be applied.
3. If the cardinality function f of the combinatorial set A that belongs to the algebra $\{\mathbb{N}, +, \times, R\}$ is obtained, then go to Step 4. Otherwise, the method for developing combinatorial generation algorithms cannot be applied.
4. Using the cardinality function f of the combinatorial set A, construct the corresponding AND/OR tree D.

5. Find a bijection $A \leftrightarrow W(D)$ between the elements of the combinatorial set A and the set of variants v of the AND/OR tree D in the form of the algorithms ObjectToVariant$(a, D) : A \to W(D)$, where $a \in A$, and VariantToObject$(v, D) : W(D) \to A$, where $v \in W(D)$.
6. Find a bijection $W(D) \leftrightarrow \mathbb{N}_{|W(D)|}$ between the elements of the set of variants v of the AND/OR tree D and the finite set of natural numbers $\mathbb{N}_{|W(D)|} = \{0, 1, \ldots, |W(D)| - 1\}$, using the algorithms RankVariant$(root, v, D) : W(D) \to \mathbb{N}_{|W(D)|}$, where $root$ is the root of an AND/OR tree D, $v \in W(D)$, and UnrankVariant$(r, D) : \mathbb{N}_{|W(D)|} \to W(D)$, where $r \in \mathbb{N}_{|W(D)|}$.

The combination of the algorithms defined in the last two steps of the modified method forms a bijection $A \leftrightarrow \mathbb{N}$ and represents the combinatorial generation algorithms Rank$(a) : A \to \mathbb{N}$ for ranking and Unrank$(r) : \mathbb{N} \to A$ for unranking elements of the combinatorial set A.

4. Application of the Modification of the Method for Developing Combinatorial Generation Algorithms

Next, we consider the process of developing ranking and unranking algorithms using the obtained method for developing combinatorial generation algorithms. We describe a combinatorial set, construct an AND/OR tree, find a bijection between the elements of the combinatorial set and the set of variants of the AND/OR tree, and develop algorithms for ranking and unranking the variants.

4.1. Combinatorial Set

Let us consider the following combinatorial object: a labeled Dyck n-path with m ascents on return steps. A Dyck n-path is a lattice path in the plane which begins at $(0,0)$, ends at $(2n, 0)$, and consists of steps $(1, 1)$ called rises or up-steps and $(1, -1)$ called falls or down-steps [36]. A return step is a down-step at level 1 (a return to the ground level 0) [37]. In a labeled Dyck n-path with m ascents on return steps, each down-step has its own label (a unique value from 1 to n). If we consider the sequence of labels of down-steps of a Dyck n-path starting from $(0, 0)$, then it has exactly m ascents.

Figure 3 shows all possible variants of the considered labeled Dyck paths for $n = 3$ and $m = 1$.

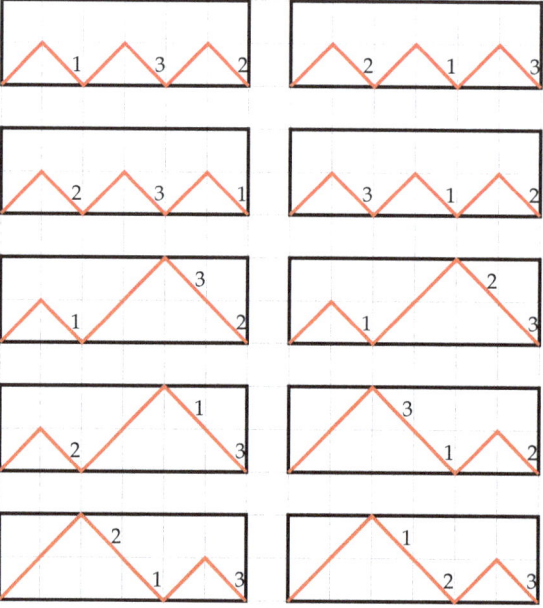

Figure 3. All labeled Dyck paths of size 3 with 1 ascent on return steps.

The total number of labeled Dyck n-paths with m ascents on return steps is defined by the elements of the Euler–Catalan number triangle (the sequence $A316773$ in [38]). That is, the cardinality function of this combinatorial set is equal to the Euler–Catalan numbers, which are denoted by EC_n^m [39].

Applying the method of compositae for obtaining explicit expressions for the coefficients of generating functions, we have found a generating function and the following explicit formula [39]:

$$EC_n^m = \begin{cases} 1, & \text{for } n = m = 0; \\ \sum_{k=m+1}^{n} \frac{n!}{k!} CT_n^k E_k^m, & \text{otherwise}, \end{cases} = \begin{cases} 1, & \text{for } n = m = 0; \\ \sum_{k=m+1}^{n} CT_n^k C_n^k E_k^m P_{n-k}, & \text{otherwise}, \end{cases} \quad (1)$$

where CT_n^m is the transposed Catalan triangle, C_n^m is the number of m-combinations of n elements, E_n^m is the Euler triangle, and P_n is the number of permutations of n elements.

4.2. AND/OR Tree

Equation (1) belongs to the algebra $\{\mathbb{N}, +, \times, R\}$, and there are the well-known formulas for the components of Equation (1), which also belong to the required algebra:

- the elements of the transposed Catalan triangle (the sequence $A033184$ in [38]) show the number of Dyck n-paths with m return steps and can be calculated using the following recurrence [37]:

$$CT_n^m = CT_{n-1}^{m-1} + CT_n^{m+1}, \quad CT_n^0 = 0, \quad CT_n^n = 1; \quad (2)$$

- the number of m-combinations of n elements (the sequence $A007318$ in [38]) can be calculated using the following recurrence [40]:

$$C_n^m = C_{n-1}^m + C_{n-1}^{m-1}, \quad C_n^n = C_n^0 = 1; \quad (3)$$

- the elements of the Euler triangle (the sequence $A173018$ in [38]) show the number of permutations of n elements with m ascents and can be calculated using the following recurrence [41]:

$$E_n^m = (m+1)E_{n-1}^m + (n-m)E_{n-1}^{m-1}, \quad E_n^{n-1} = E_n^0 = 1; \quad (4)$$

- the number of permutations of n elements (the sequence $A000142$ in [38]) can be calculated using the following recurrence [40]:

$$P_n = nP_{n-1}, \quad P_0 = 1. \quad (5)$$

Since Equation (1) and all the formulas for its components belong to the algebra $\{\mathbb{N}, +, \times, R\}$, we can construct the AND/OR tree structure for EC_n^m, which is presented in Figure 4. A bijection between the labeled Dyck n-paths with m ascents on return steps and the variants of the AND/OR tree is defined by the following rules:

- the number of return steps in a Dyck path is determined by the value of k (the selected son of the OR node labeled EC_n^m in a variant of the AND/OR tree);
- the subtree of the node labeled CT_n^k determines the version of a Dyck n-path with k return steps;
- the subtree of the node labeled C_n^k determines k values given from the set of n values, which are used as the labels for the return steps (the remaining $n - k$ values are used as the labels for the remaining $n - k$ down-steps);
- the subtree of the node labeled E_k^m determines the version of a permutation of the labels for k return steps, which form exactly m ascents;
- the subtree of the node labeled P_{n-k} determines the version of a permutation of the labels for the remaining $n - k$ down-steps.

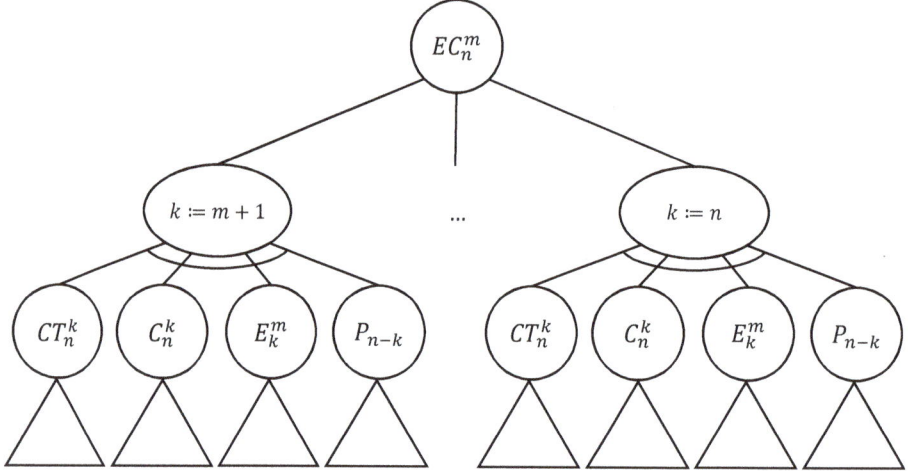

Figure 4. An AND/OR tree for EC_n^m.

Next, we need to construct AND/OR tree structures for all the subtrees of the AND/OR tree for EC_n^m. We also need to find bijections between the variants of these AND/OR trees and the corresponding combinatorial objects.

Since Equation (2) belongs to the algebra $\{\mathbb{N}, +, \times, R\}$, we can construct the AND/OR tree structure for CT_n^m, which is presented in Figure 5. A bijection between the Dyck n-paths with m return steps and the variants of the AND/OR tree is defined by the following rules:

- if a Dyck n-path is not completely filled and it is at the ground level 0, then it is necessary to add an up-step;
- each selected left son of the AND/OR tree (the node labeled CT_{n-1}^{m-1}) corresponds to a down-step in a Dyck n-path;
- each selected right son of the AND/OR tree (the node labeled CT_n^{m+1}) corresponds to an up-step in a Dyck n-path;
- if a leaf of the AND/OR tree is reached (when $m = n$), then it is necessary to add n down-steps.

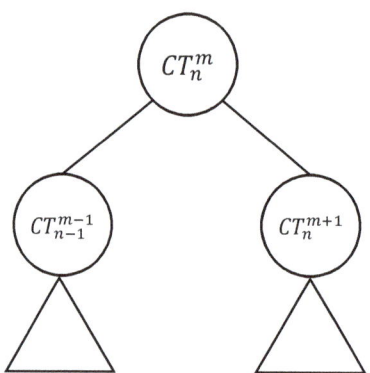

Figure 5. An AND/OR tree for CT_n^m.

For a compact representation, we encode a Dyck n-path by a sequence $a = (a_1, \ldots, a_{2n})$, where $a_i = $ 'u' encodes an up-step and $a_i = $ 'd' encodes a down-step. We also encode a variant of an AND/OR tree by a sequence $v = (v_1, v_2, \ldots)$ of the selected sons of the OR nodes in this tree (the left son corresponds to $v_i = 0$ and the right son corresponds to $v_i = 1$). An example of applying the obtained bijection for Dyck paths with return steps is presented in Table A1.

Since Equation (3) belongs to the algebra $\{\mathbb{N}, +, \times, R\}$, we can construct the AND/OR tree structure for C_n^m, which is presented in Figure 6. A bijection between the m-combinations of n elements and the variants of the AND/OR tree is defined by the following rules:

- each selected left son of the AND/OR tree (the node labeled C_{n-1}^m) determines that the element n is not selected in an m-combination of n elements;
- each selected right son of the AND/OR tree (the node labeled C_{n-1}^{m-1}) determines that the element n is selected in an m-combination of n elements;
- if a leaf of the AND/OR tree is reached (when $m = 0$), then it is necessary to not select all n elements in an m-combination of n elements;
- if a leaf of the AND/OR tree is reached (when $m = n$), then it is necessary to select all n elements in an m-combination of n elements.

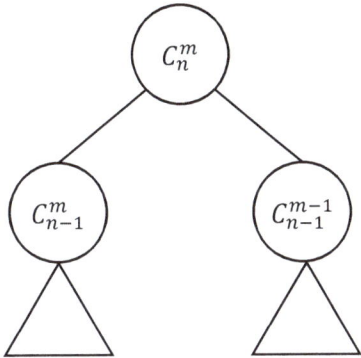

Figure 6. An AND/OR tree for C_n^m.

For a compact representation, we encode an m-combination of n elements by a sequence $a = (a_1, \ldots, a_n)$, where $a_i = 0$ encodes that the i-th element is not selected and $a_i = 1$ encodes that the i-th element is selected. We also encode a variant of an AND/OR tree by a sequence $v = (v_1, v_2, \ldots)$ of the selected sons of the OR nodes in this tree (the left son corresponds to $v_i = 0$ and the right son corresponds to $v_i = 1$). An example of applying the obtained bijection for combinations is presented in Table A2.

Since Equation (4) belongs to the algebra $\{\mathbb{N}, +, \times, R\}$, we can construct the AND/OR tree structure for E_n^m, which is presented in Figure 7. A bijection between the permutations of n elements with m ascents and the variants of the AND/OR tree is defined by the following rules:

- each selected left son of the OR node labeled E_n^m determines that the element n does not add an ascent in a permutation of n elements, and the selected son of the OR node labeled $m + 1$ determines the position of the element n in the permutation (there are exactly $m + 1$ possible positions);
- each selected right son of the OR node labeled E_n^m determines that the element n adds an ascent in a permutation of n elements, and the selected son of the OR node labeled $n - m$ determines the position of the element n in the permutation (there are exactly $n - m$ possible positions);

- if a leaf of the AND/OR tree is reached (when $m = 0$), then it is necessary to arrange all n elements in a permutation of n elements in decreasing order;
- if a leaf of the AND/OR tree is reached (when $m = n - 1$), then it is necessary to arrange all n elements in a permutation of n elements in increasing order.

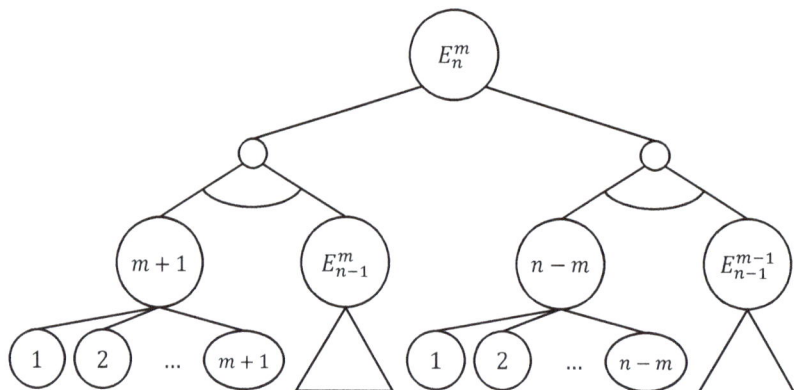

Figure 7. An AND/OR tree for E_n^m.

For a compact representation, we encode a variant of an AND/OR tree by a sequence $v = (v_1, v_2, \ldots)$ of the selected sons of the OR nodes in this tree, where each v_i is represented as a pair $(v_{i,1}, v_{i,2})$. In this pair: $v_{i,1} = 0$ corresponds to the left son of the OR node labeled E_n^m and $v_{i,2}$ determines the selected son of the OR node labeled $m + 1$; $v_{i,1} = 1$ corresponds to the right son of the OR node labeled E_n^m, $v_{i,2}$ determines the selected son of the OR node labeled $n - m$. An example of applying the obtained bijection for permutations with ascents is presented in Table A3.

Since Equation (5) belongs to the algebra $\{\mathbb{N}, +, \times, R\}$, we can construct the AND/OR tree structure for P_n, which is presented in Figure 8. A bijection between the permutations of n elements and the variants of the AND/OR tree is defined by the following rules:

- each selected son of the OR node labeled n determines the position of the element n in a permutation of n elements;
- if a leaf of the AND/OR tree is reached (when $n = 0$), then it is necessary to form an empty permutation.

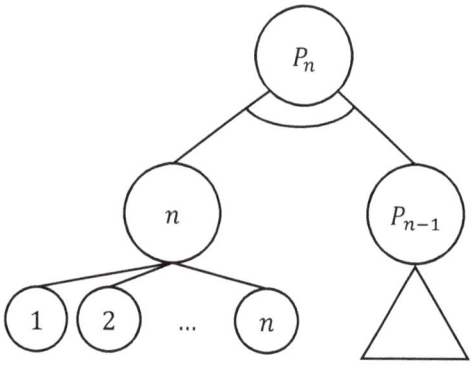

Figure 8. An AND/OR tree for P_n.

For a compact representation, we encode a variant of an AND/OR tree by a sequence $v = (v_1, v_2, \ldots)$ of the selected sons of the OR nodes in this tree. An example of applying the obtained bijection for permutations is presented in Table A4.

We have constructed AND/OR trees for all the combinatorial sets presented in this paper. Hence, we can develop algorithms for ranking and unranking the variants of the AND/OR trees.

4.3. Ranking and Unranking Algorithms

Based on Algorithms 1 and 2, we can develop algorithms for ranking and unranking the variants of the constructed AND/OR trees for EC_n^m, CT_n^m, C_n^m, E_n^m, and P_n.

For the AND/OR tree for CT_n^m, which is presented in Figure 5, we develop an algorithm for ranking its variants (Algorithm 3) and an algorithm for unranking its variants (Algorithm 4). Combining the developed algorithms with the derived rules for the bijection, we get algorithms for ranking and unranking the combinatorial set of Dyck n-paths with m return steps.

Algorithm 3: An algorithm for ranking the variants of the AND/OR tree for CT_n^m.

1 RankVariant_CT $(v = (v_1, v_2, \ldots), n, m)$
2 **begin**
3   **if** $m = n$ **then** $r = 0$
4   **else**
5   **if** $v_1 = 0$ **then** $r :=$ RankVariant_CT $((v_2, \ldots), n-1, m-1)$
6   **else** $r :=$ RankVariant_CT $((v_2, \ldots), n, m+1) + CT_{n-1}^{m-1}$
7   **end**
8   **return** r
9 **end**

Algorithm 4: An algorithm for unranking the variants of the AND/OR tree for CT_n^m.

1 UnrankVariant_CT (r, n, m)
2 **begin**
3   **if** $m = n$ **then** $v := ()$
4   **else**
5   **if** $r < CT_{n-1}^{m-1}$ **then** $v :=$ concat $((0),$ UnrankVariant_CT $(r, n-1, m-1))$
6   **else** $v :=$ concat $((1),$ UnrankVariant_CT $(r - CT_{n-1}^{m-1}, n, m+1))$
7   **end**
8   **return** v
9 **end**

In these algorithms, () denotes an empty sequence and a function concat denotes merging sequences. That is, if we have a sequence $a = (a_1, \ldots, a_n)$ and a sequence $b = (b_1, \ldots, b_m)$, then we can get the following sequence:

$$\text{concat}(a, b) = (a_1, \ldots, a_n, b_1, \ldots, b_m).$$

For the AND/OR tree for C_n^m, which is presented in Figure 6, we develop an algorithm for ranking its variants (Algorithm 5) and an algorithm for unranking its variants (Algorithm 6). Combining the developed algorithms with the derived rules for the bijection, we get algorithms for ranking and unranking the combinatorial set of m-combinations of n elements.

Algorithm 5: An algorithm for ranking the variants of the AND/OR tree for C_n^m.

1 RankVariant_C $(v = (v_1, v_2, \ldots), n, m)$
2 **begin**
3   **if** $m = 0$ or $m = n$ **then** $r = 0$
4   **else**
5   **if** $v_1 = 0$ **then** $r :=$ RankVariant_C $((v_2, \ldots), n-1, m)$
6   **else** $r :=$ RankVariant_C $((v_2, \ldots), n-1, m-1) + C_{n-1}^m$
7   **end**
8   **return** r
9 **end**

Algorithm 6: An algorithm for unranking the variants of the AND/OR tree for C_n^m.

1 UnrankVariant_C (r, n, m)
2 **begin**
3   **if** $m = 0$ or $m = n$ **then** $v := ()$
4   **else**
5   **if** $r < C_{n-1}^m$ **then** $v :=$ concat $((0),$ UnrankVariant_C $(r, n-1, m))$
6   **else** $v :=$ concat $((1),$ UnrankVariant_C $(r - C_{n-1}^m, n-1, m-1))$
7   **end**
8   **return** v
9 **end**

For the AND/OR tree for E_n^m, which is presented in Figure 7, we develop an algorithm for ranking its variants (Algorithm 7) and an algorithm for unranking its variants (Algorithm 8). Combining the developed algorithms with the derived rules for the bijection, we get algorithms for ranking and unranking the combinatorial set of permutations of n elements with m ascents.

Algorithm 7: An algorithm for ranking the variants of the AND/OR tree for E_n^m.

1 RankVariant_E $(v = ((v_{1,1}, v_{1,2}), (v_{2,1}, v_{2,2}), \ldots), n, m)$
2 **begin**
3   **if** $m = 0$ or $m = n - 1$ **then** $r = 0$
4   **else**
5   **if** $v_{1,1} = 0$ **then**
6             $l_1 := v_{1,2} - 1$
7             $l_2 :=$ RankVariant_E $(((v_{2,1}, v_{2,2}), \ldots), n-1, m)$
8             $r := l_1 + (m+1)l_2$
9   **end**
10   **else**
11            $l_1 := v_{1,2} - 1$
12            $l_2 :=$ RankVariant_E $(((v_{2,1}, v_{2,2}), \ldots), n-1, m-1)$
13            $r := l_1 + (n-m)l_2 + (m+1)E_{n-1}^m$
14   **end**
15   **end**
16   **return** r
17 **end**

Algorithm 8: An algorithm for unranking the variants of the AND/OR tree for E_n^m.

```
1  UnrankVariant_E (r, n, m)
2  begin
3      if m = 0 or m = n then v := ()
4      else
5          if r < (m+1)E_{n-1}^m then
6              l_1 := r mod m + 1
7              l_2 := ⌊r/(m+1)⌋
8              v := concat(((0, l_1 + 1)), UnrankVariant_E(l_2, n − 1, m))
9          end
10         else
11             r := r − (m+1)E_{n-1}^m
12             l_1 := r mod n − m
13             l_2 := ⌊r/(n−m)⌋
14             v := concat(((1, l_1 + 1)), UnrankVariant_E(l_2, n − 1, m − 1))
15         end
16     end
17     return v
18 end
```

For the AND/OR tree for P_n, which is presented in Figure 8, we develop an algorithm for ranking its variants (Algorithm 9) and an algorithm for unranking its variants (Algorithm 10). Combining the developed algorithms with the derived rules for the bijection, we get algorithms for ranking and unranking the combinatorial set of permutations of n elements.

Algorithm 9: An algorithm for ranking the variants of the AND/OR tree for P_n.

```
1  RankVariant_P (v = (v_1, v_2, ...), n)
2  begin
3      if n = 0 then r = 0
4      else
5          l_1 := v_1 − 1
6          l_2 := RankVariant_P((v_2, ...), n − 1)
7          r := l_1 + nl_2
8      end
9      return r
10 end
```

Algorithm 10: An algorithm for unranking the variants of the AND/OR tree for P_n.

```
1  UnrankVariant_P (r, n)
2  begin
3      if n = 0 then v := ()
4      else
5          l_1 := r mod n
6          l_2 := ⌊r/n⌋
7          v := concat((l_1 + 1), UnrankVariant_P(l_2, n − 1))
8      end
9      return v
10 end
```

For the AND/OR tree for EC_n^m, which is presented in Figure 4, we develop an algorithm for ranking its variants (Algorithm 11) and an algorithm for unranking its variants (Algorithm 12). Combining the developed algorithms with the derived rules for the bijection, we get algorithms for ranking and unranking the combinatorial set of labeled Dyck n-paths with m ascents on return steps.

Algorithm 11: An algorithm for ranking the variants of the AND/OR tree for EC_n^m.

1 RankVariant_EC ($v = (k, v_1, v_2, v_3, v_4), n, m$)
2 **begin**
3     $l_1 :=$ RankVariant_CT (v_1, n, k)
4     $l_2 :=$ RankVariant_C (v_2, n, k)
5     $l_3 :=$ RankVariant_E (v_3, k, m)
6     $l_4 :=$ RankVariant_P ($v_4, n - k$)
7     $r := l_1 + CT_n^k(l_2 + C_n^k(l_3 + E_k^m l_4)) + \sum_{i=m+1}^{k-1} CT_n^i C_n^i E_i^m P_{n-i}$
8   **return** r
9 **end**

Algorithm 12: An algorithm for unranking the variants of the AND/OR tree for EC_n^m.

1 UnrankVariant_EC (r, n, m)
2 **begin**
3     $k := m + 1$
4     $sum := 0$
5   **while** $sum + CT_n^k C_n^k E_k^m P_{n-k} \leq r$ **do**
6        $sum := sum + CT_n^k C_n^k E_k^m P_{n-k}$
7        $k := k + 1$
8   **end**
9     $r := r - sum$
10     $l_1 := r \bmod CT_n^k$
11     $r := \left\lfloor \frac{r}{CT_n^k} \right\rfloor$
12     $l_2 := r \bmod C_n^k$
13     $r := \left\lfloor \frac{r}{C_n^k} \right\rfloor$
14     $l_3 := r \bmod E_k^m$
15     $l_4 := \left\lfloor \frac{r}{E_k^m} \right\rfloor$
16     $v_1 :=$ UnrankVariant_CT (l_1, n, k)
17     $v_2 :=$ UnrankVariant_C (l_2, n, k)
18     $v_3 :=$ UnrankVariant_E (l_3, k, m)
19     $v_4 :=$ UnrankVariant_P ($l_4, n - k$)
20     $v = (k, v_1, v_2, v_3, v_4)$
21   **return** v
22 **end**

In these algorithms, we use all the above mentioned algorithms for ranking and unranking the variants of the AND/OR trees for CT_n^m, C_n^m, E_n^m, and P_n. For a compact representation, a variant of the AND/OR tree for EC_n^m is encoded by a sequence $v = (k, v_1, v_2, v_3, v_4)$, where:

- k is the label of the selected son of the OR node labeled EC_n^m in a variant of the AND/OR tree;
- v_1 corresponds to the variant of the subtree of the node labeled CT_n^k;
- v_2 corresponds to the variant of the subtree of the node labeled C_n^k;

- v_3 corresponds to the variant of the subtree of the node labeled E_k^m;
- v_4 corresponds to the variant of the subtree of the node labeled P_{n-k}.

5. Conclusions

In this paper, we study methods for developing combinatorial generation algorithms and present basic general methods for solving this task. We consider one of these methods, which is based on AND/OR trees, and extend it by using the mathematical apparatus of the theory of generating functions.

Using an AND/OR tree structure, it is possible to develop listing, ranking, and unranking algorithms for a given combinatorial set. However, the use of the method for developing combinatorial generation algorithms based on AND/OR trees has the following restriction: the cardinality function of a combinatorial set must belong to the algebra $\{\mathbb{N}, +, \times, R\}$. For solving this problem, we propose to apply the method of compositae for obtaining explicit expression of the coefficients of generating functions, since the theory of generating functions is one of the basic approaches in combinatorics. The limitation of this method is that it can be applied only for a combinatorial set for which a generating function is known.

As a result, we formalize the proposed idea in our modification of the original method for developing combinatorial generation algorithms. In addition, one of the main contributions of the paper is the application of this method. To confirm the effectiveness of using the proposed method, we develop new ranking and unranking algorithms for the following combinatorial sets: labeled Dyck n-paths with m ascents on return steps, Dyck n-paths with m return steps, m-combinations of n elements, permutations of n elements with m ascents, permutations of n elements. For each of them, we construct an AND/OR tree, find a bijection between the elements of the combinatorial set and the set of variants of the AND/OR tree, and develop algorithms for ranking and unranking the variants of the AND/OR tree.

All the developed algorithms have been realized in the computer algebra system "Maxima" and validated by exhaustive generation for fixed values of combinatorial set parameters. It also has shown that all the developed algorithms have polynomial time complexity. Several examples of applying the obtained results can be found in Appendix A.

As further research, we will consider the development of new and effective combinatorial generation algorithms in the field of applied mathematics. For example, it can be done for combinatorial sets that represent different types of chemical compounds [42], molecular structures such as RNA and DNA [43–45], etc. We also plan further improvements to the presented method for developing combinatorial generation algorithms, for example, through the usage of other types of trees [46].

Author Contributions: Investigation, Y.S., V.K., and D.K.; methodology, V.K.; writing—original draft preparation, Y.S.; writing—review and editing, D.K. All authors have read and agreed to the published version of the manuscript.

Funding: The reported study was supported by the Russian Science Foundation (project no. 18-71-00059).

Acknowledgments: The authors would like to thank the referees for their helpful comments and suggestions.

Conflicts of Interest: The authors declare no conflict of interest.

Appendix A. Examples of Ranking the Elements of Combinatorial Sets

Table A1. Ranking the combinatorial set of Dyck n-paths with m return steps for $n = 5$ and $m = 2$.

Dyck path	Variant of AND/OR Tree	Rank
u d u u d u d u d d	(0, 1, 0, 1, 0, 1)	0
u d u u d u u d d d	(0, 1, 0, 1, 1)	1
u d u u u d d u d d	(0, 1, 1, 0, 0, 1)	2
u d u u u d u d d d	(0, 1, 1, 0, 1)	3

Table A1. *Cont.*

Dyck path	Variant of AND/OR Tree	Rank
uduuuudddd	(0,1,1,1)	4
uuddduududd	(1,0,0,1,0,1)	5
uudduuuddd	(1,0,0,1,1)	6
uududduudd	(1,0,1,0,0,1)	7
uudududdud	(1,0,1,0,1)	8
uuduudddud	(1,0,1,1)	9
uuudddduudd	(1,1,0,0,0,1)	10
uuuddudddud	(1,1,0,0,1)	11
uuududddud	(1,1,0,1)	12
uuuudddddud	(1,1,1)	13

Table A2. Ranking the combinatorial set of m-combinations of n elements for $n=5$ and $m=2$.

Combination	Variant of AND/OR Tree	Rank
(1,1,0,0,0)	(0,0,0)	0
(1,0,1,0,0)	(0,0,1,0)	1
(0,1,1,0,0)	(0,0,1,1)	2
(1,0,0,1,0)	(0,1,0,0)	3
(0,1,0,1,0)	(0,1,0,1)	4
(0,0,1,1,0)	(0,1,1)	5
(1,0,0,0,1)	(1,0,0,0)	6
(0,1,0,0,1)	(1,0,0,1)	7
(0,0,1,0,1)	(1,0,1)	8
(0,0,0,1,1)	(1,1)	9

Table A3. Ranking the combinatorial set of permutations of n elements with m ascents for $n=4$ and $m=2$.

Permutation	Variant of AND/OR Tree	Rank
(4,1,2,3)	((0,1))	0
(1,4,2,3)	((0,2))	1
(1,2,4,3)	((0,3))	2
(3,4,1,2)	((1,1),(0,1))	3
(3,1,2,4)	((1,2),(0,1))	4
(1,3,4,2)	((1,1),(0,2))	5
(1,3,2,4)	((1,2),(0,2))	6
(2,3,4,1)	((1,1),(1,1))	7
(2,3,1,4)	((1,2),(1,1))	8
(2,4,1,3)	((1,1),(1,2))	9
(2,1,3,4)	((1,2),(1,2))	10

Table A4. Ranking the combinatorial set of permutations of n elements for $n=4$.

Permutation	Variant of AND/OR Tree	Rank
(4,3,2,1)	(1,1,1,1)	0
(3,4,2,1)	(2,1,1,1)	1
(3,2,4,1)	(3,1,1,1)	2
(3,2,1,4)	(4,1,1,1)	3
(4,2,3,1)	(1,2,1,1)	4
(2,4,3,1)	(2,2,1,1)	5
(2,3,4,1)	(3,2,1,1)	6
(2,3,1,4)	(4,2,1,1)	7
(4,2,1,3)	(1,3,1,1)	8
(2,4,1,3)	(2,3,1,1)	9
(2,1,4,3)	(3,3,1,1)	10

Table A4. *Cont.*

Permutation	Variant of AND/OR Tree	Rank
(2, 1, 3, 4)	(4, 3, 1, 1)	11
(4, 3, 1, 2)	(1, 1, 2, 1)	12
(3, 4, 1, 2)	(2, 1, 2, 1)	13
(3, 1, 4, 2)	(3, 1, 2, 1)	14
(3, 1, 2, 4)	(4, 1, 2, 1)	15
(4, 1, 3, 2)	(1, 2, 2, 1)	16
(1, 4, 3, 2)	(2, 2, 2, 1)	17
(1, 3, 4, 2)	(3, 2, 2, 1)	18
(1, 3, 2, 4)	(4, 2, 2, 1)	19
(4, 1, 2, 3)	(1, 3, 2, 1)	20
(1, 4, 2, 3)	(2, 3, 2, 1)	21
(1, 2, 4, 3)	(3, 3, 2, 1)	22
(1, 2, 3, 4)	(4, 3, 2, 1)	23

References

1. Knuth, D.E. *The Art of Computer Programming, Volume 4A: Combinatorial Algorithms, Part 1*; Addison-Wesley Professional: Boston, MA, USA, 2011.
2. Ruskey, F. Combinatorial generation. Working Version (1j-CSC 425/520). Available online: http://page.math.tu-berlin.de/~felsner/SemWS17-18/Ruskey-Comb-Gen.pdf (accessed on 1 May 2020).
3. Reingold, E.M.; Nievergelt, J.; Deo, N. *Combinatorial Algorithms: Theory and Practice*; Prentice-Hall: Upper Saddle River, NJ, USA, 1977.
4. Kreher, D.L.; Stinson, D.R. *Combinatorial Algorithms: Generation, Enumeration, and Search*; CRC Press: Boca Raton, FL, USA, 1999.
5. Barcucci, E.; Del Lungo, A.; Pergola, E.; Pinzani, R. ECO: A methodology for the enumeration of combinatorial objects. *J. Differ. Equ. Appl.* **1999**, *5*, 435–490. [CrossRef]
6. Barcucci, E.; Del Lungo, A.; Pergola, E. Random generation of trees and other combinatorial objects. *Theoret. Comput. Sci.* **1999**, *218*, 219–232. [CrossRef]
7. Bacchelli, S.; Barcucci, E.; Grazzini, E.; Pergola, E. Exhaustive generation of combinatorial objects by ECO. *Acta Inform.* **2004**, *40*, 585–602. [CrossRef]
8. Bacchelli, S.; Ferrari, L.; Pinzani, R.; Sprugnoli, R. Mixed succession rules: The commutative case. *J. Combin. Theory Ser. A* **2010**, *117*, 568–582.
9. Del Lungo, A.; Frosini, A.; Rinaldi, S. ECO method and the exhaustive generation of convex polyominoes. *Lect. Notes Comput. Sci. Discret. Math. Theor. Comput. Sci.* **2003**, *2731*, 129–140.
10. Del Lungo, A.; Duchi, E.; Frosini, A.; Rinaldi, S. On the generation and enumeration of some classes of convex polyominoes. *Electron. J. Combin.* **2004**, *11*, 1–46. [CrossRef]
11. Vajnovszki, V. Generating involutions, derangements, and relatives by ECO. *Discrete Math. Theor. Comput. Sci.* **2010**, *12*, 109–122.
12. Vajnovszki, V. An efficient Gray code algorithm for generating all permutations with a given major index. *J. Discret. Algorithms* **2014**, *26*, 77–88. [CrossRef]
13. Do, P.T.; Tran, T.T.H.; Vajnovszki, V. Exhaustive generation for permutations avoiding (colored) regular sets of patterns. *Discret. Appl. Math.* **2019**, *268*, 44–53. [CrossRef]
14. Flajolet, P.; Zimmerman, P.; Cutsem, B. A calculus for the random generation of combinatorial structures. *Theoret. Comput. Sci.* **1994**, *132*, 1–35. [CrossRef]
15. Sedgewick, R.; Flajolet, P. *An Introduction to the Analysis of Algorithms*, 2nd ed.; Addison-Wesley: Boston, MA, USA, 2013.
16. Martinez, C.; Molinero, X. A generic approach for the unranking of labeled combinatorial classes. *Random Struct. Algorithms* **2001**, *19*, 472–497. [CrossRef]
17. Martinez, C.; Molinero, X. Generic algorithms for the generation of combinatorial objects. *Lect. Notes Comput. Sci. Math. Found. Comput. Sci.* **2003**, *2747*, 572–581.
18. Martinez, C.; Molinero, X. An experimental study of unranking algorithms. *Lect. Notes Comput. Sci. Exp. Effic. Algorithms* **2004**, *3059*, 326–340.

19. Martinez, C.; Molinero, X. Efficient iteration in admissible combinatorial classes. *Theoret. Comput. Sci.* **2005**, *346*, 388–417. [CrossRef]
20. Molinero, X.; Vives, J. Unranking algorithms for combinatorial structures. *Int. J. Appl. Math. Inf.* **2015**, *9*, 110–115.
21. Ryabko, B.Y. Fast enumeration of combinatorial objects. *Discret. Math. Appl.* **1998**, *8*, 163–182. [CrossRef]
22. Medvedeva, Y.S.; Ryabko, B.Y. Fast enumeration algorithm for words with given constraints on run lengths of ones. *Probl. Inf. Transm.* **2010**, *46*, 390–399. [CrossRef]
23. Medvedeva, Y. Fast enumeration of words generated by Dyck grammars. *Math. Notes* **2014**, *96*, 68–83. [CrossRef]
24. Kruchinin, V.V. *Methods for Developing Algorithms for Ranking and Unranking Combinatorial Objects Based on AND/OR Trees*; V-Spektr: Tomsk, Russia, 2007. (In Russian)
25. Shablya, Y.; Kruchinin, D.; Kruchinin, V. Application of the method of compositae in combinatorial generation. In Proceedings of the Book of the 2nd Mediterranean International Conference of Pure and Applied Mathematics and Related Areas (MICOPAM2019), Paris, France, 28–31 August 2019; pp. 91–94.
26. Hartman, P.; Sawada, J. Ranking and unranking fixed-density necklaces and Lyndon words. *Theoret. Comput. Sci.* **2019**, *791*, 36–47. [CrossRef]
27. Pai, K.J.; Chang, J.M.; Wu, R.Y.; Chang, S.C. Amortized efficiency of generation, ranking and unranking left-child sequences in lexicographic order. *Discrete Appl. Math.* **2019**, *268*, 223–236. [CrossRef]
28. Amani, M.; Nowzari-Dalini, A. Efficient generation, ranking, and unranking of (k,m)-ary trees in B-order. *Bull. Iranian Math. Soc.* **2019**, *45*, 1145–1158. [CrossRef]
29. Stanley, R.P. *Enumerative Combinatorics*, 2nd ed.; Cambridge University Press: Cambridge, MA, USA, 2011; Volume 1.
30. Kruchinin, D.V.; Kruchinin, V.V. A method for obtaining generating functions for central coefficients of triangles. *J. Integer Seq.* **2012**, *15*, 12.9.3.
31. Kruchinin, D.V.; Kruchinin, V.V. Explicit formulas for some generalized polynomials. *Appl. Math. Inf. Sci.* **2013**, *7*, 2083–2088. [CrossRef]
32. Kruchinin, V.V.; Kruchinin, D.V. Composita and its properties. *J. Anal. Number Theory* **2014**, *2*, 37–44.
33. Kruchinin, D.V.; Shablya, Y.V. Explicit formulas for Meixner polynomials. *Int. J. Math. Math. Sci.* **2015**, *2015*, 620569. [CrossRef]
34. Kruchinin, D.V.; Shablya, Y.V.; Kruchinin, V.V.; Shelupanov, A.A. A method for obtaining coefficients of compositional inverse generating functions. In Proceedings of the International Conference of Numerical Analysis and Applied Mathematics (ICNAAM 2015), Rhodes, Greece, 22–28 September 2015; Volume 1738, p. 130003.
35. Kruchinin, D.V.; Kruchinin, V.V. Explicit formula for reciprocal generating function and its application. *Adv. Stud. Contemp. Math. Kyungshang* **2019**, *29*, 365–372.
36. Banderier, C.; Krattenthaler, C.; Krinik, A.; Kruchinin, D.; Kruchinin, V.; Nguyen, D.; Wallner, M. Explicit formulas for enumeration of lattice paths: Basketball and the kernel method. In *Lattice Path Combinatorics and Applications*; Springer: Cham, Switzerland, 2019; pp. 78–118.
37. Deutsch, E. Dyck path enumeration. *Discret. Math.* **1999**, *204*, 167–202. [CrossRef]
38. Sloane, N.J.A. The On-Line Encyclopedia of Integer Sequences. Available online: https://oeis.org (accessed on 1 May 2020).
39. Shablya, Y.; Kruchinin, D. Euler–Catalan's number triangle and its application. *Symmetry* **2020**, *12*, 600. [CrossRef]
40. Graham, R.L.; Knuth, D.E.; Patashnik, O. *Concrete Mathematics*, 2nd ed.; Addison-Wesley: Boston, MA, USA, 1994.
41. Petersen, T.K. *Eulerian Numbers*; Birkhauser Advanced Texts; Basler Lehrbucher, Birkhauser/Springer: New York, NY, USA, 2015.
42. Shimizu, T.; Fukunaga, T.; Nagamochi, H. Unranking of small combinations from large sets. *J. Discret. Algorithms* **2014**, *29*, 8–20. [CrossRef]
43. Seyedi-Tabari, E.; Ahrabian, H.; Nowzari-Dalini, A. A new algorithm for generation of different types of RNA. *Int. J. Comput. Math.* **2010**, *87*, 1197–1207. [CrossRef]
44. Nebel, M.E.; Scheid, A.; Weinberg, F. Random generation of RNA secondary structures according to native distributions. *Algorithms Mol. Biol.* **2011**, *6*, 24. [CrossRef] [PubMed]

45. Chee, Y.M.; Chrisnata, J.; Kiah, H.M.; Nguyen, T.T. Efficient encoding/decoding of irreducible words for codes correcting tandem duplications. In Proceedings of the 2018 IEEE International Symposium on Information Theory (ISIT), Vail, CO, USA, 17–22 June 2018.
46. Pop, P.C. The generalized minimum spanning tree problem: An overview of formulations, solution procedures and latest advances. *Eur. J. Oper. Res.* **2020**, *283*, 1–15. [CrossRef]

© 2020 by the authors. Licensee MDPI, Basel, Switzerland. This article is an open access article distributed under the terms and conditions of the Creative Commons Attribution (CC BY) license (http://creativecommons.org/licenses/by/4.0/).

Article

Efficient Dynamic Flow Algorithms for Evacuation Planning Problems with Partial Lane Reversal

Urmila Pyakurel [1,†]**, Hari Nandan Nath** [1,*]**, Stephan Dempe** [2] **and Tanka Nath Dhamala** [1]

1. Central Department of Mathematics, Tribhuvan University, P.O. Box: 13143, Kathmandu, Nepal; urmilapyakurel@gmail.com (U.P.); amb.dhamala@daadindia.org (T.N.D.)
2. TU Bergakademie, Fakultät für Mathematik und Informatik, 09596 Freiberg, Germany; dempe@math.tu-freiberg.de
* Correspondence: hari672@gmail.com
† Current address: TU Bergakademie, Fakultät für Mathematik und Informatik, 09596 Freiberg, Germany.

Received: 9 September 2019; Accepted: 16 October 2019; Published: 19 October 2019

Abstract: Contraflow technique has gained a considerable focus in evacuation planning research over the past several years. In this work, we design efficient algorithms to solve the maximum, lex-maximum, earliest arrival, and quickest dynamic flow problems having constant attributes and their generalizations with partial contraflow reconfiguration in the context of evacuation planning. The partial static contraflow problems, that are foundations to the dynamic flows, are also studied. Moreover, the contraflow model with inflow-dependent transit time on arcs is introduced. A strongly polynomial time algorithm to compute approximate solution of the quickest partial contraflow problem on two terminal networks is presented, which is substantiated by numerical computations considering Kathmandu road network as an evacuation network. Our results show that the quickest time to evacuate a flow of value 100,000 units is reduced by more than 42% using the partial contraflow technique, and the difference is more with the increase in the flow value. Moreover, the technique keeps the record of the portions of the road network not used by the evacuees.

Keywords: network optimization; dynamic flow; evacuation planning; contraflow configuration; partial lane reversals, algorithms and complexity; logistic supports

1. Introduction

Because of the significant occurrences of many predictable and unpredictable large-scale disasters worldwide, regardless of various discoveries and urbanization, an efficient, implementable, and reliable evacuation planning is indispensable for saving life and supporting humanitarian relief with optimal use and equitable distribution of available resources. Among prevalent disasters, e.g., earthquakes, volcanic eruptions, landslides, floods, tsunamis, hurricanes, typhoons, chemical explosions, and terrorist attacks, the most remarkable losses are noted in earthquakes in Nepal (April 2015), Japan (March 2011), Haiti (January 2010), Chichi (Taiwan, September 1999), Bam (Iran, December 2003), Kashmir (Pakistan, October 2005), and Chile (May 1960); various tsunamis in Japan and the Indian Ocean; the major hurricanes Katrina, Rita, and Sandy in USA; and the September 11 attacks in USA. The threat of disasters always persists, e.g., there is a prediction of earthquakes of more than 8.4 in Richter scale around the capital of Nepal in the near future (Pyakurel et al. [1]). Therefore, there is always a need for an effective emergency plan to cope with disasters worldwide, including Nepal.

Evacuating people from disastrous areas to safe places is one of the important aspect of emergency planning. Among the diversified fields (e.g., traffic simulation, fluid dynamics, control theory, variational inequalities, and network flow) of mathematical research in evacuation planning, network flow methodologies are the most efficient [2]. An evacuation optimizer looks after a plan on evacuation network for an efficient transfer of maximum evacuees from the dangerous (sources) to safer (sinks) locations as quickly as possible [3,4]. An optimal shelter location and support of humanitarian logistics within these emergency scenarios are equally demanding but challenging issues. The aim of this paper is to look at the transportation planning based on strong mathematical modeling with applications, not only in emergency evacuations, but also in the heavy traffic hours of a city. A comprehensive explanation on diversified theories and applications can be found in the survey papers of Hamacher and Tjandra [4], Cova and Johnson [5], Altay and Green III [6], Pascoal et al. [7], Moriarty et al. [8], Chen and Miller-Hooks [9], Yusoff et al. [10], Dhamala et al. [11], Kotsireas et al. [12], and the literature therein. The theoretical background needed in this paper, in particular, is given in Section 2 and inside other sections, wherever necessary.

The evacuation network, which corresponds to a region (or a building, shopping mall, etc.), is represented by a dynamic network in which nodes represent the street intersections (or rooms), and arcs represent the connections (streets, doors, etc.) between the nodes. The hazardous locations of evacuees are termed as the source nodes and the safe locations where the evacuees are to be transferred are termed as sink nodes. The nodes and arcs have capacities. Further, each arc has a transit time or a cost associated with it. The evacuees or the vehicles carrying evacuees traveling through the network are modeled as a flow. An evacuation plan largely depends on the number of sources, number of sinks, and the parameters associated with nodes and arcs, which may be dependent on time or the amount of the flow, along with other constraints. The time variable which is continuous may be discretized. The discrete time steps approximate the computationally heavy continuous models at the cost of solution approximations. Also the constant time probably approximated by free flow speeds or certain queuing rules and constant capacity settings mostly realize the evacuation problems to be linear, at least more tractable, in contrast to the more general and realistic flow-dependent real-world evacuation scenarios.

Following the pioneers of Ford and Fulkerson [13], with an objective to maximize the flow from a source to a sink at the end of given discrete time period, Gale [14] shows an existence of the maximum flow from the very beginning in discrete time setting. Two pseudo-polynomial time algorithms are presented by Wilkinson [15] and Minieka [16] for the latter problem with constant arc transit times. An upward scaling approximation algorithm of Hoppe [17] polynomially solves this problem within a factor of $1+\epsilon$, for every $\epsilon > 0$. For the special series-parallel networks, Ruzika et al. [18] solve this problem, applying a minimum cost circulation flow algorithm by exploiting a property that every cycle of its residual network is of non-negative length. Minieka [16], and Hoppe and Tardos [19] maximize the flow in priority ordering, which is important in some scenarios of evacuation planning. Burkard et al. [20], and Hoppe and Tardos [19] present efficient algorithms for shifting the already fixed evacuees in minimum time. However, the general multiterminal evacuation problems, with variable number of evacuees at sources, are computationally hard even with constant attributes on arcs. Likewise, the earliest arrival transshipment solutions that fulfill the specific demands at sinks by the specific supplies from sources maximizing the flow at each point of time are also not solved in polynomial time yet. The multisource single-sink (cf. Baumann and Skutella [21]) and zero transit times, in either one source or one sink (cf. Fleischer [22]) earliest arrival transshipment problems, have polynomial time solutions. Gross et al. [23] propose efficient algorithms to calculate the approximate earliest arrival flows on arbitrary networks. Several works have been done with continuous time settings as well (see, Pyakurel and Dhamala [24] for the references). Several polynomial time algorithms by natural transformations are obtained by Fleischer and Tardos [25].

In urban road networks, the vehicles in a particular road segment are allowed to move in a specified direction only. In case of disasters, the evacuation planners discourage people to move towards risk areas from safer places. As a result, the road segments leading towards the risk areas become unoccupied and those towards safer places become overoccupied. If the direction of the traffic flow in unoccupied segments is reversed, then the vehicles moving towards the safe areas can use those segments as well. Reversing the traffic flow in a particular road segment is also known as lane reversal. The optimal lane reversal strategy makes the traffic systematic and smooth by removing the traffic jams caused, not only in different natural and human-created disasters of large scale, but also in busy office hours, special events and street demonstrations. The contraflow reconfiguration, by means of various operations research models (cf. Section 2), heuristics, optimization techniques and simulation, reverses the usual direction of empty lanes towards sinks satisfying the given constraints that increase the value of the flow, and decrease the average time of evacuation. However, an efficient and universally acceptable solution approach that meets the macroscopic and microscopic behavioral characteristics of the evacuees (e.g., threat conditions, community context, and preparedness) is still lacking (see [4,26]). Because of the large size of the problem (a large number of variables and parameters involved), designing algorithms to calculate optimal solution within a desired time is challenging. Heuristics and approximation methods provide solutions within an acceptable time but the optimality of the solution is not guaranteed. A trade-off between computational costs and solution quality should always be desired when designing solution algorithms.

As the computational costs of the exact mathematical solutions for general contraflow techniques are quite high, a series of heuristic procedures are approached in literature that are computationally manageable. Kim et al. [27] present two greedy and bottleneck heuristics for possible numerical approximate solutions to the quickest contraflow problem, and they show that at least 40% evacuation time can be reduced by reverting at most 30% arcs in their case study. They model the problem of lane reversals mathematically as an integer programming problem by means of flows on network and also prove that it is $\mathcal{NP}$-hard. All contraflow evacuation problems are at least harder than the corresponding problems without contraflow. We recommend Dhamala et al. [11] and the references therein for different approaches of contraflow heuristics.

Though comparatively less, recent interest also includes analytical techniques, after Rebennack et al. [28] solved the two-terminal maximum contraflow and quickest contraflow problems optimally in polynomial times. The earliest arrival and the maximum contraflow problems are solved with the temporally repeated solutions in Dhamala and Pyakurel [29]. Its solution with continuous time is obtained in [30]. The authors of [31] solve the earliest arrival contraflow on two-terminal network in pseudo-polynomial time. They also introduce the lex-maximum dynamic contraflow problem in which flow is maximized in given priority ordering and solved with polynomial time complexity. These problems are also solved in continuous time setting by using natural transformation (cf. Section 2.2) of flow in discrete times to continuous time intervals in [24,30]. With the given supplies at the sources and demands of the sink, the earliest arrival transshipment contraflow problem is modeled in discrete time [32] and solved on multisource network with polynomial algorithm. The problem with zero transit time on arcs is also solved on multi-sink network with a polynomial time complexity. For the multiterminal network, they present approximation algorithms to solve the earliest arrival transshipment contraflow problem. The discrete solutions are extended into continuous time in [1,24]. The problems with similar objectives, in what is known as an abstract network, are solved in [33].

In the present work, we propose algorithms to reverse the road segments up to the necessary capacity only, to record segments with unused capacities so that they can be used for other purposes of facilitating evacuation. Our proposed algorithms also summarize the earlier results on contraflow in compact form. To the best of our knowledge, this is the first attempt to address the issues on different partial contraflow problems with constant transit times and inflow-dependent transit times associated to the arcs. These node–arc partial contraflow models contribute in saving the unnecessary arc reversals improving the complete contraflow approaches.

The organization of this paper is as follows. Section 2 presents the basic terminology necessary in the paper and different flow models. All the previously solved dynamic contraflow problems with constant transit time are extended in the partial contraflow configuration and solved with efficient algorithms in Section 4, after presenting the fundamental static partial contraflow solutions in Section 3. Section 5 introduces the partial contraflow approach to solve the quickest flow problem with inflow-dependent transit times presenting efficient algorithms. Section 6 presents numerical computations related to the quickest partial contraflow, with inflow-dependent transit times and taking a case of the Kathmandu road network. The paper is concluded in Section 7.

2. Basic Terminology

An evacuation network is represented by $\mathcal{N} = (V, A, b, \tau, S, D, T)$, where V is the set of n nodes, $A \subseteq V \times V$ is the set of m arcs with a set of source nodes S and that of sink nodes D. For each $v \in V$, we define $A_v = \{e : e = (v, u) \in A\}$, the set of arcs outgoing from v, and $B_v = \{e : e = (u, v) \in A\}$, the set of arcs incoming to v. The network is assumed to be without parallel arcs between the nodes as they can be combined to a single arc with added capacity. It is connected with m arcs, so that $n - 1 \leq m$, and therefore $n + m = O(m)$. The nodes may be equipped with the initial occupancy $o : V \to R^+$. The predefined parameter T denotes a permissible time window within which the whole evacuation process has to be completed. It may be discretized into discrete time steps $\mathbf{T} = \{0, 1, \ldots, T\}$ or can be considered a continuous one as $\mathbf{T} = [0, T]$.

On arcs, the upper capacity (bound) function $b : A \times T \to R^+$ limits the flow rate passing along the arcs for each point in time. The transit time function $\tau : A \times T \to R^+$ measures the time the flow units take to travel along the arcs. We frame this work with constant and inflow-dependent transit times on arcs. Smith and Cruz [34] give various approaches of travel time estimation on arterial links, free and high ways. With inflow-dependent transit times, the transit time $\tau_e(x_e(\theta))$ is a function of inflow rate $x_e(\theta)$ on the arc e at given time point θ, so that at a time flow units enter an arc with the uniform speed and remain with the uniform speed traveling through this arc.

The flow rate function is defined by $x : A \times T \to R^+$, where $x_e(\theta)$ denotes the flow rate on e at time θ. It may be taken as an inflow, outflow, and intermediate flow rate that measure the flow at entry, exit, and intermediate points on an arc, respectively. For $\theta \in \{0, 1, \ldots, T\}$ and constant function τ, the amount of flow sent at time θ into e arrives to its end at time $\theta + \tau_e$. Whereas, for continuous time $\theta \in [0, T]$ and constant function τ, the amount of flow per time unit enters at this rate e at time θ and proceed continuously.

One may introduce an additional parameter $\lambda_e \in R^+$ on arc e, e.g., a gain factor, to model a generalized dynamic flow when only λ_e units of flow leave from w at time $\theta + \tau_e$, by entering a unit of flow on $e = (v, w)$ at time θ. If the flow is not gained, practically, along any arc, then $\lambda_e \leq 1$ holds for each arc $e \in A$, and we call the network as a lossy network [35], which is denoted by $\mathcal{N} = (V, A, b, \tau, \lambda, S, D, T)$.

2.1. Flow Models

For a source node, s, and a sink node, d, a static s-d flow with value $\text{val}(y)$ is a function $y : \to \mathcal{R}^+$ satisfying

$$\text{val}(y) = \sum_{e \in B_d} y_e = \sum_{e \in A_s} y_e \tag{1}$$

$$\sum_{e \in B_v} y_e - \sum_{e \in A_v} y_e = 0, \quad \forall\, v \in V \setminus \{s, d\} \tag{2}$$

$$b_e \geq y_e \geq 0, \quad \forall\, e \in A \tag{3}$$

The constraints in (2) are flow-conservation constraints and the constraints in (3) are capacity constraints. The maximum static flow problem seeks to maximize the objective (1), and we denote the value of the maximum static flow by $\text{val}_{\max}(y)$. If the flow conservation constraints (2) are satisfied for each $v \in V$, then corresponding flow y is also known as a circulation. If we add an arc (d, s) with capacity $\text{val}(y)$ and set $y(d, s) = \text{val}(y)$, then the value of such a flow is zero, and the resulting flow is a circulation. Given a fixed flow value $\text{val}(y)$ and the cost c_e per unit of flow for each $e \in A$, the minimum cost static flow problem seeks to minimize the total cost $\sum_{e \in A} c_e y_e$ of shifting $\text{val}(y)$ from s

to d. Adding an arc (d,s) with capacity $val(y)$ and cost 0, the minimum cost static flow problem can be turned into a minimum cost circulation problem.

Let us assume that the arc transit times and capacities are constant over the time. With an amount of inflow $x_e(\theta)$ on arc e at discrete time $\theta = 0, 1, \ldots, T$ that may change over the planning horizon T, the generalized dynamic flow $x : A \times T \to R^+$ for given time T satisfies constraints (4–6).

$$\sum_{\sigma=\tau_e}^{T} \sum_{e \in B_v} \lambda_e x_e(\sigma - \tau_e) = \sum_{\sigma=0}^{T} \sum_{e \in A_v} x_e(\sigma), \ \forall v \notin \{s,d\} \tag{4}$$

$$\sum_{\sigma=\tau_e}^{\theta} \sum_{e \in B_v} \lambda_e x_e(\sigma - \tau_e) \geq \sum_{\sigma=0}^{\theta} \sum_{e \in A_v} x_e(\sigma), \ \forall v \notin \{s,d\}, \ \theta \in \mathbf{T} \tag{5}$$

$$b_e(\theta) \geq \lambda_e x_e(\theta) \geq 0, \ \forall e \in A, \ \theta \in \mathbf{T} \tag{6}$$

The generalized earliest arrival flow problem is to find a generalized dynamic flow of value $val_{\max}(x_e, \theta)$ for each time unit $\theta \in \mathbf{T}$ defined by objective function (7). It is defined as a generalized maximum dynamic flow problem if the maximization is considered for $\theta = T$ only:

$$val(x, \theta) = \sum_{e \in B_d} \sum_{\sigma=\tau_e}^{\theta} \lambda_e x_e(\sigma - \tau_e), \ \theta \in \mathbf{T} \tag{7}$$

Note that, besides the sink, no flow units remain in the dynamic network after time T. It is ensured by assuming that $x_e(\theta) = 0$ for all $\theta \geq T - \tau_e$.

For the following models we assume that gain factor $\lambda = 1$. Then, the generalized dynamic flow reduces to the dynamic flow and the generalized earliest arrival flow reduces to the earliest arrival flow with objective function (8).

$$val(x, \theta) = \sum_{\sigma=0}^{\theta} \sum_{e \in A_s} x_e(\sigma) = \sum_{\sigma=\tau_e}^{\theta} \sum_{e \in B_d} x_e(\sigma - \tau_e) \tag{8}$$

Given a time horizon, T, and and a set of terminals with a given priority order, the lexicographic maximum (lex-maximum) dynamic flow problem seeks to identify a feasible dynamic flow that maximizes the amount leaving (entering) a terminal in the given order. For a given value Q_0 (number of flow units representing evacuees), the quickest flow problem minimizes $T = T(Q_0)$ such that the value of the dynamic flow not less than Q_0, satisfying the constraints (4)–(6) with equality in (5) and $\lambda = 1$.

Let $\mathcal{N} = (V, A, b, \tau, S, D, \mu(s), \mu(d))$ be a multiterminal network with source-supply and sink-demand vectors $\mu(s)$ and $\mu(d)$, respectively, such that $\mu(S \cup D) = \sum_{v \in S \cup D} \mu(v) = 0$. The multiterminal earliest arrival flow problem seeks to find the dynamic flow, so that the total supply $\mu(S) = \sum_{s \in S} \mu(s)$ is sent from S to meet the total demand $\mu(D) = \sum_{d \in D} \mu(d)$ in D with maximum value at each $\theta \geq 0$. If all demands are to be fulfilled with supplies by shifting them to the sinks within given time T, then the problem is known as a transshipment problem. The earliest arrival transshipment problem maximizes $val(x, \theta)$ in objective function (9) satisfying multiterminal constraints (4)–(6) for all time points $\theta \in \{0, 1, \ldots, T\}$ with $\lambda = 1$.

$$val(x, \theta) = \sum_{\sigma=0}^{\theta} \sum_{e \in A_s : s \in S} x_e(\sigma) = \sum_{\sigma=\tau_e}^{\theta} \sum_{e \in B_d : d \in D} x_e(\sigma - \tau_e) \tag{9}$$

If the transshipment from S to D is done in the minimum time $\min T = T(\mu(S \cup D))$, then the problem is called quickest transshipment problem.

Let $e' = (w,v)$ be the reverse of an arc $e = (v,w)$. The residual network $\mathcal{N}(y)$ for a static flow y is given by $(V, A(y))$, where $A(y) = A_F(y) \cup A_B(y)$ with $A_F(y) = \{e \in A : y_e < b_e\}$ and $A_B(y) = \{e' \in A : y_e > 0\}$ with arc length τ_e for $e \in A_F(y)$ and $-\tau_e$ for $e' \in A_B(y)$. The residual capacity $b(y) : A(y) \to \mathcal{R}^+$ is defined as $b_e(y) = b_e - y_e$ for $e \in A_F(y)$ and $b_e(y) = y_{e'}$ for $e \in A_B(y)$.

Given a multiterminal S-D network, we construct an extended network by adding two extra nodes: s^* (called super-source) and d^*(called super-sink). For each $s \in S, d \in D$, we construct arcs $(s^*, s), (d, d^*)$ with zero transit time and problem-dependent capacities.

2.2. Natural Transformation

With the continuous time settings set as $\mathbf{T} = [0, T]$, all of the above models described in the above subsection can be remodeled by replacing the summation over time with respective integrals. The amount of flow entry, x_e, on the arcs considered above in discrete models naturally transfer to the entry flow rates in this continuous approach.

Fleischer and Tardos [25] connect the continuous and discrete flow models by the following natural transformation, as defined in (10), that deals with the same computational complexity to both.

$$x_e^c(\psi) = x_e(\theta), \text{ for all } \theta \text{ and } \psi \text{ with } \theta \leq \psi < \theta + 1 \tag{10}$$

where $x_e(\theta)$ is the amount of discrete dynamic flow that enters arc e at time $\theta = 0, 1, \ldots, T$ with constant capacities on the arcs. For static flow y_e on arc e, the amount of discrete dynamic flow with transit time τ_e on arc e is

$$x_e(\theta) = \sum_{\sigma=0}^{\tau(e)-1} y_e(\theta - \sigma), \text{ for all } \theta = 0, 1, \ldots, T - 1 \tag{11}$$

Notice that the flow entering an arc $e = (v, w)$ at time $\theta - \tau_e$ arrives at w at time θ in discrete time, but at time $[\theta + 1)$ in continuous time. The flow x^c is feasible and will be same for both settings at any interval $[\theta, \theta + k)$, for $\theta = 0, 1, \ldots, T, k \in \mathbb{N}$.

With standard chain decomposition, the static flow y is decomposed into a set of chain flows $\Gamma = \{\gamma_1, \ldots, \gamma_r\}$ with $r \leq m$ that satisfies $y = \sum_{k=1}^{r} \gamma_k$. Each chain in Γ starts at a source node and ends at a sink node using arcs in the same direction as y does. The travel time on each chain γ_k is such that $\tau(\gamma_k) \leq T$. The feasible dynamic flow can be obtained by summing the dynamic flows induced by each chain flow.

A nonstandard chain decomposition of feasible y, e.g., $\Gamma = \{\gamma_1, \ldots, \gamma_{r'}\}$, allows for an arc in the opposite direction also for the flow. If $e = (v, w)$ has transit time τ_e, then for its reverse arc $e' = (w, v), \tau_{e'} = -\tau_e$. A unit of flow starting from w at time $\theta + \tau_e$ and reaching v at time θ cancels the unit of flow starting from v at time θ and reaching w at time $\theta + \tau_e$, and thus it is equivalent to sending a negative unit of flow from v at time θ to w at time $\theta + \tau_e$.

Using the concept of natural transformation discussed above, Fleischer and Tardos [25] solved the continuous versions of maximum flow, universal dynamic flow, lexicographically maximum flow, quickest flow, and dynamic transshipment problems by solving their discrete counter parts with time-invariant attributes. The computational complexities for both approaches remain the same. The generalized dynamic cut capacity is defined to show the equivalent maximum flow solutions.

2.3. Models for Arc Reversals

Let $\mathcal{N} = (V, A, b, \tau, S, D, T)$ be an evacuation network. For $e = (v, w) \in A$, we denote its reverse arc (w, v) by e'. To solve a network flow problem with arc reversals, a common approach is to solve the corresponding problem in, what is known as, an auxiliary network. We denote the auxiliary network of $\mathcal{N}$ by $\overline{\mathcal{N}} = (V, E, \overline{b}, \overline{\tau}, S, D, T)$, in which

$$\overline{b}_{\overline{e}} = b_e + b_{e'}, \text{ and } \overline{\tau}_{\overline{e}} = \begin{cases} \tau_e & \text{if } e \in A \\ \tau_{e'} & \text{otherwise} \end{cases}$$

and an edge $\overline{e} \in E$ in auxiliary $\overline{\mathcal{N}}$ if $e \in A$ or $e' \in A$. While working with the auxiliary network for reconfiguration, one is allowed to redirect the edge in any direction with the modified increased capacity and same transit time in either direction. The remaining graph topology and data structures in reconfigured network are unaltered. By discarding the time factor for flow passing through the network, a static contraflow configuration will be defined analogously.

The core idea behind contraflow reconfiguration technique is to increase outbound flow with reduced time on the evacuation network. Numerous dynamic contraflow heuristics have been presented and implemented during the past few years. However, recently many analytical approaches have been investigated, and polynomial time algorithms are also presented in a few cases, though the general multiterminal problem is still $\mathcal{NP}$-hard because of a conflict with reverting intermediate arcs. We recommend a complete survey [11] for details.

Example 1. *Let us consider a single-source single-sink evacuation network as shown in Figure 1(i), where s is the source node and d is the sink node. The arcs, for example, (x, y) and (y, x), represent the two-way road segments between nodes x and y. Each arc contains capacity and transit time (cost) associated to it. For example, arc (s, w) has capacity 2 and transit time 3, that means, assuming a time unit of 5 min and a flow unit of 10 cars, a maximum of 20 cars can reach w from s in 15 min. The auxiliary network for reconfiguration is as shown in Figure 1(ii), in which the capacity of each edge $\overline{e}$ is obtained by adding the capacities of e, e' and the transit time $\overline{\tau}_{\overline{e}} = \tau_e = \tau_{e'}$.*

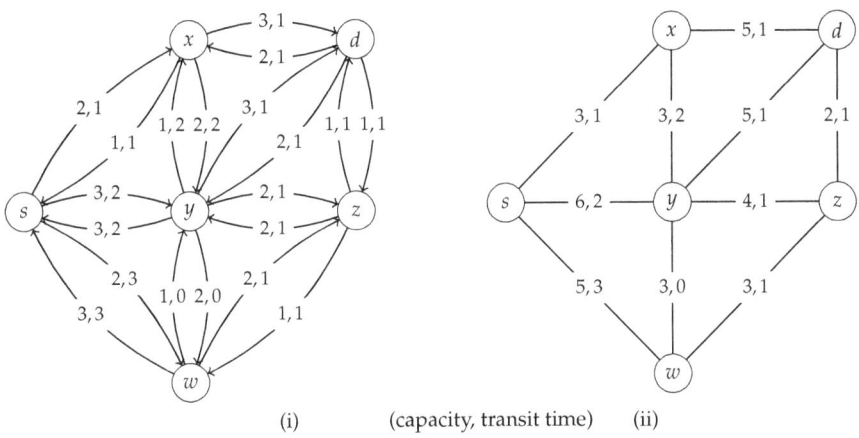

Figure 1. (i) Evacuation network $\mathcal{N}$. (ii) Auxiliary network $\overline{\mathcal{N}}$ of $\mathcal{N}$.

3. Static Partial Contraflow

In this section, we introduce the maximum static partial contraflow problem (MSPCFP) (cf. Problem 1) and the lex-maximum static partial contraflow problem (LMSPCFP) (cf. Problem 2). Thereafter, polynomial time algorithms are presented to solve these problems. These algorithms work as a foundation of solving the dynamic versions of the corresponding problems in the subsequent sections.

3.1. Maximum Static Partial Contraflow

Problem 1. *Given a static network $\mathcal{N} = (V, A, b, \tau, S, D)$, the maximum static partial contraflow problem (MSPCFP) is to maximize a static S-D flow, saving the unused arc capacity, if the direction of arcs can be reversed partially.*

Algorithm 1 is designed for a single-source single-sink static network.

Algorithm 1: The maximum static partial contraflow algorithm (MSPCFA).

Input: A static network $\mathcal{N} = (V, A, b, s, d)$
Output: A maximum static partial contraflow (MSPCF) on $\mathcal{N}$ with saved capacities of arcs

1. Construct the auxiliary network $\overline{\mathcal{N}} = (V, E, \overline{b}, s, d)$.
2. Run the maximum static flow algorithm in $\overline{\mathcal{N}}$ with capacity $\overline{b}$, that is, $\overline{b}_{\overline{e}}$ for each $\overline{e} \in E$ to compute the maximum static flow y.
3. Decompose the flow into paths and cycles and remove the flow in cycles.
4. Reverse $e' \in A$ up to the capacity $y_e - b_e$ iff $y_e > b_e$, b_e replaced by 0 whenever $e \notin A$.
5. For each $e \in A$, if e is reversed, $r(e) = \overline{b}_{\overline{e}} - y_{e'}$ and $r(e') = 0$. If neither e nor e' is reversed, $r(e) = b_e - y_e$, where $r(e)$ is the saved capacity of e.

Step 2 of Algorithm 1 relies on any of the best polynomial time maximum flow algorithms that have a long history but are still under study. Their optimal solutions are guaranteed by the fundamental max-flow min-cut theorem, which states that the maximum flow value equals to the minimum cut capacity. Following the flow augmenting path pseudo-polynomial, that is, $O(nm \log b_{max})$ time, algorithm of Ford and Fulkerson [13], which ensures a maximum static flow y if and only if the corresponding residual network does not contain an augmenting path, there exist several advancements on its improvements. By scaling the capacities, the running time can be improved to $O(m^2 \log b_{max})$, where b_{max} represents the integer valued maximum arc capacity. Furthermore, the shortest augmenting path algorithm that uses the unit path length function, the blocking flow algorithm that augments along a maximal set of shortest paths with respect to a blocking flow, and the push-relabel algorithm that functions with nonconservation of flows, except at the source, and sink nodes turn into strongly polynomial time algorithms with complexity, $O(nm^2)$, $O(n^2m)$ and $O(nm \log(n^2/m))$, respectively.

The flow decomposition of Step 3 into paths and cycle removal is less costly with $O(nm)$ running time. In a few special case, like unit capacities, some of these algorithms can be implemented with nearly linear time bounds. Using advanced data structures and dynamic trees, though much theoretical, even more strongly polynomial time algorithms are developed, however, not much implementable in usual practice (see Goldberg and Tarjan [36] for a brief review). For example, with the dynamic-tree data structure, the binary blocking flow algorithm requires $O(\min(n^{2/3}, \sqrt{m})m \log(n^2/m) \log b_{max})$ time within a factor $\log(n^2/m) \log b_{max}$ of the best algorithm for the unit arc capacity problem without parallel arcs.

In Step 4 of Algorithm 1, reversal of traffic flow is allowed only in the necessary amount on the road segment, i.e, partial reversal of the arcs in evacuation network. Unlike in the previous investigations, the remaining capacity of the segment is not reversed, which can be used for other purposes, for example, facility location–allocation and emergency logistic supports.

Based on maximum static contraflow solution with complete contraflow configuration presented in [28], Step 2 of Algorithm 1, the maximum flow algorithm, computes the maximum static flow optimally on auxiliary network $\overline{\mathcal{N}}$ and Step 5 comes as a direct consequence of it. This leads to:

Lemma 1. *Algorithm 1 computes the maximum static flow with partial arc reversal on $\mathcal{N}$ by saving capacities $r(e)$ of arc $e \in A$ correctly.*

Theorem 1. *Algorithm 1 solves the MSPCFP in strongly polynomial time with partial arc reversal capability.*

Using a maximum flow algorithm, Rebennack et al. [28] obtained a maximum contraflow solution in auxiliary network. However, their algorithm applies complete arc reversal strategy, such that an arc $e' \in A$ is reversed if and only $y_e > b_e$ or $y_e > 0$ for $e \notin A$ in the maximum static contraflow solution y. Here, we use partial lane reversal strategy, where the lanes of arcs not required by the contraflow solution are not reversed.

Example 2. *We illustrate Algorithm 1 on the network constructed in Example 1. First, we solve the maximum static flow problem on auxiliary network in Figure 1(ii). As shown in Figure 2(i), paths $P_1 = s - x - d$, $P_2 = s - y - d$, $P_3 = s - y - z - d$, $P_4 = s - w - z - d$, and $P_5 = s - w - y - x - d$ carry 3, 5, 1, 1, and 2 flow units, respectively. Thus, the obtained maximum static flow 12 units on $\overline{\mathcal{N}}$ is equivalent to the maximum static contraflow on $\mathcal{N}$ with arc reversals. According to Algorithm 1, arcs $(x,s), (y,s), (d,x), (d,y), (d,z)$ are reversed completely; arcs $(w,s), (x,y), (y,w)$ are partially reversed each up to the capacity 1; each of the arcs $(y,z), (w,z), (z,w), (y,w)$ has a saved capacity of one unit; and each of $(z,y), (w,s)$ has that of two units, as shown in Figure 2(ii).*

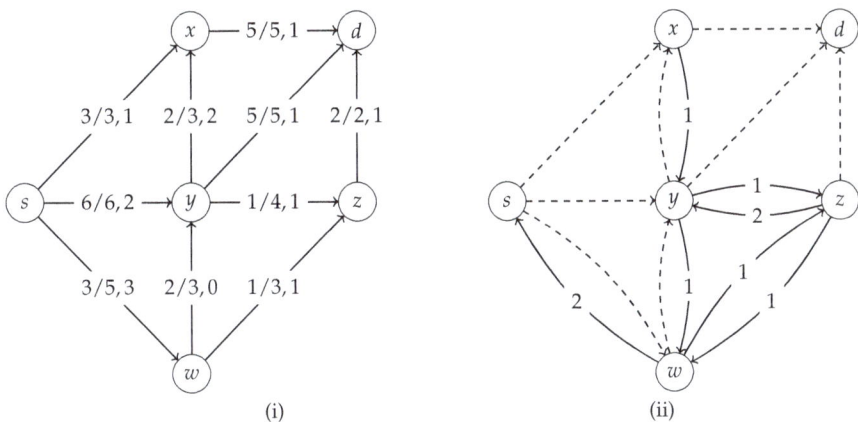

Figure 2. (i) Solution in Figure 1(ii) with (flow/capacity, transit time). (ii) Saved capacities to part (i).

3.2. Lex-Maximum Static Partial Contraflow

Assume that, in a multiterminal evacuation network, the risk zones and the destination areas are prioritized subject to specific reasons such as risk levels of disastrous areas, disabilities or urgency of the evacuees, and service requirements to them. To meet these necessities, the sets of sources and sinks are categorized as priority sets.

Let $S_1 \subseteq \cdots \subseteq S_q \subseteq S$ and $D_1 \subseteq \cdots \subseteq D_r \subseteq D$, respectively, be the priority sets of the sources and sinks, where X_i gets higher priority to X_j whenever $X_i \subseteq X_j$ holds. Considered with a maximal flow, let the greatest number of units that can enter the sink D_* be $\mathrm{val}_m(D_*)$. Then, a maximal flow that delivers $\mathrm{val}_m(D_k)$ units into each D_k is a lexicographically (lex-) maximal static flow on the sinks. A maximal flow that sends maximum number of units, $\mathrm{val}_m(S_k)$, out of each S_k is a lexicographically (lex-) maximal flow on the sources.

Problem 2. *Let $\mathcal{N} = (V, A, b, S, D)$ be a given multiterminal network with ordered sets of sources and sinks. The lex-maximum static partial contraflow problem (LMSPCFP) at sources (sinks) is used to determine a feasible flow that lexicographically maximizes the amounts leaving (entering) the terminals in given priority orders, if the partial reversal of arcs is allowed.*

To solve the lex-maximum static partial contraflow problem (LMSPCFP; Problem 2), we design Algorithm 2. An illustration is given in Figure 3.

Algorithm 2: The lex-maximum static partial contraflow algorithm (LMSPCFA).

Input: A multiterminal static network $\mathcal{N} = (V, A, b, S, D, T)$ with given orders of sources and sinks priorities
Output: A LMSPCF on $\mathcal{N}$ with saved capacities of arcs

1. Construct the auxiliary network $\overline{\mathcal{N}} = (V, E, \overline{b}, S, D)$.
2. Run the lex-maximum static algorithm of [16] on $\overline{\mathcal{N}}$ with capacity $\overline{b}$ to find the flow y.
3. Decompose y into paths and cycles, and remove flows in cycles.
4. Reverse $e' \in A$ up to the capacity $y_e - b_e$ iff $y_e > b_e$, b_e replaced by 0 whenever $e \notin A$.
5. For each $e \in A$, if e is reversed, $r(e) = \overline{b}_e - y_{e'}$ and $r(e') = 0$. If neither e nor e' is reversed, $r(e) = b_e - y_e$, where $r(e)$ is the saved capacity of e.

The authors of [31] presented a polynomial time algorithm to find contraflow reconfiguration with complete arc reversals. We establish the following result that leaves the unused parts of arcs unturned, solving a lex-maximum partial contraflow problem in polynomial time.

Theorem 2. *The LMSPCFP can be solved using Algorithm 2 in polynomial time with arc reversals partially.*

Proof. For the given priority ordering of multiterminals, the maximum static flow problem is solved iteratively on auxiliary network $\overline{\mathcal{N}} = (V, E, \overline{b}, S, D)$ that gives an optimal solution to the lexicographically (lex-) maximum static flow problem, as in [16]. The obtained solution is equivalent to the lex-maximum static contraflow solution on $\mathcal{N}$, [31]. Step 5 of Algorithm 2 saves all unused capacities of arcs in $O(m)$ time, leading to the solution of LMSPCFP in polynomial time. □

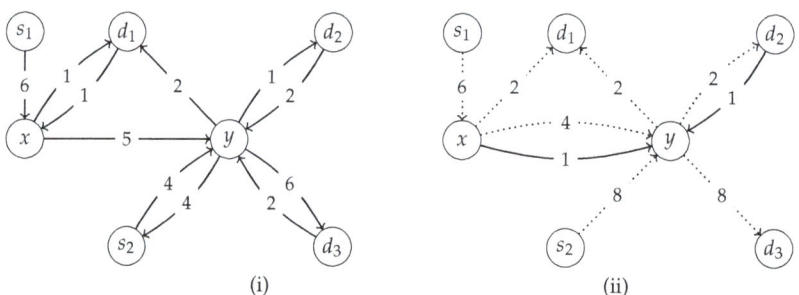

Figure 3. On the capacitated static network (i) with 2-sources and 3-sinks having priority ordering of sinks d_1, d_3, and d_2, the lex-maximum static partial contraflow is obtained as in (ii). Flow entering sinks d_1, d_3, and d_2 are maximized through paths $s_1 - x - d_1$, $s_2 - y - d_1$, $s_2 - y - d_3$, and $s_1 - x - y - d_3$, and $s_1 - x - y - d_2$, respectively, by saving arc unit capacity of arcs (x, y) and (d_2, y).

4. Partial Lane Reversals for Time-Invariant Attributes

This section deals with the concepts of dynamic partial contraflow and contra-transshipment problems with constant transit times and capacities on arcs. Some efficient algorithms are presented for their solutions. As the evacuation issues with dynamic environment have been categorized into different problem types, such as maximum flow, quickest flow, lexicographically maximum flow, earliest arrival flow, quickest transshipment, and earliest arrival transshipment, each has to be dealt to separately with contraflow and, thereby, partial contraflow configuration techniques. The maximum dynamic, earliest arrival, lex-maximum dynamic and quickest, dynamic partial contra-transshipment problems are abbreviated by MDPCFP, EAPCFP, LMDPCFP, QPCFP, and DPCTP, respectively.

4.1. Dynamic Contraflow Problems

As there is no algorithm to find temporally repeated flows on general S-D dynamic networks, an exact optimal dynamic contraflow solution for them have not been found in polynomial time yet. In particular, on the networks, like two-terminal s-d; priority-based S-D; and transshipments S-d, s-D, and S-D with different constraints, the optimal static flow can be decomposed into chains (paths) that are temporally repeated over the time horizon, T, returning the temporally repeated dynamic flows in such cases. From the series of literature on analytical contraflow [1,24,28,30–32,37], it is established that the optimal dynamic contraflow for the given network $\mathcal{N} = (V, E, b, \tau, s, d, T)$ is equivalent to the optimal dynamic flow on corresponding reconfigured network $\overline{\mathcal{N}} = (V, E, \overline{b}, \overline{\tau}, s, d, T)$. The input network may be a super-source super-sink added extended network if one is considering a multiterminal network, whenever a temporally repeated solution is possible.

Previous contraflow approaches do not apply the remaining arc capacities, which result after contraflow reconfiguration. In this subsection, we redefine a series of partial contraflow problems with time-invariant attributes and present appropriate efficient algorithms to solve the corresponding problems saving the unnecessary arc capacities. Once a dynamic flow is obtained, it should be subtracted from reconfigured capacities of arcs to record the maximum unused arc capacities. These algorithms have great benefit as the remaining unused arc capacities can be used for emergency vehicles and logistics. The following problem is more general, addressing respective partial contraflow problems in a compact form.

Problem 3. *Given a network, $\mathcal{N} = (V, A, b, \tau, S, D, T)$, with integer inputs, the dynamic partial contraflow problem (DPCFP) with objective function (G) is to find a dynamic S-D flow optimizing (G) for all time $\theta \in T$ with arc reversals partially.*

Problem 3 is stated in an abstract form for a general objective function G without its explicit nature. As per the requirement of the specific problem, we will state it explicitly in the subsequent sections.

Applying Step 2 of Algorithm 3, any technique that computes a temporally repeated flow on reconfigured network $\overline{\mathcal{N}}$ is applicable to find an optimal solution to the corresponding contraflow for the original network $\mathcal{N}$. During the computation of temporally repeated flows, Step 3 removes the cycle flows, if they exist, so that the simultaneous flows in both directions are not possible. Saving of the unused capacities are recorded in Step 5. Thus, a flow is either along arc e or arc e', and its value is not greater than the added arc capacities at all time units. Therefore, the condition of feasibility is satisfied by the algorithm. The optimality depends on the specific objective function.

Algorithm 3: The dynamic partial contraflow algorithm (DPCFA).

Input: A Dynamic network $\mathcal{N} = (V, A, b, \tau, s, d, T)$ with constant and symmetric transit time, i.e., $\tau_e = \tau_{e'}$

Output: A dynamic partial contraflow x with the partial arc reversals

1. Construct the auxiliary network $\overline{\mathcal{N}} = (V, E, \overline{b}, \overline{\tau}, s, d, T)$ of $\mathcal{N}$ for contraflow reconfiguration.
2. Use a temporally repeated flow algorithm to solve DPCFP(G) on reconfigured network $\overline{\mathcal{N}}$ with capacity $\overline{b}_{\bar{e}}$ and transit time $\overline{\tau}_{\bar{e}}$ for each $\bar{e} \in E$.
3. Decompose the flow y into paths and cycles and remove the flow in cycles.
4. For each $\theta \in T$, reverse $e' \in A$ up to the capacity $y_e - b_e$ iff $y_e > b_e$, b_e replaced by 0 whenever $e \notin A$.
5. For each $\theta \in T$ and $e \in A$, if e is reversed, $r(e) = \overline{b}_{\bar{e}} - y_{e'}$ and $r(e') = 0$. If neither e nor e' is reversed, $r(e) = b_e - y_e$, where $r(e)$ is the saved capacity of e.

4.1.1. Maximum Dynamic Contraflow

If the flow has to be maximized for given time period, T, without considering earlier periods, then Problem 3 with the arcs permissible to be reversible only at time zero is the MDPCF problem. The objective (8) is then max $G = \max \mathrm{val}(x, T)$ for $\theta = T$ subject to constraints (4)–(6).

Polynomial time algorithms to solve the S-D MDCFP has not been investigated yet. However, it can be solved in pseudo-polynomial time by reducing it into extended s-d network of its time-expanded network. The s-d MDCFP is polynomially solvable in discrete time [28] and, in continuous time, [30] reverses an arc completely whenever it is to be reversed.

Theorem 3. *The s-d MDPCFP with $G = \mathrm{val}(x, T)$ can be solved in $O(h_1(n, m) + h_2(n, m))$ time in which flow decomposition and minimum cost flow problems are solved in $h_1(n, m) = O(n.m)$ and $h_2(n, m) = O(n^2.m^3. \log n)$ times, respectively.*

Proof. Theorem 3 is proved in three steps. First, we show that the solution Algorithm 3 yields is feasible. Step 2 uses maximum dynamic flow algorithm of [13]. After the removal of the positive flow in cycles in Step 3, there is a flow either along arc e or e' but never in both arcs and the flow is not greater than the modified capacities in the auxiliary network on each arc at each time unit. Second, we show the optimality. We use the temporally repeated flow algorithm of [13] after we obtain the feasible flow that gives a maximum dynamic flow solution on reconfigured network $\overline{\mathcal{N}}$, which is equivalent to the maximum dynamic contraflow solution on $\mathcal{N}$ with the arcs reversed up to the necessary capacity in Step 4. In Step 5, we record the capacities of the arcs not used by the flow after necessary (partial) arc reversals, thereby obtaining a MDPCF solution.

The time complexity of the algorithm is dominated by the time complexity of the maximum dynamic flow algorithm of [13] in Step 2, which is equal to the complexity of a minimum cost flow computation, and that of flow decomposition in Step 3. This completes the proof. □

Example 3. *We compute the maximum dynamic flow on auxiliary network (cf. Figure 1(ii)) of evacuation network in Figure 1(i)) having the aforementioned capacities and transit times on each arc with the repetition of path flows computed in Example 2 within given time horizon $T = 6$. Using the algorithm in [13], we get the static flow y corresponding to the temporally repeated dynamic flow, which is the same as that in Figure 2(i). The temporally repeated maximum dynamic flow after partial contraflow configuration is as given in Table 1.*

Table 1. Maximum dynamic flow computation after partial contraflow configuration (cf. Figure 1).

Path P	y(P)	Repeated for $\theta =$	Dynamic Flow Value
s − x − d	3	2, 3, 4, 5, 6	15
s − y − d	5	3, 4, 5, 6	20
s − y − z − d	1	4, 5, 6	3
s − w − z − d	1	5, 6	2
s − w − y − x − d	2	6	2
Total			42

The arc reversals and saved capacities are similar to those in Example 2.

We observe another algorithm for the partial contraflow configuration that is based on the minimum cut problem. The formation of reconfigured network is from weighted undirected network on which a minimum s-d cut can be obtained in $O(n^2 \log^3 n)$ time complexity that is independent with maximum flow computations (Karger and Stein [38]). The partial contraflow configuration can be achieved by reversing only the capacities of arcs that are equal to the minimum cut capacities. However, it is difficult to identify the used capacities of other arcs that are not contained on minimum cut set. We can say that maximum flow is equal to the minimum cut but there is not any technique developed to decompose the maximum flow value in different *s-d* paths in undirected network given a minimum cut solution. If any such technique exists, then solving partial contraflow configuration problem is not harder than finding the distribution of maximum flow value in different arcs and paths of the network.

4.1.2. Earliest Arrival Contraflow

The earliest arrival (also known as the universal maximum) partial contraflow problem (EAPCFP/UMPCFP) on the given dynamic network $\mathcal{N} = (V, A, b, \tau, S, D, T)$ is to find a dynamic *S-D* flow that is maximum for all time steps $\theta = 0, 1, \ldots, T$ with arc reversal capability partially. In Step 2 of Algorithm 3, the objective function (8) is then $\max G = \max \text{val}(x, \theta)$ for all $\theta \in \mathbf{T}$ subject to constraints (4)–(6).

The earliest arrival contraflow problem for *s-d* series-parallel network is solved in polynomial time $O(nm + m \log m)$, being identical with maximum flow optimal solution in which arcs are reversed only at time zero [29,30]. The two solutions match for this case since each cycle in the residual network has nonnegative length [18]. However, for other general networks, as there does not exist any exact earliest arrival maximum flow solution with standard chain decomposition of [13], the nonstandard chain decomposition of [19] has to be applied in order to decide contraflow reconfiguration which demands arc reversals time to time, [24,31]. This results in a pseudo-polynomial time algorithm, using the successive shortest path algorithms as in [15,16]. A complication for the *s-d* earliest arrival contraflow solution arises because of the flipping requirements of intermediate arcs with respect to the time.

As the *S-D* maximum dynamic contraflow problem is $\mathcal{NP}$-hard, the corresponding *S-D* earliest arrival contraflow problem is also $\mathcal{NP}$-hard. However, the authors of [30] obtain an approximate contraflow solution within the factor $(1 - \epsilon)$ of the optimal earliest arrival contraflow in polynomial time. For this they run, the fully polynomial time approximation algorithm of [17] and obtain the approximate *s-d* earliest arrival contraflow in $O(m\epsilon^{-1}(m + n \log n) \log \bar{b}_{\max})$ time are used, where $\bar{b}_{\max} = \max_{\bar{e} \in E} \bar{b}_{\bar{e}}$.

The authors of [39] extend the results on earliest arrival contraflow problem to the partial lane reversal reconfiguration by saving unused arc capacity. Their algorithms have similar times complexities as without contraflow.

4.1.3. Generalization of Dynamic Contraflow

Given a generalized dynamic lossy network $\mathcal{N} = (V, A, b, \tau, \lambda, S, D, T)$ with integer inputs, the generalized earliest arrival partial contraflow problem (GEAPCFP) is to find a generalized maximum flow $\max G = \max \text{val}(x, \theta)$ for all $\theta \in \mathbf{T}$ defined in Equation (7), subject to the constraints (4)–(6) with partial arc reversal capability at time zero. If flow is maximized for a given time horizon T only, then the problem is a generalized maximum dynamic partial contraflow problem (GMDPCFP).

As the corresponding contraflow problems on general *S-D* network are $\mathcal{NP}$-hard, an additional gain factor on each arc make the partial contraflow problems also $\mathcal{NP}$-hard on general *S-D* lossy network, too. However, considering *s-d* lossy network, the contraflow problems can be solved computing corresponding generalized flows on the auxiliary network in pseudo-polynomial time complexity [40]. Moreover, with the same complexity, we can solve the partial contraflow problems using Algorithm 4 saving all unused arc capacities that can be used for other purposes.

There are different factors that make the flow to be lost during evacuation process. However, we consider a special case in which it is assumed that in each time unit the same percentage of the remaining flow value is lost. Thus, we consider a special case $\lambda \equiv 2^{c \cdot \tau}$ for some constant $c < 0$.

In the reconfigured network, we compute a maximum flow by calculating flow along shortest *s-d* paths, augmenting this flow, and repeating the process successively until no *s-d* path exists in the residual network. Then, such a maximum flow constructs an optimal maximum dynamic flow in pseudo-polynomial time with the standard temporally repeated flow technique in time-expanded network.

Algorithm 4: The generalized dynamic partial contraflow algorithm (GDPCFA).

Input: Given a lossy network $\mathcal{N} = (V, A, b, \tau, \lambda, s, d)$ with integer inputs
Output: The resulting flow is GMDPCF and GEAPCF with the arc reversals by saving unused arc capacities

1. Compute the generalized maximum dynamic contraflow and generalized earliest arrival contraflow on $\overline{\mathcal{N}} = (V, E, \overline{b}, \overline{\tau}, \lambda, s, d)$ with respective $\overline{b}_{\overline{e}}$ and $\overline{\tau}_{\overline{e}}$ calculated in Step 1 of Algorithm 3, and with additional gain factor $\lambda_{\overline{e}} \equiv 2^{c \cdot \overline{\tau}_{\overline{e}}}$, $c < 0$ using algorithms of [40,41]. Let y be the corresponding flow.
2. Reverse $e' \in A$ up to the capacity $y_e - b_e$ iff $y_e > b_e$, b_e replaced by 0 whenever $e \notin A$.
3. For each $e \in A$, if e is reversed, $r(e) = \overline{b}_{\overline{e}} - y_{e'}$ and $r(e') = 0$. If neither e nor e' is reversed, $r(e) = b_e - y_e$, where $r(e)$ is the saved capacity of e.

Theorem 4. *The generalized maximum dynamic partial contraflow problem (GMDPCFP) can be solved with pseudo-polynomial time complexity.*

Proof. The highest gain path is computed in $h_3(n, m) = O(mn)$ time, in which transit time $\tau_{\overline{e}} = \frac{1}{c} \log \lambda_{\overline{e}}$ is considered as cost function. In auxiliary network $\overline{\mathcal{N}}$ the generalized maximum flow using the highest gain path is computed by a standard maximum flow algorithm. For given time horizon T, there are at most T iteration, i.e., a maximum flow is computed at each iterations in $O(nm)$, and thus time complexity is $h_3(n, n) = O(nm.T)$ [41]. As Steps 2 and 3 are solved in linear time, the complexity of Algorithm 4 is dominated by Step 1. Therefore, the GMDPCFP is solved in $O(h_3(n, m) + h_3(n, n))$ time complexity. Moreover, the GMDPCF solution has the earliest arrival property maximizing the flow at each point of time and thus, the GEAPCFP is also solved in the same complexity. □

4.1.4. Lexicographically Maximum Dynamic Contraflow

With given priority ordering at terminals of the dynamic network $\mathcal{N} = (V, A, b, \tau, S, D, T)$, the LMDPCFP is to find a lexicographically maximum dynamic flow at each priority terminal sets with arc reversal capability partially at any time point.

Fixing the supplies and demands at sources and sinks, the LMDPCFP problem has been solved with polynomial time complexity [24,31]. However, it is also solvable for unknown supplies and demands on terminals with the same complexity, because there is a priority ordering in terminals but not in supplies and demands. In the reconfigured network $\overline{\mathcal{N}} = (V, E, \overline{b}, \overline{\tau}, S, D, T)$ of Algorithm 3, if we calculate the minimum cost flow in Step 2 at each iteration as in [19], the LMDPCF solution is obtained after δ (number of terminals) iterations within time horizon T by saving unused arc capacities. However, it uses so-called nonstandard flow decomposition in which backward arcs are allowed. The consequence is Theorem 5.

Theorem 5. *The LMDPCFP with partial reversals of arc capacities can be solved in $O(\delta \times MCF(m, n))$ time, where $MCF(m, n) = O(m \log n(m + n \log n))$ is the time complexity of minimum cost flow solution.*

4.1.5. Quickest Contraflow Problem

Problem 4. *For a given dynamic network $\mathcal{N} = (V, A, b, \tau, S, D, Q_0)$ with integer inputs and fixed flow value Q_0, the quickest partial contraflow problem (QPCFP) is to find a minimum time T to transship the flow value Q_0 from the sources S to the sinks D with arc reversal capability partially.*

The authors of [24,28,30,37] investigate the *s-d* network quickest contraflow problem and *S-D* network quickest contra-transshipment problem and present polynomial-time algorithms for their solution. On the *S-D* network, the quickest contraflow problem is not easier than *3-SAT* and *PARTITION* [28]. However, for an *s-d* network, they present a strongly polynomial time algorithm with discrete **T** based on the parametric search technique of Megiddo [42] and Burkard et al. [20]). They find an upper bound for the quickest time by computing *s-d* paths in polynomial time and then use parametric search to find the minimum time before which given flow value is sent to the sink resulting in a strongly polynomial time complexity of $O(m^2(\log n)^3(m + n\log n))$. In same time complexity, the quickest contraflow problem is solved in continuous times in [30].

Pyakurel et al. [37] presented the first polynomial algorithm with a time-complexity of a minimum cost flow algorithm to solve the *s-d* quickest contraflow problem. The *s-d* quickest contraflow solution has been computed by solving the parametric minimum cost flow problem using the cost scaling algorithm of Lin and Jaillet [43]. It takes $O(nm \log(n^2/m) \log(n\tau_{max}))$ time to solve this problem, where τ_{max} is the maximum transit time over all arcs. All the algorithms are presented with complete contraflow configuration.

Replacing the cost scaling algorithm of Lin and Jaillet [43], if we use the cancel and tighten algorithm of Saho and Shigeno [44] in Step 2 of Algorithm 3 that computes the quickest flow by solving the parametric minimum cost flow problem in strongly polynomial time complexity, we get what is stated in Theorem 6 without detailed proof. With this, we not only improve the complexity of algorithm to solve the *s-d* quickest contraflow problem, but also reverse necessary parts of the road segments saving all unused arc capacities, obtaining the QPCF solution in strongly polynomial time.

Theorem 6. *The quickest contraflow problem on s-d network can be solved in $O(nm^2(\log n)^2)$ time complexity with partial reversals of arc capacities.*

4.2. Dynamic Contra-Transshipment Problems

Problem 5. *Given a network $\mathcal{N} = (V, A, b, \tau, S, D, \mu(S), \mu(D), T)$ with integer inputs, the dynamic partial contra-transshipment problem (DPCTP) with objective function (H) is to find a feasible dynamic S-D flow for (G) that fulfills the supply-demand shipments with partial arc reversals.*

Problem 5 is stated in an abstract form for a general objective function G without its explicit nature. As per the requirement of the specific problem, we will state it explicitly in the subsequent sections.

If the fixed source-sink supply-demand amounts should be shifted within given time horizon T by maximizing G at every time point from the beginning with partial arc reversals, then the problem is earliest arrival partial contra-transshipment (EAPCTP).

The authors of [1,24,32] investigate the earliest arrival contra-transshipment problem and present polynomial time algorithms on multisource or multi-sink networks for specific arc transit times. Moreover, a pseudo-polynomial time algorithm has been presented and its approximation solution is computed for arbitrary transit times on each arc. If the transit time of each arc is zero, then the approximation solution is obtained in polynomial time. For an urban evacuation scenarios including life boats or pick-up bus stations, the concept of zero transit time is very important and applicable [45].

Based on the previous results from the literature, the EAPCTP can be solved in different conditions using Algorithm 5 in which all unused arc capacities are saved in contraflow configuration. However, this problem is not solved on general *S-D* network yet. For the *S-d* network $\mathcal{N} = (V, A, b, \tau, S, d, \mu(S), \mu(d), T)$ with arbitrary arc transit times, a solution of the earliest arrival contra-transshipment problem can be found in polynomial time reversing the arc in the time intervals whenever necessary. Moreover, the algorithm records all unused arc capacities.

By constructing extended network of reconfigured *S-d* network, we can compute with super-source s^* the s^*-*d* minimum cost flow circulations according to [21] and can save arc capacity using Step 3 of Algorithm 5 as follows.

First, the S-d network $\overline{\mathcal{N}}$ is converted into extended network s^*-d network, making all nodes in S intermediate and the total of the supply $\mu(S)$ is assigned to s^*. Moreover, node s^* is connected from d with a dummy arc (d, s^*) having infinite capacity. On s^*-d auxiliary network, we obtain a feasible dynamic flow by computing the minimum cost circulation to the dynamic flow on S-d auxiliary network, where $\mu(S)$ units of flow are being sent from the sources in S to d in time T. In this procedure, to overcome from the violation of individual supplies at the source nodes, an earliest arrival flow pattern $p(\theta)$, i.e., the maximum flow $\text{val}_S(x, \theta)$ in which $p(\theta) \leq \text{val}_S(x, \theta)$ for every $\theta \geq 0$, is defined on s^*-d network. If $p(\theta) = \text{val}_S(x, \theta)$, for all $\theta \geq 0$, the process is complete. The pattern is obtained polynomially in the input size plus the number of breakpoints. For given pattern $p(\theta)$ with k breakpoints on the S-d network $\overline{\mathcal{N}}$, an earliest arrival transshipment can be obtained by computing a transshipment dynamic flow in s^*-d network with k additional nodes and arcs. Thus, the obtained earliest arrival transshipment is equivalent to the earliest arrival contra-transshipment on S-d network $\overline{\mathcal{N}}$. Moreover, we can save all the unused arc capacities within the same complexity.

Algorithm 5: The dynamic partial contra-transshipment algorithm (DPCTPA).

Input: A dynamic network $\mathcal{N} = (V, A, b, \tau, S, D, \mu(S), \mu(D), T)$ with constant and symmetric transit times, i.e., $\tau_e = \tau_{e'}$

Output: A dynamic contra-transshipment with the partial arc reversals

1. Construct the reconfigured auxiliary network $\overline{\mathcal{N}} = (V, E, \overline{b}, \overline{\tau}, S, D, \mu(S), \mu(D))$ of $\mathcal{N}$ for contraflow reconfiguration.
2. Use a respective transshipment algorithm to solve DPCTP(G) on reconfigured network $\overline{\mathcal{N}}$ with capacity $\overline{b}_{\tilde{e}}$ and transit time $\overline{\tau}_{\tilde{e}}$ for each $\tilde{e} \in E$. Let y be the corresponding flow.
3. For each $\theta \in T$ and reverse $e' \in A$ up to the capacity $y_e - b_e$ iff $y_e > b_e$, b_e replaced by 0 whenever $e \notin A$.
4. For each $\theta \in T$ and $e \in A$, if e is reversed, $r(e) = \overline{b}_{\tilde{e}} - y_{e'}$ and $r(e') = 0$. If neither e nor e' is reversed, $r(e) = b_e - y_e$, where $r(e)$ is the saved capacity of e.

Theorem 7. *On the S-d network, the EAPCTP can be solved in polynomial time in the input plus output size.*

If the network is s-D with arbitrary transit times, its solution does not exist, because there is always conflict of which s-D path should be used first to make earliest possible flows. However, if transit times are assumed to be zero, every s-D path has same length yielding optimal earliest arrival contra-transshipment solution in polynomial time. Different networks can be categorized as in [45], wherein the s-D network EAPCTP can be solved polynomially using Algorithm 5 with reversing the partial capacities of arcs.

Theorem 8. *The EAPCTP problem can be solved polynomially on multi-sink networks with transit time zero on each arc, saving the unused arc capacities.*

Even with the zero transit times, for the S-D network, an earliest arrival transshipment solution is not possible. Consider a network $\mathcal{N}$ with two sources, s_1 and s_2; two sinks, d_1 and d_2; and arcs $(s_1, d_1), (s_1, d_2)$, and (s_2, d_2). Each source and sink have supply 2 and demand 2, and each arc has unit capacity. If we use all paths at time zero, we can transship three units of flows. But leaving the path s_1-d_2 empty, we can transship only two units of flow at time zero violating the maximality at every time point.

Thus, we investigate for an approximate solution for the S-D network earliest arrival partial contra-transshipment problem. For the solution, the reconfigured S-D transshipment network is transformed into time-expanded network. Then, the extended time-expanded network is constructed adding supper source s^* and super sink d^* with enough time bound, T, in which we can apply the algorithm of Gross et al. [23] that computes 2-value approximate earliest arrival transshipment solution. The optimal earliest arrival transshipment solution is bounded by 2 times the earliest arrival transshipment solution, called the 2-value approximate earliest arrival transshipment. It is equivalent to the 2-value approximate earliest arrival contra-transshipment on given network with arbitrary transit time on each arc. As it works on a time-expanded network, its time complexity is pseudo-polynomial. However, if the transit time is reduced to be zero, polynomial time 2-value approximation EAPCT can be obtained using the algorithm in [46] on Step 2 of Algorithm 5, thereby reversing the necessary parts of the segments and saving unused arc capacities in Steps 4 and 5.

Theorem 9. *The 2-value approximated EAPCT on S-D network can be solved efficiently by saving all unused arc capacities of arcs.*

5. Lane Reversals with Variable Attributes

We consider problems with variable transit times as a nonlinear function of probable congestion due to the current situation of flow in arcs. The transit times are flow dependent if they depend upon the density, speed and flow rate along the arcs. The inflow-dependent transit time $\tau_e(x_e(\theta))$ depends on inflow rate $x_e(\theta)$ at given time point θ so that, at a time, the flow units enter an arc e with uniform speed which remains uniform throughout this arc.

Contraflow with Inflow Dependent Transit Times

To introduce the inflow-dependent quickest partial contraflow problem (IFDQPCFP), the function τ in Problem 4 is replaced by inflow-dependent transit time function $\tau(x)$, which comprises functions $\tau_e(x_e)$, which denote the transit time on arc e if the inflow rate is x_e, for each $e \in A$ (see also [47]). In what follows, we model the problem and present a strongly polynomial time algorithm for an approximate solution of IFDQPCFP.

In oder to model the inflow-dependent flow over time problem, assume that at any moment of time the transit time function on an arc is given as a piecewise constant, nondecreasing, left-continuous function of inflow rate, Köhler et al. [2]. Note that this function can be restricted to be only integral values as it can be easily relaxed to allow arbitrary rational values by scaling the time with a proper way. Moreover, any general non-negative, nondecreasing, left-continuous function has been approximated by a step function within arbitrary precision.

Along with the inflow rate, the transit time functions generally depend on the free flow transit time and capacity of the arc. If the capacity of an arc is increased, more flow can be sent along the arc and the units of flow take less time to travel the same arc. In a contraflow configuration, the auxiliary network is constructed by adding the capacities of the opposite arcs. Therefore the same amount of flow may take less time to reach from one end of the arc to the other end in comparison to the one without contraflow configuration. We assume that the free flow transit time in the two opposite arcs and the arc with which they are replaced with in the contraflow configuration are identical. The value of the transit time function on the arc in the auxiliary network is the result of the free flow transit time and the enhanced capacity. We assume that the transit time τ_e on an arc e is a function of the inflow rate $x_e(\theta)$, the free flow transit time τ_e^0, and the capacity b_e. Our approach is to find the quickest flow in the form of a temporally repeated static flow y, we assume that τ_e is given as a function of the static flow rate y_e, τ_e^0, b_e. Let

$$\tau_e(y_e) = f(y_e, \tau_e^0, b_e).$$

Then, for some $y_e = \zeta$, assuming that the free flow transit times on the opposite arcs e and e' are equal, we have,

$$\tau_e(\zeta) = f(\zeta, \tau_e^0, b_e),$$
$$\tau_{e'}(\zeta) = f(\zeta, \tau_e^0, b_{e'}),$$

and on the auxiliary network
$$\tau_{\bar{e}}(\zeta) = f(\zeta, \tau_e^0, b_e + b'_e).$$

We present Algorithm 6 to solve the single-source single-sink IFDQPCF Problem 6.

Problem 6. *Given a network $\mathcal{N} = (V, A, b, \tau, s, d, Q_0)$ with inflow-dependent τ, integer inputs, and fixed flow value Q_0, the s-d inflow-dependent quickest partial contraflow problem (IFDQPCFP) is to find a minimum time T to transship the flow value Q_0 allowing partial arc reversals.*

Algorithm 6: Inflow dependent quickest partial contraflow algorithm (IFDQPCFA).

Input: Given a dynamic network $\mathcal{N} = (V, A, b, \tau, s, d, Q_0)$
Output: An inflow-dependent quickest contraflow allowing partial arc reversals

1. Consider the reconfigured network $\overline{\mathcal{N}} = (V, E, \overline{b}, \tau, s, d, Q_0)$, where
$$b_{\bar{e}} = b_e + b_{e'} \text{ and } \tau_{\bar{e}}^0 = \begin{cases} \tau_e^0 & \text{if } e \in A \\ \tau_{e'}^0 & \text{otherwise} \end{cases}$$
for $\bar{e} \in E$.

2. Compute the static flow y corresponding to the quickest flow on $\overline{\mathcal{N}}$ using algorithm of Köhler et al. [2].
3. Decompose y into paths and cycles and remove flows in cycles.
4. Reverse $e' \in A$ up to the capacity $y_e - b_e$ iff $y_e > b_e$, b_e replaced by 0 whenever $e \notin A$.
5. For each $e \in A$, if e is reversed, $r(e) = \bar{b}_{\bar{e}} - y_{e'}$ and $r(e') = 0$. If neither e nor e' is reversed, $r(e) = b_e - y_e$, where $r(e)$ is the saved capacity of e.

Before we realize the correctness of Algorithm 6, we show that the temporally repeated flow with inflow-dependent transit times can be computed using a bow network in Step 2 as in Köhler et al. [2]. Let τ^{st} be the step function representation of τ, such that for a particular arc $e = (v, w) \in A$, $\tau_e^{st}(z) = \tau^i$, $z \in (z_{i-1}, z_i]$, $i = 1 \cdots k$, where $0 = z_0 < z_1 < \cdots < z_k = b_e$ and z_i, τ^i are non-negative integers. To construct the bow graph, we introduce

(i) regulating arcs $\rho_i (i = 1 \cdots k)$ with capacity z_i and transit time 0, such that the tail of ρ_i is the head of ρ_{i+1} for $i = 1 \cdots k - 1$ and tail of ρ_k is v, and
(ii) bow arcs $\beta^i (i = 1 \cdots k)$ with infinite capacity and transit time τ^i such that the tail of β^i is the head of ρ_i and the head of β^i is w.

Figure 4 shows the bow graph representation of $e = (v, w)$ in which $\tau_e^{st}(z) = 2, z \in (0, z_1]; \tau_e^{st}(z) = 4, z \in (z_1, z_2 = b_e]$.

We denote the bow network corresponding to the network $\mathcal{N} = (V, A, b, \tau, s, d)$ by $\mathcal{N}^B = (V^B, A^B, b^B, \tau^B, s, d)$, where V^B, A^B consist of vertices and arcs constructed as a result of bow graph representation of each arc $e \in A$; b^B and τ^B represent the capacity and transit time of each $e \in A^B$ as defined above. With this, every flow over time with inflow-dependent transit times in $\overline{\mathcal{N}}$ can be considered as a flow over time with constant transit times in $\mathcal{N}^B$, but not conversely. The problem on bow network is certainly a relaxation of the original with inflow-dependent transit times flow over time problem. Lemma 2 assumes inflow-dependent nondecreasing piecewise constant transit time functions.

Lemma 2. *For given dynamic s-d flow with inflow-dependent transit times sending Q_0 units in reconfigured network $\overline{\mathcal{N}}$ within time $T^* = \min T(Q_0)$, a temporally repeated flow with inflow-dependent transit times can be computed in strongly polynomial time that sends the same amount of s-d flow within at most $2T^*$ time.*

Proof. To construct the bow graph $\overrightarrow{\mathcal{N}}^B$ of $\overrightarrow{\mathcal{N}}$, we modify E by replacing each $\bar{e} \in E$ by two opposite arcs, each with the capacity $\bar{b}$ and transit time $\tau_{\bar{e}}$, which can be done in $o(m)$ times. Then we construct the bow network as mentioned earlier. As this network has constant transit time on arcs, we can use any algorithm to calculate the quickest flow for a network with constant transit time on arcs. The best-known strongly polynomial algorithm so far is the cancel-and-tighten algorithm in [44]. Let T^B be the quickest time to send a flow of value Q_0 from s to d. Thus, T^B is a lower bound on the optimal time T^* in $\overrightarrow{\mathcal{N}}$. The quickest flow computation, e.g., by cancel-and-tighten algorithm, yields a static flow y^B on $\overrightarrow{\mathcal{N}}^B$. Temporal repetition of y^B over the time horizon T^B yields a dynamic flow x^B in $\overrightarrow{\mathcal{N}}^B$, with

$$\text{val}(x^B) = T^B \text{val}(y^B) - \sum_{e \in E^B} \tau_e y_e^B = Q_0 \tag{12}$$

The dynamic flow x^B in $\overrightarrow{\mathcal{N}}^B$ may not yield a feasible dynamic flow in $\overrightarrow{\mathcal{N}}$ [2]. We overcome the difficulty by pushing the static flow from fast bow arcs to the slowest positive flow carrying bow arc (say, β^e) for each $e \in E$. This results into a modified static flow $\bar{y}^B$, with $\text{val}(\bar{y}^B) = \text{val}(y^B)$, which induces a temporally repeated dynamic flow $\bar{x}^B$ in $\overrightarrow{\mathcal{N}}^B$ with time horizon $T \geq T^B$. T can be calculated by using the equation

$$\text{val}(\bar{x}^B) = T\text{val}(\bar{y}^B) - \sum_{e \in E^B} \tau_e \bar{y}_e^B = Q_0 \tag{13}$$

One can show that

$$2T^B \text{val}(\bar{y}^B) - \sum_{e \in E^B} \tau_e \bar{y}_e^B \geq Q_0 \tag{14}$$

and as $\text{val}(\bar{x}^B)$ is an increasing function of T, it can be realized that $T \leq 2T^B$. For any $e \in E$, as $\bar{y}^B$ uses at most one bow arc β^e, we can find a feasible dynamic flow x in $\overrightarrow{\mathcal{N}}$ such that $x_e(\theta) = \bar{x}^B_{\beta^e}(\theta) \leq \bar{y}^B_{\beta^e}$. As the temporally repeated dynamic flow induced by $\bar{y}^B$ satisfies flow conservation, x also satisfies the flow conservation with storage of flow at intermediate nodes on $\overrightarrow{\mathcal{N}}$. The time horizon of x is T such that $T \leq 2T^B \leq 2T^*$. □

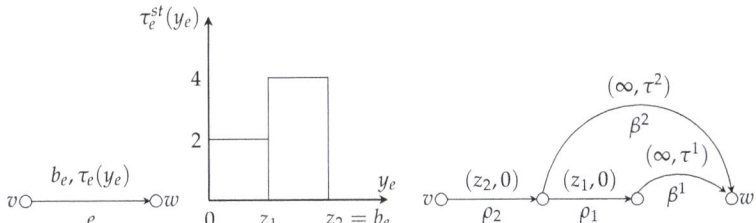

Figure 4. Expansion of a single arc $e = (v, w)$ in the bow network with transit times 2 and 4, for at most z_1 and z_2 flow units, respectively.

Theorem 10. *An approximate solution to the IFDQPCFP can be obtained using the IFDQPCFPA (Algorithm 6) in strongly polynomial time by reversing arc capacities partially.*

Proof. First, the IFDQPCFPA algorithm (cf. Algorithm 6) is feasible, as all of its steps are feasible. On auxiliary network $\overrightarrow{\mathcal{N}}$, we can compute the temporally repeated flow with inflow-dependent transit times using the algorithm of Köhler et al. [2] that gives the approximate quickest flow as in Lemma 2 as well as we can save all the unused arc capacities. From the feasibility of our algorithm, we directly conclude that every feasible quickest flow solution on the reconfigured network $\overline{\mathcal{N}}$ is equivalent to the quickest flow solution in the original network $\mathcal{N}$ as in constant transit times [28,30,32]. Thus, the obtained approximate quickest flow on $\overline{\mathcal{N}}$ is the approximate quickest partial contraflow for network $\mathcal{N}$ with inflow-dependent transit times which can be obtained in polynomial time complexity. □

6. Case Illustration

To illustrate some computational results, we consider Kathmandu road network containing major road sections (cf. Figure 5) as an evacuation network N with $n = 44$ and $m = 124$. The transit time (which we consider as the free flow transit time) in each road segment is as provided by Google Maps, and the integer capacity is assumed to be between 1 to 4 flow units per second according to the width of the segment. Related data are given in Appendix A (Tables A1 and A2).

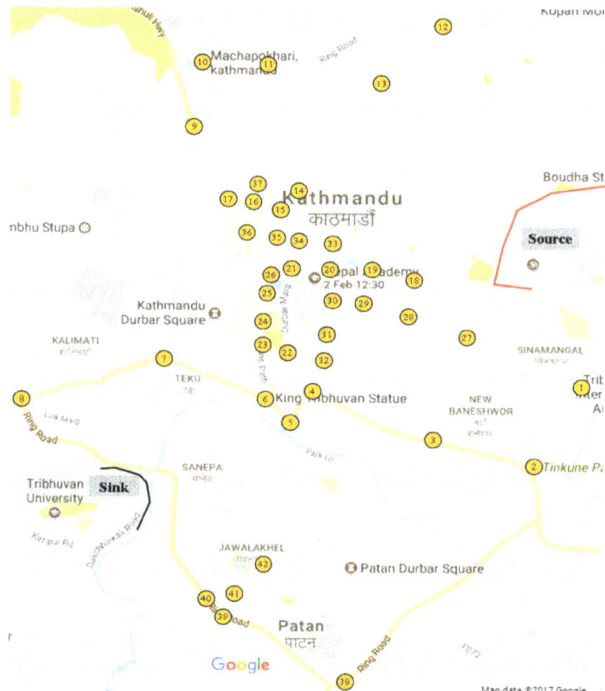

Figure 5. Kathmandu road network.

For the purpose of calculating inflow-dependent transit time on each arc e, we consider the following two functions given in [48] and present an analysis corresponding to each of them in parallel.

1. BPR function, developed by US Bureau of Public Roads:

$$\tau_e(y_e) = t_e^0 \left[1 + \alpha \left(\frac{y_e}{b_e'} \right)^\beta \right] \tag{15}$$

As an usual practice, we take $\alpha = 0.15, \beta = 4, b_e' = 0.8 b_e$.

2. Davidson's function:

$$\tau_e(y_e) = t_e^0 \left[1 + J \frac{y_e}{b_e - y_e} \right] \tag{16}$$

In our computations, we take $J = 0.1$.

Given the number of flow units Q_0 to be evacuated, to find the quickest flow allowing (partial) arc reversal, we construct the auxiliary network of the evacuation network. To solve the problem with the transit time depending on the inflow on each arc, we construct the bow graph of the auxiliary network as described in Section 5.

To construct the bow graph, measuring y_e, b_e in flow units per second and t_e^0 in seconds, we consider the transit time function as the step function

$$\tau_e^{st}(y_e) = \lfloor \tau_e(\lceil y_e \rceil - 1) \rfloor, 0 < y_e \leq b_e \qquad (17)$$

where $\lceil y_e \rceil$ represents the least integer greater than or equal to y_e and $\lfloor \tau_e(y_e) \rfloor$ is the value of $\tau_e(y_e)$ rounded to the nearest integer. As an example, the step function representation of a function on an arc with the free flow transit time 120 s and capacity four units of flow per second is given in Figure 6.

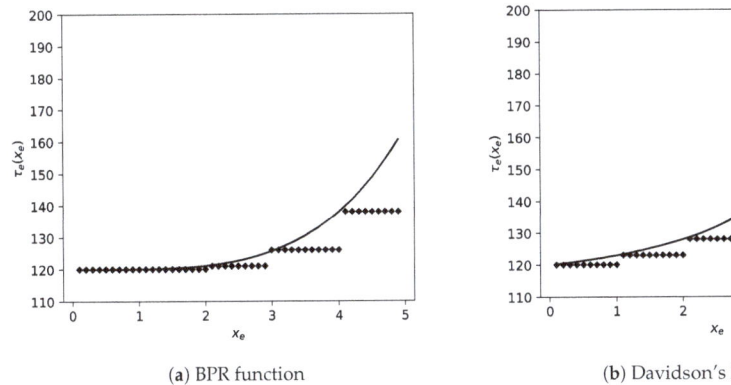

(a) BPR function   (b) Davidson's function

Figure 6. Flow-dependent transit time functions and corresponding step functions with $\tau_e^0 = 120$ s and $b_e = 4$ per second.

We find the static flow corresponding to the quickest flow in the bow graph, and we push the flow to the slowest arc (see Section 5) to find the approximate dynamic flow corresponding to the quickest flow. To compare the quickest time T^* in the bow graph and its approximate value T^{approx}, we consider Q_0 between 1 to 10,000 with a gap of 500. The results are shown in Figure 7. We find the maximum value of $\frac{T^{approx}}{T^*}$ to be 1.045 in case of BPR function and 1.098 in case of Davidson's function.

We compare the quickest times before and after allowing partial arc reversal in Figure 8. For Q_0 as small as 500, the quickest time before allowing (partial) arc reversal using the BPR function is approximately 29.5 min; whereas, after allowing arc reversal, it is 27.6 min (i.e., approximately 93.5% of the time before allowing arc reversal). With the increase in the value of Q_0, the gap increases. For Q_0 as large as 100,000, the value after allowing arc reversal is 141.7 min, 57.6% of the value before allowing arc reversal which is 246.1 min. The quickest times for some values of Q_0 before and after allowing arc reversal are listed in Table 2.

The number of arcs reversed (partially) for some values of Q_0 are given in Table 3. The observations show that increasing Q_0 beyond a sufficient large value does not increase the number of arcs reversed beyond some fixed value (e.g., 29 in this case).

The links used for the quickest flow corresponding to $Q_0 = 100,000$ allowing partial arc reversal (using BPR function and Davidson's function) are depicted in Figure 9, with appropriate direction of the flow. The road segments which need to be reversed fully are (1, Source), (12, Source), (18, Source), (27, Source), (2, 1), (13, 12), (14, 13). (15, 14). (5, 4), (7, 6), (Sink, 8), (17, 16), (16, 15), (Sink, 7), (Sink, 40). The segments which are to be reversed partially are listed in Table 4.

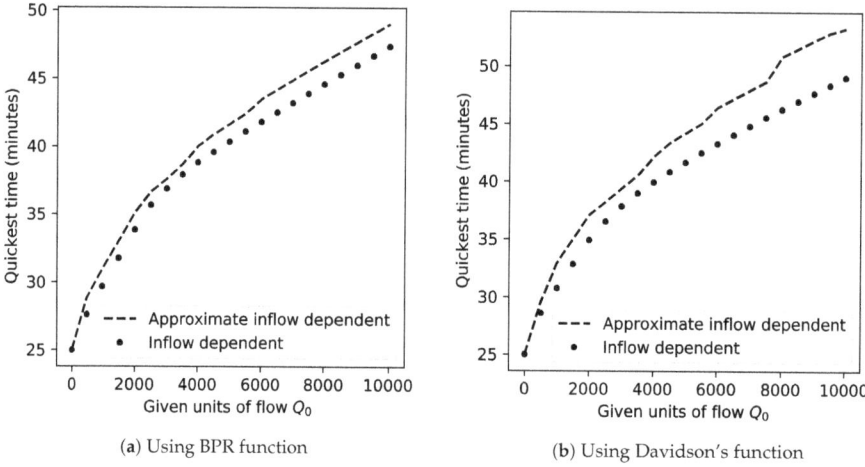

Figure 7. Quickest time in bow graph and its approximation by pushing the flow to the slowest arc.

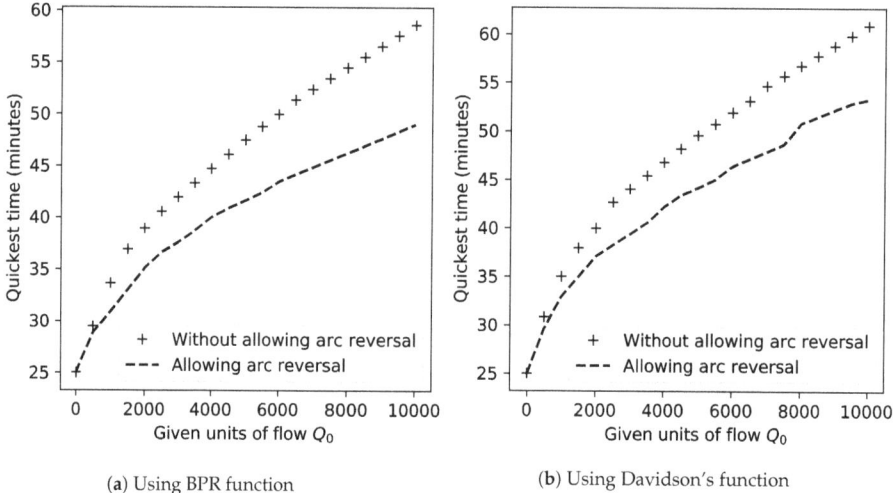

Figure 8. Quickest times before and after allowing partial arc reversal.

Table 2. Comparison of the quickest time before and after allowing arc reversal.

	BPR Function		Davidson's Function	
	Quickest Time		Quickest Time	
Q_0	before Contraflow	after Contraflow	before Contraflow	after Contraflow
500	29.5	27.6	30.8	28.6
1000	33.6	29.7	35	30.8
10,000	58.6	47.4	60.9	49
20,000	79.5	58.4	81.8	60.5
50,000	142	89.6	144.3	91.7
100,000	246.1	141.7	248.4	143.8

Table 3. Number of arcs reversed.

Q_0	Number of Arcs Reversed	
	BPR Function	Davidson's Function
500	8	5
1000	8	10
10,000	20	21
20,000	29	29
50,000	29	29
100,000	29	29

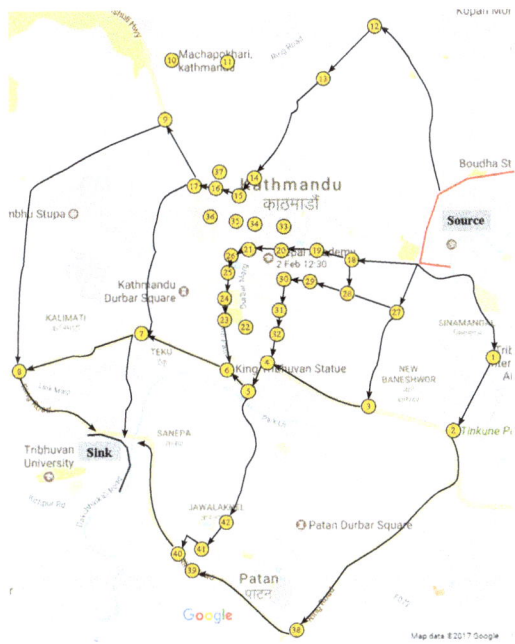

Figure 9. Direction of the approximate quickest flow allowing partial arc reversal, $Q_0 = 100{,}000$.

We also compare the quickest times with inflow-dependent transit time on arcs against the quickest times with constant transit time on arcs. For the purpose, we consider three types of constant transit time τ_e for each $e \in A$:

(i) $\tau_e = \tau_e^{st}(b_e)$, the upper bound on the step function represent of $\tau_e(y_e)$.
(ii) $\tau_e = \bar{\tau}_e^{st}(y_e) = \frac{\sum_{i=1}^{b_e} \tau_e^{st}(i)}{b_e}$, the average of the step function values.
(iii) $\tau_e = \tau_e^0$ the free flow transit time.

It is observed, in the network considered, that the quickest times corresponding to the constant time on each arc as the average of the corresponding step function are very close to the quickest time with inflow-dependent transit time (cf. Figure 10).

Table 4. Partially reversed segments.

Segment	Reversed Capacity	Capacity
(3, 27)	1	2
(38, 2)	1	3
(39, 38)	1	3
(40, 39)	1	3
(6, 5)	1	3
(4, 32)	1	2
(32, 31)	1	2
(8, 7)	1	3
(23, 24)	2	4
(24, 25)	2	4
(26, 21)	2	4
(25, 26)	2	4
(31, 30)	1	2
(19, 18) [a]	1	2
(7, 17) [b]	1	2

[a] for Davidson's function only. [b] for BPR function only.

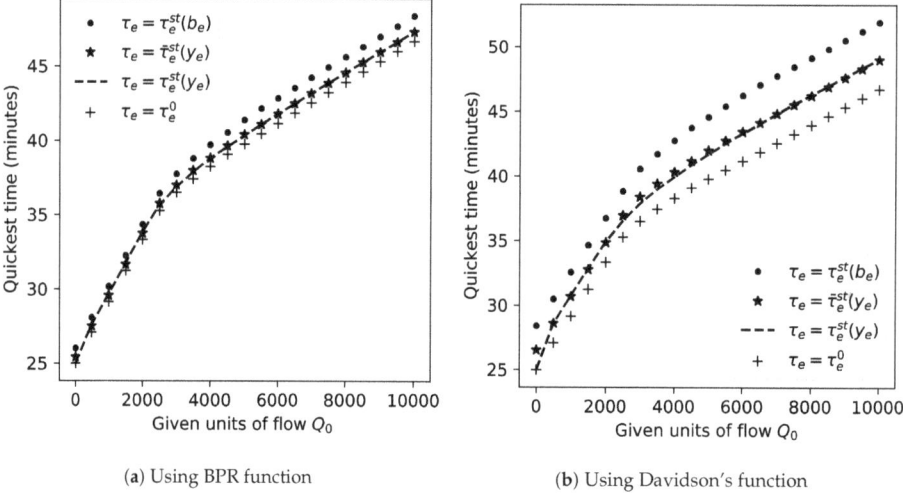

(a) Using BPR function   (b) Using Davidson's function

Figure 10. Comparison of quickest times (inflow-dependent transit time vs. constant transit time on arcs).

7. Conclusions

Highlighting the overall pros and cons of the complete contraflow models and algorithms, a new and more relevant approach—a partial lane reversal strategy—has been introduced in this paper. Using this approach, we can send maximum evacuees in minimum evacuation time recording all unused capacities of the lanes for other crucial emergency and logistic supports for the evacuees with partial reversals of lane capacities. The static partial contraflow problems and the dynamic partial contraflow problems including, the maximum dynamic, the earliest arrival, quickest, lex-maximum dynamic, generalized universally maximum, and partial contra-transshipment problems, have been solved with efficient algorithms. The maximum dynamic and earliest arrival contraflow problems are generalized on lossy networks with partial contraflow reconfiguration. Polynomial time algorithms to solve these problems with constant transit time on each arc have been proposed.

Moreover, the partial contraflow models with variable transit time on each arc have been introduced for the first time. For the inflow-dependent transit times on each arc, an algorithm with strongly polynomial time complexity has been presented that computes an approximate solution to the two terminal quickest contraflow problem with partial lane reversals, which is substantiated by numerical computations considering a case of Kathmandu road network as an evacuation network. The algorithms related to static flow are useful when one is interested to find the maximum rate of flow (evacuees) that can reach the sink(s). Within a specified time, if the maximum number of evacuees have to be evacuated, algorithms related to maximum dynamic flow are useful. The algorithms related to quickest flow are useful to identify the minimum time to evacuate a known number of evacuees.

To the best of our knowledge, the problems investigated in this work are conducted for the first time in the partial contraflow approach. Although these models provide information about the parts of the road segments not used by evacuees, they do not guarantee the existence of such a path between given nodes which may be required for movement of facilities from a node towards sources. As we have investigated only a single-source single-sink model with variable attributes to identify the quickest time, we are interested to extend these contraflow and partial contraflow models and algorithms to solve other network flow over time problems with variable attributes. In addition, we intend to implement the results for supporting logistics in emergencies using the partial contraflow techniques.

Author Contributions: Conceptualization, U.P.; Formal analysis, U.P. and H.N.N.; Investigation, U.P. and H.N.N.; Supervision, S.D. and T.N.D.; Writing–original draft, U.P.; Writing–review and editing, H.N.N.

Funding: This research received no external funding.

Acknowledgments: This research has been carried out during the stay of the first author as a research fellow with George Foster Fellowship for Post-doctoral Research and the stay of the third the author under the *AvH Research Group Linkage Program* at TU Bergakedemie Freiberg, Germany. The authors acknowledge the support of the Alexander von Humboldt Foundation, and thank the anonymous reviewers for their worthwhile suggestions to improve the quality of this work.

Conflicts of Interest: The authors declare no conflicts of interest.

Abbreviations

The following abbreviations are used in this manuscript:

BPR	Bureau of Public Roads
DPCFA	Dynamic partial contraflow algorithm
DPCFP	Dynamic partial contraflow problem
DPCTA	Dynamic partial contra-transshipment algorithm
DPCTP	Dynamic partial contra-transshipment problem
EAPCFA	Earliest arrival partial contraflow algorithm
EAPCFP	Earliest arrival partial contraflow problem
GDPCFA	Generalized dynamic partial contraflow algorithm
GDPCFP	Generalized dynamic partial contraflow problem
GEAPCFA	Generalized earliest arrival partial contraflow algorithm
GEAPCFP	Generalized earliest arrival partial contraflow problem
GMDPCFA	Generalized maximum dynamic partial contraflow algorithm
GMDPCFP	Generalized maximum dynamic partial contraflow problem
IFDQPCFA	Inflow-dependent quickest partial contraflow algorithm
IFDQPCFP	Inflow-dependent quickest partial contraflow problem
LMDPCFA	Lexicographic maximum dynamic partial contraflow algorithm

LMDPCFP   Lexicographic maximum dynamic partial contraflow problem
LMSPCFA   Lexicographic maximum static partial contraflow algorithm
LMSPCFP   Lexicographic maximum static partial contraflow problem
MDPCFA   Maximum dynamic partial contraflow algorithm
MDPCFP   Maximum dynamic partial contraflow problem
MSPCFA   Maximum static partial contraflow algorithm
MSPCFP   Maximum static partial contraflow problem
QPCFA   Quickest partial contraflow algorithm
QPCFP   Quickest partial contraflow problem
UMPCFA   Universal maximum partial contraflow algorithm
UMPCFP   Universal maximum partial contraflow problem

Appendix A

Table A1. Network data considered in Section 6.

e	b_e (per second)	$b_{e'}$ (per second)	τ_e^0 (minutes)
(Source, 1)	2	2	6
(Source, 12)	2	2	10
(Source, 18)	2	2	3
(Source, 27)	2	2	4
(1, 2)	2	2	3
(1, 27)	2	2	5
(2, 3)	3	3	5
(2, 38)	3	3	12
(3, 4)	3	3	5
(3, 27)	2	2	6
(4, 5)	3	3	1
(4, 32)	2	2	1
(5, 6)	3	3	1
(5, 42)	2	2	7
(6, 7)	3	3	5
(6, 23)	2	2	2
(7, 8)	3	3	8
(7, 17)	2	2	10
(7, Sink)	2	2	5
(8, 9)	2	2	16
(8, Sink)	3	3	7
(9, 10)	2	2	3
(9, 17)	2	2	3
(10, 11)	2	2	5
(11, 12)	2	2	17
(11, 37)	2	2	7
(12, 13)	2	2	4
(13, 14)	2	2	6
(13, 33)	2	2	9
(14, 15)	2	2	1
(14, 33)	2	2	3
(15, 16)	2	2	1
(15, 35)	2	2	1

Table A2. Network data considered in Section 6 (contd...).

e	b_e (per second)	$b_{e'}$ (per second)	τ_e^0 (minutes)
(16, 17)	2	2	1
(16, 36)	2	2	3
(16, 37)	4	0	1
(18, 19)	2	2	2
(18, 28)	2	2	2
(19, 20)	2	2	2
(19, 29)	2	2	2
(20, 21)	2	2	2
(20, 30)	2	2	2
(20, 33)	2	2	1
(21, 26)	0	4	1
(21, 34)	2	2	1
(22, 23)	4	0	1
(22, 32)	2	2	1
(23, 24)	4	0	1
(24, 25)	4	0	2
(25, 26)	4	0	1
(26, 35)	2	2	2
(27, 28)	2	2	3
(28, 29)	4	0	2
(29, 30)	4	0	1
(30, 31)	2	2	2
(31, 32)	2	2	2
(33, 34)	2	2	2
(34, 35)	2	2	1
(35, 36)	2	2	2
(38, 39)	3	3	7
(38, 42)	2	2	8
(39, 40)	3	3	1
(39, 41)	2	2	1
(40, 41)	2	2	2
(40, Sink)	3	3	8
(41, 42)	2	2	2

References

1. Pyakurel, U.; Dhamala, T.N.; Dempe, S. Efficient continuous contraflow algorithms for evacuation planning problems. *Ann. Oper. Res.* **2017**, *254*, 335–364. [CrossRef]
2. Köhler, E.; Langkau, K.; Skutella, M. Time expanded graphs for flow depended transit times. In *European Symposium on Algorithms*; Springer: Berlin/Heidelberg, Germany, 2002; pp. 599–611.
3. Dhamala, T.N. A survey on models and algorithms for discrete evacuation planning network problems. *J. Ind. Manag. Optim.* **2015**, *11*, 265–289. [CrossRef]
4. Hamacher H.W.; Tjandra, S.A. Mathematical modeling of evacuation problems: A state of the art. In *Pedestrain and Evacuation Dynamics*; Schreckenberger, M., Sharma, S.D., Eds.; Springer: Berlin/Heidelberg, Germany, 2002; pp. 227–266.
5. Cova, T.; Johnson J.P. A network flow model for lane-based evacuation routing. *Transp. Res. Part Policy Pract.* **2003**, *37*, 579–604. [CrossRef]
6. Altay, N.; Green, W.G., III. OR/MS research in disaster operations management. *Eur. J. Oper. Res.* **2006**, *175*, 475–493. [CrossRef]
7. Pascoal, M.M.; Captivo, M.E.V.; Climaco, J.C.N. A comprehensive survey on the quickest path problem. *Ann. Oper. Res.* **2006**, *147*, 5–21. [CrossRef]
8. Moriarty, K.D.; Ni, D.; Collura, J. Modeling traffic flow under emergency evacuation situations: Current practice and future directions. In Proceedings of the 86th Transportation Research Board Annual Meeting, Washington, DC, USA, 21–25 January 2007.

9. Chen, L.; Miller-Hooks, E. The building evacuation problem with shared information. *Nav. Res. Logist.* **2008**, *55*, 363–376. [CrossRef]
10. Yusoff, M.; Ariffin, J.; Mohamed, A. Optimization approaches for macroscopic emergency evacuation planning: A survey. In Proceedings of the International Symposium on Information Technology, Kuala Lumpur, Malaysia, 26–28 August 2008; Volume 3, pp. 1–7.
11. Dhamala, T.N.; Pyakurel, U.; Dempe, S. A critical survey on the network optimization algorithms for evacuation planning problems. *Int. J. Oper. Res.* **2018**, *15*, 101–133.
12. Kotsireas, I.S.; Nagurney, A.; Pardalos, P.M. Dynamics of disasters—Key concepts, models, algorithms, and insights. In *Springer Proceedings in Mathematics & Statistics*; Springer: Berlin/Heidelberg, Germany, 2015.
13. Ford, L.R.; Fulkerson, D.R. *Flows in Networks*; Princeton University Press: Princeton, NJ, USA, 1962.
14. Gale, D. Transient flows in networks. *Mich. Math. J.* **1959**, *6*, 59–63. [CrossRef]
15. Wilkinson, W.L. An algorithm for universal maximal dynamic flows in a network. *Oper. Res.* **1971**, *19*, 1602–1612. [CrossRef]
16. Minieka, E. Maximal, lexicographic, and dynamic network flows. *Oper. Res.* **1973**, *21*, 517–527. [CrossRef]
17. Hoppe, B. Efficient Dynamic Network Flow Algorithms. Ph.D. Thesis, Cornell University, Ithaca, NY, USA, 1995.
18. Ruzika, S.; Sperber, S.; Steiner, M. Earliest arrival flows on series-parallel graphs. *Networks* **2011**, *10*, 169–173. [CrossRef]
19. Hoppe, B.; Tardos, E. The quickest transshipment problem. *Math. Oper. Res.* **2000**, *25*, 36–62. [CrossRef]
20. Burkard, R.E.; Dlaska, K.; Klinz, B. The quickest flow problem. *ZOR-Methods Model. Oper. Res.* **1993**, *37*, 31–58. [CrossRef]
21. Baumann, N.; Skutella M. Earliest arrival flows with multiple sources. *Math. Oper. Res.* **2009**, *34*, 499–512. [CrossRef]
22. Fleischer, L. Universally maximum flow with piecewise-constant capacities. *Networks* **2001**, *38*, 115–125. [CrossRef]
23. Groß, M.; Kappmeier, J.P.W.; Schmidt, D.R.; Schmidt, M. Approximating earliest arrival flows in arbitrary networks. In *European Symposium on Algorithms*; Springer: Berlin/Heidelberg, Germany, 2012; pp. 551–562.
24. Pyakurel, U.; Dhamala, T.N. Continuous dynamic contraflow approach for evacuation planning. *Ann. Oper. Res.* **2017**, *253*, 573–598. [CrossRef]
25. Fleischer, L.; Tardos, E. Efficient continuous-time dynamic network flow algorithms. *Oper. Res. Lett.* **1998**, *23*, 71–80. [CrossRef]
26. Quarantelli, E.L. *Evacuation Behavior and Problems: Findings and Implications from the Research Literature*; Ohio State University Columbus Disaster Research Center: Columbus, OH, USA, 1980.
27. Kim, S.; Shekhar, S.; Min, M. Contraflow transportation network reconfiguration for evacuation route planning. *IEEE Trans. Knowl. Data Eng.* **2008**, *20*, 1115–1129.
28. Rebennack, S.; Arulselvan, A.; Elefteriadou, L.; Pardalos, P.M. Complexity analysis for maximum flow problems with arc reversals. *J. Comb. Optim.* **2010**, *19*, 200–216. [CrossRef]
29. Dhamala, T.N.; Pyakurel, U. Earliest arrival contraflow problem on series-parallel graphs. *Int. J. Oper. Res.* **2013**, *10*, 1–13.
30. Pyakurel, U.; Dhamala, T.N. Continuous time dynamic contraflow models and algorithms. *Adv. Oper. Res. Hindawi* **2016**, *2016*, 368587. [CrossRef]
31. Pyakurel, U.; Dhamala, T.N. Models and algorithms on contraflow evacuation planning network problems. *Int. J. Oper. Res.* **2015**, *12*, 36–46. [CrossRef]
32. Pyakurel, U.; Dhamala, T.N. Evacuation planning by earliest arrival contraflow. *J. Ind. Manag. Optim.* **2017**, *13*, 489–503. [CrossRef]
33. Pyakurel, U.; Nath, H.N.; Dhamala, T.N. Partial contraflow with path reversals for evacuation planning. *Ann. Oper. Res.* **2018**. [CrossRef]
34. Smith, J.M.G.; Cruz, F.R.B. $M/G/c/c$ state dependent travel time models and properties. *Phys. Stat. Mechan. Its Appl.* **2014**, *395*, 560–579. [CrossRef]
35. Ahuja, R.K.; Magnanti, T.L.; Orlin, J.B. *Ntwork Flows: Theory, Algorithms, and Applications*; Prentice Hall: Englewood Cliffs, NJ, USA, 1993.
36. Goldberg, A.V.; Tarjan, R.E. Efficient maximum flow algorithms. *Commun. ACM* **2014**, *57*, 82–89. [CrossRef]

37. Pyakurel, U.; Nath, H.N.; Dhamala, T.N. Efficient contraflow algorithms for quickest evacuation planning. *Sci. China Math.* **2018**, *61*, 2079–2100. [CrossRef]
38. Karger, D.R.; Stein C. A new approach to the minimum cut problem. *J. ACM* **1996**, *43*, 601–640. [CrossRef]
39. Pyakurel, U.; Dempe, S. Earliest arrival flow with partial lane reversals for evacuation planning. *Int. J. Oper. Res. Nepal* **2019**, *8*, 27–37.
40. Pyakurel, U.; Hamacher, H.W.; Dhamala T.N. Generalized maximum dynamic contraflow on lossy network. *Int. J. Oper. Res. Nepal* **2014**, *3*, 27–44.
41. Groß, M.; Skutella, M. Generalized maximum flows over time. In *Approximation and Online Algorithms*; Solis-Oba, R., Persiano, G., Eds.; Springer: Berlin/Heidelberg, Germany, 2012.
42. Megiddo, N. Combinatorial optimization with rational objective functions. *Math. Oper. Res.* **1979**, *4*, 414–424. [CrossRef]
43. Lin, M.; Jaillet, P. On the quickest flow problem in dynamic networks—A parametric min-cost flow approach. In *Proceedings of the Twenty-Sixth Annual ACM-SIAM Symposium on Discrete Algorithms*; Society for Industrial and Applied Mathematics: San Diego, CA, USA, 2015; pp. 1343–1356.
44. Saho, M.; Shigeno, M. Cancel-and-tighten algorithm for quickest flow problems. *Networks* **2017**, *69*, 179–188. [CrossRef]
45. Schmidt, M.; Skutella, M. Earliest arrival flows in networks with multiple sinks. *Electron. Notes Discret. Math.* **2010**, *36*, 607–614. [CrossRef]
46. Kappmeier, J.P.W.; Matuschke, J.; Peis, B. Abstract flows over time: A first step towards solving dynamic packing problems.*Theor. Comput. Sci.* **2014**, *544*, 74–83. [CrossRef]
47. Pyakurel, U.; Dempe, S.; Dhamala, T.N. Network reconfiguration with variable transit times for evacuation planning. In Proceedings of the APORS 2018, Kathmandu, Nepal, 6–9 August 2018; pp. 25–28.
48. Sheffi, Y. *Urban Transportation Networks: Equilibrium Analysis with Mathematical Programming Techniques*; Prentice-Hall, Inc.: Englewood Cliffs, NJ, USA, 1984.

 © 2019 by the authors. Licensee MDPI, Basel, Switzerland. This article is an open access article distributed under the terms and conditions of the Creative Commons Attribution (CC BY) license (http://creativecommons.org/licenses/by/4.0/).

Article

A Deep Learning Algorithm for the Max-Cut Problem Based on Pointer Network Structure with Supervised Learning and Reinforcement Learning Strategies

Shenshen Gu *,† and Yue Yang

School of Mechatronic Engineering and Automation, Shanghai University, Shanghai 200444, China; yangyue0328@shu.edu.cn
* Correspondence: gushenshen@shu.edu.cn
† Current address: 99 Shangda Road, Shanghai 200444, China.

Received: 5 January 2020; Accepted: 20 February 2020; Published: 22 February 2020

Abstract: The Max-cut problem is a well-known combinatorial optimization problem, which has many real-world applications. However, the problem has been proven to be non-deterministic polynomial-hard (NP-hard), which means that exact solution algorithms are not suitable for large-scale situations, as it is too time-consuming to obtain a solution. Therefore, designing heuristic algorithms is a promising but challenging direction to effectively solve large-scale Max-cut problems. For this reason, we propose a unique method which combines a pointer network and two deep learning strategies (supervised learning and reinforcement learning) in this paper, in order to address this challenge. A pointer network is a sequence-to-sequence deep neural network, which can extract data features in a purely data-driven way to discover the hidden laws behind data. Combining the characteristics of the Max-cut problem, we designed the input and output mechanisms of the pointer network model, and we used supervised learning and reinforcement learning to train the model to evaluate the model performance. Through experiments, we illustrated that our model can be well applied to solve large-scale Max-cut problems. Our experimental results also revealed that the new method will further encourage broader exploration of deep neural network for large-scale combinatorial optimization problems.

Keywords: Max-cut problem; combinatorial optimization; deep learning; pointer network; supervised learning; reinforcement learning

1. Introduction

Combinatorial optimization is an important branch of operations research. It refers to solving problems of variable combinations by minimizing (or maximizing) an objective function under given constraints, and is based on the research of mathematical methods to find optimal arrangements, groupings, orderings, or screenings of discrete events. As a research hot-spot in combinatorial optimization, the Max-cut problem is one of the 21 typical non-deterministic polynomial (NP)-complete problems proposed by Richard M. Karp [1]. It refers to obtaining a maximum segmentation for a given directed graph, such that the sum of the weights across all edges of two cutsets is maximized [2]. The Max-cut problem has a wide range of applications in engineering problems, such as Very Large Scale Integration (VLSI) circuit design, statistical physics, image processing, and communications network design [3]. As a solution of the Max-cut problem can be used to measure the robustness of a network [4] and as a standard for network classification [5], it can also be applied to social networks.

It has been discovered that many classic combinatorial optimization problems derived from engineering, economics, and other fields are NP-hard. The Max-cut problem concerned in this paper is among these problems. For combinatorial optimization problems, algorithms can be roughly divided into two categories: one is represented by exact solution approaches, including enumeration methods [6] and branch and bound methods [7], etc. The other category is represented by heuristic algorithms including genetic algorithms, ant colony algorithms, simulated annealing algorithms, neural networks, Lin–Kernighan Heuristic (LKH) algorithms, and so on [8]. However, there is no polynomial time solvable algorithm to find a global optimal solution for NP-hard problems. Compared with the exact approach, heuristic algorithms are able to deal with large-scale problems efficiently. They have certain advantages in computing efficiency and can be applied to solving large-scale problems with huge amount of variables. In order to solve the Max-cut problem, a large number of heuristic algorithms have been proposed, such as evolutionary algorithms and ant colony algorithms. However, for these algorithms, the most obvious disadvantage of them is that they are easy to fall into local optima. For this reason, more and more experts have begun working on the research and innovation of some novel and effective algorithms for large-scale Max-cut problems.

Deep learning is a research field which has developed very rapidly in recent years, achieving great success in many sub-fields of artificial intelligence. From its root, deep learning is a sub-problem of machine learning. Its main purpose is to automatically learn effective feature representations from a large amount of data, such that it can better solve the credit assignment problem (CAP) [9]; that is, the contribution or influence of different components in a system or their parameters to the output of the final system. The emergence of deep neural networks has provided the possibility for solving large-scale combinatorial optimization problems. In recent years, with the development of the combination of deep neural networks and operations research for large-scale combinatorial optimization problems, scholars have explored how to apply deep neural networks in these fields, and have achieved certain results. The related research has mainly focused on the algorithm design for combinatorial optimization problems based on pointer networks. Vinyals used the attention mechanism [10] to integrate a pointer structure into the sequence-to-sequence model, thus creating the pointer network. Bello improved the pointer network structure and used a strategy gradient algorithm combined with time-series difference learning to train the pointer network in reinforcement learning to solve the combinatorial optimization problem [11]. Mirhoseini removed the coding part of a recurrent neural network (RNN) and used the embedded input to replace the hidden state of the RNN. With this modification, the computational complexity was greatly reduced and the efficiency of the model was improved [12]. In Reference [13], a purely data-driven method to obtain approximate solutions of NP-hard problems was proposed. In Reference [14], a pointer network was used to establish a flight decision prediction model. Khalil solved classical combinatorial optimization problems by Q-learning [15]. The pointer network model has also been used, in Reference [16], to solve the weightless Max-cut problem. Similarly, solved the unconstrained boolean quadratic programming problem (UBQP) through the pointer network [17].

The section arrangement of this paper is as follows. Section 2 mainly introduces the Max-cut problem and the method for generating its benchmark. Section 3 demonstrates the pointer network model, including the Long Short-Term Memory network and Encoder–Decoder. Section 4 introduces two ways to train the pointer network model to solve the Max-cut problem, namely supervised learning and reinforcement learning. Section 5 illustrates the details of the experimental procedure and the results. Section 6 provides the conclusions.

2. Motivation and Data Set Structure

2.1. Unified Model of the Max-Cut Problem

The definition of the Max-cut problem is given as follows.

An undirected graph $G = (V, E)$ consists of a set of vertices V and a set of edges E, where $V = \{1, 2, ..., n\}$ is its set of vertices, and $E \subseteq V \times V$ is its set of edges, and $w_{i,j}$ is the weight on the edge connecting vertex i and vertex j. For any proper subset S of the vertex set V, let:

$$\delta(S) = \{e_{i,j} \in E; i \in S, j \in V - S\}, \tag{1}$$

where $\delta(S)$ is a set of edges, one end of which belongs to S and the other end belongs to $V-S$. Then, the cut $cut(S)$ determined by S is:

$$cut(S) = \sum_{e_{i,j} \in \delta(S)} w_{i,j}. \tag{2}$$

In simple terms, the Max-cut problem is to find a segmentation $(S, V-S)$ of a vertex set V, where the maximum weight of the edges is segmented.

2.2. Benchmark Generator of the Max-Cut Problem

When applying deep learning to train and solve the Max-cut problem, whether supervised learning or reinforcement learning, a large number of training samples are necessary. The method of data set generation introduced here is to transform the $\{-1,1\}$ quadratic programming problem into the Max-cut problem.

First of all, the benchmark generator method for the boolean quadratic programming (BQP) problem, proposed by Michael X. Zhou [18], is used to generate random $\{-1,1\}$ quadratic programming problems, which can be solved in polynomial time. Next, inspired by [19], we transform the results of the previous step into solutions of the Max-cut problem. The specific implementation is described below.

Michael X. Zhou transformed the quadratic programming problem shown by Equation (3) into the dual problem shown by Equation (4) through the Lagrangian dual method.

$$\min \{f(x) = \frac{1}{2}x^T Q x - c^T x \,|\, x \in \{-1,1\}^n\}, \tag{3}$$

where $Q = Q^T \in \mathbb{R}^{n \times n}$ is a given indefinite matrix, and $c \in \mathbb{R}^n$ is a given non-zero vector.

The dual problem is described as follows:

$$\begin{array}{ll} \text{find} & Q, c, x, \lambda \\ \text{s.t.} & (Q + diag(\lambda)) = c \\ & Q + diag(\lambda) > 0 \\ & x \in \{-1,1\}^n \end{array} \tag{4}$$

Then, according to the paper [19], the solution of the $\{-1,1\}$ quadratic programming problem can be transformed into the solution of the Max-cut problem.

The integer programming for the Max-cut problem is given by:

$$\begin{array}{ll} \max & \frac{1}{2} \sum_{i<j} w_{i,j}(1 - x_i \cdot x_j) \\ \text{s.t.} & x_i \in \{-1,1\}, \, \forall i = 1, \cdots, n, \end{array} \tag{5}$$

where i in $x_i \in \{-1,1\}$ represents the vertex i, and -1 and 1 represent the values of the two sets. If $x_i \cdot x_j$ is equal to 1 and the vertices of edge (i, j) are in the same set, then $(i, j) \in E$ is not a cut edge; if $x_i \cdot x_j$ is equal to -1 and the vertices of the edge (i, j) are not in the same set, then $(i, j) \in E$ is the cut edge. If $(i, j) \in E$ is a cut edge, $(1 - x_i \cdot x_j)/2$ is equal to 1; if $(i, j) \in E$ is not a cut edge, $(1 - x_i \cdot x_j)/2$ is equal to 0. Thus, the objective function represents the sum of the weights of the cut edges of the Max-cut. Define $S = \{i : x_i = 1\}$, $\overline{S} = \{i : x_i = -1\}$, and the weight of the cut is $w(S, \overline{S}) = \sum_{i<j} w_{i,j}(1 - x_i \cdot x_j)/2$.

The pseudocode for generating the benchmark of the Max-cut problem is shown in Algorithm 1, where the parameter *base* is used to control the value range of the elements in matrix Q.

Algorithm 1 A benchmark generator for the Max-cut problem

Input: Dimension: n; base = 10;

Output: Matrix: Q; Vector: x

1: Randomly generate an n-dimensional matrix that conforms to the standard normal distribution to obtain Q;
2: $Q = base \times Q$;
3: Convert Q to a symmetric matrix with $\frac{Q+Q^T}{2}$;
4: Generate random numbers in the range (0,1) of n rows and 1 column as a vector x;
5: $x = 2x - 1$;
6: Take the absolute value of Q and sum it over the rows, assigning the result to λ;
7: Place the value of the vector λ on the main diagonal of the square matrix Q' in order, and let the values of Q' (except the main diagonal) be zero.
8: $c = (Q + Q') \times x$;
9: Set an additional variable w_{oj}, $w_{oj} = \frac{1}{4}(\sum_{j=1}^{i-1} q_{ji} + \sum_{j=i+1}^{n} q_{ij}) + \frac{1}{2}c_i$, $1 \leq j \leq n$, and $w_{ij} = \frac{1}{4}q_{ij}$, $1 \leq i < j \leq n$;
10: Update Q: $Q = (q_{1j}^T, q_{2j}^T, ..., q_{(n+1)j}^T) \leftarrow (w_{0j}^T, w_{1j}^T, ..., w_{nj}^T)$;
11: Set an additional variable $x_0 = 1$, and let $x = 2x + 1$;
12: Update x: $x = (x_1, x_2, ..., x_{n+1}) \leftarrow (x_0, x_1, ..., x_n)$.

This method for obtaining Max-cut benchmark data sets effectively solves the difficulty in training the network to solve the Max-cut problem model when lacking a large number of training samples. However, there is a common defect in this method: in the training set obtained using the dual problem to deduce the solution of the original problem, its data samples obey certain rules. This may lead to difficulty in learning the general rule of the Max-cut problem when training with the method by deep learning.

Therefore, in addition to the above method, we consider using the benchmark generator in the Biq Mac Library to solve the Max-cut problem. The Biq Mac Library offers a collection of Max-cut instances. Biq Mac is a branch and bound code based on semi-definite programming (SDP). The dimension of the problems (i.e., number of variables or number of vertices in the graph) ranges from 60–100. These instances are mainly used to test the pointer network model for the Max-cut problem.

3. Models

3.1. Long Short-Term Memory

It is difficult for traditional neural networks to classify subsequent events by using previous event information. However, an RNN can continuously operate information in a cyclic manner to ensure that the information persists, thereby effectively processing time-series data of any length. Given an input sequence $x_{1:T} = (x_1, x_2, ..., x_t, ..., x_T)$, the RNN updates the activity value h_t of the hidden layer with feedback and calculates the output sequence $y_{1:T} = (y_1, y_2, ..., y_t, ..., y_T)$ using the following equations:

$$h_t = \text{sigmoid}(M^{hx}x_t + M^{hh}h_{t-1}), \tag{6}$$

$$y_t = M^{yh}h_t. \tag{7}$$

As long as the alignment between input and output is known in advance, an RNN can easily map sequences to sequences. However, the RNN cannot solve the problem when the input and output sequences have different lengths or have complex and non-monotonic relationships [20]. In addition, when the input sequence is long, the problem of gradient explosion and disappearance will occur [21]; which is also known as the long-range dependence problem. In order to solve these problems, many improvements have been made to RNNs; the most effective way, thus far, is to use a gating mechanism.

A long short-term memory (LSTM) network [22] is a variant of RNN, which is an outstanding embodiment of RNN based on the gating mechanism. Figure 1 shows the structure of the loop unit of a LSTM. By applying the LSTM loop unit of the gating mechanism, the entire network can establish long-term timing dependencies to better control the path of information transmission. The equations of the LSTM model can be briefly described as:

$$\begin{bmatrix} \tilde{c}_t \\ o_t \\ i_t \\ f_t \end{bmatrix} = \begin{bmatrix} \tanh \\ \sigma \\ \sigma \\ \sigma \end{bmatrix} \left(M \begin{bmatrix} x_t \\ h_{t-1} \end{bmatrix} + b \right), \tag{8}$$

$$c_t = f_t \odot c_{t-1} + i_t \odot \tilde{c}_t, \tag{9}$$

$$h_t = o_t \odot \tanh(c_t), \tag{10}$$

where $x_t \in \mathbb{R}^e$ is the input at the current time; $M \in \mathbb{R}^{4d \times (d+e)}$ and $b \in \mathbb{R}^{4d}$ are the network parameters; $\sigma(\cdot)$ is the Logistic function, with output interval $(0,1)$; h_{t-1} is the external state at the previous time; $\odot$ is the product of vector elements; c_{t-1} is the memory unit at the previous moment; and $\tilde{c}_t$ is the candidate state obtained by the non-linear function. At each time t, the internal state c_t of the LSTM records historical information up to the current time. The three gates used to control the path of information transmission are f_t, i_t, and o_t. The functions of three gates are:

- The forget gate f_t controls how much information the previous state c_{t-1} needs to forget;
- The input gate i_t controls how much information the candidate state $\tilde{c}_t$ needs to be saved at the current moment; and
- The output gate o_t controls how much information the internal state c_{t-1} of the current moment needs to be output to the external state h_{t-1}.

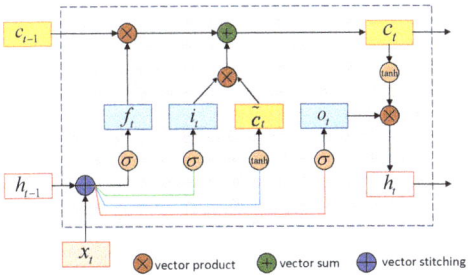

Figure 1. Long short-term memory (LSTM) loop unit structure.

In our algorithm, the purpose of the LSTM is to estimate the conditional probability $p(y_1, ..., y_{T'} | x_1, ..., x_T)$, where $(x_1, ..., x_T)$ is the input sequence, $y_1, ..., y_{T'}$ is the corresponding output sequence, and the length T' may be different from T. The LSTM first obtains a fixed dimension representation X of the input sequence $(x_1, ..., x_T)$ (given by the last hidden state of the LSTM), then calculates $y_1, ..., y_{T'}$, whose initial hidden state is set to $x_1, ..., x_T$:

$$p(y_1, ..., y_{T'}|x_1, ..., x_T) = \prod_{t=1}^{T'} p(y_t|X, y_1, ..., y_{t-1}), \tag{11}$$

where each $p(y_t|X, y_1, ..., y_{t-1})$ distribution is represented by the softmax of all variables in the input Max-cut problem matrix.

3.2. Encoder–Decoder Model

The encoder–decoder model is also called the asynchronous sequence-to-sequence model; that is, the input sequence and the output sequence neither need to have a strict correspondence relationship, nor do they need to maintain the same length. Compared with traditional structures, it greatly expands the application scope of the model. It can directly model sequence problems in a pure data-driven manner and can train the model using an end-to-end method. It can be seen that it is very suitable for solving combinatorial optimization problems.

In the encoder–decoder model (shown in Figure 2), the input is a sequence $x_{1:T} = (x_1, ..., x_T)$ of length T, and the output is a sequence $y_{1:T'} = (y_1, ..., y_{T'})$ of length T'. The implementation process is realized by first encoding and then decoding. Firstly, a sample x is input into an RNN (encoder) at different times to obtain its encoding h_T. Secondly, another RNN (decoder) is used to obtain the output sequence $\hat{y}_{1:T'}$. In order to establish the dependence between the output sequences, a non-linear autoregressive model is usually used in the decoder:

$$h_t = f_1(h_{t-1}, x_t), \forall t \in [1, T], \tag{12}$$

$$h_{T+t} = f_2(h_{T+t-1}, \hat{y}_{t-1}), \forall t \in [1, T'], \tag{13}$$

$$y_t = g(h_{T+t}), \forall t \in [1, T'], \tag{14}$$

where $f_1(\cdot)$ and $f_2(\cdot)$ are RNNs used as encoder and decoder, respectively; $g(\cdot)$ is a classifier; and $\hat{y}_t$ are vector representations used to predict the output.

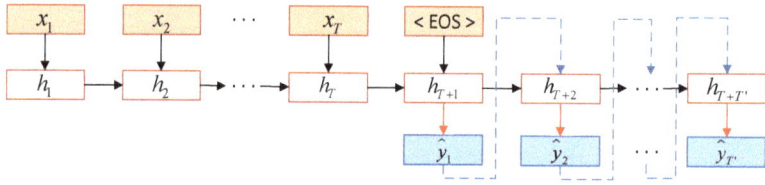

Figure 2. Encoder–decoder model.

3.3. Pointer Network

The amount of information that can be stored in a neural network is called the network capacity. Generally speaking, if more information needs to be stored, then more neurons are needed or the network must be more complicated, which will cause the number of necessary parameters of the neural network to increase exponentially. Although general RNNs have strong capabilities, when dealing with complex tasks, such as processing large amounts of input information or complex computing processes, the computing power of computers is still a bottleneck that limits the development of neural networks.

In order to reduce the computational complexity, we use the mechanisms of the human brain to solve the information overload problem. In such a way, we add an attention mechanism to the RNN.

When the computing power is limited, it is used as a resource allocation scheme to allocate computing resources to more important tasks.

A pointer network is a typical application for combining an attention mechanism and a neural network. We use the attention distribution as a soft pointer to indicate the location of relevant information. In order to save computing resources, it is not necessary to input all the information into the neural network, only the information related to the task needs to be selected from the input sequence X. A pointer network [9] is also an asynchronous sequence-to-sequence model. The input is a sequence $X = x_1, ..., x_T$ of length T, and the output is a sequence $y_{1:T'} = y_1, y_2, ..., y_{T'}$. Unlike general sequence-to-sequence tasks, the output sequence here is the index of the input sequence. For example, when the input is a group of out-of-order numbers, the output is the index of the input number sequence sorted by size (e.g., if the input is 20, 5, 10, then the output is 1, 3, 2).

The conditional probability $p(y_{1:T'} | x_{1:T})$ can be written as:

$$p(y_{1:T'} | x_{1:T}) = \prod_{i=1}^{m} p(y_i | y_{1:i-1}, x_{1:T}) \approx \prod_{i=1}^{m} p(y_i | x_{y_1}, ..., x_{y_{i-1}}, x_{1:T}), \quad (15)$$

where the conditional probability $p(y_i | x_{y_1}, ..., x_{y_{i-1}}, x_{1:T})$ can be calculated using the attention distribution. Suppose that an RNN is used to encode $x_{y_1}, ..., x_{y_{i-1}}, x_{1:T}$ to obtain the vector h_i, then

$$p(y_i | y_{1:i-1}, x_{1:T}) = \text{softmax}(s_{i,j}), \quad (16)$$

where $s_{i,j}$ is the unnormalized attention distribution of each input vector at the ith step of the encoding process,

$$s_{i,j} = v^T \tanh(U_1 x_j + U_2 h_i), \forall j \in [1, T], \quad (17)$$

where v, U_1, and U_2 are learnable parameters.

Figure 3 shows an example of a pointer network.

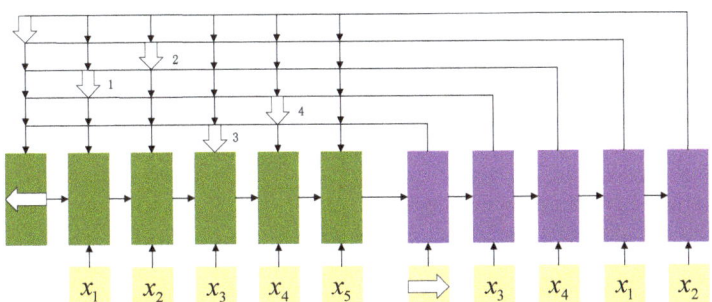

Figure 3. The architecture of pointer network (encoder in green, decoder in purple).

4. Learning Mechanism

Machine learning methods can be classified according to different criteria. Generally speaking, according to the information provided by the training samples and different feedback mechanisms, we classify machine learning algorithms into three categories: supervised learning, unsupervised learning, and reinforcement learning. Our algorithm uses supervised learning (SL) and reinforcement learning (RL) to train the pointer network model to obtain the solution of the Max-cut problem, which will be described in detail below.

4.1. Supervised Learning

4.1.1. Input and Output Design

The feature of the Max-cut problem is that its variable is either 0 or 1, such the problem is equivalent to selecting a set of variables from all variables with a value of 1 to maximize the objective function. This is a typical choice problem in combinatorial optimization problems. The goal of supervised learning is to learn the relationship between the input x and the output y by modeling $y = f(x;\theta)$ or $p(y|x;\theta)$. For the Max-cut problem, the pointer network uses an $n \times n$ symmetric matrix Q to represent the input sequence of the n nodes, where q_{ij} is an element in the symmetric matrix, which represents the weight of the connection between vertex i vertex and vertex j ($q_{ij} \geq 0$, $q_{ij} = 0$ means there is no connection between vertex i and vertex j). The output sequence of the pointer network is represented by $X = x_1, x_2, ..., x_n$, which contains two variables; that is 0 and 1. Vertices with 0 and vertices with 1 are divided into two different sets. The result of summing weights with all edges across the two cut sets is the solution to the Max-cut problem.

The following example is used to explain the input and output design of the pointer network to solve the Max-cut problem.

Example 1.

$$f(x) = 3x_1x_2 + 4x_1x_4 + 5x_2x_3 + 2x_2x_4 + x_3x_4 \tag{18}$$
$$x_i \in \{0,1\}, (i = 1, ..., 4)$$

The symmetric matrix Q of the above problem can be expressed as:

$$Q = \begin{pmatrix} 0 & 3 & 0 & 4 \\ 3 & 0 & 5 & 2 \\ 0 & 5 & 0 & 1 \\ 4 & 2 & 1 & 0 \end{pmatrix},$$

and the characteristics of the variables x_1, x_2, x_3, and x_4 are represented by the vectors $q_1 = (0,3,0,4)^T$, $q_2 = (3,0,5,2)^T$, $q_3 = (0,5,0,1)^T$, and $q_4 = (4,2,1,0)^T$, respectively.

For the Max-cut problem, the optimal solution of the above example is x. The sequence (q_1, q_2, q_3, q_4) is the input of the pointer network, and the known optimal solution is used to train the network model and guide the model to select q_1 and q_3. The input vector selected by the decoder represents the corresponding variable value of 1, while the corresponding variable value of the unselected vector is 0.

For the output part of the pointer network model, for the $n \times n$ matrix, we design a matrix of dimension $(n+1)$ to represent the network output. Exactly as in Example 1, the output result is a label that be described by the matrix O_{label}:

$$O_{label} = \begin{pmatrix} 0 & 1 & 0 & 0 & 0 \\ 0 & 0 & 0 & 1 & 0 \\ 1 & 0 & 0 & 0 & 0 \\ 1 & 0 & 0 & 0 & 0 \\ 1 & 0 & 0 & 0 & 0 \end{pmatrix}.$$

The relationship between O_{label} and the variable x is:

$$x_j = \begin{cases} 1, & \text{if } o_{ij} = 1, j \neq 0; \\ \text{EOS}, & \text{if } o_{ij} = 1, j = 0; \\ 0, & \text{others}. \end{cases} \tag{19}$$

We use $\text{EOS} = (1, 0, \cdots, 0)^T$ to indicate the end of the pointer network solution process. After the model training is completed, the probability distribution of the softmax of the output matrix is obtained.

The corresponding result may be as described by the matrix $O_{predict}$. In the solution phase, we select the one with the highest probability in the output probability distribution and set it to 1, and the rest of the positions to 0. According to the result of $O_{predict}$, the pointer network selects the variables x_1 and x_3 with a value of 1, and the remaining variables have a value of 0—which is consistent with the result selected by O_{label}:

$$O_{predict} = \begin{pmatrix} 0.03 & 0.8 & 0.02 & 0.1 & 0.05 \\ 0.1 & 0 & 0.2 & 0.7 & 0 \\ 0.9 & 0.03 & 0.03 & 0.01 & 0 \\ 1 & 0 & 0 & 0 & 0 \\ 1 & 0 & 0 & 0 & 0 \end{pmatrix}.$$

4.1.2. Algorithm Design

When training deep neural networks, for N given training samples $\left\{\left(\mathbf{x}^{(n)}, y^{(n)}\right)\right\}_{n=1}^{N}$, the softmax regression in supervised learning uses cross entropy as a loss function and uses gradient descent to optimize the parameter matrix W. The goal of neural network training is to learn the parameters which minimize the value of the cross-entropy loss function. In practical applications, the mini-batch stochastic gradient descent (SGD) method has the advantages of fast convergence and small computational overhead, so it has gradually become the main optimization algorithm used in large-scale machine learning [23]. Therefore, during the training process, we use mini-batch SGD. At each iteration, we randomly select a small number of training samples to calculate the gradient and update the parameters. Assuming that the number of samples per mini-batch is K, the training process of softmax regression is: initialize $W_0 \leftarrow 0$, and then iteratively update by the following equation

$$W_{t+1} \leftarrow W_t + \alpha \left(\frac{1}{K} \sum_{n=1}^{N} x^{(n)} \left(y^{(n)} - \widehat{y}_{W_t}^{(n)} \right)^T \right), \tag{20}$$

where α is the learning rate and $\widehat{y}_{W_t}^{(n)}$ is the output of the softmax regression model when the parameter is W_t.

The training process of mini-batch SGD is shown in Algorithm 2.

Algorithm 2 Mini-batch SGD of pointer network

Input: training set: $D = \{(x^{(n)}, y^{(n)})\}_{n=1}^{N}$; mini-batch size: K; number of training steps: L; learning rate: α;

Output: optimal: W

1: random initial W;
2: **repeat**
3:     randomly reorder the samples in training set D;
4:     **for** $t = 1, ..., L$ **do**
5:         select samples $(x^{(n)}, y^{(n)})$ from the training set D;
6:         update parameters: $W_{t+1} \leftarrow W_t + \alpha(\frac{1}{K} \sum_{n=1}^{N} x^{(n)} (y^{(n)} - \widehat{y}_{W_t}^{(n)})^T)$;
7:     **end for**
8: **until** the error rate of model $f(x; W)$ no longer decreases.

4.2. Reinforcement Learning

Reinforcement learning is a very attractive method in machine learning. It can be described as an agent continuously learning from interaction with the environment to achieve a specific goal (such as obtaining the maximum reward value). The difference between reinforcement learning and supervised learning is that reinforcement learning does not need to give the "correct" strategy as supervised information, it only needs to give the return of the strategy and then adjust the strategy to achieve the maximum expected return. Reinforcement learning is closer to the nature of biological learning and can cope with a variety of complex scenarios, thus coming closer to the goal of general artificial intelligence systems.

The basic elements in reinforcement learning include:

- The *agent* can sense the state of the external environment and the reward of feedback, then make decisions;
- The *environment* is everything outside the *agent*, which is affected by the actions of the *agent* by changing its state and feeding the corresponding reward back to the *agent*;
- s is a description of the *environment*, which can be discrete or continuous, and its state space is S;
- a is a description of the behavior of the *agent*, which can be discrete or continuous, and its action space is A;
- The reward $r(s, a, s')$ is a scalar function—that is, after the agent makes an action a based on the current state s, the environment will give a reward to the agent. This reward is related to the state s' at the next moment.

For simplicity, we consider the interactions between *agent* and *environment* as a discrete time-series in this paper. Figure 4 shows the interaction between an *agent* and an *environment*.

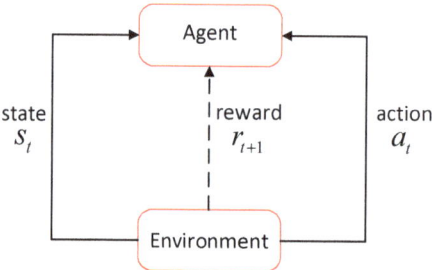

Figure 4. Agent–environment interaction.

4.2.1. Input and Output Design

The pointer network input under reinforcement learning is similar to that under supervised learning. The only difference is that, when applying reinforcement learning, a special symbol $Split$ needs to be added, as reinforcement learning only focuses on those variables selected before the variable $Split$. $Split$ is a separator that divides a variable into two types. We use the following rules: when inputting into the pointer network, all variables before $Split$ are set to 1, and all variables after $Split$ are set to 0. We use the zero vector to represent the $Split$. Therefore, in order to change the n-dimensional matrix Q into $n + 1$ dimensions, we add a row and a column of 0 to the last row and the last column of matrix Q. Under this rule, taking Example 1 as an example, to convert the matrix Q into the matrix P, the input sequence of the pointer network is $(p_1, p_2, p_3, p_4, Split)$:

$$P = \begin{pmatrix} 0 & 3 & 0 & 4 & 0 \\ 3 & 0 & 5 & 2 & 0 \\ 0 & 5 & 0 & 1 & 0 \\ 4 & 2 & 1 & 0 & 0 \\ 0 & 0 & 0 & 0 & 0 \end{pmatrix}.$$

Similar to supervised learning, at the output of the pointer network, a symbol EOS is added to divide the set of output vertices. As in Example 1, the output of the pointer network is (1, 3, EOS, 2, 4), which means that the four vertices are divided into two sets, which are (1, 3) and (2, 4). The numbers in front of EOS indicate that the value at these vertex positions is 1, and the numbers after EOS indicate that the value at these positions is 0. Thus, the max-cut value can be calculated according to the divided sets, and it is this value that is used as the reward in reinforcement learning.

4.2.2. Actor–Critic Algorithm

The actor–critic algorithm is a reinforcement learning method which combines a policy gradient and temporal difference learning. We combine the input–output structure characteristics of the Max-cut problem with the actor–critic algorithm in reinforcement learning to train the pointer network model. The actor–critic algorithm used for solving such combinatorial optimization problems uses the same pointer network encoder for both the actor network and the critic network. First, the actor network encodes the input sequence. Next, the decoder part selects the variable with value 1, according to the probability. The critic network encodes the input sequence, then predicts the optimal value of the Max-cut problem using a value function.

In the actor–critic algorithm, $\phi(s)$ is the input to the actor network, which corresponds to the given symmetric matrix Q in the Max-cut problem; that is, Q is used as the input sequence of the actor network. The actor refers to the policy function $\pi_\theta(s, a)$, which can learn a strategy to obtain the highest possible reward. For the Max-cut problem, $\pi_\theta(s, a)$ represents the strategy scheme in which variables are selected as 1. The critic refers to the value function $V_\phi(s)$, which estimates the value function of the current strategy. With the help of the value function, the actor–critic algorithm can update the parameters in a single step, without having to wait until the end of the round to update. In the actor–critic algorithm, the policy function $\pi_\theta(s, a)$ and the value function $V_\phi(s)$ are both functions that need to be learned simultaneously during the training process.

Assuming the return $G(\tau_{t:T})$ from time t, we use Equation (21) to approximate it:

$$\hat{G}(\tau_{t:T}) = r_{t+1} + \gamma V_\phi(s_{t+1}), \tag{21}$$

where s_{t+1} is the state at $t+1$ and r_{t+1} is the instant reward.

In each step of the update, the strategy function $\pi_\theta(s, a)$ and the value function $V_\phi(s)$ are learned. On one hand, the parameter ϕ is updated, such that the value function $V_\phi(s_t)$ is close to the estimated real return $\hat{G}(\tau_{t:T})$:

$$\min_\phi \left(\hat{G}(\tau_{t:T}) - V_\phi(s_t)\right)^2. \tag{22}$$

On the other hand, the value function $V_\phi(s_t)$ is used as a basis function to update the parameter, in order to reduce the variance of the policy gradient:

$$\theta \leftarrow \theta + \alpha \gamma^t \left(\hat{G}(\tau_{t:T}) - V_\phi(s_t)\right) \frac{\partial}{\partial \theta} \log \pi_\theta(a_t|s_t). \tag{23}$$

In each update step, the actor performs an action a, according to the current environment state s and the strategy $\pi_\theta(a|s)$; the environment state becomes s' and the actor obtains an instant reward r. The critic (value function $V_\phi(s)$) adjusts its own scoring standard, according to the real reward given by the environment and the previous score $(r + \gamma V_\phi(s'))$, such that its own score is closer to the real return of the environment. The actor adjusts its strategy π_θ according to the critic's score, and strives to do better next time. At the beginning of the training, actors performs randomly and critic gives random marks. Through continuous learning, the critic's ratings become more and more accurate, and the actor's movements become better and better.

Algorithm 3 shows the training process of the actor–critic algorithm.

Algorithm 3 Actor–critic algorithm

Input: state space: S; action space: A; differentiable strategy function: $\pi_\theta(a|s)$; differentiable state value function: $V_\phi(s)$; discount rate: γ; learning rate: $\alpha > 0, \beta > 0$;

Output: strategy: π_θ

1: random initial θ, ϕ;
2: **repeat**
3:     initial starting state s;
4:     $\lambda = 1$;
5:     **repeat**
6:         In state s, select an action $a = \pi_\theta(a|s)$;
7:         perform the action a to get an instant reward r and a new state s';
8:         $\delta \leftarrow r + \gamma V_\phi(s') - V_\phi(s)$;
9:         $\phi \leftarrow \phi + \beta \delta \frac{\partial}{\partial \phi} V_\phi(s)$;
10:        $\theta \leftarrow \theta + \alpha \lambda \delta \frac{\partial}{\partial \theta} \log \pi_\theta(a|s)$;
11:        $\lambda \leftarrow \gamma \lambda$;
12:        $s \leftarrow s'$;
13:     **until** s is the termination state;
14: **until** θ converges.

---

5. Experimental Results and Analysis

Based on the TensorFlow framework, this paper uses two learning strategies (supervised learning and reinforcement learning) to train and predict the Max-cut problem with a pointer network. The model is trained on a deep learning server platform consisting of two NVIDIA TITAN Xp GPUs and an Intel Core i9-7960X CPU.

The initial parameters in the pointer network are randomly generated by a uniform distribution in $[-0.08, 0.08]$, and the initial learning rate is 0.001. During the training process, when supervised learning is applied to train the pointer network, the model uses a single-layer LSTM with 256 hidden units and is trained with mini-batch SGD. When applying reinforcement learning to train the pointer network, the model uses three layers of LSTMs, with each layer consisting of 128 hidden units, and is trained with the actor–critic algorithm. In the prediction stage, the heat parameter in the pointer network is set to 3, and the initial reward baseline is set to 100. The model tested during the prediction phase is the last iteration of the training phase. For the Max-cut problem of different dimensions, except for increasing the sequence length, the other hyperparameter settings are the same. In the implementation, we use the Adam algorithm to adjust the learning rate. Adam algorithm can make effective dynamic adjustments to the model to make the changes in hyperparameters relatively stable [24].

We constructed a data set based on the method mentioned in Section 2.2 (using the $\{-1,1\}$ quadratic programming problem transformed into the Max-cut problem), which we refer to as the Zhou data set. In order not to lose generality, we also used the Binary quadratic and Max cut Libraries

(Biq Mac Library), which are the most commonly used benchmark generators for the Max-cut problem. We performed experiments on the Zhou data set and the Biq Mac Library data set, respectively.

5.1. Experiments on Zhou Data Set

According to the method for randomly generating Max-cut problems in Section 2.2, a large number of Max-cut problems with known exact solutions were obtained, which we formed into data sets with specified input and output formats. The training set and the test set are both data generated from the same probability distribution, and the density of the input matrix Q in the data sample is 94.6%. Each sample in the training and test sets is unique. Then, we divided the data sets randomly into training and test sets according to the ratio of 10:1. For different dimensions, the training set contained 1000 samples and the test set contained 100 samples. The maximum number of training iterations was set to 100,000. The accuracy of the solution trained by the model is defined as:

$$\text{Accuracy} = \frac{v\,(\text{Ptr-Net})}{v\,(\text{Opt})} \times 100\%, \tag{24}$$

where $v\,(\text{Ptr-Net})$ is the solution of trained pointer network model, and $v\,(\text{Opt})$ is the optimal value of the Max-cut problem.

We first used supervised learning to train the pointer network on the 10-, 30-, 50-, 60-, 70-, 80-, 90-, 100-, 150-, and 200-dimensional Max-cut problems, respectively. Table 1 shows average accuracy of the Max-cut problem of the above dimensions. And the detailed experimental results are listed in Table 2.

Table 1. Average accuracies and training times for Max-cut problems with different dimensions by SL.

Dimensions	Average Accuracy	Average Training Time
10	100%	1:13:20
30	98.78%	2:12:35
50	97.56%	5:29:16
60	94.51%	6:37:10
70	90.95%	7:33:42
80	88.64%	8:39:08
90	86.35%	9:58:06
100	80.50%	11:14:57
150	74.94%	14:52:46
200	71.95%	19:28:25

Table 2. Detailed solutions and accuracies for Max-cut problems with different dimensions by supervised learning (SL).

Sample	Optimum	Solution	Accuracy	Sample	Optimum	Solution	Accuracy
S10.1	330	330	100%	S80.1	85,462	74,634	87.33%
S10.2	281	281	100%	S80.2	94,552	83,357	88.16%
S10.3	240	240	100%	S80.3	100,512	82,782	82.36%
S10.4	236	236	100%	S80.4	92,108	83,735	90.91%
S10.5	171	171	100%	S80.5	89,311	79,299	88.79%
S10.6	124	124	100%	S80.6	100,862	97,624	97.79%
S10.7	208	208	100%	S80.7	88,919	81,996	92.21%
S10.8	230	230	100%	S80.8	91,045	79,974	87.84%
S10.9	245	245	100%	S80.9	87,873	75,870	86.34%
S10.10	257	257	100%	S80.10	111,327	95,374	85.67%
S30.1	4861	4861	100%	S90.1	136,959	114,251	83.42%
S30.2	5820	5698	97.90%	S90.2	134,022	124,033	92.55%
S30.3	4708	4617	98.07%	S90.3	145,727	115,448	79.22%
S30.4	6123	6123	100%	S90.4	132,287	117,391	88.74%
S30.5	6033	6008	99.59%	S90.5	134,420	126,341	93.99%
S30.6	5380	5342	99.29%	S90.6	133,817	116,491	87.05%
S30.7	6927	6799	98.15%	S90.7	142,957	123,292	86.24%
S30.8	4914	4741	96.48%	S90.8	120,026	104,207	86.82%
S30.9	6401	6340	99.05%	S90.9	145,635	120,749	82.91%
S30.10	5185	5147	99.27%	S90.10	141,741	117,019	82.56%
S50.1	24,468	23,386	95.58%	S100.1	174,947	144,963	82.86%
S50.2	22,462	21,646	96.37%	S100.2	199,441	181,966	91.24%
S50.3	23,246	23,246	100%	S100.3	166,682	130,995	78.59%
S50.4	19,776	19,273	97.46%	S100.4	179,885	146,426	81.40%
S50.5	25,057	23,947	95.57%	S100.5	184,363	146,653	79.55%
S50.6	27,510	27,037	98.28%	S100.6	191,636	165,283	86.25%
S50.7	26,698	26,368	98.76%	S100.7	189,959	136,784	72.01%
S50.8	20,627	20,261	98.23%	S100.8	177,545	144,373	81.32%
S50.9	20,493	20,213	98.63%	S100.9	181,022	145,642	80.46%
S50.10	22,130	21,404	96.72%	S100.10	189,239	134,965	71.32%
S60.1	43,173	40,062	92.79%	S150.1	542,081	453,348	83.63%
S60.2	42,057	39,280	93.40%	S150.2	571,793	390,333	68.26%
S60.3	43,190	40,360	93.45%	S150.3	678,393	551,327	81.27%
S60.4	54,174	53,138	98.09%	S150.4	574,523	481,574	83.82%
S60.5	43,638	40,180	92.08%	S150.5	545,008	412,718	75.73%
S60.6	38,255	37,333	97.59%	S150.6	613,130	467,820	76.30%
S60.7	52,689	49,260	93.49%	S150.7	545,500	354,314	64.95%
S60.8	43,902	40,741	92.80%	S150.8	632,578	521,155	82.39%
S60.9	39,098	37,980	97.14%	S150.9	612,560	349,406	57.04%
S60.10	41,005	38,655	94.27%	S150.10	630,733	479,420	76.01%
S70.1	64,914	59,371	91.46%	S200.1	1,444,264	1,080,545	74.82%
S70.2	63,306	62,872	99.31%	S200.2	1,488,701	1,006,851	67.63%
S70.3	71,127	62,855	88.37%	S200.3	1,368,359	1,052,517	76.92%
S70.4	65,673	57,286	87.23%	S200.4	1,352,301	1,102,923	81.56%
S70.5	59,045	54,864	92.92%	S200.5	1,309,815	1,152,570	87.99%
S70.6	60,016	50,337	83.87%	S200.6	1,338,423	928,162	69.35%
S70.7	63,158	57,117	90.44%	S200.7	1,311,058	805,257	61.42%
S70.8	63,478	59,211	93.28%	S200.8	1,462,304	822,691	56.26%
S70.9	67,019	60,871	90.83%	S200.9	1,350,077	1,114,423	82.55%
S70.10	70,616	64,818	91.79%	S200.10	1,347,381	822,037	61.01%

Then the pointer network based on reinforcement learning was also trained with the Zhou data set, on 10-, 50-, 150-, 200-, 250-, and 300-dimensional Max-cut problems. Table 3 shows the average accuracy of the Max-cut problem for the above dimensions. And the detailed experimental results are listed in Table 4.

Table 3. Average accuracies and training times for Max-cut problems with different dimensions by RL.

Dimensions	Average Accuracy	Average Training Time
10	100%	0:10:07
50	98.28%	0:23:03
100	96.32%	0:42:33
150	95.06%	1:03:57
200	92.38%	1:27:18
250	89.88%	1:53:28
300	87.64%	2:21:30

Table 4. Detailed solutions and accuracies for Max-cut problems with different dimensions by reinforcement learning (RL).

Sample	Optimum	Solution	Accuracy	Sample	Optimum	Solution	Accuracy
R10.1	233	233	100%	R150.6	584,968	572,689	97.90%
R10.2	248	248	100%	R150.7	553,878	453,361	81.86%
R10.3	193	193	100%	R150.8	618,615	583,545	94.33%
R10.4	192	192	100%	R150.9	529,739	522,427	98.62%
R10.5	302	302	100%	R150.10	559,414	513,463	91.79%
R10.6	187	187	100%	R200.1	1,274,866	1,267,884	99.45%
R10.7	341	341	100%	R200.2	1,392,200	1,165,174	83.69%
R10.8	133	133	100%	R200.3	1,358,320	1,345,870	99.07%
R10.9	301	301	100%	R200.4	1,320,006	1,118,705	84.75%
R10.10	272	272	100%	R200.5	1,368,199	1,020,056	74.55%
R50.1	25,565	25,302	98.97%	R200.6	1,397,432	1,292,628	92.50%
R50.2	22,528	22,441	99.61%	R200.7	1,421,061	1,420,172	99.94%
R50.3	25,426	24,783	97.47%	R200.8	1,376,229	1,357,875	98.67%
R50.4	25,787	25,425	98.60%	R200.9	1,344,436	1,266,442	94.20%
R50.5	21,030	19,755	93.94%	R200.10	1,388,152	1,332,225	95.97%
R50.6	25,079	24,614	98.15%	R250.1	2,590,918	2,242,263	86.54%
R50.7	22,077	21,820	98.84%	R250.2	2,700,294	2,503,768	92.72%
R50.8	28,899	28,715	99.36%	R250.3	2,542,443	2,230,460	87.73%
R50.9	29,101	28,614	98.33%	R250.4	2,542,413	2,357,060	92.71%
R50.10	25,729	25,607	99.53%	R250.5	2,702,833	2,547,463	94.25%
R100.1	187,805	182,369	97.02%	R250.6	2,764,901	2,750,197	99.47%
R100.2	197,470	193,171	97.82%	R250.7	2,777,948	2,027,296	72.98%
R100.3	187,495	185,929	99.16%	R250.8	2,671,835	2,530,103	94.70%
R100.4	216,339	211,431	97.73%	R250.9	2,593,131	2,031,284	78.33%
R100.5	161,961	161,067	99.45%	R250.10	2,596,843	2,579,798	99.34%
R100.6	178,737	176,723	98.87%	R300.1	4,430,263	3,858,903	87.10%
R100.7	183,560	183,315	99.87%	R300.2	4,363,482	2,344,354	53.73%
R100.8	155,038	154,292	99.52%	R300.3	4,459,682	4,248,761	95.27%
R100.9	191,120	142,059	74.33%	R300.4	4,562,319	2,369,015	51.93%
R100.10	174,202	173,729	99.73%	R300.5	4,404,895	4,113,113	93.38%
R150.1	568,452	554,128	97.48%	R300.6	4,497,912	4,483,644	99.68%
R150.2	549,303	542,731	98.80%	R300.7	4,364,640	4,298,760	98.49%
R150.3	672,601	628,655	93.47%	R300.8	4,589,744	4,372,106	95.26%
R150.4	590,417	553,860	93.81%	R300.9	4,655,631	4,652,613	99.94%
R150.5	563,674	561,554	99.62%	R300.10	4,956,332	4,944,887	99.77%

It can be seen, from Tables 1 and 3 that, regardless of whether supervised learning or reinforcement learning was used, the average accuracy of the pointer network solution decreased as the number of dimensions increased. However, the average accuracy of reinforcement learning decreased very slightly. Secondly, by comparing the two tables, we find that the pointer network model obtained through reinforcement learning was more accurate than that obtained by supervised learning. Finally, it can be seen that the time taken to train the model with reinforcement learning was faster than that for supervised learning. Figure 5 shows the accuracy of the solution for the Max-cut problem samples trained with supervised learning and reinforcement learning.

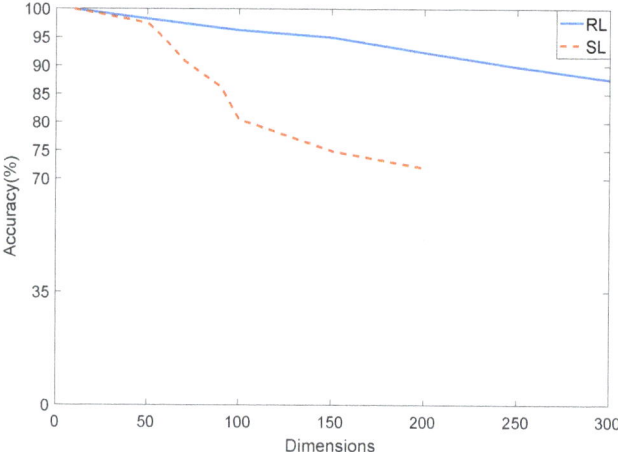

Figure 5. Average accuracies of SL and RL.

5.2. Experiments on Biq Mac Library

In order to further verify the generalization ability of the pointer network model, ten groups of 60-, 80-, and 100-dimensional Max-cut samples were selected from the Biq Mac Library (http://biqmac.uni-klu.ac.at/biqmaclib.html). As the Max-cut problem data set of each dimension in the Biq Mac Library only has ten groups of data, the amount of data was not enough to train the pointer network (training the pointer network model requires at least 100 groups of data), so we only used the Biq Mac Library as the test set; the Zhou data set was still used as the training set.

Table 5 shows the detailed experimental results of the Max-cut problem with 60, 80, and 100 dimensions using the Biq Mac Library by reinforcement learning.

Table 5. Solution and accuracy on Biq Mac Library data set by RL.

Sample	Optimum	Solution	Accuracy	Sample	Optimum	Solution	Accuracy
R60.1	536	441	82.28%	R80.6	926	817	88.23%
R60.2	532	478	89.85%	R80.7	929	773	83.21%
R60.3	529	463	87.52%	R80.8	929	785	84.50%
R60.4	538	478	88.85%	R80.9	925	830	89.73%
R60.5	527	486	92.22%	R80.10	923	640	69.34%
R60.6	533	479	89.87%	R100.1	2019	1530	75.78%
R60.7	531	438	82.49%	R100.2	2060	1507	73.16%
R60.8	535	473	88.41%	R100.3	2032	1461	71.90%
R60.9	530	468	88.30%	R100.4	2067	1573	76.10%
R60.10	533	483	90.62%	R100.5	2039	1433	70.28%
R80.1	929	829	89.24%	R100.6	2108	1483	70.35%
R80.2	941	753	80.02%	R100.7	2032	1464	72.04%
R80.3	934	824	88.22%	R100.8	2074	1585	76.42%
R80.4	923	819	88.73%	R100.9	2022	1477	73.05%
R80.5	932	805	86.37%	R100.10	2005	1446	72.20%

The average prediction results are shown in Table 6.

Table 6. Average accuracies of different dimensional Max-cut problems using the Biq Mac Library by RL.

Dimensions	Average Accuracy
60	88.05%
80	84.76%
100	73.09%

It can be seen, from Tables 3 and 6, that the average accuracies when predicting the Biq Mac Library using reinforcement learning were lower than the accuracies on Zhou dataset. This is because the Biq Mac Library is composed of data samples with different distributions, which can better characterize the essential characteristics of the Max-cut problem. We believe that, in future research, if the model can be trained on a larger training set with the distribution of the Biq Mac Library, its performance can be definitely improved.

6. Conclusions

In this paper, we proposed an effective deep learning method based on a pointer network for the Max-cut problem. We first analyzed the structural characteristics of the Max-cut problem and introduced a method to generate a large data set of Max-cut problems. Then, the algorithmic frameworks for training the pointer network model under two learning strategies (supervised learning and reinforcement learning) were introduced in detail. We applied supervised learning and reinforcement learning strategies separately to train the pointer network model, and experimented with Max-cut problems with different dimensions. The experimental results revealed that, for the low-dimensional Max-cut problem (below 50 dimensions), the models trained by supervised learning and reinforcement learning both have high accuracy and that the accuracies are basically consistent. For high-dimensional cases (above 50 dimensions), the accuracy of the solution in the training mode using reinforcement learning was significantly better than that with supervised learning. This illustrates that reinforcement learning can better discover the essential characteristics behind the Max-cut problem and can mine better optimal solutions from the data. This important finding will instruct us to further improve the performance and potential of pointer networks as a deep learning method for Max-cut problems and other combinatorial optimization problems in future research.

Author Contributions: S.G. put forward the idea and algorithms. S.G. investigated and supervised the project. Y.Y. and S.G. simulated the results. Y.Y. validated and summarized the results in tables. S.G. and Y.Y. prepared and wrote the article. All authors have read and agreed to the published version of the manuscript.

Funding: The work described in the paper was supported by the National Science Foundation of China under Grant 61876105.

Conflicts of Interest: The authors declare no conflict of interest.

Abbreviations

The following abbreviations are used in this manuscript:

NP-hard	Non-deterministic Polynomial-hard
NP	Non-deterministic Polynomial
VLSI	Very Large Scale Integration
LKH	Lin–Kernighan Heuristic
CAP	Credit Assignment Problem
RNN	Recurrent Neural Network
UBQP	Unconstrained Boolean Quadratic Programming
BQP	Boolean Quadratic Programming
SDP	Semi-Definite Programming
LSTM	Long Short-Term Memory
SL	Supervised Learning
RL	Reinforcement Learning
SGD	Stochastic Gradient Descent

Variables

The following variables are used in this manuscript:

D	Training set
E	Edge set, $E \subseteq V \times V$
G	Undirected graph of the Max-cut problem, $G = (V, E)$
$G(\tau_{t:T})$	The return from time t in actor–critic
K	The mini-batch size
L	Number of training steps
M	Network parameter in LSTM, $M \in \mathbb{R}^{4d \times (d+e)}$
O_{label}	Label matrix of the output
$O_{predict}$	The probability distribution of the output matrix
P	The transformed reinforcement learning input matrix
Q	Adjacency matrix, $Q = Q^T = (q_{ij})_{n \times n}$ and $q_{ij}(i=j)$ are zero
S	Subset of vertex set
U	Learnable parameter in attention mechanism
V	Vertex set, $V = \{1, 2, ..., n\}$
V_ϕ	Value function of actor–critic algorithm
W	Parameter matrix to be updated in mini-batch SGD
X	Input sequence, $X = x_{1:T}$
Y	Output sequence, $Y = y_{1:T}$
a	Action in agent–environment interaction
b	Network parameter in LSTM, $b \in \mathbb{R}^{4d}$
c	Non-zero vector, $c \in \mathbb{R}^n$
c_t	Memory unit for the current moment
f_t	Forget gate for the current moment
$f(x)$	$f(x) = \frac{1}{2} x^T Q x - c^T x$
h	Hidden layer
i_t	Input gate for the current moment
n	Dimensions of the Max-cut problem
o_t	Output gate for the current moment
p	Conditional probability
r	Reward of agent-environment interaction
s	State of agent-environment interaction
$s_{i,j}$	Non-normalized attention distribution of each input vector
v	Learnable parameter in attention mechanism
$w_{i,j}$	The weight on the edge connecting vertex i and vertex j
α	Learning rate in mini-batch SGD
β	Learning rate in actor–critic algorithm
γ	Discount rate in actor–critic algorithm
θ	Parameters to be updated in strategy function
λ	Lagrange multiplier, $\lambda \in \mathbb{R}^n$
π_θ	Strategy function of actor–critic algorithm
σ	Logistic function in LSTM
ϕ	Parameters to be updated in value function

References

1. Goemans, M.X. Improved approximation algorithms for maximum cut and satisfiability problems using semidefinite programming. *J. ACM* **1995**, *42*, 1115–1145. [CrossRef]
2. Mcmahan, H.B.; Holt, G.; Sculley, D.; Young, M.; Kubica, J. Ad click prediction: A view from the trenches. In Proceedings of the 19th ACM SIGKDD international conference on Knowledge discovery and data mining, Chicago, IL, USA, 14 August 2013.
3. Barahona, F.; Grötschel, M.; Junger, M.; Reinelt, G. An application of combinatorial optimization to statistical physics and circuit layout design. *Oper. Res.* **1988**, *36*, 493–513. [CrossRef]
4. Xi, Y.J.; Dang, Y.Z. The method to analyze the robustness of knowledge network based on the weighted supernetwork model and its application. *Syst. Eng. Theory Pract.* **2007**, *27*, 134–140. [CrossRef]
5. Dreiseitla, S.; Ohno-Machadob, L. Logistic regression and artificial neural network classification models: A methodology review. *J. Biomed. Inf.* **2002**, *35*, 352–359. [CrossRef]
6. Croce, F.D.; Kaminski, M.J.; Paschos, V.T. An exact algorithm for max-cut in sparse graphs. *Oper. Res. Lett.* **2007**, *35*, 403–408. [CrossRef]
7. Krishnan, K.; Mitchell, J.E. A semidefinite programming based polyhedral cut and price approach for the maxcut problem. *Comput. Optim. Appl.* **2006**, *33*, 51–71. [CrossRef]
8. Funabiki, N.; Kitamichi, J.; Nishikawa, S. An evolutionary neural network algorithm for max cut problems. In Proceedings of the International Conference on Neural Networks (ICNN'97), Houston, TX, USA, 12 June 1997.
9. Denrell, J.; Fang, C.; Levinthal, D.A. From t-mazes to labyrinths: Learning from model-based feedback. *Manag. Sci.* **2004**, *50*, 1366–1378. [CrossRef]
10. Vinyals, O.; Fortunato, M.; Jaitly, N. Pointer networks. In *Advances in Neural Information Processing Systems*; MIT Press: Cambridge, MA, USA, 2014; pp. 2692–2700.
11. Bello, I.; Pham, H.; Le, Q.V.; Norouzi, M.; Bengio, S. Neural combinatorial optimization with reinforcement learning. *arXiv* **2016**, arXiv:1611.09940.
12. Mirhoseini, A.; Pham, H.; Le, Q.V.; Steiner, B.; Larsen, R.; Zhou, Y. Device placement optimization with reinforcement learning. In Proceedings of the 34th International Conference on Machine Learning, Sydney, Australia, 11 August 2017; pp. 2430–2439.
13. Milan, A.; Rezatofighi, S.H.; Garg, R.; Dick, A.; Reid, I. Data-Driven Approximations to NP-Hard Problems. In Proceedings of the Thirty-First AAAI Conference on Artificial Intelligence, San Francisco, CA, USA, 4 February 2017; pp. 1453–1459.
14. Mottini, A.; Acuna-Agost, R. Deep Choice Model Using Pointer Networks for Airline Itinerary Prediction. In Proceedings of the 23rd ACM SIGKDD International Conference, Halifax, NS, Canada, 13–17 August 2017; pp. 1575–1583.
15. Bahdanau, D.; Cho, K.; Bengio, Y. Neural machine translation by jointly learning to align and translate. In Proceedings of the 3rd International Conference on Learning Representations (ICLR), San Diego, CA, USA, 7–9 May 2015.
16. Gu, S.; Yang, Y. A Pointer Network Based Deep Learning Algorithm for the Max-Cut Problem. *ICONIP* **2018**, *LNCS 11301*, 238–248.
17. Gu, S.; Hao, T.; Yao, H. A Pointer Network Based Deep Learning Algorithm for Unconstrained Binary Quadratic Programming Problem. *Neurocomputing* **2020**, (accepted).
18. Zhou, M.X. A benchmark generator for boolean quadratic programming. *Comput. Sci.* **2015**, arXiv:1406.4812.
19. Barahona, F.; Michael, J.; Reinelt, G. Experiments in quadratic 0-1 programming. *Math. Program.* **1989**, *44*, 127–137. [CrossRef]
20. Sutskever, I.; Vinyals, O.; Le, Q.V. Sequence to Sequence Learning with Neural Networks. In *Advances in Neural Information Processing Systems*; MIT Press: Cambridge, MA, USA, 2014; pp. 3104–3112.
21. Jing, L.; Shen, Y.; Dubček, T.; Peurifoy, J.; Skirlo, S.; Lecun, Y.; Tegmark, M. Tunable efficient unitary neural networks (eunn) and their application to rnns. In Proceedings of the 34th International Conference on Machine Learning, Sydney, Australia, 6 August 2016.
22. Hochreiter, S.; Schmidhuber, J. Long short-term memory. *Neural Comput.* **1997**, *9*, 1735–1780. [CrossRef] [PubMed]

23. Konecny, J.; Liu, J.; Richtarik, P.; Takac, M. Mini-batch semi-stochastic gradient descent in the proximal setting. *IEEE J. Sel. Top. Signal Process.* **2016**, *10*, 242–255. [CrossRef]
24. Kingma, D.P.; Ba, J. Adam: A Method for Stochastic Optimization. In Proceedings of the 3rd International Conference on Learning Representations (ICLR), San Diego, CA, USA, 7–9 May 2015.

© 2020 by the authors. Licensee MDPI, Basel, Switzerland. This article is an open access article distributed under the terms and conditions of the Creative Commons Attribution (CC BY) license (http://creativecommons.org/licenses/by/4.0/).

Article

A New Formulation for the Capacitated Lot Sizing Problem with Batch Ordering Allowing Shortages

Yajaira Cardona-Valdés [1,†]**, Samuel Nucamendi-Guillén** [2,*,†]**, Rodrigo E. Peimbert-García** [3,†,‡]**, Gustavo Macedo-Barragán** [2,†] **and Eduardo Díaz-Medina** [2,†]

1. Centro de Investigación en Matemáticas Aplicadas, Universidad Autónoma de Coahuila, Blvd. V. Carranza s/n. Col. República Oriente, Saltillo C.P. 25280, Coahuila, Mexico; y.cardona@uadec.edu.mx
2. Facultad de Ingeniería, Universidad Panamericana, Álvaro del Portillo 49, Zapopan, Jalisco 15010, Mexico, gustavo.macedo@up.edu.mx (G.M.-B.); ediazm@up.edu.mx (E.D.-M.)
3. School of Engineering and Sciences, Tecnológico de Monterrey, Eugenio Garza Sada 2501 Sur, Monterrey, Nuevo León 64849, Mexico; rodrigo.peimbert@tec.mx
* Correspondence: snucamendi@up.edu.mx
† These authors contributed equally to this work.
‡ Current address: School of Engineering, Macquarie University. Balaclava Road, North Ryde, Sydney 2109, Australia.

Received: 23 April 2020; Accepted: 26 May 2020; Published: 1 June 2020

Abstract: This paper addresses the multi-product, multi-period capacitated lot sizing problem. In particular, this work determines the optimal lot size allowing for shortages (imposed by budget restrictions), but with a penalty cost. The developed models are well suited to the usually rather inflexible production resources found in retail industries. Two models are proposed based on mixed-integer formulations: (i) one that allows shortage and (ii) one that forces fulfilling the demand. Both models are implemented over test instances and a case study of a real industry. By investigating the properties of the obtained solutions, we can determine whether the shortage allowance will benefit the company. The experimental results indicate that, for the test instances, the fact of allowing shortages produces savings up to 17% in comparison with the model without shortages, whereas concerning the current situation of the company, these savings represent 33% of the total costs while preserving the revenue.

Keywords: capacitated lot sizing; mixed integer formulation; retail; inventory; shortages

1. Introduction

Today's companies are immersed in a globalized market crammed with considerable demands for different products. Regardless of the size or type of industry, the selling business always involves two parties: customers and sellers. Customers search for quality products that meet their requirements, whereas companies search for adaptive strategies to compete against each other and satisfy the preferences of their individual customers while minimizing any negative impact on their operations and other aspects of their businesses [1].

One of the most integral operations of a company is inventory management, because many resources are expended on goods and products that will generate profits for the company. Yet, the purchased materials must be ordered at an economic price and then stored. On the other hand, companies that sell products in large quantities or with high variability in their demand commonly face the overstock problem, wherein the optimum stocking level is exceeded, and the stored products cannot be sold beyond an estimated time. This is the problem faced by a Mexican textile and footwear distributor that inspired this research. This company handles several products in catalogues and sells them to wholesale and retail customers. Currently, the company manually gathers the data that allow the analysts and sales force to understand the products' behaviours and, consequently, plan the most appropriate purchase strategy. This data collection process is merely supported by the empirical sui generis methods of older, more experienced workers. Coupled with this problem, suppliers place product-dependent batch orders of different sizes with the company. The delivery time of each item is established by the corresponding supplier and hinders the empirical purchasing planning conducted by the company. Thus, the present work develops a couple of Mixed-Integer Linear Programming (MILP) models for inventory management. The two MILP models involve multiple items and multiple periods, including different capacity constraints related to the batch size, lead time, and budget. These models seek to improve the timing and quantity of the ordering scheme, thereby reducing the operating expenses of the company. The effectiveness of these formulations is evaluated through the case study. The remaining sections of this paper are organized as follows. Section 2 reviews the relevant literature, and Section 3 introduces the proposed MILP formulations. In Section 4, the proposed models are implemented and evaluated in the case study of a Mexican fashion retail company. Section 5 presents the conclusions and some directions for future research.

2. Literature Review

Interest in streamlining inventory processes has recently increased, particularly in tasks related to classifying products, registration methods, and re-inventory models. According to De Horatius [2], inventory management focuses on minimizing the inventory levels while ensuring stock availability. Inventory planning then is a fundamental part of the fashion retail industry, which covers the business of fashion products such as apparel, shoes, and beauty products. This fashion market is very challenging as it is characterized by short life cycles compared to other retail and service industries, high volatility, poor predictability, and critical-mass shopping behaviour [1]. As Liu et al. [3] stated, retailers' optimal management of their inventories largely depends on the accuracy of predicting future demand, which is commonly affected by uncertainty factors that are difficult to manage, such as trends, seasonality, and availability.

The history of inventory problems is rooted in the Economic Order Quantity (EOQ) model, which assumes a single item in a continuous and constant demand over an infinite planning horizon. This EOQ model is easily solved to determine the optimal order quantity, balancing the setups, and computing the inventory holding costs. However, the model becomes NP-hard when more items are considered and multiple capacity restrictions are in place [4]. Thus, this problem has been extended to considering multiple items under the imposition of different cost conditions and limited production capacities, which in turn have culminated in the known Capacitated Lot Sizing Problem (CLSP). Furthermore, the combination of these considerations, along with additional features such as the demand uncertainty, setup costs and/or time, and alternative suppliers have introduced different complexities to address the problem [5]. Hence, every firm must decide the lot sizing problem with the specific considerations that are better suited to describe its current circumstances and that will directly affect its system operation, productivity, and overall performance.

The general CLSP problem seeks to optimize the order quantity that will meet the customer demand, over a finite time horizon, while minimizing the sum of ordering, purchasing, and holding inventory costs. Although constraints in the formulation of this problem might limit the production, space, or budget capacity for any given product, the basic mathematical model is easily adapted by appropriately interpreting the model elements, such as the variables and parameters [5].

The lot sizing problem has been studied since the early Twentieth Century, and its application in real scenarios currently constitute an active area of research [6–8]. Moreover, this problem has split into a wide diversity of variants to the problem, which in turn have been solved by an even wider variety of approaches. In this regard, different reviews summarizing the literature available have been published since the mid 1990s [9–16]. Overall, this literature can be classified into three groups: (1) studies considering space or monetary budgets; (2) studies on lead time; and (3) studies considering batch size.

First, among the studies considering space or monetary budgets, Ben-Daya and Raouf [17] developed an algorithm that optimizes the ordering policy in the multi-item, single-period inventory problem, within the context of perishable commodities. In this study, the algorithm determines the setup costs and lot size decisions when items compete for budgetary and floor- or shelf-space restrictions. Any shortage is translated into a penalty cost, and the excess of products (above demand) is disposed at a reduced price. Nonetheless, the policy is fixed during the planning horizon and does not consider the possibility of replenishment. A few years later, Guan and Li [18] presented two models of lot size, one with a restricted storage capacity and a second with a constraint on the order capacity. Meanwhile, a different study was developed by Fan and Wang [19], who modelled a lot size problem that sought to reduce ordering and storing expenses in a scenario where the size of the warehouse could be altered at a cost for each change in capacity. One more study was conducted by Woarawichai et al. [20], who developed an MILP for the multi-period inventory lot size problem with supplier selection under storage and budget constraints. The authors introduced a supplier-dependent transaction cost, but the order lead time was deterministic and assumed identical for all products, which limited the contribution of the model.

Second, among the studies on lead time, Karmarkar [21] related the lead time to lot size and concluded that extending the lead time negatively affected the response to customers. The author also argued that small batches created more setups while large batches lengthened the lead time. Some years later, Ben-Daya and Raouf [22] reviewed different inventory models that considered the lead time and order quantity as decision variables. However, most of the revised works assumed an invariant lead time over the entire planning horizon. For this reason, Kuik et al. [9] defined these models as finite capacity models. Meanwhile, Hariga and Ben-Daya [23] developed a continuous-review inventory model in which the reorder point, the ordering quantity, and the lead time were the decision variables, followed by Ouyang, Chen, and Chang [24], who studied a reorder-point inventory model with a variable lead time and partial backorders in an imperfect production process. These authors similarly defined the lead time as a decision variable to be minimized, in combination with other variables such as lot size, reorder point, process quality, and setup cost. A more recent study conducted by Helber and Sahling [25] solved a dynamic multilevel CLSP with positive lead times by a mathematical programming approach. Here, the authors developed an iterative procedure that fixed the products to be purchased in each period and then optimized the lot size of each purchased item. They developed four variants of this procedure with different decomposition structures to diversify the search in the solution space.

Finally, among studies considering batch size, more recent studies have been carried out. For example, Yang et al. [26] analyzed the optimal inventory management of a single product that is purchased in batches. This method accounts for the placement cost in a limited warehouse capacity. A different algorithm was later proposed by Akbalik et al. [27], who addressed a company that buys its products in batches and varies the storage cost when supply exceeds the determined capacity. At the same time, Farhat et al. [28] studied batch purchases with the option of returning the items that were not sold to the supplier. When the products purchased in batches are perishable, they must be treated with special care, and the model must consider a series of constraints, as proposed by Broekmeulen and van Donselaar [29].

As shown in this section, there is a vast literature addressing different variants of the lot sizing problem. Yet, literature focusing on optimizing the CLSP by considering variable batch sizes and variable delivery times (lead times) in retail applications is very scarce. In addition, many articles apply discounts when buying in large volumes, which is managed differently by the company under study.

3. Problem Statement

The problem analyzes a retailer who buys products in large quantities (lot sizes) from manufacturers and then sells them in smaller quantities (possibly single units) to consumers. The ordered quantities are merely considered as integer multiples of the batch size of each item. The retailer, should decide "what products to order", "how much to order for each product", and "when to order".

Unlike common lot sizing models in manufacturing environments, in our model, the retailers depend mainly on the lead time (LT) established by the suppliers. To ensure their timely receipt, purchase orders must be placed ahead of time (i.e., in advance). This anticipation is represented by an artificial shift (ST) created over the planning horizon, where the periods to be phased out are added before the first week of demand.

The following example attempts to show how the artificial shift over the time horizon periods will work. The instance considers three products to be purchased within a planning horizon of six ($r = 6$) periods $(1, \ldots, r)$. Item 1 belongs to imported products, i.e., is provided by an external supplier. The demand for each product in each period is displayed in Table 1.

Table 1. Demands per item and period for the original planning horizon.

	1			...		r
Item/Demand	1	2	3	4	5	6
1	110	99	164	181	95	164
2	197	120	146	101	152	131
3	174	132	137	142	119	183

The lead time (LT) of items 1, 2, and 3 are 3, 1, and 2 periods, respectively. The batch sizes of Items 1, 2, and 3 are 200, 200, and 150 units, respectively.

To create the shifted horizon (ST), we must determine the maximum number of periods that must be added to the horizon planning time. We assume that the maximum lead time ($LTmax$) is the maximum LT of the imported products ($LTmax = 3$ in this example). Therefore, the end of the time horizon is shifted from the sixth (r) to the ninth period ($s = r + LT_{max}$). Table 2 shows the corresponding demands over the ST shifted horizon in the illustrative example.

Table 2. Demand per item and per period for the shifted planing horizon.

	1			...		r		...	s
Item / Period	1	2	3	4	5	6	7	8	9
1	0	0	0	110	99	164	181	95	164
2	0	0	0	197	120	146	101	152	131
3	0	0	0	174	132	137	142	119	183

As shown in Table 2, the first three periods represent the artificial shift (with zero demand values). From the fourth to the ninth period, the demands correspond to those of the original planning horizon (Table 1).

Since the company has the policy of paying the orders in advance (to seize the discount prices), we formulated the model to take into account this feature. For this reason, we considered the shifted time horizon, in which the company can place the orders LT_i periods ahead (depending on the item) of the period in which they will be sold. To avoid dealing with negative indices, the first demand period is shifted to the $LT_{max} + 1$ period.

3.1. Mathematical Formulation

In this section, we formally define the addressed problem and propose an MILP formulation. Prior to that, let us introduce the following sets and parameters:

N : Set of n items to sell, indexed by i, where $i \in \{1,\ldots,n\}$.
M : Set of m imported items, the first m items from N, $M \subset N$, indexed by j, where $j \in \{1,\ldots,m\}$.
T : Set of r original time horizon periods, indexed by h, where $h \in \{1,\ldots,r\}$.
ST : Set of s shifted periods, indexed by t, where $t \in \{LT_{max}+1,\ldots,s\}$.
D_{it} : Demand for item i at period t.
TD_i : Total demand per item i, expressed as the sum of the periodical demands over the time horizon.
O_i : Setup order cost for item i.
P_i : Unitary purchase cost for item i.
H_i : Holding cost for item i.
W_{ini} : Initial purchase budget.
W_t : Periodical budget (after the initial purchase) for period t.
LT_i : Lead time of item i.
LT_{max} : Maximum lead time over imported items.
B_i : Batch size of item i (In number of units).
S_i : Penalty shortage cost for every unit of item i.

Two particular features must be considered in the proposed model. On the one hand, the company classifies the items into two categories: imported (purchased from external suppliers) and domestic (acquired from internal manufacturers). On the other hand, imported items share a common lead time that requires the placing of future orders. Therefore, for them, only an initial order must be placed. For this reason, imported items represent the first items in the N set.

3.2. MILP Formulation with Shortages

The problem consists of determining the purchase orders, i.e., the quantity of batches to be purchased per item over the planning period, in such a way that the total cost incurred by the holding, shortages, and final inventory costs are minimized. We also determined the maximum inventory levels in order to take advantage of the availability of space in the warehouse.

Let us introduce the following decision variables to establish the mathematical model:

$$X_{it} = \text{Quantity of batches of item } i \text{ to purchase at period } t.$$
$$I_{it} = \text{Quantity of units of item } i \text{ to hold in inventory at period } t.$$
$$Z_{it} = \text{Quantity of shortage units of item } i \text{ at period } t.$$
$$R_i = \text{Maximum allowed inventory of item } i \text{ in any period.}$$
$$Y_{it} = \begin{cases} 1, & \text{if purchase order } i \text{ is placed at period } t \\ 0, & \text{otherwise.} \end{cases}$$

The objective function (1) aims to minimize three terms: the holding cost, the shortage cost, and the final inventory costs. The third term is considered only over the last period stock per item.

$$\text{Minimize} \sum_{i=1}^{n}\sum_{t=1}^{s-1} H_i I_{it} + \sum_{i=1}^{n}\sum_{t=1}^{s} S_i Z_{it} + 2\sum_{i=1}^{n} P_i I_{is}. \quad (1)$$

The balance equation for item i in period t (equals the sum of the inventory over the previous period and the purchase in the current period, minus the remaining surplus in the current period) is established by Equation (2).

$$I_{i(t-1)} + B_i X_{i(t-LT_i)} - D_{it} + Z_{it} = I_{it}, \quad \forall i \in N; t \in ST. \quad (2)$$

The maximum allowable inventory of each item in each period is ensured by Equation (3). Meanwhile, Equation (4) ensure that, for each item, no inventory is generated before the LT_{max} period.

$$I_{it} \leq R_i, \quad \forall i \in N, t \in ST, \quad (3)$$
$$I_{ih} = 0, \quad \forall i \in N, h \in T, h \leq LT_{max}. \quad (4)$$

For imported items, only one purchase order must be placed, and it should be placed at the beginning of the time horizon; this is ensured by Equation (5).

$$\sum_{t=LT_{max}+2}^{s} Y_{j(t-LT_i)} = 0, \quad \forall j \in M. \tag{5}$$

The amount of the initial purchase cannot exceed the budget assigned at the first original time period; this is ensured by Equation (6), while the total lot size based on the weekly budget is ensured by Equation (7).

$$\sum_{i=1}^{n} (B_i P_i X_{i(LT_{max}-LT_i+1)} + O_i Y_{i(LT_{max}-LT_i+1)}) \leq W_{ini}, \tag{6}$$

$$\sum_{i=m+1}^{n} (B_i P_i X_{i(t-LT_i)} + O_i Y_{i(t-LT_i)}) \leq W_t, \quad \forall t \in ST, t > LT_{max}+1. \tag{7}$$

Based on the batch size of each item, the maximum allowable quantity of purchased batches (in general) of item i is determined by Equation (8).

$$X_{i(t-LT_i)} \leq \left(1 + \frac{TD_i}{B_i}\right) Y_{i(t-LT_i)}, \quad \forall i \in N; t \in ST. \tag{8}$$

The units of shortage per item and per period is estimated by Equations (9) and (10), respectively.

$$Z_{it} \geq D_{it} - B_i X_{it} - I_{it}, \quad \forall i \in N; t \in ST, \tag{9}$$
$$Z_{it} \leq D_{it}, \quad \forall i \in N; t \in ST. \tag{10}$$

The nature of the variables is established by the group of Equations (11)–(15).

$$R_i \geq 0, \quad \forall i \in N, \tag{11}$$
$$I_{it} \geq 0, \quad \forall i \in N; t \in ST, \tag{12}$$
$$X_{it} \geq 0, \quad \forall i \in N; t \in ST, \tag{13}$$
$$Z_{it} \geq 0, \quad \forall i \in N; t \in ST, \tag{14}$$
$$Y_{it} \in \{0,1\}, \quad \forall i \in N; t \in ST. \tag{15}$$

This model will be referred to as the "AS model" in the computational experiments section.

Let us continue with the illustrative example of Section 3. The optimal solution has an objective function value of $33,895.40. Table 3 shows the purchase orders of each item over the planning period.

Table 3. Optimal purchase orders during the shifted planning horizon.

Item/Period	1	2	3	4	5	6	7	8	9
1	4	0	0	0	0	0	0	0	0
2	0	1	1	1	0	1	0	0	0
3	0	0	2	1	0	1	1	1	0

During the first three periods, the purchases listed in Table 3 satisfy the demands in the fourth period. The created inventories and shortages of each item are described in Tables 4 and 5, respectively.

Table 4. Periodical inventory held in units per item during the shifted planning horizon.

Item/Period	1	2	3	4	5	6	7	8	9
1	0	0	0	690	591	427	246	151	0
2	0	0	0	3	283	137	36	284	153
3	0	0	0	126	144	7	15	46	13

Table 5. Periodical shortage during the shifted planning horizon.

Item/Period	1	2	3	4	5	6	7	8	9
1	0	0	0	0	0	0	0	0	13
2	0	0	0	0	0	0	0	0	47
3	0	0	0	0	0	0	0	0	0

Regarding the maximum inventories per item, they were determined as follows: Item 1: 690 units, Item 2: 284 units, and Item 3: 144 units. These results could help the company to determine a strategic decision about the distribution of the warehouse so they can seize the space availability.

3.3. MILP Formulation without Shortages

To evaluate properly if allowing the model to consider shortages was beneficial, a second formulation that did not allow having shortages was developed to compare. The version without shortages is denoted as "WS model". The WS model was obtained by removing the Z_{it} variables from the AS model and also removing the equations involving them. In particular, Equations (2), (9), and (10) were reformulated in terms of the inequalities proposed in [30]. The reformulated expressions are given by (16), (17), and (18) respectively:

$$I_{i(t-1)} + B_i X_{i(t-LT_i)} - D_{it} = I_{it}, \qquad \forall\, i \in N;\, t \in ST, \tag{16}$$

$$I_{i(t-1)} \geq D_{it}(1 - Y_{it}), \qquad \forall\, i \in N;\, t \in ST, \tag{17}$$

$$B_{it} X_{it} \leq D_{it} + I_{it}, \qquad \forall\, i \in N;\, t \in ST. \tag{18}$$

Under these equations, the model must fulfill the demand of each period. In particular, Equation (16) balances the inventories to suit the demand in period t. Equation (17) imposes an inventory in weeks when no purchases will be placed. Finally, Equation (18) places an upper bound on the maximum purchase amount in a specific period.

In addition, the objective function (1) is modified by removing the component of shortage. It will be stated as shown in Equation (19). For the sake of clarity, we present the condensed MILP formulation; from the AS model, this model preserves Equations (3)–(8) and (11)–(15).

$$\text{Minimize} \sum_{i=1}^{n} \sum_{t=1}^{s-1} H_i I_{it} + 2 \sum_{i=1}^{n} P_i I_{is}, \tag{19}$$

subject to:

$$I_{i(t-1)} + B_i X_{i(t-LT_i)} - D_{it} = I_{it}, \quad \forall\, i \in N;\, t \in ST,$$

$$I_{it} \leq R_i, \quad \forall\, i \in N,\, t \in ST,$$

$$I_{ih} = 0, \quad \forall\, i \in N,\, h \in T,\, h \leq LT_{max}.$$

$$\sum_{t=LT_{max}+2}^{s} Y_{j(t-LT_i)} = 0, \quad \forall\, j \in M.$$

$$\sum_{i=1}^{n}\left(B_i P_i X_{i(LT_{max}-LT_i+1)} + O_i Y_{i(LT_{max}-LT_i+1)}\right) \leq W_{ini},$$

$$\sum_{i=m+1}^{n}\left(B_i P_i X_{i(t-LT_i)} + O_i Y_{i(t-LT_i)}\right) \leq W_t, \quad \forall\, t \in ST,\, t > LT_{max}+1.$$

$$X_{i(t-LT_i)} \leq \left(1 + \frac{TD_i}{B_i}\right) Y_{i(t-LT_i)}, \quad \forall\, i \in N;\, t \in ST.$$

$$I_{i(t-1)} \geq D_{it}(1 - Y_{it}), \quad \forall\, i \in N;\, t \in ST,$$

$$B_{it} X_{it} \leq D_{it} + I_{it}, \quad \forall\, i \in N;\, t \in ST.$$

$$R_i \geq 0, \quad \forall\, i \in N,$$

$$I_{it} \geq 0, \quad \forall\, i \in N;\, t \in ST,$$

$$X_{it} \geq 0, \quad \forall\, i \in N;\, t \in ST,$$

$$Z_{it} \geq 0, \quad \forall\, i \in N;\, t \in ST,$$

$$Y_{it} \in \{0,1\}, \quad \forall\, i \in N;\, t \in ST.$$

In the following section, the experimental results and a discussion about the advantages and disadvantages for each mathematical model will be presented.

4. Computational Experiments

This section is devoted to the implementation and experimental results of the two proposed formulations. The validation was performed over a test set of random instances. The performance assessments were over the instances' size and the CPU time. We also present the results over the instance of the case study that motivated this research.

The computational experimentation was executed on a Workstation Think Centre, ThinkStation P910 Xeon E5-2620, and a Windows 10 operating system. The mathematical model was coded using Visual Studio 2019 and solved by the commercial solver CPLEX 12.9. The time limit was set to two hours (7200 s).

4.1. Case Study

As explained in the Introduction Section, this study was motivated by a real problem in which a retailer sells clothes, shoes, and accessories by catalogue. This company has over 29 years of experience in the market and manages two sales seasons per year: the spring-summer season and the autumn-winter season. In each 21 week season, the company handles around 465 stock-keeping units (items) purchased from different suppliers. Each supplier has its own delivery time (lead time). Around 30% of these items (139 out of 465) belong to the category of imported, and the remaining 70% are domestic. In the case of the imported items, the length of the lead time (seven weeks) requires the company to place replenishment orders. For this reason, they established the policy of placing only an initial purchase order, purchasing as much as possible. To ensure competitive prices, the suppliers also stipulate batch ordering of their items, which complicates the purchase and inventory management policy. Currently, the company performs the ordering process empirically, through a purchasing department involving approximately 20 personnel. Each person is in charge of purchasing certain items and considers both the placement and the order size under individual criteria. As these criteria are non-standardized (i.e., there is no institutionalized ordering scheme for either time or order size), the fear of overfilling the inventory constantly generates an opportunity cost. For this reason, the company decided to incorporate efficient methods for determining the appropriate number of orders per week, as well as the number of orders per item type that would minimize both the costs of placing purchase orders and keeping the inventory. To fulfill this need, we developed an efficient ordering scheme policy based on MILP.

4.2. Test Instances

Based on the case study instance that considered 465 items, we created nine different instances by randomly selecting subsets of items from the case study instance (10, 20, 30, or 50 items). As in the case study, we preserved the percentages of items provided by external and internal suppliers (30% and 70%, respectively).

For each subset size, we generated four instances. Each instance was labelled by "X-Y", where X and Y denote the number of items and the instance number, respectively.

The demands, lead times, setup order costs, batch sizes, unitary purchasing costs, and holding cost of the items in a specific instance were decided from the case study instance. The initial and periodical budget were proportionally scaled regarding the vales considered for the case study instance. The time horizon was set to 21 weeks, and the penalty cost per unit of shortage was set to 2.5 times the unitary purchasing cost.

4.3. Experimental Results over the Test Instances

The AS model was solved by the commercial solver CPLEX 12.9. The results obtained for the test instances are reported in Table 6. The first column displays the instance name, while columns 2, 3, and 4 report the integer solution, the best bound obtained, and the GAPreported by the commercial solver, respectively. For optimal solutions, the GAP value reported is zero. Finally column 5 reports the elapsed CPU time.

Table 6. Performance of the AS model over the test random instances.

Instance	Objective Function	Best Bound	GAP%	CPU Time (Seconds)
10-1	$372,721	$372,721	0.00	82.03
10-2	$349,317	$349,317	0.00	528.52
10-3	$ 543,781	$543,781	0.00	94.96
10-4	$1,258,256	$1,258,256	0.00	406.59
20-1	$679,124	$679,124	0.00	8.891
20-2	$676,166	$676,166	0.00	56.49
20-3	$964,839	$964,839	0.00	19.66
20-4	$1,303,250	$1,303,250	0.00	47.562
30-1	$977,660	$965,095	1.29	7200.00
30-2	$1,385,400	$1,331,200	0.52	7200.00
30-3	$985,798	$973,146	1.28	7200.00
30-4	$1,115,070	$1,104,540	0.94	7200.00
50-1	$3,535,310	$3,519,520	0.45	7200.00
50-2	$2,966,000	$2,950,390	0.53	7200.00
50-3	$3,533,190	$3,496,460	1.04	7200.00
50-4	$3,454,710	$3,428,850	0.75	7200.00
Average			0.43	3677.794

According to the results, the model solved to optimality eight out of 16 instances. For the remaining instances, it reached the specified time limit imposed on the solver. The GAPs deviated by up to 1.30% from the best bound. Among these instances, the average GAP was only 0.43%. Based on these results, the model showed a reasonable performance by providing high-quality integer values.

In particular, from the instances of 30 and 50 items, the formulation was unable to obtain optimal solutions; for those instances, we decided to increase the time limit of 7200 s imposed on the solver in pursuit of getting optimal solutions. However, the commercial solver got stuck in a positive GAP after more than 10,000 s. We analyzed the behaviour of the solver during the solution process on a test instance, evaluating the behaviour of the GAP when the elapsed time increased. Figure 1 illustrates this behaviour for Instance 30-1. Due to this, we decided to preserve the time limit of 7200 s for solving the case study.

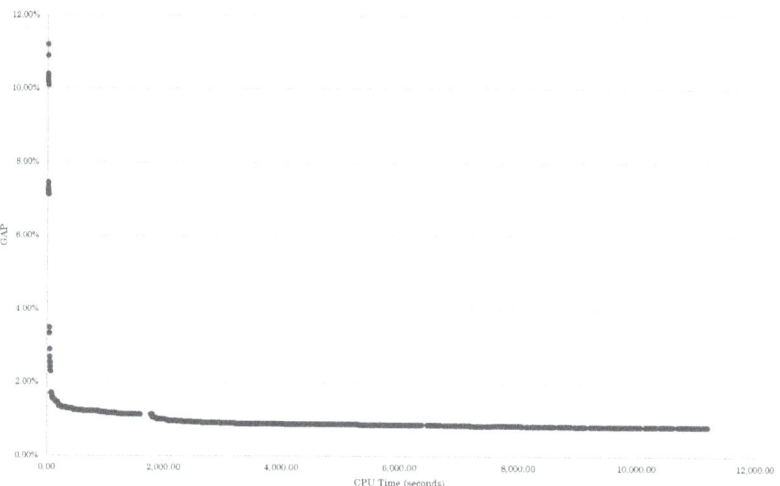

Figure 1. GAP analysis for the test instance 30-1.

Another point to highlight is that, for the instance of 50 items, the solution value was significantly higher than the rest of the cases. When revising in detail the cost segments contributing to the objective function, we realized that, for some cases, the model preferred to generate a high level of shortage, in particular for imported items. Based on this result, we could infer that the model determined when it was worth purchasing products or leaving a shortage in pursuit of minimizing costs. To validate this hypothesis, we conducted a complementary analysis by evaluating the case in which the formulation must fulfill the demand, through the WS model. The next section presents the results of this comparison.

4.4. Comparison between the AS Model and WS Model over Test Instances

The WS model was solved by the commercial solver CPLEX 12.9. The results obtained for the test instances are summarized in Table 7. In this table, column 1 indicates the instance name, whereas columns 2, and 3 display the objective function and the elapsed CPU time, respectively.

As expected, the WS model solved all of the instances to optimality and produced higher objective function values than the AS model. In particular, the AS model obtained solutions that produced savings up to 35.80% of the objective function. However, we conducted a Man–Whitney analysis with a confidence level of 95 % to determine if the the savings were statistically significant. The alternative hypothesis (H_a), "the WS model computes higher objective function values than the AS model", was evaluated against the null hypothesis (H_0) "there is no objective function value difference between the models". The obtained p-value was 0.107, which accepted the null hypothesis that, eventually, both models could produce an equivalent total cost.

Table 7. Performance of the WS model over the test random instances.

Instance	Objective Function	CPU Time (Seconds)
10-1	$528,137	0.688
10-2	$544,109	1.140
10-3	$595,973	0.625
10-4	$1,529,770	2.875
20-1	$1,008,150	0.813
20-2	$782,212	2.469
20-3	$1,120,920	0.750
20-4	$1,611,150	3.531
30-1	$1,447,230	3.969
30-2	$1,510,310	2.266
30-3	$1,340,600	3.515
30-4	$1,414,390	1.422
50-1	$4,015,590	1.953
50-2	$3,495,540	2.875
50-3	$4,135,130	2.953
50-4	$3,998,180	2.266
Average		2.220

Another important aspect to analyze is the one related to the inventories. In this case, the inventory cost of the WS model was compared with the inventory cost plus the opportunity cost of the AS model. The results are displayed in Table 8. For this table, column 1 displays the instance name, column 2 the Inventory Costs (IC) for the WS formulation, columns 3 and 4 the inventory cost, the Opportunity Cost (OC), and the Total Cost (TC) composed by the sum of the inventory cost and the opportunity cost (lost-sales costs), for the AS formulation, respectively. Column 6 estimates the savings gained by AS (monetary savings), and column 7 presents the percentage of savings of the AS model over the WS model (% of savings AS vs. WS). In the table, IC, OC, and TC are estimated by Equations (20), (21), and (22), respectively:

$$IC = \sum_{i=1}^{n} \sum_{t=LT_{max}+1}^{S} H_i I_{it} \quad (20)$$

$$OC = \sum_{i=1}^{n} \sum_{t=LT_{max}+1}^{S} P_i Z_{it} \quad (21)$$

$$TC = IC + OC \quad (22)$$

From Table 8, we observe that the AS model, in terms of inventory costs, was more cost effective than the WS model, producing savings up to 35.8%. However, the total cost, composed by the inventory and opportunity cost, presented a mixed behaviour, showing that 68.5% of the instances produced savings (values with a positive sign in the last column) when the AS model was compared with the WS model. We also observed that the WS approach produced higher values (up to 18%) of inventory costs than the AS model. Thus, the AS formulation could be useful to decrease the overstock while ensuring the availability of items (i.e., minimizing the shortage levels).

In particular, for the 20-3 instance and all 30 item instances, it was observed that the AS model produced an opportunity cost (estimated as the prices of the unsold items) that, after being added to the inventory cost, exceeded the inventory cost created by the WS model. The effect could be explained by the fact that, in order to minimize the final inventory, the AS model preferred to generate a high level of shortages in the last period, which was less expensive than buying complete batches that would be penalized at the end of the time horizon. However, in general, the AS model produced lower opportunity cost values, which also supported the fact that the model offered the advantage of determining which products should be prioritized at the time of purchase for each period.

Table 8. Comparison between the total inventory costs for the model with and without shortage for the test random instances. IC, Inventory Cost; OC, Opportunity Cost; TC, Total Cost.

| Instance | WS Model | AS Model | | | Monetary | % of Savings |
	IC	IC	OC	TC	Savings	AS vs. WS
10-1	$266,897	$217,126	$33,570	$250,696	$16,111	6.04
10-2	$257,546	$214,408	$46,332	$260,740	$ −3194	−1.24
10-3	$435,105	$408,633	$17,151	$425,784	$9321	2.14
10-4	$1,157,045	$1,100,163	$40,816	$1,140,979	$16,066	1.39
20-1	$455,757	$353,577	$76,390	$429,967	$25,790	5.66
20-2	$397,775	$358,129	$51,600	$409,729	$ −11,955	−3.01
20-3	$770,110	$739,858	$40,505	$780,363	$−10,253	−1.33
20-4	$1,069,308	$996,489	$68,938	$1,065,427	$3880	0.36
30-1	$706,880	$635,984	$82,238	$717,922	$−11,042	−1.56
30-2	$815,893	$734,699	$84,400	$819,099	$−3205	−0.39
30-3	$638,094	$536,434	$119,413	$655,847	$−17,752	−2.78
30-4	$719,120	$660,574	$89,034	$749,658	$−30,537	−4.24
50-1	$3,042,402	$2,906,978	$3640	$2,910,619	$131,783	4.33
50-2	$2,697,426	$2,322,546	$2720	$2,325,267	$372,159	1.38
50-3	$2,886,510	$2,717,491	$4391	$2,721,882	$164,628	5.70
50-4	$2,953,503	$2,821,755	$2933	$2,824.688	$128,815	4.36

In terms of CPU time, the WS model optimized the test instances within 4 s of CPU time, whereas for the AS model, the fastest instance required almost 9 s to solve to optimality. The results of the WS model could be used to obtain a "fast" purchasing plan in which all items will have overstock, and with the aim of minimizing the inventory levels, the company could create a clearance or sales strategy for the last period.

The inventory cost is commonly used to assess the efficiency of the overall purchase policy. We compared the final inventory costs of the two models. The results are reported in Table 9. Column 1 shows the instance name and columns 2 ans 3 the final inventory cost for the WS model and the AS model, respectively. Column 4 shows the percentage of saving of the AS model over the WS model.

Table 9. Comparison of the final inventory cost for the AS model and the WS model.

| | Final Inventory Cost | | |
Instance	WS Model	AS Model	% of Saving AS vs. WS
10-1	$3266.63	$888.88	72.57%
10-2	$3582.04	$238.48	93.34%
10-3	$2010.85	$11,153.38	42.64%
10-4	$4659.02	$700.67	84.96%
20-1	$6904.91	$1682	75.64%
20-2	$4809.59	$2362.96	50.83%
20-3	$4385.10	$1546.48	64.73%
20-4	$6772.98	$1680.18	75.19%
30-1	$9254.41	$1704.75	81.57%
30-2	$6180.20	$11,160.50	81.22%
30-3	$8781.33	$1885.40	78.52%
30-4	$8690.87	$2897.33	66.67%
50-1	$12,164.00	$3303.10	72.85%
50-2	$11,843.28	$4642.65	60.80%
50-3	$15,607.00	$4706.79	69.85%
50-4	$13,058.00	$4246.13	67.48%

From Table 9, the AS model saved up to 93% of the final inventory cost computed in the WS model. Even when there was not significant evidence of savings with respect to the objective value against the AS model, a statistical analysis was conducted to determine if the AS model produced significant savings regarding the final inventory cost. We analyzed the results of the test instances by performing a hypothesis test. Given the distribution of the costs, a Mann–Whitney (non-parametric test) with a confidence level of 95 % was selected again for this purpose. The alternative hypothesis (H_a), "the WS model computes higher final inventory costs than the AS model", was evaluated against the null hypothesis (H_0) "there is no final inventory cost difference between the models". The test yielded a p-value of 0.013, which supported the evidence that the WS model computed higher final inventory costs than the AS model. However, the WS model reached the optimal results in a significantly shorter CPU time than the AS model. Based on these findings, we could conclude that the WS model could be used to determine the total cost for the worst scenario (i.e., avoiding shortages). In other words, by the WS model, it was possible to obtain an upper bound for the AS model.

4.5. Experimental Results over the Case Study

To conclude the present set of computational experiments, the performances of the WS and AS models were compared on real data provided by the company. Whereas the WS model solved this instance to optimality in 15.94 s, the AS model reached the time limit of 7200 s and reported a GAP of 0.81%. Regarding the objective function values, Table 10 summarizes the results. In this table, columns 1 and 2 provide the objective function value obtained by the WS and AS models, respectively. Columns 3 and 4 display the savings of the AS model over the WS model, in monetary and percentage terms, respectively.

Table 10. Comparison between the solution cost for the AS model and the WS model for the case study instance.

Objective Function		Monetary Savings	% of Savings AS vs. WS
WS Model	AS Model		
$35,306,310	$29,532,400	$5,773,910	16.35

It is important to note that the AS model achieved significant monetary savings, which represented almost 17% of the total cost incurred by the WS model. In the real situation, these savings rose up to 33.33% with regard to the last year's operation costs of the company (their costs according to our proposed objective function were equal to $44,078,209). For the case of the administration of the warehouse capacity, it could be observed that, according to the results of the R_i variables, the AS formulation distributed the maximum inventories of the items in a way that the total stock in any period did not exceed 70% of the full capacity. In the case of the WS model, this value was not superior to 90%. Finally, the comparisons of the final inventory between both models is presented in Table 11.

Table 11. Comparison between the final inventory cost for the AS and WS model for the case study instance.

	Final Inventory Cost		
Instance	WS Model	AS Model	% of Savings AS vs. WS
Case study	$141,073	$36,201	74%

In the real situation, the company incurred a final inventory cost of $195,778.50. Therefore, both models improved this cost substantially. In particular, the AS model recovered approximately 32.29% of the inventory cost, increasing the company's profits by approximately 1.2%.

5. Conclusions and Future Work

This work addressed the capacitated lot sizing problem with lead times, batch ordering, and shortages. This problem was motivated by a retail company that sells clothes, shoes, and accessories by catalogue. The problem considered a retail environment in which items were not produced, but purchased from manufacturers (suppliers) who stipulated the batch ordering and who provided different delivery times (lead times) for each item. The problem was modelled and solved via an MILP formulation.

Two mathematical formulations were developed, one that allowed shortage (AS model) and another that did not (WS model). In the AS version, instances up to 20 items were solved to optimality, whereas the WS version solved all instances. The performance of the formulation was evaluated on different instance sizes. The AS model obtained significant savings (up to 17%) over the current situation of the company. Most relevantly, both mathematical formulations significantly improved the current situation of the company, saving up to 33% of the current final inventory costs. These results would be of interest to both academics and practitioners.

To assess whether introducing shortages benefited the company, we compared the final inventories computed by both models. The inventory costs in the AS version were cost-effective in all cases, but these savings were not statistically different from the savings gained in the WS model. In addition, the WS version optimized the test instances within 16 s of CPU time. In the case study, the best obtained solution deviated by 0.80% from the lower bound after reaching the specified time limit.

This research contributes to the body of knowledge as it optimized the CLSP by considering multiple items along with variable batch sizes and variable delivery times (lead times) in a retail environment. Still, the study addressed a comprehensive problem in the retail industry, but the approach assumed the following: first, there was a deterministic demand, which represented the first limitation of this study; and second, suppliers presented the right deliveries in terms of quality and amount and time, which represented the second limitation of this study. Any change to these assumptions would create a different scenario that should be addressed in a very particular way.

Further research should consider stochastic environments, in particular the uncertainty in the demand and/or lead times. Moreover, considerations such as the quality and timely deliveries by the suppliers can be questioned since these scenarios might not reflect 100% of the real-life cases. Including these variants will bring value to the lot sizing problem literature. Other considerations may include not only the lost-sales costs, but the cost of losing future customers in order to improve the estimation of opportunity costs. Finally, high-quality solutions within short computational time frames can be obtained by heuristics and metaheuristic algorithms. The models presented in this study will benefit companies by providing a good starting point to facilitate their decision-making process.

Author Contributions: Conceptualization, Y.C.-V., S.N.-G.; methodology, S.N.-G., Y.C.-V. and G.M.-B.; software, G.M.-B. and E.D.-M.; validation, S.N.-G. and Y.C.-V.; formal analysis, S.N.-G., Y.C.-V. and R.E.P.-G.; investigation, S.N.-G., Y.C.-V. and R.E.P.-G.; resources, Y.C.-V., R.E.P.-G. and E.D.-M.; data curation, G.M.-B., S.N.-G. and E.D.-M.; writing–original draft preparation, S.N.-G., Y.C.-V., G.M.-B., R.E.P.-G. and E.D.-M.; writing–review and editing, S.N.-G., Y.C.-V. and R.E.P.-G.; visualization, Y.C.-V. and S.N.-G.; supervision, S.N.-G.; project administration, S.N.-G. All authors have read and agreed to the published version of the manuscript.

Funding: This work was supported by Universidad Panamericana through the grant "Fomento a la Investigación 2019", under Project Code UP-CI-2019-ING-GDL-08, and by the FORDECYT-CONACYT, under Project Code 265667.

Conflicts of Interest: The authors declare no conflict of interest.

Abbreviations

The following abbreviations are used in this manuscript:

MILP   Mixed-Integer Linear Program
CLSP   Capacitated Lot Sizing Problem
AS   Allowing Shortage model
WS   Without Shortage model
EOQ   Economic Order Quantity

References

1. Christopher, M.; Lowson, R.; Peck, H. Creating agile supply chains in the fashion industry. *Int. J. Retail. Distrib. Manag.* **2004**, *32*, 367–376. [CrossRef]
2. DeHoratius, N.; Mersereau, A.J.; Schrage, L. Retail inventory management when records are inaccurate. *Manuf. Serv. Oper. Manag.* **2008**, *10*, 257–277. [CrossRef]
3. Liu, N.; Ren, S.; Choi, T.M.; Hui, C.L.; Ng, S.F. Sales forecasting for fashion retailing service industry: A review. *Math. Probl. Eng.* **2013**, *2013*, 1–9. [CrossRef]
4. Hsu, W.L. On the general feasibility test of scheduling lot sizes for several products on one machine. *Manag. Sci.* **1983**, *29*, 93–105. [CrossRef]
5. Bruno, G.; Genovese, A.; Piccolo, C. The capacitated Lot Sizing model: A powerful tool for logistics decision making. *Int. J. Prod. Econ.* **2014**, *155*, 380–390. [CrossRef]
6. Rezaei, J.; Davoodi, M. Multi-objective models for lot sizing with supplier selection. *Int. J. Prod. Econ.* **2011**, *130*, 77–86. [CrossRef]
7. Ferreira, D.; Clark, A.R.; Almada-Lobo, B.; Morabito, R. Single-stage formulations for synchronised two-stage lot sizing and scheduling in soft drink production. *Int. J. Prod. Econ.* **2012**, *136*, 255–265. [CrossRef]
8. Liao, J.J.; Huang, K.N.; Chung, K.J. Lot-sizing decisions for deteriorating items with two warehouses under an order-size-dependent trade credit. *Int. J. Prod. Econ.* **2012**, *137*, 102–115. [CrossRef]
9. Kuik, R.; Salomon, M.; Van Wassenhove, L.N. Batching decisions: Structure and models. *Eur. J. Oper. Res.* **1994**, *75*, 243–263. [CrossRef]
10. Drexl, A.; Kimms, A. Lot sizing and scheduling—Survey and extensions. *Eur. J. Oper. Res.* **1997**, *99*, 221–235. [CrossRef]
11. Karimi, B.; Ghomi, S.F.; Wilson, J. The capacitated lot sizing problem: A review of models and algorithms. *Omega* **2003**, *31*, 365–378. [CrossRef]
12. Minner, S. Multiple-supplier inventory models in supply chain management: A review. *Int. J. Prod. Econ.* **2003**, *81*, 265–279. [CrossRef]
13. Quadt, D.; Kuhn, H. Capacitated lot sizing with extensions: A review. *4OR* **2008**, *6*, 61–83. [CrossRef]
14. Gicquel, C.; Minoux, M.; Dallery, Y. *Capacitated Lot Sizing Models: A Literature Review*; Working Paper or Preprint. Available online: https://hal.archives-ouvertes.fr/hal-00255830 (accessed on 1 March 2008).
15. Jans, R.; Degraeve, Z. Modeling industrial lot sizing problems: A review. *Int. J. Prod. Res.* **2008**, *46*, 1619–1643. [CrossRef]
16. Buschkühl, L.; Sahling, F.; Helber, S.; Tempelmeier, H. Dynamic capacitated lot sizing problems: A classification and review of solution approaches. *Or Spectr.* **2010**, *32*, 231–261. [CrossRef]
17. Ben-Daya, M.; Raouf, A. On the constrained multi-item single-period inventory problem. *Int. J. Oper. Prod. Manag.* **1993**, *13*, 104–112. [CrossRef]
18. Guan, Y.; Liu, T. Stochastic lot sizing problem with inventory-bounds and constant order-capacities. *Eur. J. Oper. Res.* **2010**, *207*, 1398–1409. [CrossRef]
19. Fan, J.; Wang, G. Joint optimization of dynamic lot and warehouse sizing problems. *Eur. J. Oper. Res.* **2018**, *267*, 849–854. [CrossRef]
20. Woarawichai, C.; Kullpattaranirun, T.; Rungreunganun, V. Inventory lot sizing problem with supplier selection under storage space and budget constraints. *Int. J. Comput. Sci. Issues (IJCSI)* **2011**, *8*, 250.
21. Karmarkar, U.S. Lot sizes, lead times and in-process inventories. *Manag. Sci.* **1987**, *33*, 409–418. [CrossRef]
22. Ben-Daya, M.A.; Raouf, A. Inventory models involving lead time as a decision variable. *J. Oper. Res. Soc.* **1994**, *45*, 579–582. [CrossRef]
23. Hariga, M.; Ben-Daya, M. Some stochastic inventory models with deterministic variable lead time. *Eur. J. Oper. Res.* **1999**, *113*, 42–51. [CrossRef]
24. Ouyang, L.Y.; Chen, C.K.; Chang, H.C. Quality improvement, setup cost and lead-time reductions in lot size reorder point models with an imperfect production process. *Comput. Oper. Res.* **2002**, *29*, 1701–1717. [CrossRef]
25. Helber, S.; Sahling, F. A fix-and-optimize approach for the multi-level capacitated lot sizing problem. *Int. J. Prod. Econ.* **2010**, *123*, 247–256. [CrossRef]
26. Yang, Y.; Yuan, Q.; Xue, W.; Zhou, Y. Analysis of batch ordering inventory models with setup cost and capacity constraint. *Int. J. Prod. Econ.* **2014**, *155*, 340–350. [CrossRef]

27. Akbalik, A.; Hadj-Alouane, A.B.; Sauer, N.; Ghribi, H. NP-hard and polynomial cases for the single-item lot sizing problem with batch ordering under capacity reservation contract. *Eur. J. Oper. Res.* **2017**, *257*, 483–493. [CrossRef]
28. Farhat, M.; Akbalik, A.; Sauer, N.; Hadj-Alouane, A. Procurement planning with batch ordering under periodic buyback contract. *IFAC-PapersOnLine* **2017**, *50*, 13982–13986. [CrossRef]
29. Broekmeulen, R.A.; Van Donselaar, K.H. A heuristic to manage perishable inventory with batch ordering, positive lead-times, and time-varying demand. *Comput. Oper. Res.* **2009**, *36*, 3013–3018. [CrossRef]
30. Pochet, Y.; Wolsey, L.A. *Production Planning by Mixed Integer Programming*; Springer Science & Business Media: Berlin/Heidelberg, Germany, 2006.

© 2020 by the authors. Licensee MDPI, Basel, Switzerland. This article is an open access article distributed under the terms and conditions of the Creative Commons Attribution (CC BY) license (http://creativecommons.org/licenses/by/4.0/).

Article

Packing Oblique 3D Objects

Alexander Pankratov [1,2], Tatiana Romanova [1,2,*] and Igor Litvinchev [3]

[1] Department of Mathematical Modeling and Optimal Design, Institute for Mechanical Engineering Problems of the National Academy of Sciences of Ukraine, 2/10, Pozharsky str., 61046 Kharkiv, Ukraine; pankratov2001@yahoo.com
[2] Department of Department of Systems Engineering of Kharkiv National University of Radio Electronics, Nauky Ave. 14, 61166 Kharkiv, Ukraine
[3] Faculty of Mechanical and Electrical Engineering, Graduate Program in Systems Engineering, Nuevo Leon State University (UANL), 66450 Monterrey, Mexico; igorlitvinchev@gmail.com
* Correspondence: tetiana.romanova@nure.ua

Received: 23 May 2020; Accepted: 3 July 2020; Published: 10 July 2020

Abstract: Packing irregular 3D objects in a cuboid of minimum volume is considered. Each object is composed of a number of convex shapes, such as oblique and right circular cylinders, cones and truncated cones. New analytical tools are introduced to state placement constraints for oblique shapes. Using the phi-function technique, optimized packing is reduced to a nonlinear programming problem. Novel solution approach is provided and illustrated by numerical examples.

Keywords: packing; irregular 3D objects; quasi-phi-function s; nonlinear optimization

---

1. Introduction

Packing problems aim to allocate a set of objects in a container subject to placement constraints. The latter typically stipulates non-overlapping between the objects and the boundary of the container. Additional placement constraints may include weight distribution and stacking, cargo stability, balance constraints, loading and unloading preferences, etc. [1]. In optimized packing certain criteria have to be optimized, e.g., maximizing the number of the packed objects, minimizing the waste or optimizing characteristics of the container, its volume or shape [2]. Packing problems are proved to be NP-hard [3].

Packing issues traditionally are important in logistics, e.g., in maritime transportation and container loading, cutting industrial materials in furniture and glass manufacturing [4]. Packing problems also arise in modelling liquid and glass structures, in analyses of powder and granular materials in mineral industry, in molecular and nanotechnologies [5–9].

Different classifications for packing problems are proposed (see, e.g., [2,4,10] and the references therein). Focusing on the shapes involved, packing problems can be divided into two large groups: regular and irregular. While regular 3D packing deals with relatively simple shapes (spheres, ellipsoids, convex polyhedrons), irregular packing focuses basically on nonconvex figures (see, e.g., [2,11–19]).

Various modelling and solution approaches, exact and approximated, are known for the regular packing (see, e.g., [20–22] and the references therein). For irregular 3D packing, heuristics are widely used [2]. A large group of heuristics is based on representing complex irregular shapes by corresponding collections of simpler (regular) figures thus reducing the problem approximately to a regular case [23–27]. Techniques in the other group combine a local search with simple decision rules, such as the deepest bottom-left approach or random allocation [28–33]. An alternative methodology is using genetic algorithms, directly or in combination with the first two approaches [34–37].

In this paper, packing irregular 3D objects in a cuboid of minimum volume is considered. Each complex object is composed of a number of convex shapes, such as oblique and right circular cylinders, cones and truncated cones. Studying composed objects requires new modelling tools, different from those used previously for simpler objects (see, e.g., [20–22] and the references therein). In this paper, the phi-function approach is applied to represent analytically containment and non-overlapping conditions. Using the concept of quasi phi-functions [38], an exact mathematical model is formulated and a corresponding nonlinear programming problem is stated. A solution algorithm is proposed and computational results are presented to illustrate the approach. To the best of our knowledge, exact mathematical models for packing complex objects composed of a mixture of oblique and right convex shapes nether were considered before.

Nonspherical particles presented by the superquadric equations are widely used in different industrial production, and significantly affect the macro- and microcharacteristics of granular materials, see, e.g., [39] and the references therein. However, the particle shapes constructed by the superquadric equations are geometrically symmetrical and strictly convex, which significantly limits their further engineering applications [40]. In recent years, the composed element method has been successfully used. In this approach a complex nonconvex object is composed of basic convex elements, e.g., spheres, cylinders, super-quadratic elements and other convex (irregular) shapes [41,42].

Another source of packing complex composed shapes is additive manufacturing (AM), also known as 3D printing. AM refers to technologies for producing complex parts in a layer-by-layer material deposition process. The process takes place inside the machine in an enclosed build container or a "build volume". AM does not use any conventional physical tooling such as moulds, cutting implements or dies. Using AM, products previously designed and manufactured as assemblies of multiple components can now be manufactured as single items [43]. As a parallel manufacturing process, AM permits producing various complex parts in a single build volume. This gives rise to a build volume packing problem arising during the machine setup process. Thus, packing complex objects composed of different shapes plays an important role 3D printing.

Our interest in studying oblique objects is motivated by modelling particulate systems of nonspherical shapes [44,45] and by build volume packing problems in 3D printing [10].

The main contributions of the paper are as follows:

1. New tools of mathematical modelling are presented to describe analytically non-overlapping and containment constraints for packing irregular 3D objects composed by the union of oblique and right basic shapes (circular cylinders, cones, truncated cones and spheres). The objects can be freely translated and rotated.
2. An exact mathematical model for the irregular packing problem is formulated in the form of nonlinear continuous programming problem.
3. A solution algorithm for the irregular packing problem is developed.
4. New benchmark instances are provided to illustrate the efficiency of the approach.

The paper is organized as follows. Section 2 provides the general formulation for the irregular packing problem. Quasi-phi-function s for the composed 3D objects are defined to describe analytically placement constraints in Section 3. The nonlinear programming model for the irregular packing problem is presented and solved by an algorithm described in Section 4. Computational results are given in Section 5, while Section 6 concludes. Definitions of the phi-functions and quasi-phi-function s are provided in Appendix A.

2. Problem Formulation

The packing problem is considered in the following setting. Denote a cuboid of variable length l, width w and height h by Ω (see Figure 1). Let a set of objects $T_q \subset \mathbb{R}^3$, $q \in J_N = \{1, 2, \ldots, N\}$ be given.

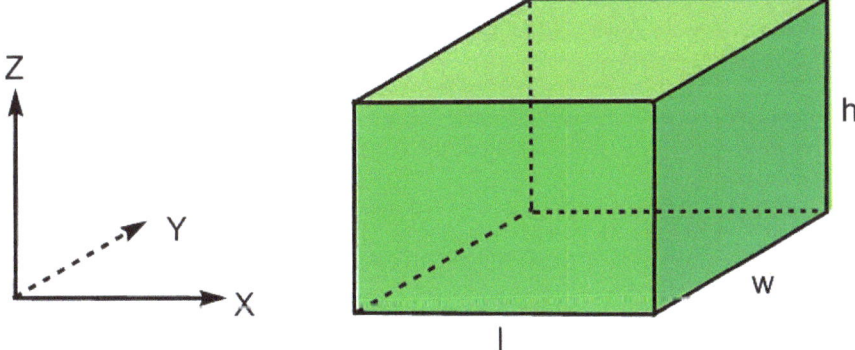

Figure 1. Container Ω.

The location and orientation of each 3D object T_q is defined by a vector $u_q = (v_q, \theta_q)$ of its (variable) placement parameters in the fixed coordinate system $OXYZ$. Here $v_q = (x_q, y_q, z_q)$ is a translation vector and $\theta_q = (\theta_q^1, \theta_q^2, \theta_q^3)$ is a vector of rotation parameters, where $\theta_q^1, \theta_q^2, \theta_q^3$ are Euler angles.

The notation $T_q(u_q) = \{\tilde{p} \in \mathbb{R}^3 : \tilde{p} = v_q + \Theta(\theta_q) \cdot (p)^T, \forall p \in T_q\}$ is used for translated and rotated object T_q, where $\Theta(\theta_q) = \Theta(\theta_q^1, \theta_q^2, \theta_q^3)$ is a rotation matrix of the form:

$$\Theta(\theta_q) = \begin{pmatrix} \cos\theta_q^1 \cos\theta_q^3 - \sin\theta_q^1 \cos\theta_q^2 \sin\theta_q^3 & -\cos\theta_q^1 \sin\theta_q^3 - \sin\theta_q^1 \cos\theta_q^2 \cos\theta_q^3 & \sin\theta_q^1 \sin\theta_q^2 \\ \sin\theta_q^1 \cos\theta_q^3 + \cos\theta_q^1 \cos\theta_q^2 \sin\theta_q^3 & -\sin\theta_q^1 \sin\theta_q^3 + \cos\theta_q^1 \cos\theta_q^2 \cos\theta_q^3 & -\cos\theta_q^1 \sin\theta_q^2 \\ \sin\theta_q^2 \sin\theta_q^3 & \sin\theta_q^2 \cos\theta_q^3 & \cos\theta_q^2 \end{pmatrix}.$$

Assume that $T_q(u_q) = \bigcup_{i=1}^{n_q} T_i^q(u_q)$, where T_i^q denotes a basic object from a family of oblique and right circular cylinders, cones, truncated cones denoted by $\Im$ and spheres (Figure 2a). An object T_q for $n_q \geq 2$ is referred to the composed object (Figure 2b).

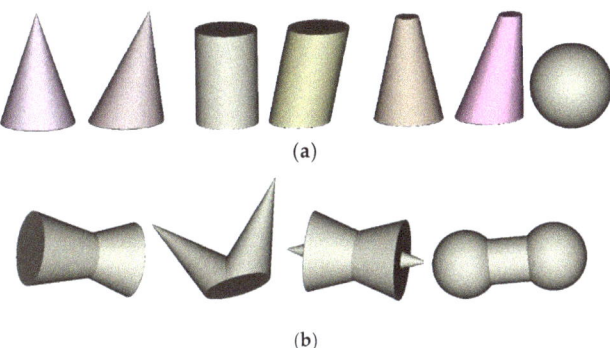

Figure 2. Placement objects: (**a**) basic; (**b**) composed.

Let a sphere be defined by its centre $p_i^q = (x_i^q, y_i^q, z_i^q)$ and a radius r_i^q. Each basic object from the family $\Im$ is defined by three vectors $p_{i1}^q = (x_{i1}^q, y_{i1}^q, z_{i1}^q)$, $p_{i2}^q = (x_{i2}^q, y_{i2}^q, z_{i2}^q)$ and $\mathbf{n}_i^q = (n_i^x, n_i^y, n_i^z)$, as well as a pair of parameters r_{i1}^q and r_{i2}^q. Here p_{i1}^q, p_{i2}^q are the centres and r_{i1}^q, r_{i2}^q are the radii of the bottom and top bases of T_i^q, $\mathbf{n}_i^q$ is the unit normal vector to the bottom (top) base of T_i^q.

Note that if $r_{i1}^q = r_{i2}^q > 0$ then T_i^q is a circular cylinder; if $r_{i1}^q \neq r_{i2}^q$ and $r_{i1}^q > 0, r_{i2}^q > 0$ then T_i^q is a circular truncated cone; if $r_{i1}^q > 0, r_{i2}^q = 0$ or $r_{i1}^q = 0, r_{i2}^q > 0$ then T_i^q is a circular cone. The height of each object $T_i^q \in \mathfrak{I}$ is denoted by h_i^q.

Let $p_i^q = (x_i^q, y_i^q, z_i^q)$ be a reference point of the basic object $T_i^q \subset T_q$: the centre point of a sphere or the central point for a circular base of the objects from the family $\mathfrak{I}$. In what follows, notation $\widetilde{p}_i^q = (\widetilde{x}_i^q, \widetilde{y}_i^q, \widetilde{z}_i^q) = v_q + \Theta(\theta_q) \cdot (p_i^q)^T$ is used, where v_q is the translation vector and $\Theta(\theta_q)$ is the rotation matrix of the object T_q.

The *placement constraints* can be stated in the following form:

$$\text{int} T_q(u_q) \cap \text{int} T_g(u_g) = O \text{ for } q > g \in J_N, \tag{1}$$

$$T_q(u_q) \subset \Omega \text{ for } q \in J_N. \tag{2}$$

Conditions (1) describe non-overlapping for all pairs of objects $T_q(u_q)$ and $T_g(u_g)$ for $q > g \in J_N$ (further, *non-overlapping constraints*), while conditions (2) assure containing $T_q(u_q)$ in the container Ω for $q \in J_N$ (further, containment constraints).

Irregular packing problem. Pack the set of objects T_q, $q \in J_N$, within a cuboidal container Ω of minimal volume $\kappa = l \cdot w \cdot h$, taking into account the placement constraints (1)–(2).

3. Analytical Tools

In this section geometric tools for the mathematical modelling of the placement constraints (1) and (2) are presented. To describe analytically the relations between a pair of objects considered in the placement constraints, the phi-functions [38] and quasi-phi-functions [18] are used (see Appendix A for the details. These functions for irregular objects composed of oblique and right shapes are introduced in this paper for the first time.

3.1. Modeling Non-Overlapping Constraints

Consider a pair of the composed objects $T_q(u_q) = \bigcup_{i=1}^{n_q} T_i^q(u_q)$ and $T_g(u_g) = \bigcup_{j=1}^{n_g} T_j^g(u_g)$.

To describe non-overlapping constraints (1) a phi-function for two composed objects $T_q(u_q)$ and $T_g(u_g)$ is introduced. It can be written in the form

$$\Phi'_{qg}(u_q, u_g, \tau_{qg}) = \min_{i=1,\dots,n_q, j=1,\dots,n_g} \Phi'^{qg}_{ij}(u_q, u_g, \tau_{ij}^{qg}) \tag{3}$$

where $\tau_{qg} = (\tau_{ij}^{qg}, i = 1, \dots, n_q, j = 1, \dots, n_g)$, $\Phi'^{qg}_{ij}(u_q, u_g, \tau_{ij}^{qg})$ is an adjusted quasi-phi-function for convex objects $T_i^q(u_q)$ and $T_j^g(u_g)$.

Let $P_{ij}^{qg} = \{(x, y, z) : x \le 0\}$ be a half space. Denote the half-space P translated along OX on value μ_{ij}^{qg} and rotated by angles θ^1, θ^2, by $\widetilde{P}_{ij}^{qg} = \{p : \Psi_{ij}^{qg}(p, \tau_{ij}^{qg}) \le 0\}$, where

$$\Psi_{ij}^{qg}(p, \tau_{ij}^{qg}) = \cos\theta_{ij}^{1qg} \cdot \cos\theta_{ij}^{2qg} \cdot x - \sin\theta_{ij}^{2qg} \cdot y + \sin\theta_{ij}^{1qg} \cdot \cos\theta_{ij}^{2qg} \cdot z + \mu_{ij}^{qg} \tag{4}$$

$p = (x, y, z)$, $\tau_{ij}^{qg} = (\theta_{ij}^{1qg}, \theta_{ij}^{2qg}, \mu_{ij}^{qg})$, θ_{ij}^{1qg} and θ_{ij}^{2qg} are rotation angles of the half space P_{ij}^{qg} around the axes OY and OZ in the fixed coordinate system. Define the plane $\widetilde{L}_{ij}^{qg} = \{(x, y, z) : \Psi_{ij}^{qg}(p, \tau) = 0\}$.

A quasi-phi-function for convex basic objects $T_i^q(u_q)$ and $T_j^g(u_g)$ can be defined in the form

$$\Phi'^{qg}_{ij}(u_q, u_g, \tau_{ij}^{qg}) = \min\{\Phi_i^q(u_q, \tau_{ij}^{qg}), \Phi_j^{*g}(u_g, \tau_{ij}^{qg})\} \tag{5}$$

where $\Phi_i^q(u_q, \tau_{ij}^{qg})$ is a phi-function for the object $T_i^q(u_q)$ and the half space $\widetilde{P}_{ij}^{qg}$, $\Phi_j^{*g}(u_g, \tau_{ij}^{qg})$ is a phi-function for the object $T_j^g(u_g)$ and the half space $\widetilde{P}_{ij}^{*qg} = \mathbb{R}^3 \backslash \text{int} \widetilde{P}_{ij}^{qg}$. Here $\tau_{ij}^{qg} = (\theta_{ij}^{1qg}, \theta_{ij}^{2qg}, \mu_{ij}^{qg})$ is the vector of auxiliary variables of the quasi-phi-function Φ'^{qg}_{ij}.

Therefore to define the *quasi-phi-functions* (5) *for each pair of convex basic objects* the phi-functions $\Phi_i^q(u_q, \tau_{ij}^{qg})$ and $\Phi_j^{*g}(u_g, \tau_{ij}^{qg})$ have to be derived.

First, define the phi-function $\Phi_i^q(u_q, \tau_{ij}^{qg})$ for the basic object $T_i^q(u_q)$ and the half space $\widetilde{P}_{ij}^{qg}$.

Let $T_i^q(u_q)$ be a sphere centred at the point $\widetilde{p}_{i1}^q = (\widetilde{x}_i^q, \widetilde{y}_i^q, \widetilde{z}_i^q)$ and having its radius r_i^q.

The phi-function for the sphere $T_i^q(u_q)$ and a half space $\widetilde{P}_{ij}^{qg}$ has the form

$$\Phi_i^q(u_q, \tau_{ij}^{qg}) = -\cos\theta_{ij}^{1qg} \cdot \cos\theta_{ij}^{2qg} \cdot \widetilde{x}_{i1}^q - \sin\theta_{ij}^{2qg} \cdot \widetilde{y}_{i1}^q + \sin\theta_{ij}^{1qg} \cdot \cos\theta_{ij}^{2qg} \cdot \widetilde{z}_{i1}^q + \mu_{ij}^{qg} - r_i^q =$$
$$\widetilde{\mathbf{n}}_{ij}^{qg} \cdot \widetilde{p}_{i1}^q + \mu_{ij}^{qg} - r_i^q,$$

while the phi-function for the sphere $T_j^g(u_g)$ and the half space $\widetilde{P}_{ij}^{*qg}$ can be defined as follows

$$\Phi_j^{*g}(u_g, \tau_{ij}^{qg}) = -\cos\theta_{ij}^{1qg} \cdot \cos\theta_{ij}^{2qg} \cdot \widetilde{x}_{i1}^q + \sin\theta_{ij}^{2qg} \cdot \widetilde{y}_{i1}^q - \sin\theta_{ij}^{1qg} \cdot \cos\theta_{ij}^{2qg} \cdot \widetilde{z}_{i1}^q - \mu_{ij}^{qg} - r_i^q =$$
$$-\widetilde{\mathbf{n}}_{ij}^{qg} \cdot \widetilde{p}_{i1}^q - \mu_{ij}^{qg} - r_i^q.$$

For $T_i^q(u_q)$ from the family $\mathfrak{I}$, let the bottom base of $T_i^q(u_q)$ is a circle centred at the point $\widetilde{p}_{i1}^q = (\widetilde{x}_i^q, \widetilde{y}_i^q, \widetilde{z}_i^q)$ and having its radius r_{i1}^q, while the top circular base of $T_i^q(u_q)$ is centred at $\widetilde{p}_{i2}^q = (\widetilde{x}_i^q, \widetilde{y}_i^q, \widetilde{z}_i^q + h_i^q)$ and has its radius r_{i2}^q (the cone corresponds to $r_{i2}^q = 0$).

The phi-function for the object $T_i^q(u_q) \in \mathfrak{I}$ and a half space $\widetilde{P}_{ij}^{qg}$ has the form

$$\Phi_i^q(u_q, \tau_{ij}^{qg}) = \min\left\{f_1(u_q, \tau_{ij}^{qg}), f_2(u_q, \tau_{ij}^{qg})\right\},$$

$$f_1(u_q, \tau_{ij}^{qg}) = \widetilde{\mathbf{n}}_{ij}^{qg} \cdot \widetilde{p}_{i1}^q + \mu_{ij}^{qg} - r_{i1}^q \sqrt{1 - (\widetilde{\mathbf{n}}_{ij}^{qg} \cdot \widetilde{\mathbf{n}}_i^q)^2},$$

$$f_2(u_q, \tau_{ij}^{qg}) = \widetilde{\mathbf{n}}_{ij}^{qg} \cdot \widetilde{p}_{i2}^q + \mu_{ij}^{qg} - r_{i2}^q \sqrt{1 - (\widetilde{\mathbf{n}}_{ij}^{qg} \cdot \widetilde{\mathbf{n}}_i^q)^2},$$

where $\widetilde{\mathbf{n}}_{ij}^{qg} = (\widetilde{\mathbf{n}}_{ij}^x, \widetilde{\mathbf{n}}_{ij}^y, \widetilde{\mathbf{n}}_{ij}^z)$ denotes a unit vector of the external normal to the half space $\widetilde{P}_{ij}^{qg}$ and $\widetilde{\mathbf{n}}_i^q = (\widetilde{\mathbf{n}}_i^x, \widetilde{\mathbf{n}}_i^y, \widetilde{\mathbf{n}}_i^z)$ stands for a unit vector of the external normal to the object $T_i^q(u_q)$ (see Figure 3).

The phi-function for the object $T_j^g(u_g)$ and a half space $\widetilde{P}_{ij}^{*qg}$ has the form

$$\Phi_j^{*g}(u_g, \tau_{ij}^{qg}) = \min\{f_1(u_g, \tau_{ij}^{qg}), f_2(u_g, \tau_{ij}^{qg})\},$$

$$f_1(u_g, \tau_{ij}^{qg}) = -\widetilde{\mathbf{n}}_{ij}^{qg} \cdot \widetilde{p}_{j1}^g - \mu_{ij}^{qg} - r_{j1}^g \sqrt{1 - (\widetilde{\mathbf{n}}_{ij}^{qg} \cdot \widetilde{\mathbf{n}}_i^q)^2},$$

$$f_2(u_g, \tau_{ij}^{qg}) = -\widetilde{\mathbf{n}}_{ij}^{qg} \cdot \widetilde{p}_{j2}^g - \mu_{ij}^{qg} - r_{j2}^g \sqrt{1 - (\widetilde{\mathbf{n}}_{ij}^{qg} \cdot \widetilde{\mathbf{n}}_i^q)^2}.$$

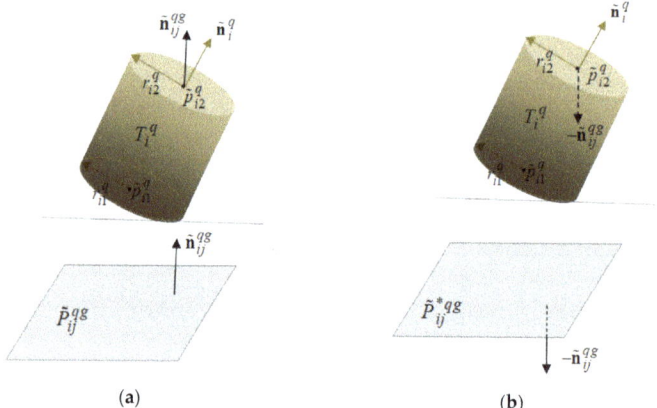

Figure 3. Non-overlapping: (**a**) convex basic objects T_i^q and the half space $\widetilde{P}_{ij}^{qg}$; (**b**) convex basic objects T_i^q and the half space $\widetilde{P}_{ij}^{*qg}$.

The inequality $\Phi'_{qg}(u_q, u_g, \tau_{qg}) \geq 0$ assures the non-overlapping condition $\text{int}\, T_q(u_q) \cap \text{int}\, T_g(u_g) = \varnothing$ in (1).

3.2. Modeling the Containment Constraints

Let us express the containment constraint $T_q(u_q) \subset \Omega$ in the equivalent form as $\text{int}\, T_q(u_q) \cap \Omega^* = \varnothing$, where $\Omega^* = R^3 \setminus \text{int}\, \Omega$.

The phi-function for the objects $T_q(u_q) = \bigcup_{i=1}^{n_q} T_i^q(u_q)$ and Ω^* can be defined in the form:

$$\Phi_q(u_q) = \min_{i=1,\dots,n_q} \Phi_i^q(u_q), \tag{6}$$

where $\Phi_i^q(u_q)$ is the phi-function for the basic convex object $T_i^q(u_q)$ and the object Ω^*.

The inequality $\Phi_q(u_q) \geq 0$ implies fulfilling the condition $\text{int}\, T_q(u_q) \cap \Omega^* = \varnothing$ that describes the containment condition (2).

Containment of a sphere $T_i^q(u_q)$ in a cuboid Ω. The phi-function for the sphere $T_i^q(u_q)$ and the object Ω^* can be defined by the equation

$$\Phi_i^q(u_q) = \min\{l - \widetilde{p}_{i1}^{qx} - r_{i1}^q, w - \widetilde{p}_{i1}^{qy} - r_{i1}^q, h - \widetilde{p}_{i1}^{qz} - r_{i1}^q, \widetilde{p}_{i1}^{qx} - r_{i1}^q, \widetilde{p}_{i1}^{qy} - r_{i1}^q, \widetilde{p}_{i1}^{qz} - r_{i1}^q\}.$$

Containment of the object $T_i^q(u_q)$ from the family $\Im$ in a cuboid Ω. The phi-function for the objects $T_i^q(u_q)$ and Ω^* has the form

$$\Phi_i^q(u_q) = \min\{\varphi_k(u_q), k = 1, \dots, 6\} \tag{7}$$

where

$$\varphi_1(u_q) = \min\{\widetilde{p}_{i1}^{qx} - r_{i1}^q \sqrt{1 - (\widetilde{n}_i^{qx})^2}, \widetilde{p}_{i2}^{qx} - r_{i2}^q \sqrt{1 - (\widetilde{n}_i^{qx})^2}\},$$

$$\varphi_2(u_q) = \min\{\widetilde{p}_{i1}^{qy} - r_{i1}^q \sqrt{1 - (\widetilde{n}_i^{qy})^2}, \widetilde{p}_{i2}^{qy} - r_{i2}^q \sqrt{1 - (\widetilde{n}_i^{qy})^2}\},$$

$$\varphi_3(u_q) = \min\{\widetilde{p}_{i1}^{qz} - r_{i1}^q \sqrt{1 - (\widetilde{n}_i^{qz})^2}, \widetilde{p}_{i2}^{qz} - r_{i2}^q \sqrt{1 - (\widetilde{n}_i^{qz})^2}\},$$

$$\varphi_4(u_q) = \min\{l - \widetilde{p}_{i1}^{qx} - r_{i1}^q \sqrt{1 - (\widetilde{n}_i^{qx})^2}, l - \widetilde{p}_{i2}^{qx} - r_{i2}^q \sqrt{1 - (\widetilde{n}_i^{qx})^2}\},$$

$$\varphi_5(u_q) = \min\{w - \widetilde{p}_{i1}^{qy} - r_{i1}^q \sqrt{1 - (\widetilde{n}_i^{qy})^2}, w - \widetilde{p}_{i2}^{qy} - r_{i2}^q \sqrt{1 - (\widetilde{n}_i^{qy})^2}\},$$

$$\varphi_6(u_q) = \min\{h - \widetilde{p}_{i1}^{qz} - r_{i1}^q \sqrt{1 - (\widetilde{n}_i^{qz})^2}, h - \widetilde{p}_{i2}^{qz} - r_{i2}^q \sqrt{1 - (\widetilde{n}_i^{qz})^2}\}.$$

The inequality $\Phi_i^q(u_q) \geq 0$ implies the condition $\text{int} T_i^q(u_q) \cap \Omega^* = \varnothing$.

4. Mathematical Model and Solution Algorithm

4.1. Mathematical Model

Now the irregular packing problem can be formulated as the following nonlinear optimization problem:

$$\min \kappa \text{ s.t. } (u, \tau) \in W, \quad (8)$$

$$W = \left\{ (u, \tau) : \Phi'_{qg}(u_q, u_g, \tau_{qg}) \geq 0, q > g \in \mathbf{I}_N, \Phi_q(u_q) \geq 0, q \in \mathbf{I}_N \right\}, \quad (9)$$

where $\tau_{qg} = (\tau_{ij}^{qg}, i = 1, \ldots, n_q, j = 1, \ldots, n_g)$ is a vector of the auxiliary variables for the quasi-phi-function $\Phi'_{qg}(u_q, u_g, \tau_{qg})$ (3)–(5) for the objects $T_q(u_q)$ and $T_g(u_g)$, $\Phi_q(u_q)$ is the phi-function (6)–(7) for the objects $T_q(u_q)$ and Ω^*, $q \in \mathbf{I}_N$.

The feasible region W in (9) is defined by a system of nonsmooth inequalities that can be reduced to a system of inequalities with differentiable functions.

The model (8)–(9) is a nonconvex and continuous nonlinear programming problem. This is an exact formulation in the sense that it gives all optimal solutions to the irregular packing problem.

The number of the problem variables is $\sigma = 3(1 + 2N + m)$, where $m = \sum_{q>g \in \mathbf{I}_N} n_q n_g$. The model (8)–(9) involves $O(N^2)$ nonlinear inequalities and $O(N^2)$ variables.

The model (8)–(9) represents all globally optimal solutions to the original irregular packing problem. It can be solved by any available global solver, e.g., BARON or LGO included in AMPL [46]. However, due to large number of variables and constraints, a direct solution to this problem may be time consuming and complicated. In the next section a solution approach is proposed to search for a local minimum of the problem (8)–(9). This solution can be used either as a reasonable approximation to the original global solution, or as a starting point for a global solver or heuristics.

4.2. Solution Algorithm

The following multistart strategy is used to solve the problem (8)–(9). A number of feasible starting points is generated. Then a local maximum of the problem (8)–(9) is obtained starting from each feasible point generated at the first stage. Finally, the best solution is selected from those obtained at the second stage. This result is considered as the solution of the problem (8)–(9).

4.2.1. Feasible Starting Points

To find feasible starting points for the problem (8)–(9) an algorithm based on the homothetic (scaling) transformation of objects is applied. The basic steps of the algorithm are as follows.

Step 1. Circumscribe spheres $S_q, q \in J_N$ around the objects $T_q(u_q), q \in J_N$.

Step 2. Construct a container Ω^0 with sufficiently large starting length l^0, width w^0 and height h^0 allowing placement of all the spheres $S_q, q \in J_N$.

Step 3. Generate a set of N randomly chosen centre points (x_q^0, y_q^0, z_q^0) for the spheres $S_q, q \in J_N$ inside the container Ω^0.

Step 4. Grow up the spheres S_q of the radii $\lambda r_q, q \in J_N$, starting from $\lambda = 0$ to the full size ($\lambda = 1$). Here the decision variables are the centres of S_q and the homothetic coefficient λ, where $0 \leq \lambda \leq 1$.

Step 5. Form a vector of feasible parameters of our objects $T_q(u_q) \subset S_q(v_q)$.

Now proceed with a more detailed description of the algorithm.

First, fix $l = l^0$, $w = w^0$, $h = h^0$ and then starting from the point $\omega^0 = (x_1^0, y_1^0, z_1^0, \ldots, x_N^0, y_N^0, z_N^0, \lambda^0 = 0)$ solve the following nonlinear programming subproblem:

$$\max \lambda \text{ s.t. } \omega \in W_\lambda \subset R^{3N+1},$$

$$W_\lambda = \{\omega : \Phi^{S_q S_g}(\omega) \geq 0, \Phi^{S_q \Omega^*}(\omega) \geq 0, q < g \in I_N, 1 - \lambda \geq 0, \lambda \geq 0\}.$$

Here $\omega = (x_1, y_1, z_1, \ldots, x_N, y_N, z_N, \lambda)$,

$$\Phi^{S_q S_g}(\omega) = (x_q - x_g)^2 + (y_q - y_g)^2 + (z_q - z_g)^2 - (\lambda r_q + \lambda r_g)^2$$

is the phi-function for the sphere S_q of the radius λr_q and the sphere S_g of the radius λr_g,

$$\Phi^{S_q \Omega^*}(\omega) = \min\{\varphi(\omega)_{kq}, k = 1, \ldots, 6\}$$

is the phi-function for the sphere S_q of the radius λr_q and the object Ω^*, where

$$\varphi_{1q}(\omega) = l^0 - x_q - \lambda r_q, \quad \varphi_{2q}(\omega) = x_q - \lambda r_q,$$
$$\varphi_{3q}(\omega) = w^0 - y_q - \lambda r_q, \quad \varphi_{4q}(\omega) = y_q - \lambda r_q,$$
$$\varphi_{5q}(\omega) = h^0 - z_q - \lambda r_q, \quad \varphi_{6q}(\omega) = z_q - \lambda r_q.$$

Denote the global maximum point of the above subproblem by $(x_1^*, y_1^*, z_1^*, \ldots, x_N^*, y_N^*, z_N^*, \lambda^* = 1)$ and form a vector of feasible parameters $\varsigma^0 = (l^0, w^0, h^0, u_1^0, \ldots, u_q^0, \ldots, u_N^0)$. Here $u_q^0 = (x_q^*, y_q^*, z_q^*, \theta_q^0)$ and θ_q^0 is a vector of randomly generated rotation parameters of the objects $T_q, q \in J_N$.

To generate a starting point $u^0 = (\varsigma^0, \tau^0)$ for a subsequent search for a local minimum of the problem (8)–(9), define a vector $\tau^0 = (\tau_{ij}^{qg}, i = 1, \ldots, n_q, j = 1, \ldots, n_q, q > g \in I_N)$ by solving the following optimization subproblems: $\max_{\tau_{ij}^{qg}} \Phi'^{qg}_{ij}(u_q, u_g, \tau_{ij}^{qg})$ for $i = 1, \ldots, n_q, j = 1, \ldots, n_g, q > g \in I_N$.

4.2.2. Local Optimization

The definition of the feasible set W in (9) involves a large number $O(N^2)$ of inequalities and variables. To cope with this large-scale problem the decomposition algorithm [47] is used that reduces the problem (8)–(9) to a sequence of nonlinear programming subproblems with a smaller number $O(N)$ of inequalities and variables. The key idea of the algorithm is as follows. For each vector of feasible placement parameters of the objects, fixed individual cubic containers are constructed containing spheres that circumscribe the appropriate convex basic object. Each sphere is allowed to move within the appropriate individual container. The motion of each sphere is described by a system of six linear inequalities. Then a subregion of the feasible region W is formed as follows. For all spheres, $O(N)$ inequalities are added to the system (9) and $O(N^2)$ phi-inequalities corresponding to the pairs of basic objects with individual containers non-overlapping each other are deleted. Moreover, some redundant containment constraints are also deleted. This auxiliary local minimization subproblem has $O(N)$ variables and nonlinear constraints. The solution to this problem is used as a starting feasible point for the next iteration. On the last iteration of the algorithm a local minimum to the problem (8)–(9) is obtained.

5. Computational Results

In this section, five new benchmark instances are provided to demonstrate the efficiency of the proposed methodology. All experiments were running on an AMD FX(tm)-6100, 3.30 GHz computer (Ultra A0313). Programming Language C++, Windows 7. For the local optimisation, the IPOPT code (https://projects.coin-or.org/Ipopt) reported in [48] was used under default options. The multistart approach was used for the problem (8)–(9) as follows. For each problem instance, 10 starting points were generated using the algorithm of SubSection 4.2.1. Then 10 corresponding local minima were obtained by the algorithm described in SubSection 4.2.2 and the best local mimimum was selected as an approximate solution to the problem (8)–(9). The CPU time indicated for each problem instance is the total time for all 10 runs.

Example 1. *The optimized packings of $N = 25$ irregular objects. Each object $T_q(u_q), q = 1, \ldots, 25$ is composed by the union of two cones $T_i^q(u_q), i = 1, 2$ given by the following parameters: $p_{11}^q = (0,0,0)$, $p_{12}^q = (9,0,0)$, $\mathbf{n}_1^q = (1,0,0)$, $r_{11}^q = 3$, $r_{12}^q = 0$ and $p_{21}^q = (7,0,0)$, $p_{22}^q = (-2,0,0)$, $\mathbf{n}_2^q = (1,0,0)$, $r_{21}^q = 3$, $r_{22}^q = 0$ respectively.*

The best local minimum obtained for 2857.99 sec is $\kappa^* = l^* \cdot w^* \cdot h^* = 17.673918 \cdot 20.065788 \cdot 24.972631 = 8856.3211208954$.

The corresponding packing is shown in Figure 4a.

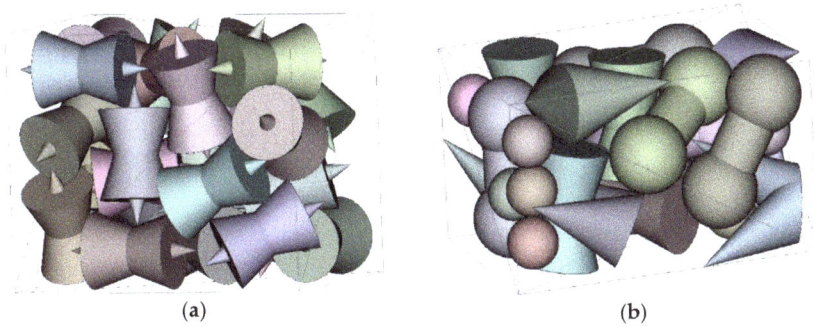

Figure 4. Optimized packings of composed objects: (**a**) Example 1; (**b**) Example 2.

Example 2. *The optimized packings of $N = 25$ basic and irregular objects, including:*

- spheres $T_q(u_q), q = 1, \ldots, 10$ of radii $r^q = 2$ centred at $p^q = (0,0,0)$;
- irregular objects $T_q(u_q), q = 11, \ldots, 15$ composed by the union of three basic objects $T_i^q(u_q), i = 1, 2, 3$, where $T_1^q(u_q)$ is the cylinder with $p_{11}^q = (0,0,0)$, $p_{12}^q = (8,0,0)$, $\mathbf{n}_1^q = (1, 0, 0)$, $r_{11}^q = 2$, $r_{12}^q = 2$; $T_2^q(u_q)$ and $T_3^q(u_q)$ are the spheres of radii $r_2^q = r_3^q = 3$, centred at the points $p_2^q = p_3^q = (0,0, 0)$;
- irregular objects $T_q(u_q), q = 16, \ldots, 20$ composed by the union of two basic objects $T_i^q(u_q), i = 1, 2$, where $T_1^q(u_q)$ is the truncated cone with $p_{11}^q = (0,0,0)$, $p_{12}^q = (9,0,0)$, $\mathbf{n}_1^q = (1,0,0)$, $r_{11}^q = 1$, $r_{12}^q = 3$; $T_2^q(u_q)$ is the cone with $p_{21}^q = (0,0,0)$, $p_{22}^q = (9,0,0)$, $\mathbf{n}_2^q = (1,0,0)$, $r_{21}^q = 3$, $r_{22}^q = 0$;
- irregular objects $T_q(u_q), q = 21, \ldots, 25$ composed by the union of two cones $T_i^q(u_q), i = 1, 2$ with $p_{11}^q = (0,0,0)$, $p_{12}^q = (8,6,0)$, $\mathbf{n}_1^q = (1,0,0)$, $r_{11}^q = 3$, $r_{12}^q = 0$; $p_{21}^q = (0,0,0)$, $p_{22}^q = (8,6,0)$, $\mathbf{n}_2^q = (1,0,0)$, $r_{21}^q = 3$, $r_{22}^q = 0$ respectively.

The best local minimum obtained for 3478.23 sec. is $\kappa^* = l^* \cdot w^* \cdot h^* = 14.889393 \cdot 24.925430 \cdot 17.165791 = 6370.6459961746$.

The corresponding packing is shown in Figure 4b.

Example 3. *The optimized packings of $N = 2, 3, 4, 5$ objects from Example 1:*

(1) for $N = 2$ the best local minimum

$$\kappa^* = l^* \cdot w^* \cdot h^* = 8.085071 \cdot 10.392305 \cdot 6.000000 = 504.135155$$

was obtained for 11.638 sec. starting from the feasible solution

$$\kappa^0 = l^0 \cdot w^0 \cdot h^0 = 26.147623 \cdot 14.547834 \cdot 14.550066 = 5534.7182137988.$$

The corresponding packings are shown in Figure 5.

(2) for $N = 3$ the best local minimum

$$\kappa^* = l^* \cdot w^* \cdot h^* = 12.682284 \cdot 11.050980 \cdot 6.000000 = 840.910031$$

was obtained for 26.146 sec. starting from the feasible solution

$$\kappa^0 = l^0 \cdot w^0 \cdot h^0 = 27.617932 \cdot 26.392891 \cdot 13.953370 = 10170.849561976.$$

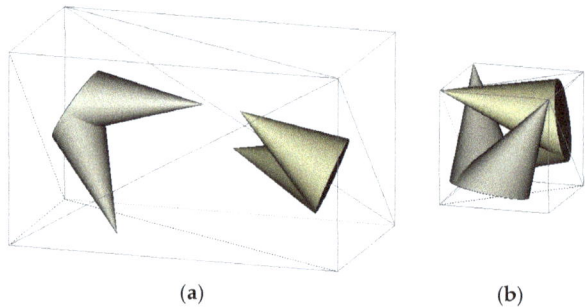

(a)   (b)

Figure 5. Packings of two composed objects: (**a**) the feasible starting point; (**b**) the corresponding local optimal packing.

The corresponding packings are shown in Figure 6.

(3) for $N = 4$ the best local minimum

$$\kappa^* = l^* \cdot w^* \cdot h^* = 8.091432 \cdot 13.454112 \cdot 10.099594 = 1099.4724285295$$

was obtained for 53.602 sec. starting from the feasible solution

$$\kappa^0 = l^0 \cdot w^0 \cdot h^0 = 14.771808 \cdot 27.984482 \cdot 24.620742 = 10177.75667595.$$

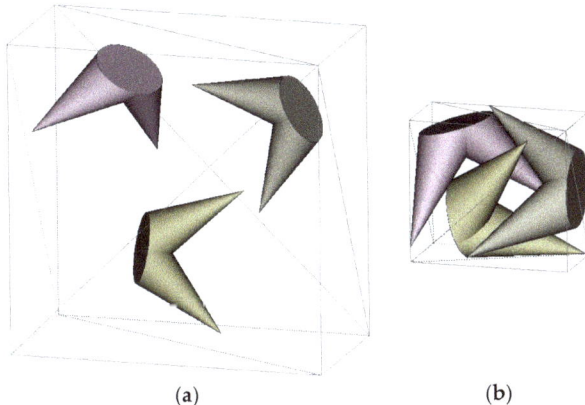

Figure 6. Packings of three composed objects: (**a**) the feasible starting point; (**b**) the corresponding local optimal packing.

The corresponding packings are shown in Figure 7.

(4) *for N = 5 the best local minimum*

$$\kappa^* = l^* \cdot w^* \cdot h^* = 11.492738 \cdot 11.267728 \cdot 10.652737 = 1379.4979707504$$

was obtained for 79.186 sec. starting from the feasible solution

$$\kappa^0 = l^0 \cdot w^0 \cdot h^0 = 26.126900 \cdot 26.524042 \cdot 23.378807 = 16201.302676444.$$

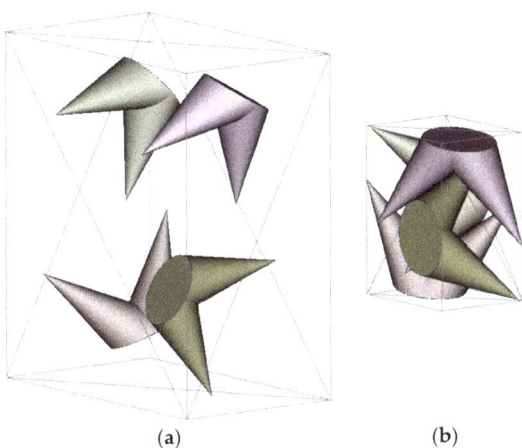

Figure 7. Packings of four composed objects: (**a**) the feasible starting point; (**b**) the corresponding local optimal packing.

The corresponding packings are shown in Figure 8.

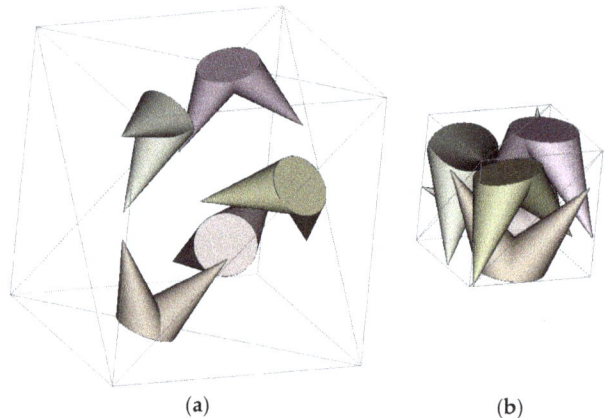

(a)   (b)

Figure 8. Packings of five composed objects: (**a**) the feasible starting point; (**b**) the corresponding local optimal packing.

Example 4. *The optimized packings of $N = 12$ objects (three objects of each type from Example 2). The best local minimum*

$$\kappa^* = l^* \cdot w^* \cdot h^* = 15.084109 \cdot 15.073729 \cdot 16.129947 = 3667.5268798149$$

was obtained for 934.103 sec. starting from the feasible solution

$$\kappa^0 = l^0 \cdot w^0 \cdot h^0 = 20.944159 \cdot 24.547251 \cdot 32.847595 = 16887.65573111.$$

The corresponding packings are shown in Figure 9.

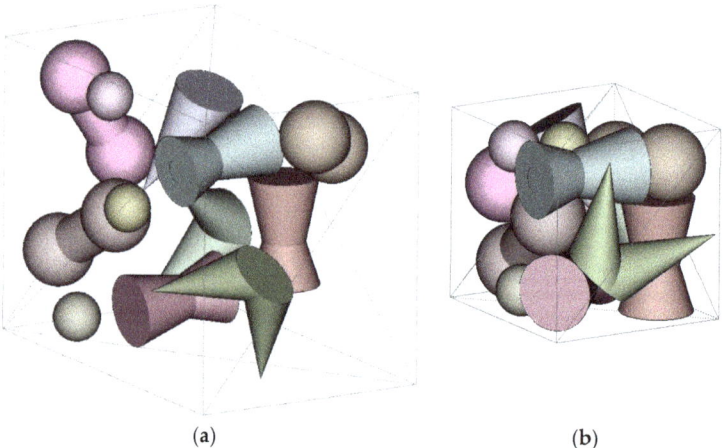

(a)   (b)

Figure 9. Packings of twelve composed objects: (**a**) the feasible starting point; (**b**) the corresponding local optimal packing.

Table 1 below provides objective function values for the feasible starting point (κ^0) and the corresponding local optimal solution (κ^*) for all 10 starting points. These values are indicated for N = 2, 3, 4, 5 (Example 3) and N = 12 (Example 4). The best local optimal solutions are highlighted in bold.

Table 1. Objective values for feasible starting points and corresponding local optima.

	N = 2		N = 3		N = 4		N = 5		N = 12	
	κ^0	κ^*	κ^0	κ^*	κ^0	κ^*	κ^0	κ^*	κ^0	κ^*
1	5337.97	816.30	10,170.84	**840.91**	11,077.98	1565.10	15,543.36	1536.67	18,810.97	4132.31
2	5215.23	835.79	8733.59	1118.40	12,479.28	1372.18	15,564.86	1639.94	18,886.64	4039.23
3	5580.59	1002.95	9085.77	1319.98	10,011.8	1668.54	15,415.77	1507.72	16,951.12	4014.97
4	5457.55	670.08	9949.29	1145.73	10,177.75	**1099.47**	16,201.30	**1379.50**	16,708.20	4478.20
5	5534.71	**504.13**	9029.34	1063.02	12,348.33	1214.28	15,253.70	1920.94	16,726.72	4160.61
6	5260.31	670.08	9801.35	1409.52	17,169.69	1365.29	16,085.74	1733.08	16,480.82	4055.20
7	5395.85	767.15	10,148.29	1152.08	11,556.14	1826.22	15,634.75	1398.68	16,887.65	**3667.53**
8	5403.27	1091.65	12,206.66	1134.85	10,317.50	1597.13	17,578.42	1670.72	17,137.79	4102.02
9	5332.57	761.25	12,003.17	1094.66	12,295.11	1264.55	17,556.03	1682.84	17,311.19	3809.62
10	4787.59	896.41	9634.21	1426.20	12,860.21	1637.15	15,228.21	1800.38	15,868.61	3879.69

As can be seen from Table 1, different starting points lead to different local minima.

6. Conclusions

Packing irregular 3D objects in a cuboid of minimum volume is considered. Each irregular object is composed by convex shapes from the family of oblique and right circular cylinders, cones, truncated cones and spheres. Continuous translations and rotations for all objects are allowed. The optimized packing problem is formulated for 3D regular and irregular objects. New analytical tools (quasi-phi-functions and phi-functions) are defined for the first time to describe non-overlapping and containment constraints for irregular objects composed by oblique shapes.

The phi-function technique is used to state the irregular packing in the form of nonlinear programming problem. The solution approach is proposed and illustrated by the numerical examples. The problem instances were selected to demonstrate the ability of the proposed modelling techniques to work with complex objects composed by different convex shapes used in applications.

The multistart algorithm used in this paper consists of two stages: constructing a number of initial feasible solutions and using local minimization (compaction) procedure to improve starting points. A simple and fast heuristic was implemented to get a starting solution. Using more advanced heuristics to construct better starting points may result in improving the overall optimization scheme. Some results in this direction are on the way.

The model (8)–(9) provides all global solutions to the original irregular packing problem. It can be solved by global solvers, e.g., BARON or LGO included in AMPL [46]. However, due to a large number of variables and constraints, the direct solution to this problem is time consuming and this approach was not used in the paper. Instead, the decomposition technique was used for the large-scale problem (8)–(9) and combined with IPOPT for solving NLP subproblems. An interesting direction for the future research is using alternative decomposition techniques [49] or constructing Lagrangian relaxations with respect to binding constraints (see, e.g., [50] and the references therein).

Author Contributions: Investigation, A.P., T.R., I.L.; Methodology, A.P., T.R., I.L.; Programming Algorithms, A.P.; Project administration, T.R., I.L.; Writing—original draft, T.R., I.L.; Writing—review and editing, T.R., I.L. All authors have read and agreed to the published version of the manuscript.

Funding: This research received no external funding.

Acknowledgments: A. Pankratov and T. Romanova were partially supported by the "Program for the State Priority Scientific Research and Technological (Experimental) Development of the Department of Physical and Technical Problems of Energy of the National Academy of Sciences of Ukraine" (#6541230).

Conflicts of Interest: The authors declare no conflict of interest.

Appendix A

To place feasibly two objects within a container, an analytical description of the relations between the pair of objects is required. In this paper the phi-function technique is used to express this relation. Let A and B be three-dimensional objects. The position of the object A is defined by a vector of placement parameters $u_A = (v_A, \theta_A)$, where: $v_A = (x_A, y_A, z_A)$ is a translation vector and θ_A is a vector of rotation parametrs. The object A, rotated by θ_A and translated by v_A is denoted by $A(u_A)$. Phi-functions allow us to distinguish the following three cases: A and B are intersecting so that A and B have common interior points; A and B do not intersect, i.e., A and B do not have common points; A and B are in contact, i.e., A and B have only common frontier points.

A continuous function $\Phi^{AB}(u_A, u_B)$ is called a phi-function of the objects $A(u_A)$ and $B(u_B)$ if the following conditions are fulfilled [38]: $\Phi^{AB}(u_A, u_B) > 0$, for $A(u_A) \cap B(u_B) = \emptyset$ (see Figure A1a); $\Phi^{AB}(u_A, u_B) = 0$, for $\text{int}A(u_A) \cap \text{int}B(u_B) = \emptyset$ and $frA(u_A) \cap frB(u_B) \neq \emptyset$ (see Figure A1b); $\Phi^{AB}(u_A, u_B) < 0$, for $\text{int}A(u_A) \cap \text{int}B(u_B) \neq \emptyset$ (see Figure A1c).

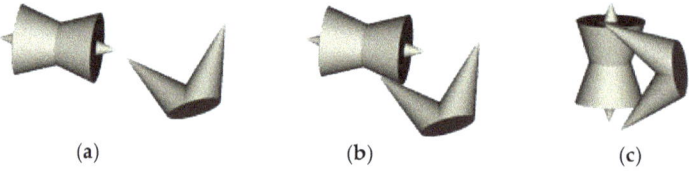

(a)　　　　　　　(b)　　　　　　　(c)

Figure A1. Attributes of a phi-function: (**a**) non-overlapping, $\Phi^{AB}(u_A, u_B) > 0$; (**b**) touching, $\Phi^{AB}(u_A, u_B) = 0$; (**c**) interior overlapping, $\Phi^{AB}(u_A, u_B) < 0$.

Here frA denotes the boundary (frontier) of the object A, while $\text{int}A$ stands for its interior. Thus, $\Phi^{AB}(u_A, u_B) \geq 0 \Leftrightarrow \text{int}A(u_A) \cap \text{int}B(u_B) = \emptyset$.

A function $\Phi'^{AB}(u_A, u_B, u')$ is called a *quasi-phi-function* for two objects $A(u_A)$ and $B(u_B)$ if $\max_{u' \in U} \Phi'^{AB}(u_A, u_B, u')$ is a phi-function for the objects [12]. Here u' denotes a vector of auxiliary variables depending on the object shapes. This function is defined for all values of u_A, u_B and has to be continuous in all its variables.

The main property of the quasi-phi-function for two objects $A(u_A)$ and $B(u_B)$ is as follows: if $\Phi'^{AB}(u_A, u_B, u') \geq 0$ for some u', then $\text{int}A(u_A) \cap \text{int}B(u_B) = \emptyset$.

References

1. Bortfeldt, A.; Wäscher, G. Constraints in container loading—A state-of-the-art review. *Eur. J. Oper. Res.* **2013**, *229*, 1–20. [CrossRef]
2. Leao, A.A.; Toledo, F.M.; Oliveira, J.F.; Carravilla, M.A.; Alvarez-Valdés, R. Irregular packing problems: A review of mathematical models. *Eur. J. Oper. Res.* **2020**, *282*, 803–822. [CrossRef]
3. Chazelle, B.; Edelsbrunner, H.; Guibas, L.J. The complexity of cutting complexes. *Discret. Comput. Geom.* **1989**, *4*, 139–181. [CrossRef]
4. Wäscher, G.; Haußner, H.; Schumann, H. An improved typology of cutting and packing problems. *Eur. J. Oper. Res.* **2007**, *183*, 1109–1130. [CrossRef]
5. Burtseva, L.; Salas, B.V.; Romero, R.; Werner, F. *Multi-Sized Sphere Packings: Models and Recent Approaches [Preprint]*; Fakultät für Mathematik, Otto-von-Guericke-Universität Magdeburg: Magdeburg, Germany, 2015. [CrossRef]
6. Burtseva, L.; Salas, B.V.; Romero, R.; Werner, F. Recent advances on modelling of structures of multi-component mixtures using a sphere packing approach. *Int. J. Nanotechnol.* **2016**, *13*, 44. [CrossRef]
7. Fasano, G.; Pintér, J.D. Modeling and optimization with case studies. In *Springer Optimization and Its Applications: Space Engineering*; Springer: New York, NY, USA, 2016; Volume 114. [CrossRef]
8. Gately, R.D.; Panhuis, M.I.H. Filling of carbon nanotubes and nanofibres. *Beilstein J. Nanotechnol.* **2015**, *6*, 508–516. [CrossRef]

9. Ungson, Y.; Burtseva, L.; Garcia-Curiel, E.R.; Salas, B.V.; Flores-Rios, B.L.; Werner, F.; Petranovskii, V. Filling of irregular channels with round cross-section: Modeling aspects to study the properties of porous materials. *Materials* **2018**, *11*, 1901. [CrossRef]
10. Araújo, L.J.; Özcan, E.; Atkin, J.; Baumers, M. Analysis of irregular three-dimensional packing problems in additive manufacturing: A new taxonomy and dataset. *Int. J. Prod. Res.* **2018**, *57*, 5920–5934. [CrossRef]
11. Stoyan, Y.; Панкратов, A.B.; Romanova, T.; Butenko, S.; Pardalos, P.M.; Shylo, V. Placement problems for irregular objects: Mathematical modeling, optimization and applications. *Springer Texts Stat.* **2017**, *130*, 521–559. [CrossRef]
12. Romanova, T.; Bennell, J.; Stoyan, Y.; Панкратов, A.B. Packing of concave polyhedra with continuous rotations using nonlinear optimisation. *Eur. J. Oper. Res.* **2018**, *268*, 37–53. [CrossRef]
13. Kovalenko, A.; Romanova, T.; Stetsyuk, P. Balance Layout Problem for 3D-Objects: Mathematical Model and Solution Methods. *Cybern. Syst. Anal.* **2015**, *51*, 556–565. [CrossRef]
14. Edelkamp, S.; Wichern, P. Packing irregular-shaped objects for 3D Printing. In *KI: Advances in Artificial Intelligence*; Springer: Cham, Switzerland, 2015; Volume 9324, pp. 45–58.
15. Romanova, T.; Litvinchev, I.; Pankratov, A. Packing ellipsoids in an optimized cylinder. *Eur. J. Oper. Res.* **2020**. [CrossRef]
16. Romanova, T.; Панкратов, A.B.; Litvinchev, I.; Plankovskyy, S.; Tsegelnyk, Y.; Shypul, O. Sparsest packing of two-dimensional objects. *Int. J. Prod. Res.* **2020**, 1–16. [CrossRef]
17. Stoyan, Y.; Grebennik, I.; Romanova, T.; Kovalenko, A. Optimized packings in space engineering applications: Part II. In *Modeling and Optimization in Space Engineering*; Springer International Publishing: New York, NY, USA, 2019; Volume 144, pp. 439–457.
18. Stoyan, Y.; Pankratov, A.; Romanova, T.; Fasano, G.; Pintér, J.D.; Stoian, Y.E.; Chugay, A. Optimized packings in space engineering applications: Part I. In *Modeling and Optimization in Space Engineering*; Springer International Publishing: New York, NY, USA, 2019; Volume 144, pp. 395–437. [CrossRef]
19. Fasano, G. A Modeling-based approach for non-standard packing problems. In *Springer Optimization and Its Applications: Space Engineering*; Springer: New York, NY, USA, 2015; Volume 105, pp. 67–85. [CrossRef]
20. Hifi, M.; Yousef, L. A local search-based method for sphere packing problems. *Eur. J. Oper. Res.* **2019**, *274*, 482–500. [CrossRef]
21. Pintér, J.D.; Kampas, F.J.; Castillo, I. Globally optimized packings of non-uniform size spheres in R^d: A computational study. *Optim. Lett.* **2017**, *12*, 585–613. [CrossRef]
22. Litvinchev, I.; Infante, L.; Ozuna Espinosa, E.L. Approximate Circle Packing in a Rectangular Container: Integer Programming Formulations and Valid Inequalities. In Proceedings of the International Conference on Computational Logistics, ICCL, Valparaíso, Chile, 24–26 September 2014; González-Ramírez, R.G., Schulte, F., Voß, S., Ceroni Díaz, J.A., Eds.; Springer: Cham, Switzerland, 2014; Volume 8760, pp. 47–60.
23. Depriester, D.; Kubler, R. Radical Voronoï tessellation from random pack of polydisperse spheres: Prediction of the cells' size distribution. *Comput. Des.* **2019**, *107*, 37–49. [CrossRef]
24. Lucarini, V. Three-Dimensional Random Voronoi Tessellations: From Cubic Crystal Lattices to Poisson Point Processes. *J. Stat. Phys.* **2009**, *134*, 185–206. [CrossRef]
25. Pankratov, A.; Romanova, T.; Litvinchev, I.; Marmolejo-Saucedo, J.A. An Optimized Covering Spheroids by Spheres. *Appl. Sci.* **2020**, *10*, 1846. [CrossRef]
26. Zhao, B.; An, X.; Wang, Y.; Qian, Q.; Yang, X.; Sun, X. DEM dynamic simulation of tetrahedral particle packing under 3D mechanical vibration. *Powder Technol.* **2017**, *317*, 171–180. [CrossRef]
27. Wang, X.; Zhao, L.; Fuh, J.Y.H.; Lee, H.P. Lee effect of porosity on mechanical properties of 3D Printed polymers: Experiments and micromechanical modeling based on X-Ray computed tomography analysis. *Polymers* **2019**, *11*, 1154. [CrossRef]
28. Egeblad, J.; Nielsen, B.K.; Brazil, M. Translational packing of arbitrary polytopes. *Comput. Geom.* **2009**, *42*, 269–288. [CrossRef]
29. Gogate, A.S.; Pande, S.S. Intelligent layout planning for rapid prototyping. *Int. J. Prod. Res.* **2008**, *46*, 5607–5631. [CrossRef]
30. Joung, Y.-K.; Noh, S.D. Intelligent 3D packing using a grouping algorithm for automotive container engineering. *J. Comput. Des. Eng.* **2014**, *1*, 140–151. [CrossRef]
31. Mack, D.; Bortfeldt, A.; Gehring, H. A parallel hybrid local search algorithm for the container loading problem. *Int. Trans. Oper. Res.* **2004**, *11*, 511–533. [CrossRef]

32. Liu, X.; Liu, J.-M.; Cao, A.-X.; Yao, Z.-L. HAPE3D—A new constructive algorithm for the 3D irregular packing problem. *Front. Inf. Technol. Electron. Eng.* **2015**, *16*, 380–390. [CrossRef]
33. Lutters, E.; Dam, D.T.; Faneker, T. 3D nesting of complex shapes. *Procedia CIRP* **2012**, *3*, 26–31. [CrossRef]
34. Bortfeldt, A.; Gehring, H. A hybrid genetic algorithm for the container loading problem. *Eur. J. Oper. Res.* **2001**, *131*, 143–161. [CrossRef]
35. Gehring, H.; Bortfeldt, A. A Parallel genetic algorithm for solving the container loading problem. *Int. Trans. Oper. Res.* **2002**, *9*, 497–511. [CrossRef]
36. Ma, Y.; Chen, Z.; Hu, W.; Wang, W. Packing irregular objects in 3D space via hybrid optimization. *Comput. Graph. Forum* **2018**, *37*, 49–59. [CrossRef]
37. Zhao, C.; Jiang, L.; Teo, K.L. A hybrid chaos firefly algorithm for three-dimensional irregular packing problem. *J. Ind. Manag. Optim.* **2020**, *16*, 409–429. [CrossRef]
38. Chernov, N.; Stoyan, Y.; Romanova, T. Mathematical model and efficient algorithms for object packing problem. *Comput. Geom.* **2010**, *43*, 535–553. [CrossRef]
39. You, Y.; Zhao, Y. Discrete element modelling of ellipsoidal particles using super-ellipsoids and multi-spheres: A comparative study. *Powder Technol.* **2018**, *331*, 179–191. [CrossRef]
40. Garboczi, E.; Bullard, J. 3D analytical mathematical models of random star-shape particles via a combination of X-ray computed microtomography and spherical harmonic analysis. *Adv. Powder Technol.* **2017**, *28*, 325–339. [CrossRef]
41. Wang, S.; Marmysh, D.; Ji, S. Construction of irregular particles with superquadric equation in DEM. *Theor. Appl. Mech. Lett.* **2020**, *10*, 68–73. [CrossRef]
42. Rakotonirina, A.D.; Delenne, J.-Y.; Radjaï, F.; Wachs, A. Grains3D, a flexible DEM approach for particles of arbitrary convex shape—Part III: Extension to non-convex particles modelled as glued convex particles. *Comput. Part. Mech.* **2018**, *6*, 55–84. [CrossRef]
43. Gibson, I.; Rosen, D.; Stucker, B. *Additive Manufacturing Technologies, 3D Printing, Rapid Prototyping, and Direct Digital Manufacturing*; Springer Science + Business Media: New York, NY, USA, 2015; ISBN 978-1-4939-2113-3.
44. Yuan, Y.; Liu, L.; Deng, W.; Li, S. Random-packing properties of spheropolyhedra. *Powder Technol.* **2019**, *351*, 186–194. [CrossRef]
45. Wei, G.; Zhang, H.; An, X.; Jiang, S. Influence of particle shape on microstructure and heat transfer characteristics in blast furnace raceway with CFD-DEM approach. *Powder Technol.* **2020**, *361*, 283–296. [CrossRef]
46. Robert, F.; Gay, D.M.; Brian, W. *Kernighan, AMPL: A Modeling Language for Mathematical Programming*, 2nd ed.; Pacific Grove: New York, NY, USA, 2003.
47. Romanova, T.; Stoyan, Y.G.; Pankratov, A.V.; Litvinchev, I.; Marmolejo-Saucedo, J.A. Decomposition algorithm for irregular placement problems. In *Advances in Intelligent Systems and Computing*; Springer Nature Switzerland AG: Cham, Switzerland, 2019; pp. 214–221.
48. Wächter, A.; Biegler, L.T. On the implementation of an interior-point filter line-search algorithm for large-scale nonlinear programming. *Math. Program.* **2006**, *106*, 25–57. [CrossRef]
49. Litvinchev, I. Decomposition-aggregation method for convex programming problems. *Optimization* **1991**, *22*, 47–56. [CrossRef]
50. Litvinchev, I.S. Refinement of Lagrangian bounds in optimization problems. *Comput. Math. Math. Phys.* **2007**, *47*, 1101–1107. [CrossRef]

© 2020 by the authors. Licensee MDPI, Basel, Switzerland. This article is an open access article distributed under the terms and conditions of the Creative Commons Attribution (CC BY) license (http://creativecommons.org/licenses/by/4.0/).

Article

Optimization Methodologies and Testing on Standard Benchmark Functions of Load Frequency Control for Interconnected Multi Area Power System in Smart Grids

Krishan Arora [1,2], Ashok Kumar [2], Vikram Kumar Kamboj [1], Deepak Prashar [3], Sudan Jha [3], Bhanu Shrestha [4,*] and Gyanendra Prasad Joshi [5,*]

[1] School of Electronics and Electrical Engineering, Lovely Professional University, Phagwara, Punjab 144411, India; krishan.12252@lpu.co.in (K.A.); kamboj.vikram@gmail.com (V.K.K.)
[2] Department of Electrical Engineering, Maharishi Markandeshwar University Mullana, Haryana 133207, India; ashok1234arora@gmail.com
[3] School of Computer Science and Engineering, Lovely Professional University, Phagwara, Punjab 144411, India; deepak.prashar@lpu.co.in (D.P.); sudhan.25850@lpu.co.in (S.J.)
[4] Department of Electronic Engineering, Kwangwoon University, Seoul 01897, Korea
[5] Department of Computer Science and Engineering, Sejong University, Seoul 05006, Korea
* Correspondence: bnu@kw.ac.kr (B.S.); joshi@sejong.ac.kr (G.P.J.); Tel.: +82-10-4590-4460 (B.S.); +82-2-69352481 (G.P.J.)

Received: 1 June 2020; Accepted: 10 June 2020; Published: 16 June 2020

Abstract: In the recent era, the need for modern smart grid system leads to the selection of optimized analysis and planning for power generation and management. Renewable sources like wind energy play a vital role to support the modern smart grid system. However, it requires a proper commitment for scheduling of generating units, which needs proper load frequency control and unit commitment problem. In this research area, a novel methodology has been suggested, named Harris hawks optimizer (HHO), to solve the frequency constraint issues. The suggested algorithm was tested and examined for several regular benchmark functions like unimodal, multi-modal, and fixed dimension to solve the numerical optimization problem. The comparison was carried out for various existing models and simulation results demonstrate that the projected algorithm illustrates better results towards load frequency control problem of smart grid arrangement as compared with existing optimization models.

Keywords: Harris hawks optimizer; load frequency control; sensitivity analysis; smart grid; particle swarm optimization; genetic algorithm; meta-heuristics

1. Introduction

Optimization shows a critical role in various regions of science and technology. This is the method through which the optimal solution can be found with the help of a wide range of search mechanisms like primary, secondary, and tertiary controls [1]. With recent advancement in technology, novel optimization methodologies are identified as meta-heuristic with concern of mathematical culture. Meta-heuristic algorithms (MA) is a typical technique to get the best outcomes for the issue. It plays a fictional role to find good specifications in an optimization matter [2].

Each real-life optimization problem required procedures which observe the examination zones effectively to find most operative explanations. Moth-flame optimizer (MFO) is newly projected meta-heuristics search algorithmic rule that is inspired by the direction-finding environment of lepidopteron and its convergence in the direction of lightweight. However, like alternative similar strategies, MFO contributes to being stuck into sub-optimal segments, which is mirrored within the procedure effort needed to search out the most effective rate. This case happens due to the developer used for research not performing well to research the find house. In addition, no free lunch theorem encourages planners to promote a new algorithmic rule or to boost the prevailing algorithmic rule.

The modern technology that balances the two-way communication between energy production and consumption and sense the critical behavior of voltage, current, and frequency which makes an electric grid as a smart grid. Smart grid is an opportunity in the growth of the country's economy and environmental health due to efficient electricity transmission, quicker restoration, reduced power cost, and enhanced integration with renewable energy sources, which is possible through optimal gain scheduling and the load frequency control method.

In earlier days, the load frequency control (LFC) problem was explained with respect to conventional dispatching [3], whose objective was to maintain voltages and frequency within prescribed limits. Today, LFC uses advanced numerical optimization techniques to solve constrained combinatorial and diverse number optimization issues. The type of controller [4], its architecture and choice of objective function play a very important role in enhancing achievement of the power system.

In the current scenario, the integral of time multiplied absolute error (ITAE) criteria is observed as an impartial task which is stated as [5]:

$$J = \int_0^{t_{sim}} (|\Delta F_1| + |\Delta F_2| + |\Delta P_{tie}|).t.dt, \qquad (1)$$

where, ΔF_1, ΔF_2 indicate deviation of the frequency in both areas and the total simulation time (in seconds) is denoted by 't_{sim}' and tie-line interchange [6] assessment is characterized by ΔP_{tie}.

The ITAE is implemented as a detached role to enhance gain of the PI controller in the present investigation. The reduction of the ITAE index with the binary moth flame optimizer (BMFO) algorithm offers augmented constraints of PI controllers which can be subjected to the following restraints [7–9]: Minimize J,

$$K_i^P \min \leq K^P \leq K_i^P \max, \text{ and } K_i^{Int.} \min \leq K^{Int.} \leq K_i^{Int.} \max,$$

where, $K_i^{Int.}$ and K_i^P symbolize fundamental and comparative gain of PI controller of i_{th} ($i = 1, 2$) area.

Our contributions in this work are as follows: First, we propose the two variants of binary moth flame optimizers to solve the frequency constraint issues. We implemented two different binary variants for improving performance of the moth flame optimizer (MFO) for discrete optimization problems. In the first variant, i.e., binary moth flame optimizer (BMFO1), coin flipping-based selection probability of binary numbers is used. We used the improved Sigmoid transformation in the second variant called BMFO2. These binary MFO algorithms along with the Harris hawks optimizer (HHO) algorithms are tested and analyzed for various unimodal, multi-modal, and fixed dimension numerical optimization problem. Secondly, Section 2 explores various optimization methodologies, including classical artificial intelligence techniques, modern intelligence techniques, hybrid artificial intelligence techniques, and smart grid technologies which are tested using standard benchmarks and compared with various algorithms. Lastly, in Section 3, all the latest used algorithms are evaluated and compared in terms of standard testing benchmarks in which the proposed HHO model is having improved results in terms of average and standard deviation. Finally, Section 4 concludes the paper.

2. Optimization Methodologies

In order to discover the mathematical design of load frequency control, numerous optimization methodologies are classified into three foremost groups like traditional techniques [10], recent techniques [11], and hybrid techniques [12].

2.1. Traditional Techniques

The traditional methods may be further classified into artificial neural network, fuzzy logic technique [13], and genetic algorithm.

2.1.1. Artificial Neural Network

The architecture of Artificial Neural Network (ANN) as shown in Figure 1 is promptly the emerging zone of investigation, producing attention of predictors from a noble type of scientific field, which gives a deviation of desired output and actual output as an error signal. An error signal acts like a feedback to the neural network, which balances the desired and actual output.

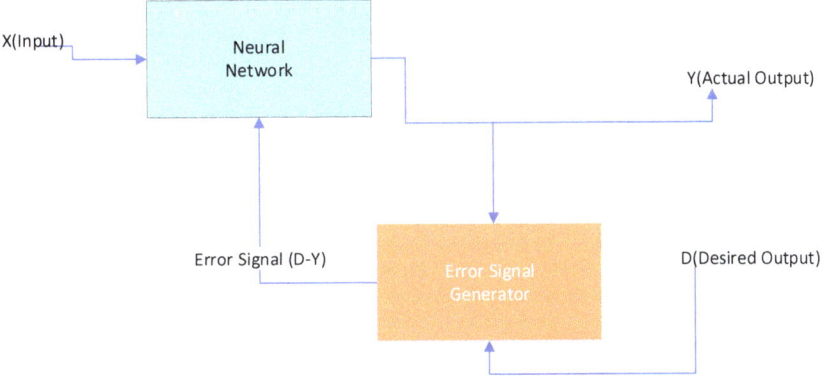

Figure 1. Artificial neural network architecture.

2.1.2. Fuzzy Logic Technique

The essential configuration of the scientific reasoning scheme in which the fuzzification [14] boundary recreates the additional contribution into a fuzzy verbal input, and likewise shows an significant character in the mathematical coherent [15] procedure as actual principles, which are delivered from current sensors, are a forever crisp analytical equivalent as shown in Figure 2.

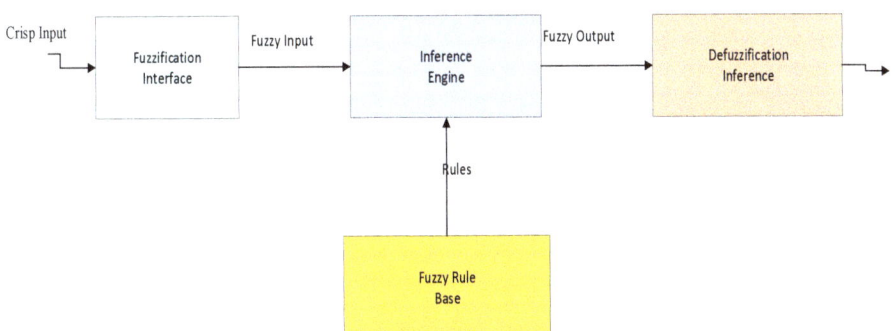

Figure 2. Fuzzy logic system architecture.

2.1.3. Genetic Algorithm

The overall thoughts were conceived by a European country [16], whereas practicality of persecution of exhausting it to untie innovative concerns was indisputable. It may be a soft computing style, which implements strategies stimulated by usual hereditary knowledge to develop conclusions to matters [17]. Genetic Algorithm (GA) as shown in Figure 3 is refreshed by Darwin's theory concerning progression, which is useful to a vast variety of methodical and industrial problems like optimization, machine learning, and automatic software design [18].

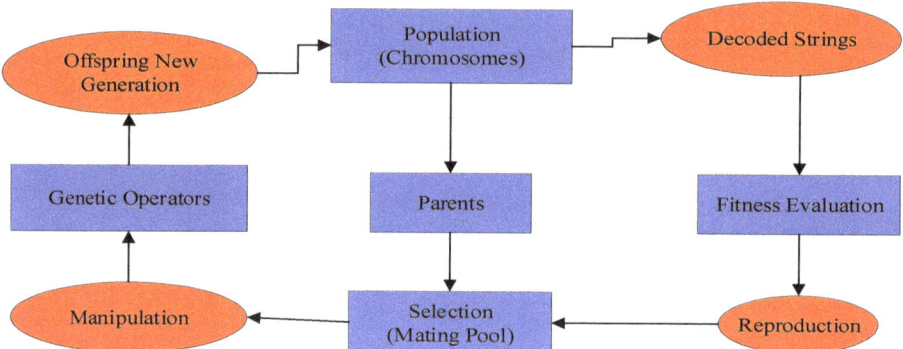

Figure 3. Genetic algorithm architecture.

2.2. Modern Intelligence Techniques

To solve the multi-disciplinary optimization problems [14], various modern practices are established by the investigators. The modern intelligence techniques are explored in the following sub-sections.

2.2.1. Differential Evolution Technique

It is a genetic-based algorithm [19] having identical operators corresponding to initialization, mutation, crossover, and selection. In this method, all constraints are expressive in genetic measurable by a genuine measurement [20]. The mathematical formulation of differential evolution is given below:

- **Initialization**

Firstly, whole vector of initial population is assigned any arbitrary assessment [21] starting with its equivalent state:

$$X_{j,i}^{(0)} = X_j^{min} + \mu_j \left(X_j^{max} - X_j^{min} \right), \qquad (2)$$

where μ_j represents uniformly dispersed arbitrary numeral initialize with the array of [0, 1], generates novel for all value of X_j^{min} and X_j^{max} are representing the uppermost and lowermost limits of the jth parameter, correspondingly.

- **Mutation**

This operator [22] generates distorted vectors X'_i by disturbing a randomly chosen vector 'X_a' and dissimilarity randomly chosen vectors 'X_b' and 'X_c' as per the following equation:

$$X_i'^{(G)} = X_a'^{(G)} + a\left(X_b^{(G)} + X_c^{(G)} \right) \ i = 1,\ldots .N\,P, \qquad (3)$$

where 'X_a','X_b', and 'X_c' represent the randomly selected vectors among set of population, and 'α' represents the scaling constant of the algorithm parameter which is used to regulate the size of the mutation operator and find better results.

- **Crossover**

    Crossover operations [23] create trial vectors X_i'' with integration of the parameters of the distorted vectors X_i' with its objective or parent vectors x_i:

$$X_{j,i}''(G) = \begin{cases} X_{j,i}'^{(G)} & \text{if } p_j \leq C_R \text{ or } j = q \\ X_{j,i}^{(G)} & \text{otherwise} \end{cases}, \qquad (4)$$

where p_j represents consistently discrete unplanned integer [24] between the variety of 0 and 1 and generates an extra for every value of j. q represents the random selected indicator {1, ..., NP} of the trial vector [25] obtain one parameter as a distorted vector. C_R representing the crossover operation constant of algorithm parameters [26] that manage the variety of population and algorithm is run absent as of local minima [27].

- **Selection**

    Selection operator [28] develops the population by choosing the trial and parent vectors (precursor) which presents a best fitness [29]:

$$X_i^{(G+1)} = \begin{cases} X_i''^{(G)} & \text{if } f\left(X_i''^{(G)}\right) \leq f\left(X_i'^{(G)}\right) \\ X_i^{(G)} & \text{otherwise} \end{cases} \quad i = 1, \ldots, NP. \qquad (5)$$

This optimization procedure is replicate to the number of generations to obtain superior fitness functions because they required optimal values to explore the search space.

2.2.2. Biogeography Based Optimization

Biogeography Based Optimization (BBO) is the investigation of topographical propagation of living classes which is based on mutation and migration procedures [30].

- **Migration**

    The migration process is either leaving or entering the species from an island. Biogeography-based optimization also used a population of candidate solution for optimization similar to partial swarm optimization and another population-based search method [31]. Depiction of all candidate solutions is complete as a vector of actual statistics. Now, all real statistics is considered in the population as suitability index variable (SIV). SIV [32] is similar to the output power of generating components in load frequency control. Few best solutions are the same in the resultant iterations; the migration process arranges to avoid the best solutions from being changed. Emigration rate [33] and immigration rate [34] for habitat contain 'k' species is express as:

$$\lambda_k = I\left(1 - \frac{K}{\eta}\right), \qquad (6)$$

$$\mu_k = \frac{Ek}{\eta}, \qquad (7)$$

where E represents the emigration rates, I represents the maximum immigration rates, and η represents the maximum number of species, respectively.

- **Mutation**

    The habitat suitability index (HSI) [35] can easily be modified with resultant in the breed calculation to be different from the symmetry value, if a number of catastrophic actions occur. In biogeography-based optimization, this procedure is modeled as SIV mutation and the mutation rates of habitats may be intended to use the species add up probabilities known unexpected modification in weather of one habitat or additional occurrence will cause the unexpected modification in HSI (habitat). This condition is replica in the form of unexpected modification in the value of the suitability index variable in BBO. The probability of some organism [36] is calculated by this equation:

$$P_S = \begin{cases} -(\lambda_S + \mu_S)P_S + \mu_{S+1}P_{S+1} & S = 0 \\ -(\lambda_S + \mu_S)P_S + \lambda_{S-1}P_{S-1} + \mu_{S+1}P_{S+1} & 1 \leq S \leq S_{max-1} \\ -(\lambda_S + \mu_S)P_S + \lambda_{S-1}P_{S-1} & S = S_{max} \end{cases} \quad (8)$$

The own probability of all members is one habitat. If probability of this is too low, and after that, this result has more probability to mutilation [37]. In a similar way, if the probability of a result is more, that result has a small probability to mutate. As a result, solutions with a low suitability index variable and high suitability index variable have a small possibility to grow an improved SIV in the new iteration. Dissimilar low suitability index variable and high suitability index [38] variable solutions, middle HSI solutions have a bigger possibility to grow improved solutions after the mutation process. By the use of equation mutation, all results can be calculated easily:

$$m(s) = m_{max}\left(\frac{1 - P_S}{P_{max}}\right), \quad (9)$$

where $m(s)$ represent the mutation rate.

2.2.3. Dragonfly Algorithm (DA)

DA [39] is an exceptional optimization process planned by Seyedali. The most important purpose of swarm is durability; thus, all individual must be unfocused outward, and opponents attracted towards nourishment sources. Taking both behaviors in swarms [40], these are five major topographies in position informing procedure of individuals. The numerical model of swarms actions as shown below: The parting procedure [41] in DA informing as in the above equation:

$$S_i = -\sum_{J=1}^{N} X - X_J, \quad (10)$$

where N represents the amount of entities of neighboring, X represents the present situation of specific, X_J indicates the location of Jth specific of the adjacent [42].

The orientation procedure in this approach can be rationalized by subsequent expression [43]:

$$A_i = \frac{\sum_{J=1}^{N} V_J}{N}, \quad (11)$$

where V_J represents velocity of Jth specific of the adjacent. The unity in DA can be intended by the above evaluation:

$$C_i = \frac{\sum_{J=1}^{N} X_J}{N} - X, \quad (12)$$

where X represents the existing specific point, X_J is the spot of Jth specific of the adjacent, and N indicates the amount of areas.

2.3. Hybrid Artificial Intelligence Techniques

2.3.1. Particle Swarm Optimization (PSO) and Gravitational Search Algorithm (GSA) Hybridization

The easiest technique to mongrelize PSO and GSA is to implement the strength separately in the successive approach [44].

Particle Swarm Optimization (PSO)

PSO is provoked with keen collective activities [45] accessible by a multiplicity of creatures, such as the group of ants or net of birds. The particle position and velocity both are updated according to the equations:

$$v_i^d(t+1) = w(t)v_i^d(t) + c_1 X r_1 X(pbest_i^d - x_i^d) + c_2 + r_2, \quad (13)$$

$$x_i^d(t+1) = x_i^d(t+1) + v_i^d(t+1), \quad (14)$$

$$Wt = rand\, X \frac{t}{t_{max}} X(w_{max} - w_{min}) + w_{min}, \quad (15)$$

where $v_i^d(t+1)$ shows velocity of (d^{th}) dimension at (t) reiteration of (i^{th}) particle, $x_i^d(t+1)$ is existing position of (d^{th}) dimensional iteration (l) of (i^{th}) particle; c_1 and c_2 representing the acceleration coefficients [46] which manage the pressure of gbest and pbest on the search procedure, r_1 and r_2 representing the arbitrary statistics in variety [0, 1]; $pbest_i^d$ represents finest point of (i^{th}) element up to now.

Gravitational Search Algorithm (GSA)

GSA is meta-heuristic population-centered approach inspired with directions of attraction and quantity associations [47–49]. In this method, cause is dignified as article encompasses of unlike multitudes and the enactment of this is considered via crowds.

2.3.2. Differential Evolution and Particle Swarm Optimization Hybrids

It is a population-based optimizer [50] alike the genetic algorithm, having identical operatives corresponding to selection, mutation, and crossover. In this method, all constraints are expressive in genetic measurable by a genuine measurement [20,51].

2.3.3. Binary Moth Flame Optimizer (BMFO1)

BMFO is a newly projected meta-heuristics search algorithm proposed by Seyedali Mirjalili [52,53] which is refreshed by direction-finding behavior of moth and its converges near light. Although, moths are having a robust capability to uphold a secure approach with respect to the moon and hold a tolerable erection for nomadic in an orthodox mark for extensive distances. Besides, they are attentive in a fatal/idle curved track over simulated basis of lights.

2.3.4. Modified SIGMOID Transformation (BMFO2)

The binary calibration of constant pursuit house and places of search representatives, resolutions to binary exploration house could be the obligatory method for optimization of binary environmental issues such as LFC. In the proposed research, a modified sigmoidal transfer function is adopted, which has superior performance than another alternatives of sigmoidal transfer function as reported in [54].

2.3.5. Harris Hawks Optimizer

HHO [55] is gradient-free and populations-centered algorithm that comprises exploitative and exploratory stages, which is fortified by astonishment swoop, the fauna of examination of a victim, and diverse stratagems built on violent marvel of Harris hawks.

2.3.6. Smart Grid Applications

The modern smart grid system as shown in Figure 4 consists of various power generating units consisting of thermal, hydro, nuclear, wind, and solar-based power producing elements. The scheduling of every power producing in optimal condition is a tedious task and requires proper commitment schedule of generating units. Further, consideration of solar and wind-based energy sources requires proper load frequency control [56].

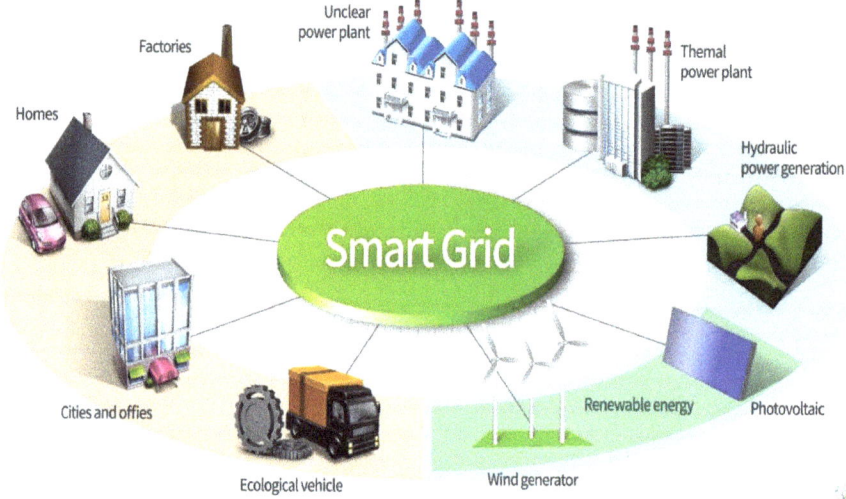

Figure 4. Modern smart grid system with Electric Vehicles (EVs) load demand.

An electric grid can easily be converted into a smart grid by balancing the voltage, current, and frequency which is possible by the load frequency control method [57]. If incoming voltage, current, and frequency is matched with the outgoing voltage, current, and frequency of an electric grid with the help of optimal gain scheduling and load frequency control approach, then steady state error will be near to zero or nil. In the proposed research, load frequency control is tested and validated with various standard benchmarks simultaneously and mathematically depicted in the following sub-sections.

3. Standard Testing Benchmarks

The consequences for various benchmark issues [58] considering the LFC situation are deliberated in the above-mentioned units.

Test System and Standard Benchmark

For confirmation of prospects of deliberate BMFO and HHO algorithms, CEC2005 benchmark functions [59] have been taken into thought, which include unimodal, multi-modal, and fixed dimensions benchmark issues and its mathematical formulation has been represented in Tables 1–3. Table 1 interprets unimodal standard performance, Table 2 portrays multi-modal standard, and Table 3 interprets fixed dimensions standard issues.

To explain the random behavior of the expected BMFO2 logarithmic rule and confirm the consequences, thirty trials were applied with all objective function check for average, variance, best and worst values for justification of output from the probable algorithmic rule, unimodal benchmark work f1, f2, f3, f4, f5, f6, and f7 are used. Table 4 (a) signifies the response of unimodal benchmark function with BMFO1 logarithmic rule, Table 4 (b) characterizes the retort of unimodal benchmark operate function by using the BMFO2 algorithmic rule and Table 4 (c) represents the answer of the fixed dimension benchmark function by using HHO algorithmic instruction.

Table 1. Unimodal benchmark.

Function	Dim	Range	f_{min}
$f_1(x) = \sum_{i=1}^{n} x_i^2$	30	[−100, 100]	0
$f_2(x) = \sum_{i=1}^{n} |x_i| + \prod_{i=1}^{n} |x_i|$	30	[−10, 10]	0
$f_3(x) = \sum_{i=1}^{n} (\sum_{j-1}^{i} x_j)^2$	30	[−100, 100]	0
$f_4(x) = max_i\{|x_i|, 1 \leq i \leq n\}$	30	[−100, 100]	0
$f_5(x) = \sum_{i=1}^{n-1} \left[100(x_{i+1} - x_i^2)^2 + (x_i - 1)^2\right]$	30	[−30, 30]	0
$f_6(x) = \sum_{i=1}^{n} ([x_i + 0.5])^2$	30	[−100, 100]	0
$f_7(x) = \sum_{i=1}^{n} ix_i^4 + random[0, 1]$	30	[−1.28, 1.28]	0

Table 2. Multimodal benchmark.

Function	Dim	Range	f_{min}
$f_8(x) = \sum_{i=1}^{n} -x_i \sin(\sqrt{|x_i|})$	30	[−500, 500]	−418.98
$f_9(x) = \sum_{i=1}^{n} [x_i^2 - 10\cos(2\Pi x_i) + 10]$	30	[−5.12, 5.12]	0
$f_{10}(x) = -20\exp(-0.2\sqrt{\frac{1}{n}\sum_{i=1}^{n} x_i^2}) - \exp(\frac{1}{n}\sum_{i=1}^{n} \cos(2\Pi x)_i + 20 + c$	30	[−32, 32]	0
$f_{11}(x) = \frac{1}{4000}\sum_{i=1}^{n} x_i^2 - \prod_{i=1}^{n} \cos(\frac{x_i}{\sqrt{i}}) + 1$	30	[−600, 600]	0
$f_{12}(x) = \frac{\Pi}{n}\left\{10\sin(\Pi y_1) + \sum_{i=1}^{n-1}(y_i - 1)^2[1 + 10\sin^2(\Pi y_{i+1}) + (y_n - 1)^2]\right\} + \sum_{i=1}^{n} u(x_i, 10, 100, 4)$ where $y_i = 1 + \frac{x_i+1}{4}$, $u(x_i, a, k, m) = \begin{cases} k(x_i - a)^m, & x_i > a \\ 0, & -a < x_i < a \\ k(-x_i - a)^m & x_i < -a \end{cases}$	30	[−50, 50]	0
$f_{13}(x) = 0.1\{\sin^2(3\Pi x_i) + \sum_{i=1}^{n}(x_i - 1)^2[1 + \sin^2(3\Pi x_i + 1] + (x_m - 1)^2[1 + \sin^2(2\Pi x_m)]\} + \sum_{i=1}^{n} u(x_i, 5, 100, 4)$	30	[−50, 50]	0
$f_{14}(x) = -\sum_{i=1}^{n} \sin(x_i) \cdot \left(\sin\left(\frac{ix_i^2}{\Pi}\right)\right)^{2m}, m = 10$	30	[0, π]	−4.687
$f_{15}(x) = \left[e - \sum_{i=1}^{n}(x_i/\beta)^{2m} - 2e - \sum_{i=1}^{n} x_i^2\right] - \prod_{i=1}^{n} \cos^2 x_i, m = 5$	30	[−20, 20]	−1
$f_{16}(x) = \left\{\sum_{i=1}^{n} \sin^2(x_i) - \exp\left(-\sum_{i=1}^{n} x_i^2\right)\right\} \cdot \exp\left[-\sum_{i=1}^{n} \sin^2 \sqrt{|x_i|}\right]$	30	[−10, 10]	−1

It is analyzed from Table 4 that the unimodel benchmark functions f1 to f7 are tested using the modern hybrid algorithms like BMFO 1, BMFO 2, and HHO, and found that Harris hawks optimizer (HHO) produces optimal outcomes in terms of mean, standard deviation, best and worst value for all functions as compared to other algorithms. The convergence curve and trial solutions for BMFO1, BMFO2, and HHO for f1 to f7 unimodal benchmark functions are presented in Figure 5.

Table 3. Fixed dimension benchmark.

Function	Dim	Range	f_{min}
$f_{14}(x) = \left(\frac{1}{500} + \sum_{j=1}^{25} \frac{1}{j+\sum_{i=1}^{2}(x_i-a_{ij})^6}\right)^{-1}$	2	[−65, 65]	1
$f_{15}(x) = \sum_{i=11}^{11} \left[a_i - \frac{x_i(b_i^2+b_ix_2)}{b_i^2+b_ix_3+x_4}\right]^2$	4	[−5, 5]	0.00030
$f_{16}(x) = 4x_1^2 - 2.1x_1^4 + \frac{1}{3}x_1^6 + x_1x_2 - 4x_2^2 + 4x_2^4$	2	[−5, 5]	−1.0316
$f_{17}(x) = \left(x_2 - \frac{5.1}{4\Pi^2}x_1^2 + \frac{5}{\Pi}x_1 - 6\right)^2 + 10\left(1 - \frac{1}{8\Pi}\right)\cos x_i + 10$	2	[−5, 5]	0.398
$f_{18}(x) = [1 + (x_1 + x_2 + 1)^2(19 - 14x_1 + 3x_1^2 - 14x_2 + 6x_1x_2 + 3x_2^2)] \times [30 + (2x_1 - 3x_2)^2 \times (18 - 32x_1 + 12x_1^2 + 48x_2 - 36x_1x_2 + 27x_2^2)]$	2	[−2, 2]	3
$f_{19}(x) = -\sum_{i=1}^{4} c_i \exp\left(-\sum_{j=1}^{3} a_{ij}(x_j - p_{ij})^2\right)$	3	[1, 3]	−3.32
$f_{20}(x) = -\sum_{i=1}^{4} c_i \exp\left(-\sum_{j=1}^{6} a_{ij}(x_j - p_{ij})^2\right)$	6	[0, 1]	−3.32
$f_{21}(x) = -\sum_{i=1}^{5}\left[(x-a_i)(x-a_i)^T + c_i\right]^{-1}$	4	[0, 10]	−10.1532
$f_{22}(x) = -\sum_{i=1}^{7}\left[(x-a_i)(x-a_i)^T + c_i\right]^{-1}$	4	[0, 10]	−10.4028
$f_{23}(x) = -\sum_{i=1}^{10}\left[(x-a_i)(x-a_i)^T + c_i\right]^{-1}$	4	[0, 10]	−10.5363

Table 4. (a) Outcomes of the BMFO1 algorithm. (b) Outcomes of the BMFO2 algorithm. (c) Outcomes of the HHO algorithm.

Benchmark Functions	Parameters				
	Mean Value	SD	Worst Value	Best Value	p-Value
(a)					
f1	5.75×10^{-34}	2.55×10^{-33}	1.40×10^{-32}	0	3.79×10^{-60}
f2	1.48×10^{-20}	2.24×10^{-20}	1.14×10^{-19}	0	3.79×10^{-60}
f3	3.87×10^{-10}	1.61×10^{-9}	8.70×10^{-9}	0	2.56×10^{-60}
f4	0.03831	0.08819	0.4401	0	2.56×10^{-60}
f5	3.14461	2.21914	6.01278	0	2.56×10^{-60}
f6	1.27×10^{32}	1.60×10^{-32}	8.32×10^{-32}	0	7.23×10^{-60}
f7	1.00564	1.00438	1.01652	0	1.74×10^{-60}
(b)					
f1	3.64×10^{-34}	1.05×10^{-33}	4.50×10^{-33}	0	2.56×10^{-60}
f2	6.08×10^{-20}	1.30×10^{-19}	6.12×10^{-19}	0	2.56×10^{-60}
f3	7.64×10^{-11}	3.00×10^{-10}	1.65×10^{-9}	9.46×10^{-15}	1.73×10^{-60}
f4	0.04709	0.09997	0.47495	0	2.56×10^{-60}
f5	3.4591	2.2489	6.2531	0.00064	1.73×10^{-60}
f6	2.85×10^{-32}	5.78×10^{-32}	3.08×10^{-31}	0	1.61×10^{-50}
f7	1.00499	1.00387	1.01831	0.00032	1.74×10^{-60}
(c)					
f1	1.0634×10^{-90}	5.82468×10^{-90}	3.19×10^{-89}	8.7×10^{-112}	1.734×10^{-6}
f2	6.9187×10^{-51}	2.46844×10^{-50}	1.31×10^{-49}	1.71×10^{-60}	1.734×10^{-6}
f3	1.251×10^{-80}	6.62663×10^{-80}	3.632×10^{-79}	8.3×10^{-99}	1.734×10^{-6}
f4	4.4615×10^{-48}	1.70307×10^{-47}	8.676×10^{-47}	2.45×10^{-59}	1.734×10^{-6}
f5	0.01500185	0.023472777	0.0874276	1×10^{-5}	1.734×10^{-6}
f6	0.00011487	0.00015409	0.0007119	4.17×10^{-7}	1.734×10^{-6}
f7	0.00015829	0.000224928	0.001202	2.87×10^{-6}	1.734×10^{-6}

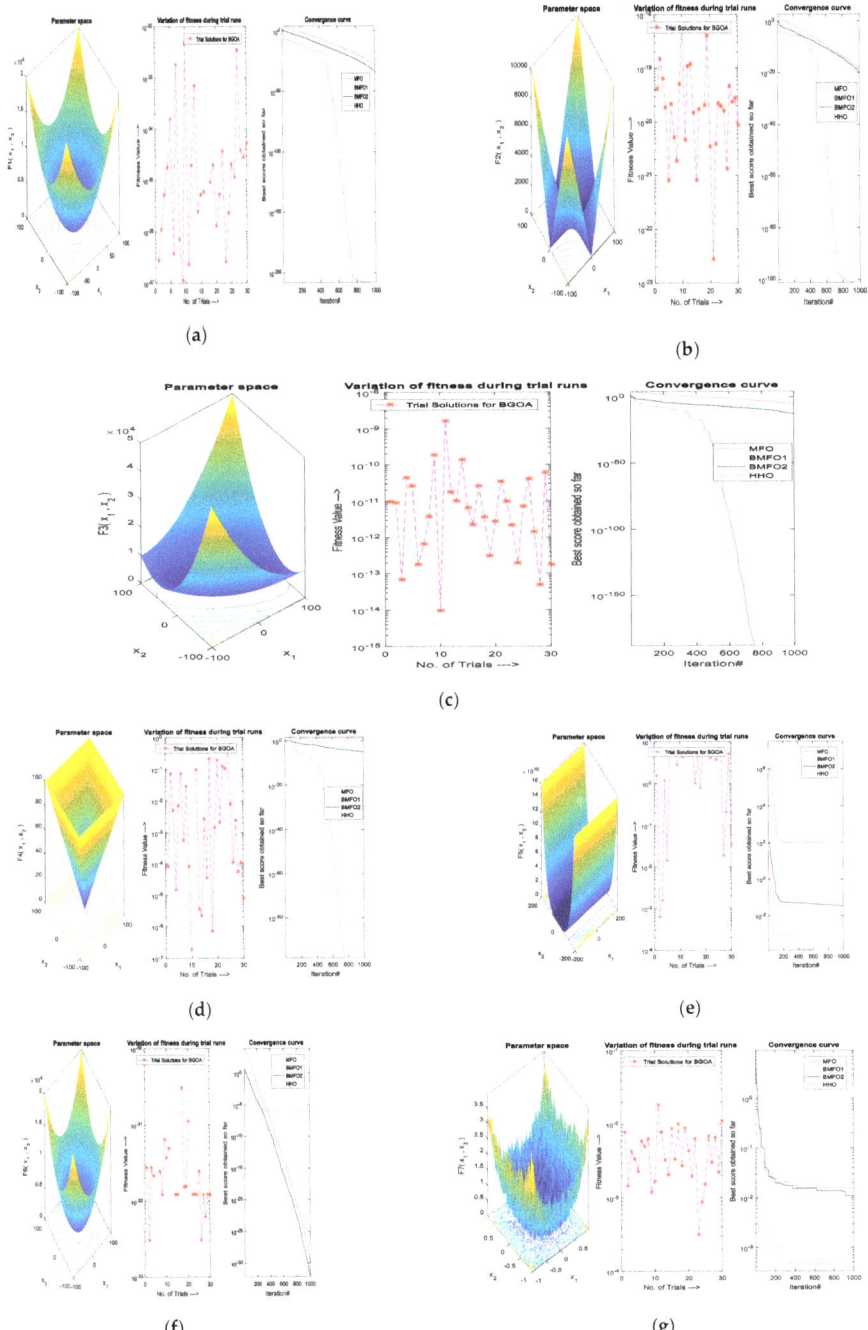

Figure 5. (a–g) Convergence curve of all algorithms for unimodal benchmark functions.

The convergence curve and trial solutions for BMFO1, BMFO2, and HHO for f1 to f7 unimodal benchmark functions are presented in Figure 5a–g.

The connected upshots for unimodal standard functions [60] have been represented in Table 5, which are correlated with various latest refined algorithms [61] grey wolf optimizer (GWO) [62], PSO [63,64], GSA [8,65], differential evolution (DE) [66,67], fruit fly optimization algorithm (FOA) [68,69], ant lion optimizer (ALO) [70,71], symbiotic organisms search (SOS) [72], bat algorithm (BA) [73], flower pollination algorithm (FPA) [74,75], cuckoo search (CS) [76], firefly algorithm (FA) [52], GA [77], grasshopper optimization algorithm (GOA) [73,78], MFO [79], multiverse optimization algorithm (MVO) [80], DA [81], binary bat optimization algorithm (BBA) [65], BBO [5,82], binary gravitational search algorithm (BGSA) [83,84], sine cosine algorithm (SCA) [85,86], FPA [74,87], salp swarm optimization algorithm (SSA) [88], and whale optimization algorithm (WOA) [89] in lieu of mean and standard deviation.

Table 5. Comparison of unimodal benchmark functions.

Algorithm	Parameter	Uni-Modal Benchmark Functions						
		f1	f2	f3	f4	f5	f6	f7
GWO [62]	Mean	0.02	0	0.01	1.02	26.81	0.82	0
	SD	0	0.03	79.15	1.32	69.9	0	0.1
PSO [63,64]	Mean	0	0.04	70.13	1.09	96.72	0	0.12
	SD	0	0.05	22.12	0.32	60.12	0	0.04
GSA [8,65]	Mean	0	0.06	896.53	7.35	67.54	0	0.09
	SD	0	0.19	318.96	1.74	62.23	0	0.04
DE [66,67]	Mean	0.01	0.01	0.01	0.01	0.01	0.01	0.01
	SD	1.01	1.01	1.01	1.01	1.01	1.01	1.01
FOA [68,69]	Mean	0.05	0.06	0.04	0.4	5.06	0.02	0.14
	SD	0.02	0.02	0.01	1.5	5.87	0	0.35
ALO [70,71]	Mean	0.01	0.01	0	0.01	0.35	0.01	0
	SD	0.01	0	0.01	0.01	0.11	0.01	0.01
SOS [72]	Mean	0.06	0.01	0.96	0.28	0.09	0.13	0
	SD	0.01	0	0.82	0.01	0.14	0.08	0
BA [73]	Mean	1.77	1.33	1.12	1.19	1.33	1.78	1.14
	SD	1.53	4.82	1.77	1.89	1.3	1.67	1.11
FPA [74,75]	Mean	0.01	0.01	0.01	0.01	0.78	0.01	0.01
	SD	0.01	0.01	0.01	0.01	0.37	0.01	0.01
CS [76]	Mean	0	1.21	1.25	0.01	0.01	0.01	0.01
	SD	0	1.04	1.02	0.01	0.01	0.01	0.01
FA [52]	Mean	0.04	0.05	0.05	0.15	2.18	0.06	0
	SD	0.01	0.01	0.02	0.03	1.45	0.01	0
GA [77]	Mean	0.12	0.15	0.14	0.16	0.71	0.17	0.01
	SD	0.13	0.05	0.12	0.86	0.97	0.87	0
GOA [73,78]	Mean	0.01	0.01	0.01	0.01	0.01	0.01	0.01
	SD	0.01	0.01	0.02	0.01	0.01	0.01	0.01
MFO [79]	Mean	0.01	0.01	0.01	0.07	27.87	3.12	0
	SD	0	0	0	0.4	0.76	0.53	0
MVO [80]	Mean	2.09	15.92	453.2	3.12	1272.1	2.29	0.05
	SD	0.65	44.75	177.1	1.58	1479.5	0.63	0.03
DA [81]	Mean	0.01	0.01	0.01	0.01	7.6	0.01	0.01
	SD	0.01	0.01	0.01	0.01	6.79	0.01	0.01
BBA [65]	Mean	1.28	1.06	15.6	1.25	24.7	1.1	1.01
	SD	1.42	1.07	23.8	1.33	35.8	1.14	1.01
BBO [5,82]	Mean	6.52	0.2	16.7	2.8	87.6	7.96	0.01
	SD	2.99	0.05	14.9	1.47	66.9	4.87	0.01
BGSA [83,84]	Mean	85	1.19	458	7.35	3110	109	0.04
	SD	48.7	0.23	275	2.25	2936	77.7	0.06
SCA [85,86]	Mean	0.01	0.01	0.06	0.1	0.01	0.01	0.01
	SD	0.01	0.01	0.14	0.58	0.01	0.01	0.01
SSA [88]	Mean	0.01	0.23	0.01	0.01	0.01	0.01	0.01
	SD	0.01	1	0.01	0.66	0.01	0.01	0.01
WOA [89]	Mean	0.01	0.01	696.73	70.69	139.15	0.01	0.09
	SD	0.01	0.01	188.53	5.28	120.26	0.01	0.05
BMFO1	Mean	0.01	0.01	0.01	0.04	3.14	0.01	0.01
	SD	0.01	0.01	0.01	0.09	2.22	0.01	0.01
BMFO2	Mean	0.01	0.01	0.01	0.05	3.46	0.01	0.01
	SD	0.01	0.01	0.01	0.1	2.25	0.01	0.01
HHO (Proposed)	Mean	1.06×10^{-90}	6.92×10^{-51}	1.25×10^{-80}	4.46×10^{-48}	0.015002	0.000115	0.000158
	SD	5.82×10^{-90}	2.47×10^{-50}	6.63×10^{-80}	1.70×10^{-47}	0.023473	0.000154	0.000225

To defend the synthesis part of the probable algorithm, multi-modal benchmark functions f8, f9, f10, f11, f12, and f13 are taken with numerous native goals with values rising violently w.r.t magnitude. Table 6 (a) presents clarification of the multimodal benchmark function with the BMFO1 algorithm and Table 6 (b) presents the explanation of the multimodal benchmark function with the BMFO2 algorithm and Table 6 (c) presents the explanation of the multimodal benchmark function with the HHO algorithm.

Table 6. (**a**) Outcomes of the BMFO1 algorithm. (**b**) Outcomes of the BMFO2 algorithm. (**c**) Results of the HHO algorithm.

Benchmark Functions	Parameters				
	Mean Value	SD	Worst Value	Best Value	p-Value
(a)					
f8	−3140.3	290.75	−2641	−4071.4	0
f9	1.63	0.96	2.98	0.01	0
f10	0.04	0.21	1.16	0.01	0
f11	0.01	0.01	0.01	0.01	1
f12	0.01	0.01	0.01	0.01	0.01
f13	0	0	0.01	0	0
(b)					
f8	−3361.2	287.325	−2879.4	−4071.4	1.73×10^{-6}
f9	1.39294	0.72032	2.98488	0	3.89×10^{-6}
f10	4.56×10^{-15}	0	4.56×10^{-15}	4.56×10^{-15}	4.33×10^{-8}
f11	0	0	0	0	1
f12	4.82×10^{-32}	8.59×10^{-34}	5.12×10^{-32}	4.71×10^{-32}	1.56×10^{-6}
f13	0.00256	0.01025	0.05478	1.35×10^{-32}	1.34×10^{-6}
(c)					
f8	−12561.4	40.82419124	−12345.3	−12569.5	1.7344×10^{-6}
f9	0.01	0.01	0.01	0.01	1
f10	8.88×10^{-161}	0.01	8.88×10^{-161}	8.88×10^{-161}	4.3205×10^{-8}
f11	0.01	0.01	0.01	0.01	1
f12	8.92×10^{-6}	1.16218×10^{-5}	4.76×10^{-5}	4.64×10^{-8}	1.7344×10^{-6}
f13	0.000101	0.000132197	0.000612	7.35×10^{-7}	1.7344×10^{-6}

It is analyzed from Table 6 that multi-model benchmark functions f8 to f13 are tested using modern hybrid algorithms like BMFO 1, BMFO 2, and HHO and found that the Harris hawks optimizer (HHO) produces optimal outcomes in terms of mean, standard deviation, best and worst value for all functions as compared to other algorithms.

The convergence curve and trial solutions for BMFO1, BMFO2, and HHO for f8 to f13 multi-modal benchmark functions are presented in Figure 6a–f.

The connected outcomes for multimodal benchmark functions has been signified in Table 7, which are associated with various latest refined meta-heuristics search algorithms like GWO [62], PSO [63,64], GSA [8,65], DE [66,67], FOA [68,69], ALO [70,71], SOS [72], BA [73], FPA [74,75], CS [76], FA [52], GA [77], GOA [73,78], MFO [79], MVO [80], DA [81], BBA [65], BBO [5,82], BGSA [83,84], SCA [85,86], FPA [74,87], SSA [88], and WOA [89] in lieu of average [90] and standard deviation.

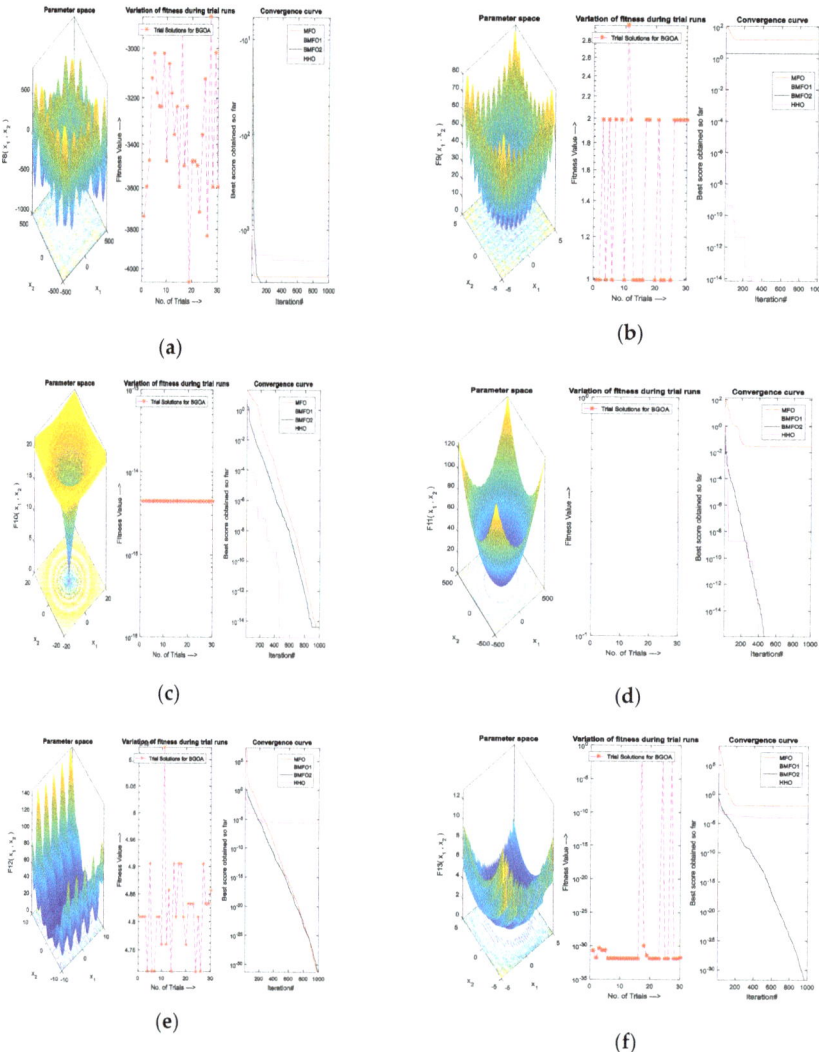

Figure 6. (a–f) Convergence curve of all algorithms for multi-modal benchmark functions.

The verified consequences for fixed dimension benchmark situations are obtainable in Table 8.

It is analyzed from Table 8 that fixed dimension benchmark functions f14 to f23 are tested using modern hybrid algorithms like BMFO 1, BMFO 2, and HHO and found that Harris hawks optimizer (HHO) produces optimal outcomes in terms of mean, standard deviation, best and worst value for all functions as compared to other algorithms.

The convergence curve and trial solutions for BMFO1, BMFO2, and HHO for f14 to f23 fixed dimension benchmark functions are presented in Figure 7a–j.

Table 7. Comparison of multi-modal benchmark functions.

Algorithms	Parameters	Multi-Modal Benchmark Functions					
		f8	f9	f10	f11	f12	f13
GWO [62]	Mean	−6120	0.31	0	0	0.05	0.65
	SD	−4090	47.4	0.08	0.01	0.02	0
PSO [63,64]	Mean	−4840	46.7	0.28	0.01	0.01	0.01
	SD	1150	11.6	0.51	0.01	0.03	0.01
GSA [8,65]	Mean	−2820	26	0.06	27.7	1.8	8.9
	SD	493	7.47	0.24	5.04	0.95	7.13
DE [66,67]	Mean	−11100	69.2	0	0	0	0
	SD	575	38.8	0	0	0	0
FOA [68,69]	Mean	−12600	0.05	0.02	0.02	0	0
	SD	52.6	0.01	0	0.02	0	0
ALO [70,71]	Mean	−1610	0	0	0.02	0	0
	SD	314	0	0	0.01	0	0
SOS [72]	Mean	−4.21	1.33	0	0.71	0.12	0.01
	SD	0	0.33	0	0.91	0.04	0
BA [73]	Mean	−1070	1.23	0.13	1.45	0.4	0.39
	SD	858	0.69	0.04	0.57	0.99	0.12
FPA [74,75]	Mean	−1840	0.27	0.01	0.09	0	0
	SD	50.4	0.07	0.01	0.04	0	0
CS [76]	Mean	−2090	0.13	0	0.12	0	0
	SD	0.01	0	0	0.05	0	0
FA [52]	Mean	−1250	0.26	0.17	0.1	0.13	0
	SD	353	0.18	0.05	0.02	0.26	0
GA [77]	Mean	−2090	0.66	0.96	0.49	0.11	0.13
	SD	2.47	0.82	0.81	0.22	0	0.07
GOA [73,78]	Mean	1	0	0.1	0	0	0
	SD	0	0	1	0	0	0
MFO [79]	Mean	−5080	0	7.4	0	0.34	1.89
	SD	696	0	9.9	0	0.22	0.27
MVO [80]	Mean	−11700	118	4.07	0.94	2.46	0.22
	SD	937	39.3	5.5	0.06	0.79	0.09
DA [81]	Mean	−2860	16	0.23	0.19	0.03	0
	SD	384	9.48	0.49	0.07	0.1	0
BBA [65]	Mean	−924	1.81	0.39	0.19	0.15	0.04
	SD	65.7	1.05	0.57	0.11	0.45	0.06
BBO [5,82]	Mean	−989	4.83	2.15	0.48	0.41	0.31
	SD	16.7	1.55	0.54	0.13	0.23	0.24
BGSA [83,84]	Mean	−861	10.3	2.79	0.79	9.53	2220
	SD	80.6	3.73	1.19	0.25	6.51	5660
SCA [85,86]	Mean	1	0.01	0.38	0.01	0.01	0.01
	SD	0.01	0.73	1	0.01	0.01	0.01
SSA [88]	Mean	0.06	0.01	0.2	0.01	0.14	0.08
	SD	0.81	0.01	0.15	0.07	0.56	0.71
MFO [79]	Mean	−8500	84.6	1.26	0.02	0.89	0.12
	SD	726	16.2	0.73	0.02	0.88	0.19
BMFO1	Mean	−3140.3	1.63	0.04	0	0	0
	SD	290.75	0.96	0.21	0	0	0
BMFO2	Mean	−3361.2	1.39	0	0	0	0
	SD	287.32	0.72	0	0	0	0.01
HHO (Proposed)	Mean	−12561.38	0	8.88×10^{-16}	0	8.92×10^{-6}	0.000101
	SD	40.82419	0	0	0	1.16×10^{-5}	0.000132

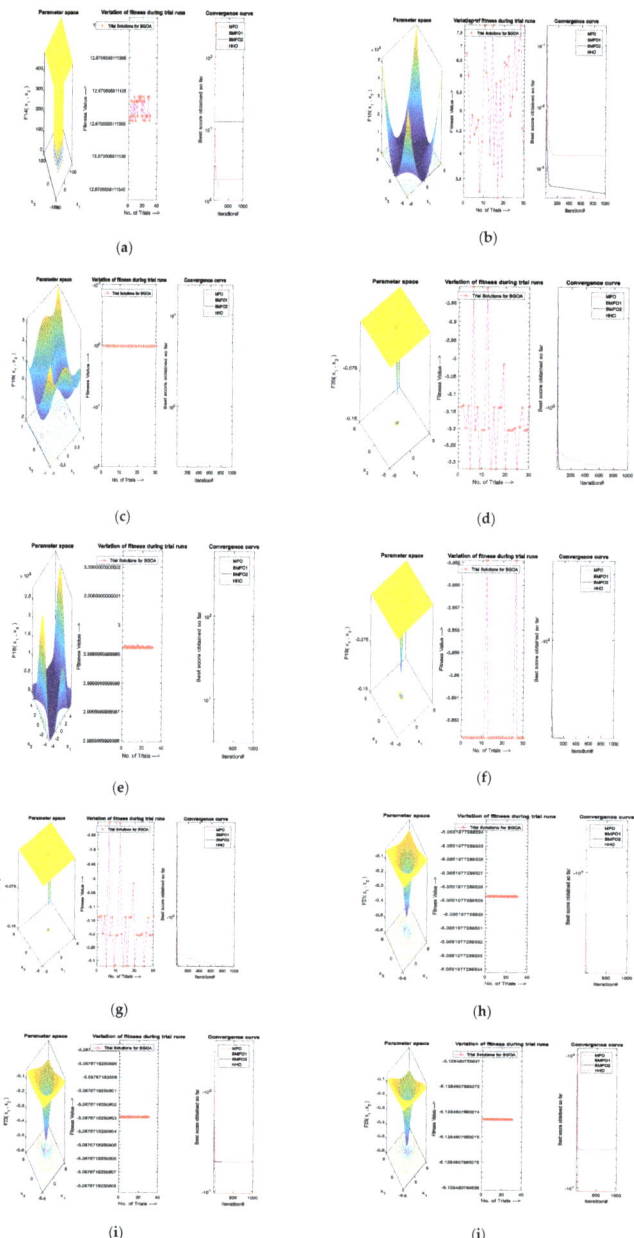

Figure 7. (a–j) Convergence curve and trial solution of BMFO2 for fixed dimension benchmark functions.

Table 8. (a) Outcomes of the BMFO1 algorithm. (b) Outcomes of the BMFO2 algorithm. (c) Outcomes of the HHO algorithm.

Benchmark Functions	Parameters				
	Mean Value	SD	Worst Value	Best Value	p-Value
(a)					
f14	12.61	1.35	12.67	10.76	0
f15	0	0	0	0	0
f16	−1.03	0	−1.03	−1.03	0
f18	3	0	3	3	0
f19	−3.86	0	−3.85	−3.86	0
f20	−3.16	0.08	−2.86	−3.32	0
f21	−5.06	0	−5.06	−5.06	0
f22	−5.09	0	−5.09	−5.09	0
f23	−5.13	0	−5.13	−5.13	0
(b)					
f14	12.67	0	12.67	12.67	0
f15	0	0	0	0	0
f16	−1.03	0	−1.03	−1.03	0
f18	3	0	3	3	0
f19	−3.86	0	−3.85	−3.86	0
f20	−3.17	0.12	−2.81	−3.32	0
f21	−5.06	0	−5.06	−5.06	0
f22	−5.09	0	−5.09	−5.09	0
f23	−5.13	0	−5.13	−5.13	0
(c)					
f14	2.361171	1.95204	5.928845	1.998004	1.73×10^{-8}
f15	1.00035	3.2×10^{-5}	0.000433	0.000309	1.73×10^{-8}
f16	−1.03162	2.86×10^{-9}	−1.03162	−1.03162	1.73×10^{-8}
f17	0.397895	1.6×10^{-5}	0.397948	0.397887	1.73×10^{-6}
f18	3.000001	4.94×10^{-6}	3.000027	2	1.73×10^{-8}
f19	−2.85977	1.005195	−3.8354	−3.86274	1.73×10^{-8}
f20	−2.06481	0.136148	−2.74389	−3.26174	1.73×10^{-8}
f21	−4.37397	1.227502	−5.0413	−10.0309	1.73×10^{-6}
f22	−5.08346	0.004672	−5.06481	−5.08765	1.73×10^{-6}
f23	−5.78398	1.712458	−5.1145	−10.3706	1.73×10^{-6}

The comparative outcomes for fixed dimension benchmark [91] functions have been represented in Tables 9 and 10, which are associated with other latest refined met heuristics search algorithms [54,92] GWO [62], PSO [63,64], GSA [8,65], DE [66,67], FOA [68,69], ALO [70,71], SOS [72], BA [73], FPA [74,75], CS [76], FA [52], GA [77], GOA [73,78], MFO [79], MVO [80], DA [81], BBA [65], BBO [5,82], BGSA [83,84], SCA [85,86], FPA [74,87], SSA [88], and WOA [89] in terms of standard deviation [93] and average.

Table 9. Comparison of fixed dimension benchmark functions.

Algorithms	Parameters	Composite Benchmark Functions					
		f14	f15	f16	f17	f18	f19
GWO [62]	Mean	3.06	0	−1.03	0.4	3	−3.86
	SD	4.25	0	−1.03	0.4	3	−3.86
PSO [63,64]	Mean	3.63	0	−1.03	0.4	3	−3.86
	SD	2.56	0	0	0	0	0
GSA [8,65]	Mean	5.86	0	−1.03	0.4	3	−3.86
	SD	3.83	0	0	0	0	0
DE [66,67]	Mean	1	0	−1.03	0.4	3	NA
	SD	0	0	0	0	0	NA
FOA [68,69]	Mean	1.22	0	−1.03	0.4	3.02	−3.86
	SD	0.56	0	0	0	0.11	0
ALO [70,71]	Mean	0	14.6	175	316	4.4	500
	SD	0	32.2	46.5	13	1.66	0.21
SOS [72]	Mean	776.48	873.8	961	899.86	741	900.5
	SD	0	9.72	67.2	0	0.79	0.84
BA [73]	Mean	182.48	487.2	588.2	756.98	542	818.5
	SD	117.02	161.4	137.8	160.1	220	152.5
FPA [74,75]	Mean	0.34	18.23	224	362.03	10.2	504
	SD	0.24	3.07	50.3	54.02	1.39	1.16
CS [76]	Mean	110	140.6	290	402	213	812
	SD	110.05	92.8	86.1	98.2	206	192
FA [52]	Mean	150.17	314.5	734.5	818.57	134	862.2
	SD	97.16	92.93	204	109.97	216	126
GA [77]	Mean	114.61	95.46	325.4	466.31	90.4	521.2
	SD	26.96	7.16	51.67	29.57	13.7	27.99
GOA [73,78]	Mean	0	0.49	0	0.82	0	0.79
	SD	0.34	0.72	0	1	0.01	0.94
MFO [79]	Mean	2.11	0	−1.03	0.4	3	−3.86
	SD	2.5	0	0	0	0	0
MVO [80]	Mean	10	30.01	50	190.3	161	440
	SD	31.62	48.31	52.7	128.67	158	51.64
DA [81]	Mean	104	193	458	596.66	230	680
	SD	91.2	80.6	165	171.06	185	199
BBA [65]	Mean	1.39	1.02	1.05	1	1.01	1
	SD	1.19	1.07	1.49	1.11	1.01	1.2
BBO [5,82]	Mean	0.06	0	0.2	0	0.14	0.08
	SD	0.81	0	0.15	0.07	0.56	0.71
MFO [79]	Mean	0	66.73	119	345.47	10.4	707
	SD	0	53.23	28.33	43.12	3.75	195
BMFO1	Mean	12.61	0	−1.03	0	3	−3.86
	SD	0.35	0	0	0	0	0
BMFO2	Mean	12.67	0	−1.03	0	3	−3.86
	SD	0	0	0	0	0	0
HHO (Proposed)	Mean	1.361171	0.00035	−1.03163	0.397895	3.000001225	−3.8597664
	SD	0.95204	3.20×10^{-5}	1.86×10^{-9}	1.60×10^{-5}	4.94×10^{-6}	0.00519467

Table 10. Comparison of results for fixed dimension functions.

Algorithms	Parameters	Benchmark Functions			
		f20	f21	f22	f23
GWO [62]	Mean	−2.79	−9.8	−9.9	−9.69
	SD	−2.84	−9.18	−7.55	−7.48
PSO [63,64]	Mean	−2.29	−7.89	−7.49	−8.99
	SD	1.06	3.07	4.08	1.76
GSA [8,65]	Mean	−2.36	−4.99	−8.64	−10.63
	SD	1.02	4.74	2.01	0
DE [66,67]	Mean	0.01	−10.2	−10.4	−10.54
	SD	0.01	0	0	0
FPA [74,75]	Mean	−4.28	−6.56	−6.57	−7.59
	SD	0.08	1.57	2.18	3.18
WOA [84]	Mean	−2.98	−7.05	−8.18	−9.34
	SD	0.38	3.63	3.83	2.41
BMFO1	Mean	−3.16	−5.06	−5.09	−5.13
	SD	0.08	0	0	0
BMFO2	Mean	−3.17	−5.06	−5.09	−5.13
	SD	0.12	0	0	0
HHO (Proposed)	Mean	−3.06481	−5.37397	−5.08346	−5.78398
	SD	0.136148	1.227502	0.004672	1.712458

4. Conclusions

The smart grid process needs a continuing matching of resource and ultimatum in accordance with recognized functioning principles of numerous algorithms. The LFC scheme delivers the consistent action of power structure by constantly balancing the resource of electricity with the response, while also confirming the accessibility of adequate supply volume in upcoming periods. In this paper, binary variations of the moth flame optimizer and HHO have been analyzed and tested to solve twenty-three benchmark problems including unimodel, multi-model, and fixed dimension functions which investigate that the proposed Harris hawks optimizer approach suggestions are offering better results as associated to substitute labeled meta-heuristics search algorithms. In upcoming work, the effectiveness of the HHO technique is deliberate for optimal matching of total generation with total consumption of electrical energy to convert an electric grid to smart grid. So, by using the Harris hawks optimizer, we can easily balance the smart grid elements by matching production and consumption of electrical energy.

Author Contributions: Conceptualization, K.A.; Data curation, K.A. and A.K.; Formal analysis, K.A., A.K. and V.K.K.; Funding acquisition, B.S. and G.P.J.; Investigation, V.K.K. and D.P.; Methodology, D.P.; Project administration, S.J.; Resources, B.S. and G.P.J.; Software, G.P.J.; Supervision, S.J.; Visualization, G.P.J.; Writing—original draft, K.A.; Writing—review & editing, G.P.J. All authors have read and agreed to the published version of the manuscript.

Funding: The present Research has been conducted by the Research Grant of Kwangwoon University in 2020.

Conflicts of Interest: The authors declare no conflict of interest.

References

1. García, J.; Martí, J.V.; Yepes, V. The Buttressed Walls Problem: An Application of a Hybrid Clustering Particle Swarm Optimization Algorithm. *Mathematics* **2020**, *8*, 862. [CrossRef]
2. Ziegler, J.G.; Nichols, N.B. Optimum settings for automatic controllers. *Trans. ASME* **1942**, *64*, 759–768. [CrossRef]

3. Elgerd, O.I.; Fosha, C.E. Optimum Megawatt-Frequency Control of Multiarea Electric Energy Systems. *IEEE Trans. Power Appar. Syst.* **1970**, *89*, 556–563. [CrossRef]
4. Mamdani, E.H. Application of fuzzy algorithms for control of simple dynamic plant. *Proc. IEEE* **1974**, *121*, 1585–1588. [CrossRef]
5. Du, D.; Simon, D.; Ergezer, M. Biogeography-Based Optimization Combined with Evolutionary Strategy and Immigration Refusal. In Proceedings of the IEEE Proc. International Conference on Systems, Man and Cybernetics, San Antonio, TX, USA, 11–14 October 2009; Volume 1, pp. 997–1002.
6. Byerly, R.T.; Sherman, D.E.; Bennon, R.J. *Frequency Domain Analysis of Low Frequency Oscillations in Large Electric Power Systems, Part-1, Basic Concepts, Mathematical Model and Computing Methods*; 727, Project 744-1; EPRI-EL: Palo Alto, CA, USA, 1979; Volume 1, pp. 1–7.
7. Meng, G.; Xiong, H.; Li, H. Power system load-frequency controller design based on discrete variable structure control theory. In Proceedings of the 2009 IEEE 6th International Power Electronics and Motion Control Conference, Wuhan, China, 17–20 May 2009; Volume 2, pp. 2591–2594.
8. Rashedi, E.; Nezamabadi-Pour, H.; Saryazdi, S. GSA: A gravitational search algorithm. *Inf. Sci.* **2009**, *179*, 2232–2248. [CrossRef]
9. Beaufays, F.; Magid, Y.A.; Widrow, B. Application of neural networks to load-frequency control in power systems. *Neural Netw.* **1994**, *7*, 183–194. [CrossRef]
10. Kothari, M.L.; Kaul, B.L.; Nanda, J. Automatic generation control of hydro-thermal system. *J. Inst. Eng. India* **1980**, *61*, 85–91.
11. Cohen, A.I.; Yoshimura, M. A Branch-and-Bound Algorithm for Unit Commitment. *IEEE Trans. Power Appar. Syst* **1983**, *102*, 444–451. [CrossRef]
12. Pan, C.T.; Liaw, C.M. An adaptive controller for power system Load-frequency control. *IEEE Trans. Power Syst.* **1989**, *4*, 122–128. [CrossRef]
13. Sugeno, M. An introductory survey of fuzzy control. *Inf. Sci.* **1985**, *36*, 59–83. [CrossRef]
14. Birch, A.P.; Sapeluk, A.T.; Ozveren, C.S. An enhanced neural network load frequency control technique. In Proceedings of the Control '94, Conference Publication, Coventry, UK, 21–24 March 1994; Volume 389, pp. 409–415.
15. Bekhouche, N.; Feliachi, A. Decentralized estimation for the Automatic Generation Control problem in power systems. In Proceedings of the [Proceedings 1992] The First IEEE Conference on Control Applications, Dayton, OH, USA, 13–16 September 1992; Volume 3, pp. 621–632.
16. Elgerd, O.J. *Electric Energy Systems Theory: An Introduction*; Tata Mcgraw Hill: New York, NY, USA, 1983.
17. Kocaarsian, I.; Cam, E. Fuzzy Logic Controller in interconnected electrical power systems for Load Frequency Control. *Electr. Power Energy Syst.* **2005**, *27*, 542–549. [CrossRef]
18. Holland, J.H. *Adaptation in Natural and Artificial Systems: An Introductory Analysis*; MIT Press: Cambridge, MA, USA, 1992.
19. Wang, Y.; Zhou, R.; Wen, C. New robust adaptive load-frequency control with system parametric uncertainties. *IEE Proc. Gener. Transm. Distrib.* **1994**, *141*, 184–190. [CrossRef]
20. Cohn, N. Techniques for improving the control of bulk power transfers on interconnected systems. *IEEE Trans. Power Appar. Syst.* **1971**, *90*, 2409–2419. [CrossRef]
21. Oneal, A.R. A Simple Method for Improving Control Area Performance: Area Control Error (ACE) Diversity Interchange. *IEEE Trans. Power Syst.* **1995**, *10*, 1071–1076. [CrossRef]
22. Kothari, D.P.; Ahmad, A. An expert system approach to the unit commitment problem. *Energy Convers. Manag.* **1995**, *36*, 257–261. [CrossRef]
23. Hsu, S.-C.; Jane, Y.-J.H.S.U.; I-Jen, C. Automatic Generation of Fuzzy Control Rules by Machine Learning Methods. In Proceedings of the IEEE International Conference on Robotics and Automation, Nagoya, Japan, 21–27 May 1995; Volume 1, pp. 287–292.
24. Indulkar, C.S.; Raj, B. Application of fuzzy controller to Automatic Generation Control. *Electr. Mach. Power Syst.* **1995**, *23*, 209–220. [CrossRef]
25. Chang, C.S.; Fu, W. Area load frequency control using fuzzy gain scheduling of PI controllers. *Electr. Power Syst. Res.* **1997**, *42*, 145–152. [CrossRef]
26. Kennedy, J.; Eberhart, R.C. A discrete binary version of the particle swarm algorithm. In Proceedings of the IEEE Conference on Systems, Man, and Cybernetics, Orlando, FL, USA, 12–15 October 1997; Volume 5, pp. 4104–4108.

27. Kumar, J. AGC simulator for price-based operation-Part I: A Model. *IEEE Trans. Power Syst.* **1997**, *12*, 527–532. [CrossRef]
28. Bakken, B.H.; Grande, O.S. Automatic Generation Control in a deregulated power system. *IEEE Trans. Power Syst.* **1998**, *13*, 1401–1406. [CrossRef]
29. Talaq, J.; AI-Basri, F. Adaptive fuzzy gain scheduling for load frequency control. *IEEE Trans. Power Syst.* **1999**, *14*, 145–150. [CrossRef]
30. Yao, X.; Liu, Y.; Lin, G. Evolutionary programming made faster. *IEEE Trans. Evolut. Comput.* **1999**, *3*, 82–102.
31. Ryu, H.S.; Jung, K.Y.; Park, J.D.; Moon, Y.H.; Rhew, H.W. Extended Integral Control for Load Frequency Control With the Consideration of Generation-Rate Constraints. In Proceedings of the 2000 Power Engineering Society Summer Meeting (Cat. No.00CH37134), Seattle, WA, USA, 16–20 July 2000; IEEE: Piscataway, NJ, USA, 2000; Volume 3, pp. 1877–1882.
32. Yukita, K.; Goto, Y.; Mizuno, K.; Miyafuji, T.; Ichiyanagi, K.; Mizutani, Y. Study of Load Frequency Control using Fuzzy Theory by Combined Cycle Power Plant. In Proceedings of the 2000 IEEE Power Engineering Society Winter Meeting (Cat. No.00CH37077), Singapore, 23–27 January 2000; IEEE: Piscataway, NJ, USA, 2000; Volume 1, pp. 422–427.
33. George, G.; Jeong, W.L. Analysis of Load Frequency Control Performance Assessment Criteria. *IEEE Trans. Power Syst.* **2001**, *16*, 520–525.
34. Wake, K.; Mizutani, Y.; Katsuura, K.; Aoki, H.; Goto, Y.; Yukita, K. A Study on Automatic Generation Control Method Decreasing Regulating Capacity. *IEEE Porto Power Tech. Proc.* **2001**, *1*, 1–6.
35. Moon, Y.H.; Ryu, H.S.; Lee, J.G.; Kim, S. Power System Load Frequency Control Using Noise-Tolerable PID Feedback. *IEEE Int. Symp. Ind. Electron.* **2001**, *1*, 1714–1718.
36. Sedghisigarchi, K.; Feliachi, A.; Davari, A. Decentralized Load Frequency Control in a Deregulated Environment using Disturbance Accommodation Control Theory. In Proceedings of the IEEE Proceedings of the Thirty-Fourth Southeastern Symposium on System Theory, Huntsville, AL, USA, 19 March 2002; Volume 1, pp. 302–306.
37. Li, P.; Zhu, H.; Li, Y. Genetic Algorithm Optimization For AGC of Multi-Area Power Systems. In Proceedings of the Proceedings of IEEE on Computers, Communications, Control and Power engineering, Beijing, China, 28–31 October 2002; Volume 1, pp. 1818–1821.
38. Bansal, R.C. Bibliography on the fuzzy set theory applications in power system. *IEEE Trans. Power Syst.* **2003**, *18*, 1291–1299. [CrossRef]
39. Suresh, V.; Sreejith, S.; Sudabattula, S.K.; Kamboj, V.K. Demand response-integrated economic dispatch incorporating renewable energy sources using ameliorated dragonfly algorithm. *Electr. Eng.* **2019**, *101*, 421–442. [CrossRef]
40. Bevrani, H.; Yasunori, M.; Kiichiro, T. A scenario on Load-Frequency Controller Design in a Deregulated Power System. In Proceedings of the SICE Annual Conference IEEE in Fukui, Fukui, Japan, 4–6 August 2003; Volume 1, pp. 3148–3153.
41. Dulpichet, R.; Amer, H.; Ali, F. Robust Load Frequency Control Using Genetic Algorithms and Linear Matrix Inequalities. *IEEE Trans. Power Syst.* **2003**, *18*, 855–861.
42. Le-Ren, C.C.; Naeb-Boon, H.; Chee-Mun, O.; Kramer, R.A. Estimation of β for adaptive frequency bias setting in load frequency control. *IEEE Trans. Power Syst.* **2003**, *18*, 904–911.
43. Ghoshal, S.P. Multi area frequency and tie-line power flow control with fuzzy logic based integral gain scheduling. *J. Inst. Eng. India* **2003**, *84*, 135–141.
44. Demiroren, A.; Yesil, E. Automatic Generation Control with fuzzy logic controllers in the power system including SMES units. *Electr. Power Energy Syst.* **2004**, *26*, 291–305. [CrossRef]
45. Kumar, P.; Ibraheem, P. Study of dynamic performance of power systems with asynchronous tie-lines considering parameter uncertainties. *J. Inst. Eng. India* **2004**, *85*, 35–42.
46. Mazinan, A.H.; Kazemi, M.F. An Efficient Solution to Load-Frequency Control Using Fuzzy-Based Predictive Scheme in a Two-Area Interconnected Power System. *IEEE Trans.* **2010**, *1*, 289–293.
47. Liu, X.; Xiaolei, Z.; Dianwei, Q. Load Frequency Control considering Generation Rate Constraints. In Proceedings of the IEEE Proceedings of the 8th World Congress on Intelligent Control and Automation, Jinan, China, 7–9 July 2010; Volume 1, pp. 1398–1401.
48. Boroujeni, S. Comparison of Artificial Intelligence Methods for Load Frequency Control Problem. *Aust. J. Basic Appl. Sci.* **2010**, *4*, 4910–4921.

49. Khodabakhshian, A.; N, G. Unified PID Design for Load Frequency Control. In Proceedings of the IEEE International Conference on Control Applications, Taipei, Taiwan, 2–4 September 2004; Volume 1, pp. 1627–1632.
50. Zeynelgil, H.L.; Demiroren, A.; Sengor, N.S. The application of ANN technique to automatic generation control for multi-area power system. *Int. J. Electr. Power Energy Syst.* **2004**, *24*, 345–354. [CrossRef]
51. Kresimir, V.; Nedjeljko, P.; Ivan, P. Applying Optimal Sliding Mode Based Load-Frequency Control in Power Systems with Controllable Hydro Power Plants. *IEEE Autom.* **2010**, *51*, 3–18.
52. Mirjalili, S. Moth-flame optimization algorithm: A novel nature-inspired heuristic paradigm. *Knowl. Based Syst.* **2015**, *89*, 228–249. [CrossRef]
53. Mohanty, B.; Acharyulu, B.V.S.; Hota, P.K. Moth-flame optimization algorithm optimized dual-mode controller for multiarea hybrid sources AGC system. *Optim. Control Appl. Methods* **2017**, *39*, 720–734. [CrossRef]
54. Swain, A.K. A simple fuzzy controller for single area hydropower system considering Generation Rate Constraints. *J. Inst. Eng. India.* **2006**, *87*, 12–17.
55. Kamboj, V.K.; Nandi, A.; Bhadoria, A.; Sehgal, S. An intensify Harris Hawks optimizer for numerical and engineering optimization problems. *Appl. Soft Comput.* **2020**, *89*, 106018. [CrossRef]
56. Xu, D.; Liu, J.; Yan, X.; Yan, W. A Novel Adaptive Neural Network Constrained Control for a Multi-Area Interconnected Power System with Hybrid Energy Storage. *IEEE Trans. Ind. Electron.* **2018**, *65*, 6625–6634. [CrossRef]
57. Xu, Y.; Li, C.; Wang, Z.; Zhang, N.; Peng, B. Load Frequency Control of a Novel Renewable Energy Integrated Micro-Grid Containing Pumped Hydropower Energy Storage. *IEEE Access* **2018**, *6*, 29067–29077. [CrossRef]
58. Giuseppe, D.; Sforna, M.; Bruno, C.; Pozzi, M. A Pluralistic LFC Scheme for Online Resolution of Power Congestions between Market Zones. *IEEE Trans. Power Syst.* **2005**, *20*, 2070–2077.
59. Nakayama, K.; G, F.; R, Y. Load Frequency Control for Utility Interaction of Wide-Area Power System Interconnection. *IEEE Trans.* **2009**, *2*, 1–4.
60. Roy, R.; Ghoshal, S.P. Evolutionary Computation Based Optimization in Fuzzy Automatic Generation Control. *IEEE Power* **2006**, *1*, 1–7.
61. Chaohua, D.; Weirong, C.; Yunfang, Z. Seeker optimization algorithm. In Proceedings of the 2006 International Conference on Computational Intelligence and Security, Guangzhou, China, 3–6 November 2006; Volume 1, pp. 225–229.
62. Mirjalili, S.; Lewis, A. The Whale Optimization Algorithm. *Adv. Eng. Softw.* **2016**, *95*, 51–67. [CrossRef]
63. Kennedy, J.; C, E.R. Particle Swarm Optimization. In Proceedings of the IEEE International Conference on Neural Networks, Perth, WA, Australia, 27 November–1 December 1995; Volume 1, pp. 1942–1948.
64. Atashpaz-Gargari, E.; Lucas, C. Imperialist competitive algorithm: An algorithm for optimization inspired by imperialistic competition. In Proceedings of the 2007 IEEE Congress on Evolutionary Computation, Singapore, 25–28 September 2007; Volume 1, pp. 4661–4667.
65. Khodabakhshian, A.; Hooshmand, R. A new PID controller design for Automatic generation Control of hydropower system. *Electr. Power Energy Syst.* **2010**, *32*, 375–382. [CrossRef]
66. Panda, G.; Sidhartha, P.; Cemal, A. Automatic Generation Control of Interconnected Power System with Generation Rate Constraints by Hybrid Neuro Fuzzy Approach. *World Acad. Sci. Eng. Technol.* **2009**, *52*, 543–548.
67. Storn, R.; Price, K. Differential Evolution—A Simple and Efficient Heuristic for global Optimization over Continuous Spaces. *J. Glob. Optim.* **1997**, *11*, 341–359. [CrossRef]
68. Soundarrajan, A.; Sumathi, S. Effect of Non-linearities in Fuzzy Based Load Frequency Control. *Int. J. Electron. Eng. Res.* **2009**, *1*, 37–51.
69. Mirjalili, S.; Gandomi, A.H.; Mirjalili, S.Z.; Saremi, S.; Faris, H.; Mirjalili, S.M. Salp Swarm Algorithm: A bio-inspired optimizer for engineering design problems. *Adv. Eng. Softw.* **2017**, *114*, 163–191. [CrossRef]
70. Tan, Y.; Tan, Y.; Zhu, Y. Fireworks Algorithm for Optimization Fireworks Algorithm for Optimization. *IEEE Trans.* **2015**, *1*, 355–364.
71. Chamnan, K.; Somyot, K. A Novel Robust Load Frequency Controller for a Two Area Interconnected Power System using LMI and Compact Genetic Algorithms. In Proceedings of the TENCON 2009—2009 IEEE Region 10 Conference, Singapore, 23–26 January 2009; Volume 1, pp. 1–6.

72. Li, M.D.; Zhao, H.; Weng, X.W.; Han, T. A novel nature-inspired algorithm for optimization: Virus colony search. *Adv. Eng. Softw.* **2016**, *92*, 65–88. [CrossRef]
73. Yang, X.-S. A New metaheuristic bat-inspired algorithm. In *Nature Inspired Cooperative Strategies for Optimization*; Springer: Heidelberg, Germany, 2010; Volume 1, pp. 65–74.
74. Mirjalili, S. SCA: A Sine Cosine Algorithm for solving optimization problems. *Knowl. Based Syst.* **2016**, *96*, 120–133. [CrossRef]
75. Hosseini, S.H.; Etemadi, A.H. Adaptive neuro-fuzzy inference system based automatic generation control. *Electr. Power Syst. Res.* **2008**, *78*, 1230–1239. [CrossRef]
76. Mirjalili, S.; Mirjalili, S.M.; Lewis, A. Grey Wolf Optimizer. *Adv. Eng. Softw.* **2014**, *69*, 46–61. [CrossRef]
77. Kazarlis, S.A.; Bakirtzis, A.G.; Petridis, V. A genetic algorithm solution to the unit commitment problem. *IEEE Trans. Power Syst.* **1996**, *11*, 83–92. [CrossRef]
78. Sreenath, A.; Atre, Y.R.; Patil, D.R. Two area load frequency control with fuzzy gain scheduling of PI controller. In Proceedings of the 2008 First International Conference on Emerging Trends in Engineering and Technology, Nagpur, Maharashtra, India, 16–18 July 2008; Volume 1, pp. 899–904.
79. Taher, A.S.; Reza, H. Robust Decentralized Load Frequency Control Using Multi Variable QFT Method in Deregulated Power Systems. *Am. J. Appl. Sci.* **2008**, *5*, 818–828. [CrossRef]
80. Erlich, I.; Venayagamoorthy, G.K.; Worawat, N. A Mean-Variance Optimization algorithm. *IEEE Congr. Evolut. Comput.* **2010**, *1*, 1–6.
81. Cheng, M.Y.; Prayogo, D. Symbiotic Organisms Search: A new metaheuristic optimization algorithm. *Comput. Struct.* **2014**, *139*, 98–112. [CrossRef]
82. Shayeghi, H.; Shayanfar, H.A.; Jalili, A. Multi stage fuzzy PID load frequency controller in a restructured power system. *J. Electr. Eng.* **2007**, *58*, 61–70.
83. Wen, T.; Zhan, X. Robust analysis and design of Load Frequency Controller for power systems. *Electr. Power Syst. Res.* **2009**, *79*, 846–853.
84. Rashedi, E.; Nezamabadi-Pour, H.; Saryazdi, S. BGSA: Binary gravitational search algorithm. *Nat. Comput.* **2010**, *9*, 727–745. [CrossRef]
85. Ghaemi, M.; Feizi-Derakhshi, M.R. Forest optimization algorithm. *Expert Syst. Appl.* **2014**, *41*, 6676–6687. [CrossRef]
86. Mathur, H.D.; Manjunath, H.V. Study of dynamic performance of thermal units with asynchronous tie-lines using fuzzy based controller. *J. Electr. Syst.* **2007**, *3*, 124–130.
87. Caliskan, F.; Genc, I. A robust fault detection and isolation method in Load Frequency Control loops. *IEEE Trans. Power Syst.* **2008**, *1*, 1756–1767. [CrossRef]
88. Nakamura, R.Y.M.; Pereira, L.A.M.; Costa, K.A.; Rodrigues, D.; Papa, J.P.; Yang, X.S. BBA: A binary bat algorithm for feature selection. In Proceedings of the Brazilian Symposium of Computer Graphic and Image Processing, Ouro Preto, Brazil, 22–25 August 2012; Volume 1, pp. 291–297.
89. Yang, X.S. Flower Pollination Algorithm for Global Optimization. *Unconv. Comput. Nat. Comput.* **2012**, *1*, 2409–2413.
90. Eusuff, M.; Lansey, K.; Pasha, F. Shuffled frog-leaping algorithm: A memetic meta-heuristic for discrete optimization. *Eng. Optim.* **2006**, *38*, 129–154. [CrossRef]
91. Elaziz, M.A.; Oliva, D.; Xiong, S. An improved Opposition-Based Sine Cosine Algorithm for global optimization. *Expert Syst. Appl.* **2017**, *90*, 484–500. [CrossRef]
92. Bansal, R.C. Overview and literature survey of Artificial Neural Network applications to power system. *J. Inst. Eng.* **2006**, *86*, 282–296.
93. Koji, A.; Satoshi, O.; Shinichi, I. New Load Frequency Control Method suitable for Large Penetration of Wind Power Generations. In Proceedings of the 2006 IEEE Power Engineering Society General Meeting, Montreal, QC, Canada, 18–22 June 2006.

© 2020 by the authors. Licensee MDPI, Basel, Switzerland. This article is an open access article distributed under the terms and conditions of the Creative Commons Attribution (CC BY) license (http://creativecommons.org/licenses/by/4.0/).

Article

RISC Conversions for LNS Arithmetic in Embedded Systems

Peter Drahoš *, Michal Kocúr, Oto Haffner, Erik Kučera and Alena Kozáková

Institute of Automotive Mechatronics, Faculty of Electrical Engineering and Information Technology, Slovak University of Technology in Bratislava, 812 19 Bratislava, Slovakia; michal.kocur@stuba.sk (M.K.); oto.haffner@stuba.sk (O.H.); erik.kucera@stuba.sk (E.K.); alena.kozakova@stuba.sk (A.K.)
* Correspondence: peter.drahos@stuba.sk

Received: 19 June 2020; Accepted: 20 July 2020; Published: 22 July 2020

Abstract: The paper presents an original methodology for the implementation of the Logarithmic Number System (LNS) arithmetic, which uses Reduced Instruction Set Computing (RISC). The core of the proposed method is a newly developed algorithm for conversion between LNS and the floating point (FLP) representations named "looping in sectors", which brings about reduced memory consumption without a loss of accuracy. The resulting effective RISC conversions use only elementary computer operations without the need to employ multiplication, division, or other functions. Verification of the new concept and related developed algorithms for conversion between the LNS and the FLP representations was realized on Field Programmable Gate Arrays (FPGA), and the conversion accuracy was evaluated via simulation. Using the proposed method, a maximum relative conversion error of less than ±0.001% was achieved with a 22-ns delay and a total of 50 slices of FPGA consumed including memory cells. Promising applications of the proposed method are in embedded systems that are expanding into increasingly demanding applications, such as camera systems, lidars and 2D/3D image processing, neural networks, car control units, autonomous control systems that require more computing power, etc. In embedded systems for real-time control, the developed conversion algorithm can appear in two forms: as RISC conversions or as a simple RISC-based logarithmic addition.

Keywords: LNS; numerical conversion; RISC; FPGA; embedded systems

1. Introduction

The Logarithmic Number System (LNS) provides comparable range and precision as the floating point (FLP) representation, however—for certain applications—it can surpass it in terms of complexity. The range of logarithmic numbers depends on the exponent's integer part, and the precision is defined by its fraction part [1]. Yet, LNS would outperform FLP only if the logarithmic addition and subtraction can be performed with at least the same speed and accuracy as FLP [2].

The long history of LNS numbers dates back to the 1970's when the "logarithmic arithmetic" for digital signal processing was introduced [3]. To avoid negative logarithms, a complementary notation for LNS was introduced in [4]. Architecture for the LNS-based processor was proposed in [5]. Implementations of basic arithmetic operations on FPGA [6] using FLP and LNS have shown that the multiplication and division operations are more effective if using LNS, as they require fewer area resources and have a significantly lower time latency. Otherwise, addition and subtraction are more suitable using FLP representation. A higher efficiency of some LNS operations was a motivation for using the LNS format for the realization of control algorithms for the autonomous electric vehicle developed within a running research project.

Nowadays, LNS representation is implemented in various applications, such as deep-learning networks [7], Cartesian to polar coordinates converters [8], or embedded model predictive control [9].

In a very interesting paper [10], it is proposed how numbers close to zero can be represented in the denormal LNS method (DLNS) using either fixed-point or LNS representations, guaranteeing constant absolute or constant relative precisions, respectively. Up to now, LNS have not been standardized.

Various methods have been developed to decrease the costs and complexity of LNS implementation, e.g., interpolation, co-transformation, multipartite tables, etc. [11–13]. Typically, there are three main categories of LNS arithmetic techniques: lookup tables, piecewise polynomial approximation, and digit–serial methods [14].

Generally, LNS addition and subtraction are carried out based on the evaluation of the transcendental functions as follows:

$$a \pm_{LNS} b = \log_2(2^a \pm 2^b) = a + \log_2(1 \pm 2^{b-a}), \qquad (1)$$

where a, b are logarithmic numbers.

Signal processing in embedded systems based on LNS has three stages: logarithmic conversions, simple operations, and antilogarithmic conversions. In the first processing stage, logarithmic conversions are applied to convert binary numbers into logarithmic ones. In the second stage, simple operations are used to perform corresponding calculations, such as addition and subtraction. In the last stage, logarithmic numbers are converted back to binary ones. There are many approaches to solving logarithmic conversion that can be classified into three categories: memory-based methods [15,16], mathematical approximations [6], and shift-and-add-based methods [11,13,17–20]. Very fast conversions (e.g., shift-and-add) allow us to combine calculations in LNS and FLP systems and design hybrid LNS/FLP processors [13,21,22]. In hybrid systems, the conversions are carried out several times during the calculation, not only the first and last phases.

Using memory-based methods, fast and more accurate conversions are achieved; however, memory size costs may increase significantly while the bit-width of the inputs increases. On the other hand, using polynomial approximations will reduce the area costs, while sacrificing the accuracy and speed. Approximation-based methods almost always use a multiplication operation, for example, an antilogarithmic converter [6] uses 20 multipliers and achieves a latency of more than 70 ns. Compared with these two kinds of implementation, shift-and-add methods can be used to achieve better design tradeoffs between accuracy, memory costs, and speed. All the above-mentioned "shift-and-add" methods achieve a latency of less than 1 ns at the cost of a low accuracy (above 1% relative error [20]) except for [23], where the attained accuracy of the LNS/FLP conversions is 0.138%. Using the proposed looping-in-sectors method in combination with a very simple approximation based on bit manipulations, a radical increase in accuracy and an acceptably low latency can be achieved.

The logarithmic and antilogarithmic conversions are a gateway to LNS algorithms. Yet, the conversions are not freely available from FPGA producers [24] and, thus, must be implemented by our own means. Furthermore, the above methods available in the literature do not meet our requirements in that the accuracy of modern industrial sensors is better than 0.1%, and sampling periods in embedded systems for motion control are less than 1 ms, which places high demands on conversion speed and application calculations. Therefore, our motivation was to develop a simple and efficient FLP/LNS conversion guaranteeing sufficient accuracy and speed, which will be a "golden mean" between the accurate but complex approximation methods and the very efficient and fast (up to 1ns) but inaccurate (relative error higher than 1%) shift-and-add methods.

Embedded systems are expanding into increasingly demanding applications, such as camera systems, lidars and 2D/3D image processing, neural networks, car control units, autonomous control systems that require more computing power, etc. A necessary reliability and functional safety are often based on redundancy of two/three channel technologies. Therefore, alternative calculations (one option is LNS) on independent HW and SW (hardware and software) solutions are needed; outputs of the independent channels are then compared according to the principles of fault tolerant systems (e.g., two out of three).

However, embedded control systems based on LNS arithmetic that operate in real-time necessitate an efficient conversion in every sampling period. Input data from sensors, counters, A/D converters, etc., are typically fixed-point (FXP) numbers, thus they have to be converted to LNS, and, the other way round, the LNS arithmetic results are to be converted back to the FXP format conventionally required by typical output devices (actuators). In this paper, the focus is on conversions between LNS and FLP, conversions from FXP to FLP and back are supposed to be resolved.

The Xilinx's industry-leading tool suite natively supports different FLP precisions including half (FLP16), single (FLP32), and double (FLP64) precisions, as well as fixed-point data types. The added flexibility of custom precision is also available in MATLAB System Generator for DSP toolbox. The FLP to FXP conversion is dealt with in [24], however the LNS data type is not yet officially supported and conversions from LNS to FLP and back are not available in FPGA libraries.

This paper presents an application of the proposed RISC conversions for logarithmic addition using the Reduced Instruction Set Computing (RISC) realizable just by means of simple operations without using multiplication, etc. Herein, RISC indicates a set of simple computer operations (add, minus, shift by 2, i.e., multiplication/division, logical operations, and bit manipulations). The proposed approach has the ambition to apply just the above-mentioned RISC operations fully excluding multiplication, division, and all other functions (log, square, ...). Using the unified format of LNS and FLP, conversion between them can be realized only by dealing with the mantissa and the fraction. To reduce memory requirements for conversions, a novel method called "looping in sectors" was developed.

The paper presents a novel effective RISC-based method, which uses the so-called "looping-in-sectors" procedure and a simple interpolation in the conversion between FLP and LNS number representations. The novel algorithm of logarithmic addition based on the developed conversions performs differently from previously known approaches. The partial results on the development of RISC conversions and algorithms for LNS [25] are completed by the conversion algorithm from LNS to FLP and its realization on FPGA.

The paper is organized as follows. In Section 2, an overview of FLP and LNS number representations is provided. Section 3 presents two developed algorithms of the RISC conversion between both systems. A simple interpolation method along with accuracy analysis are dealt with in Section 4. Principle of the RISC-based LNS addition is explained in Section 5. FPGA realization of the RISC conversion is demonstrated on a simple example in Section 6. Discussion on obtained results, their potential, and future research concludes the paper.

2. Number Systems

Let us briefly revisit the FLP and LNS number representations. According to Table 1, a floating point (FLP) number is expressed as follows:

$$FLP = (-1)^S \times 2^E \times (1+m) = (-1)^S \times 2^E \times \left(1 + \frac{N}{M}\right), \qquad (2)$$

where m is a mantissa, and N is an integer or a real number from the intervals $\langle 0, M-1 \rangle$ or $\langle 0, M \rangle$, respectively. M is the maximum of the mantissa (fractional part) with t bits.

$$M = 2^t \qquad (3)$$

Table 1. Floating point (FLP) number representation.

Sign Bit	Exponent: e-Bits	Mantissa: t-Bits
S	E	$m = N/M$

Table 2 shows the principle of an LNS number representation; according to it:

$$LNS = (-1)^S \times 2^{E_f} = (-1)^S \times 2^E \times 2^f. \quad (4)$$

The logarithmic fraction f can be expressed as follows:

$$f = F/M, \quad (5)$$

where F is an integer in the range $\langle 1 - M, M - 1 \rangle$ or a real number in the range $(-M, M)$.

Table 2. Logarithmic number (LNS) representation.

Sign Bit	Integer: e-Bits	Fractional: t-Bits
S	E	$f = F/M$

In terms of individual bits, the whole exponent E_f consists of "integer bits" (i_x) and "fractional bits" (f_y), placed next to each other. S_E denotes the sign of the exponent.

$$E_f = S_E i_{e-2} ... i_1 i_0 \quad f_{t-1} ... f_1 f_0, \quad (6)$$

For both numerical systems, the number of bits of the exponent E corresponds to the range of the numbers, and the number of the fractional part bits reflects the accuracy.

2.1. Two Possible Representations of LNS Numbers

The mantissa m is always a positive number ($0 \leq m < 1$), but the logarithmic fraction depends on the sign of the exponent S_E. Still, there is also another possibility to represent LNS fraction as always positive, similar to mantissas.

Let a number $X < 1$, $E \leq 0$ and a fraction $f < 0$. F_{SE} and F_{AP} are positive real numbers from $\langle 0, M \rangle$.

$$X = (-1)^S \times 2^E \times 2^{-\frac{F_{SE}}{M}} = (-1)^S \times 2^{E-1} \times 2^{\frac{F_{AP}}{M}}, \quad (7)$$

where F_{AP} is a complement of F_{SE} to the range of the fraction M, i.e.,

$$F_{AP} = M - F_{SE} \quad (8)$$

For numbers $X > 1$, $F_{AP} = F_{SE}$ and the integer E is unchanged. F_{AP} is an always-positive fraction.

The sign of F_{SE} is the same as the sign S_E. For the sake of completeness note that for $X = 1$ there are two possible ways (i.e., possible codes) to represent zero. In the F_{SE} representation, $E = \pm 0$ and the fraction $F_{SE} = 0$ (the same as for $X = -1$). In the F_{AP} representation there is no such anomaly; the conversion between F_{AP} and F_{SE} proceeds (7) and (8).

2.2. Equivalence between FLP and LNS

The FLP (2) and LNS (4) representations are equivalent if integer parts of both representations are equal numbers of e-bits, and both the fraction and the mantissa are equal numbers of t-bits. It is also necessary to use an always positive fraction F_{AP}. The sign S and the exponent E are matching:

$$X = (-1)^S \times 2^E \times 2^{\frac{L_X}{M}} = (-1)^S \times 2^E \times \left(1 + \frac{N_X}{M}\right). \quad (9)$$

Let N_X denote the mantissa (FLP) and $L_X = F_{AP}$ is the always positive fraction (LNS). The subscript "x" specifies that they represent (code) the equivalent number X in diverse number systems. Using the following conversion between the mantissa and the fraction, equivalence of FLP and LNS can be attained:

$$Z = 2^{\frac{L_Z}{M}} = \left(1 + \frac{N_Z}{M}\right), \tag{10}$$

where $Z \in \langle 1, 2 \rangle$ and L_Z, N_Z are positive integers from the interval $\langle 0, M-1 \rangle$ or positive real numbers within the interval $\langle 0, M \rangle$. In the same range, sequences of integers for L and N are geometric and arithmetic, respectively. The integer form of L and N is used to address the look up table (LUT) memory, while their real form is needed to attain a required accuracy. By extending the number of bits of the lower fraction and mantissa to $t + r$ bits, the accuracy can be improved. Using the following corrections, the mutual number conversions over the interval $\langle 0, M \rangle$ can be defined as follows:

$$L_Z = N_Z + C_{NZ}(N_Z), \tag{11}$$

$$N_Z = L_Z - C_{LZ}(L_Z), \tag{12}$$

where C_{NZ} and C_{LZ} are correction functions for conversions in both directions:

$$C_{NZ}(N_Z) = M \times \log_2\left(1 + \frac{N_Z}{M}\right) - N_Z, \tag{13}$$

$$C_{LZ}(L_Z) = -M \times \left(2^{\frac{L_Z}{M}} - 1\right) + L_Z. \tag{14}$$

From the corresponding diagrams in Figure 1 it is evident that both functions have the same maximum, however at various arguments:

$$\max\nolimits_{NZ} C_{NZ}(453.319721) = \max\nolimits_{LZ} C_{LZ}(541.456765) = 88.137044.$$

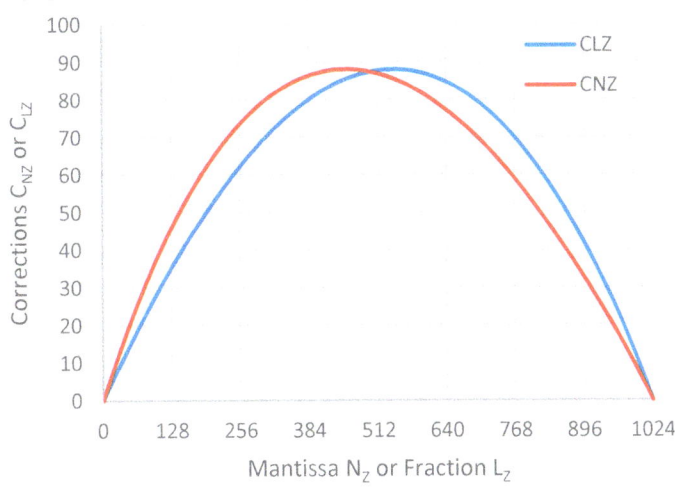

Figure 1. Diagram of correction functions from N_Z to L_Z (red plot) and from L_Z to N_Z (blue plot) for $M = 1024$.

3. RISC-Based Conversion between FLP and LNS

If we assume equivalence of the FLP and LNS numbers, the conversion can be completed between the fraction and mantissa. The proposed algorithm aims to use RISC-type computing operations to reduce memory consumption and costs, while achieving as high accuracy as possible. In the Reduced

Instruction Set Computing (RISC), operations of multiplication, division, square, square root, logarithm, etc., are not used.

3.1. Conversion from LNS to FLP

The conversion from LNS to FLP is carried out using (12) and (14). When converting L_Z it is required to cover the whole range of the fraction, guarantee a sufficient accuracy, and avoid a large memory consumption. According to the proposed approach, L_Z is split in two parts:

$$L_Z = L_S + L_B, \qquad (15)$$

where L_S is relevant for each sector and L_B for the common base. The fraction is split into sectors (e.g., 32 sectors in our case). In (15), L_S represents a relocation of the converted number to the base L_B —a part of the C_{LZ} function diagram placed in the close vicinity of the extreme (Figure 2).

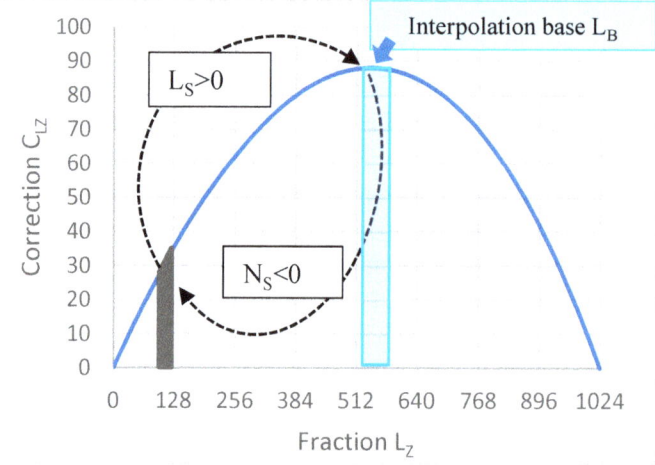

Figure 2. Looping in sectors: conversion from L_Z to N_Z for $M = 1024$.

To obtain N_B from L_B, a simple interpolation is performed (as described in Section 5). The calculation loop closes by applying N_S in (18). This procedure was named "looping in sectors" (LiS) by the first author. Mathematically, the LiS method is expressed by the Equations (16)–(18):

$$Z = 2^{\frac{L_Z}{M}} = 2^{\frac{L_S}{M}} \times 2^{\frac{L_B}{M}}. \qquad (16)$$

The equivalent FLP representation in the mantissa form is as follows:

$$Z = \left(1 + \frac{N_Z}{M}\right) = \left(1 + \frac{N_S}{M}\right) \times \left(1 + \frac{N_B}{M}\right). \qquad (17)$$

From (17) results:

$$N_Z = N_B + N_S + \frac{N_S}{M} \times N_B. \qquad (18)$$

By appropriately choosing N_S and calculating corresponding L_S according to (11), a conversion look-up table LUT1LS was generated using 10 bits, i.e., $M = 1024$ (Table 3). For each sector, the numbers N_S and L_S form a pair when using the looping-in-sectors method.

Table 3. Look-up table LUT1LS for 32 sectors.

Min Lz	Max Lz	Address	Ns	Ls
0	31	0	−304	520.3424
32	63	1	−290	491.8925
...	...	...	...	...
480	511	15	−32	46.9030
512	543	16	−8	11.5869
544	575	17	12	−17.2117
576	607	18	36	−51.0449
...	...	...	...	...
960	991	30	352	−436.4951
992	1023	31	380	−466.2551

The procedure for selecting the sector number Ns is as follows. Choose the interval for the "interpolation base L_B" in the vicinity of the argument (541.456765) of the extreme of the correction function in Figure 2. In our case, the minimum interval is min L_B = 520. Each sector has a minimum, denoted "min Lz" in Table 3. We choose Ns with three active bits (Table 4) under the condition Ls > (520-minLz) where Ls is calculated according to (11) with the required accuracy. The choice of Ns is specific in each sector; as a result the interval for the interpolation base is not 32 but 40 (wider) where $L_B \in (520, 560)$.

The RISC conversion procedure from L to N is described in the Algorithm 1 (input variable is the fraction L_Z; the sign S and the exponent E do not change):

Algorithm 1 RISC conversion from L to N
---
1. Input: LNS fraction LZ
2. Output: FLP mantissa N_Z
3. Round L_Z to the upper 5 bits = address (cut the lower bits).
4. Read L_S and N_S from the memory LUT1LS (Table 3), where 5 MSB of L_Z represent the LUT1S address.
5. $L_B = L_Z + L_S$.
6. Read N_{B0} and N_{B1} from the memory LUT2NB, where the 9 MSB of L_B represent the LUT2NB address.
7. Calculate N_B using the interpolation between N_{B0} and N_{B1} (see the next Section).
8. Calculate N_Z using bit shifting (19).

In the above Algorithm 1, N_Z is calculated as follows:

$$N_z = N_B + N_S \pm shift1\,(N_B) \pm shift2\,(N_B) \pm shift3\,(N_B) \tag{19}$$

where *shift1, shift2, shift3* denote shifting bits of N_B according to Table 4, which is suitable for FPGA implementation (it substitutes the original product $N_B * N_S/M$). In (18), we modified the expression N_S/M to 3 "shift operations" according to the weights (*w* bits) of the 3 active bits (Table 4). $M = 2^{10}$ represents a shift by 10 places to the right.

Table 4. Generating Ns using 3 active bits in 32 sectors.

|Ns|\w Bits	8	7	6	5	4	3	2	1
304	1	0	1	0	−1	0	0	0
290	1	0	0	1	0	0	0	1
...	...	...	...	...	...	...	...	...
32	0	0	0	1	0	0	0	0
8	0	0	0	0	0	1	0	0
12	0	0	0	0	0	1	1	0
36	0	0	0	1	0	0	1	0
...	...	...	...	...	...	...	...	...
352	1	1	0	−1	0	0	0	0
380	1	1	0	0	0	0	−1	0

The variable N_S was chosen to include just 3 "active bits", i.e., bits with nonzero (+1 or −1) values, see Table 4. The algorithm results that the conversion can be performed as RISC, i.e., using just elementary computer operations: addition, subtraction, shifting (multiplication and division by 2), and addressing/reading from the memory.

3.2. Conversion from FLP to LNS

Conversion from FLP to LNS is based on Equations (11) and (13) and modified Equations (16)–(18). Details are described in [25].

The RISC conversion from N to L (input variable is the mantissa N_Z; the sign S and the exponent E do not change) is described in Algorithm 2:

Algorithm 2 RISC conversion from N to L

1. Input: FLP mantissa NZ
2. Output: LNS fraction L_Z
3. Round N_Z to the upper 5 bits = address (cut the lower bits).
4. Read L_S and N_S from the memory LUT1LS (Table 3), where 5 MSB of N_Z are the memory address.
5. Calculate N_B according to (20).
6. Read L_{B0} and L_{B1} from the memory.
7. Calculate L_B using the interpolation between L_{B0} and L_{B1}.
8. Calculate $L_Z = L_B + L_S$.

In the above Algorithm 2, N_B is calculated as follows:

$$N_B = N_Z + N_S \pm shift1\,(N_Z) \pm shift2\,(N_Z) \pm shift3\,(N_Z). \qquad (20)$$

Note that (20) is a modification of (18). Again, N_S has to be chosen to include just 3 "active bits". It can be concluded that both conversions from FLP to LNS and vice versa can be performed as RISC, using just elementary computer operations.

4. Interpolation and Accuracy

For both FLP and LNS, the accuracy is given by the number of bits of the mantissa and the fraction; it usually decreases due to approximation or interpolation. In the effective RISC conversion design, a simple interpolation is based on bit manipulation. The interpolation is demonstrated on the LNS to FLP conversion. The base generated according to (12) and (14) and the corresponding look-up table LUT2NB were chosen so that the base falls within the plateau in the close vicinity of the C_{LZ} function maximum (Figure 2).

The plateau region was intentionally chosen as the interpolation base because differences between the "adjacent" logarithmic values of L_B and the arithmetic sequence values of N_B are approximately matching; this allows us to interpolate between two points in the table in a very simple way (Figure 3).

ADDRESS			Number/MEMORY
Higher bits	Lower bits	Operation	All bits
$L_{B1} = L_{B0}+1$	000...000	MEM ⇨	N_{B1}
$L_B = L_{B0} +$	$+ LL_B$	APROX.	$N_B = N_{B0} + LL_B + ...$
L_{B0}	000...000	MEM ⇨	N_{B0}

Figure 3. Principle of approximating N_B using a simple bit manipulation.

Approximating the mantissa N_B (denoted as the interpolation base L_B in Figure 2) using the look-up table LUT2NB is demonstrated in Figure 3:

- The fraction L_B is rounded to the higher n bits (e.g., n = 10), thus obtaining the address L_{B0}. Then, N_{B0} is read from the memory.
- Lower bits of L_B are denoted as LL_B:

$$L_B = L_{B0} + LL_B. \tag{21}$$

- The simplest approximation is the "quasilinear" interpolation as follows:

$$N_B = N_{B0} + LL_B. \tag{22}$$

According to this method, the error dN_{B01} is calculated as a difference of adjacent memory locations in the considered memory LUT2NB:

$$dN_{B01} = N_{B1} - N_{B0} - 1. \tag{23}$$

In case of a 10-bit memory quantization in the range $L_B \in \langle 520, 560 \rangle$, the maximum positive error is 1.1988×10^{-5} ($dN_{B01} = 1.2275 \times 10^{-2}$) at the right limit of the interval, and the maximum negative error is -1.3742×10^{-5} ($dN_{B01} = -1.4072 \times 10^{-2}$) at the left limit, which corresponds to an accuracy of 16 bits (10 + 6). These errors of the variable N_B will then be influenced by a factor $(1 + N_S/M)$ in the range $\langle 0.70312, 1.37109 \rangle$, according to (18). The resulting error analysis is in Section 6. The accuracy can be improved using a finer memory quantization in the plateau region. Note that the above accuracy levels are attained using RISC, i.e., multiplication is not used.

If a higher accuracy is needed, the following linear interpolation can be applied:

$$N_B = N_{B0} + LL_B + dN_{B01} \times LL_B. \tag{24}$$

In the memory region LUT2NB with 40 cells where $L_B \in \langle 520, 560 \rangle$, the approximation error occurs due to nonlinearity. Local extremes between adjacent memory cells (approximately in the middle of them) were examined. For a 10-bit memory quantization, the related accuracy is 23 bits (10 + 13). The proposed linear interpolation is a RISC extended by one multiplication operation.

Finally, it has to be noted that the FLP to LNS interpolation procedure is similar, only based on Equations (11) and (13). Details are described in [25]. Accuracy assessment in terms of memory quantization is the same.

5. Application of RISC Conversions for Logarithmic Addition

The principle of the logarithmic addition based on the developed RISC conversions is briefly presented in this section. The aim is to show the possibility of applying the developed conversions for RISC-based LNS addition without additional memory requirements. Similar algorithms developed for data conversion at the input and output of the embedded system can be used for the LNS adder.

Consider two real numbers A, B, where $A > B$ represented in LNS using integer exponents E_A, E_B, and fractions L_A, L_B, respectively:

$$A + B = 2^{E_A} \times 2^{\frac{L_A}{M}} + 2^{E_B} \times 2^{\frac{L_B}{M}}. \tag{25}$$

The proposed operation of logarithmic addition will be demonstrated under the assumptions $E_A \geq E_B$ and $L_A \geq L_B$. Applying the distributive law, we obtain:

$$A + B = 2^{E_A} \times 2^{\frac{L_B}{M}} \times \left(2^{\frac{L_A - L_B}{M}} + \frac{2^t}{M} \times 2^{-(E_A - E_B)} \right). \tag{26}$$

Assume $E_A > E_B$ and $L_A < L_B$, we obtain:

$$A + B = 2^{E_A-1} \times 2^{\frac{L_B}{M}} \times \left(2^{\frac{M+L_A-L_B}{M}} + \frac{2^t}{M} \times 2^{-(E_A-1-E_B)}\right). \tag{27}$$

Denote L_{AB}, d and E_{AS} as follows:

$$L_{AB} = L_A - L_B \quad \text{or} \quad L_{AB} = M + L_A - L_B \tag{28}$$

$$d = E_A - E_B \quad \text{or} \quad d = E_A - E_B - 1 \tag{29}$$

$$E_{AS} = E_A \quad \text{or} \quad E_{AS} = E_A - 1. \tag{30}$$

Then from both assumptions results:

$$A + B = 2^{E_{AS}} \times 2^{\frac{L_B}{M}} \times \left(2^{\frac{L_{AB}}{M}} + \frac{2^{t-d}}{M}\right). \tag{31}$$

The RISC-based conversion of the fraction $L_{AB} = L_A - L_B$ to the mantissa N_{AB} is carried out using (12). Then, $1 + N_{AB}/M$ has a mantissa format; $M = 2^t$ can be considered as a mantissa or a fraction, as needed. For M, the corrections (13), (14) are zero, i.e., no conversion is required. Dividing M by powers of 2^d, i.e., applying the shifting by d, we obtain a result, which has the range and character of a mantissa; let us denote it $N_E = 2^{t-d}$, then

$$A + B = 2^{E_{AS}} \times 2^{\frac{L_B}{M}} \times \left(1 + \frac{N_{AB}}{M} + \frac{N_E}{M}\right), \tag{32}$$

where the two rightmost terms in the expression in parentheses have a mantissa format and can be simply summed:

$$N_{ABE} = N_{AB} + N_E. \tag{33}$$

If is $N_{ABE} \geq M$ (overflow), then

$$N_{ABE} = (N_{ABE} - M)/2 \quad \text{and} \quad E_{AS} = E_{AS} + 1. \tag{34}$$

Applying the RISC conversion of N_{ABE} to L_{ABE} according to (11) we obtain:

$$A + B = 2^{E_{AS}} \times 2^{\frac{L_B}{M}} \times 2^{\frac{L_{ABE}}{M}} = 2^{E_{SUM}} \times 2^{\frac{L_{SUM}}{M}}, \tag{35}$$

which is a logarithmic sum of the original numbers A, B.

When implemented, the overflow $L_{ABE} + L_B$ of the range $\langle 0, M-1 \rangle$ by the fraction L has to be treated. The integer exponent of the sum E_{SUM} can take two values: either E_{AS} or $E_{AS} + 1$ (under overflown).

$$L_{SUM} = L_{ABE} + L_B \quad \text{or} \quad L_{SUM} = L_{ABE} + L_B - M \tag{36}$$

The above-presented original procedure of LNS addition is based solely on RISC-type operations including two RISC conversions that determine the accuracy of the adder. The LNS adder can be implemented using the six standard additions or subtractions (28)–(30), (33), (34) and (36), by comparing four pairs of numbers, two shifting operations, and two RISC conversions. The developed approach is promising for applications in embedded control systems realized, e.g., on FPGA [26]. More details on LNS addition and subtraction via RISC computing can be found in [25].

6. Implementation on FPGA

To verify the proposed conversion algorithm, a hardware realization of a simple converter from LNS to FLP number representations was realized. The provided example demonstrates the feasibility of the proposed approach and the solution accuracy.

6.1. Design of an LNS to FLP Converter

The implemented LNS input of the converter represented according to Table 2 has a 32-bit data width. The most significant bit is assigned to the sign, the next 8 bits belong to the integer part, and the last 23 significant bits are the fractional part of the number. The convertor output is a 32-bit FLP number format compatible with the IEEE 754 standard (Table 1).

The LNS to FLP converter is realized as a digital logic circuit and contains only combinational logic. The converter design was developed in VHDL (Very High-Speed Integrated Circuit Hardware Description Language) using Vivado IDE and was targeted for the FPGA (Field Programmable Gate Array) chip Xilinx Artix-7 XC7A100T-1CSG324C mounted on Nexys 4 trainer board [27]. FPGAs are a powerful tool for prototyping and testing hardware designs based on digital logic. FPGAs have broad resources for implementation of combinational and sequential logic, which allows us to implement even more complex and tailored digital designs.

The structure of the LNS to FLP convertor represented by RTL (Register Transfer Level) is shown in Figure 4. The design is based on the algorithm of RISC conversion from LNS to FLP as described in Section 3. The sign bit and the integer part of the LNS input are directly connected to the sign and exponent parts of the FPL output. The fractional part of the input signal is marked as L_Z. The five MSB bits of L_Z are the input for Table 3, LUT1LS (Address column). L_B described in (15) is computed as a sum of LUT1LS output L_S and the signal L_Z. In this converter design, only a simple approximation according to (22) is realized. L_B is calculated as the sum of N_{B0} and LL_B. The mantissa N_Z (18) as a part of the output FLP number is computed as the sum of three signals: N_B, N_S, and output product of N_B and N_S. Division in (18) is realized by a proper choice of the output product bits.

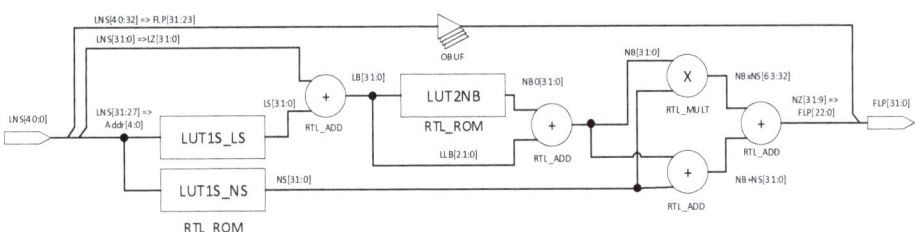

Figure 4. Illustration of the principle of approximating N_B using a simple interpolation.

Both tables LUT1LS and LUT2NB at the RTL level are implemented as an ROM memory. After performing the synthesis and implementation steps in Vivado IDE, the tables are assigned to the LUT tables [28] of FPGA. In FPGAs, LUT tables are part of configuration blocks and are the main resource for implementation of combinational logic. Multiplication and summation operations are automatically assigned to DSP cores [29]. The overall converter design needs only 50 slices representing just 0.32% of the used FPGA chip capacity. In Table 5, the obtained results are compared with the reference conventional approximation methods [6] and very fast shift-and-add methods [20]. Latency of the circuit, or the time from input to output as a performance measure, is very circuit-dependent. In the considered reference design, the FPGA Virtex II was used, which is an older product line of the Xilinx FPGA chips.

Table 5. Comparison of area, latency, and accuracy of a conversion from LNS to FLP.

	Conventional Design [6]	Proposed Design	Shift-and-Add Design [20]
Slices	236	50	NA
Multipliers	20	1	0
Memory	4 × 18 k BRAM	0	0
Latency	72 ns	22/12 ns	<1 ns
Accuracy	High	<0.001%	>1%

Previous converters are realized on 65 nm full customizable CMOS technology with latency under 1 ns [20]. For our implementation, a 28-nm prefabricated structure of FPGA Artix-7 was used. Even with this significant limitation, the achieved latency was 22.198 ns (12.083-ns logic delay and 10.115 net delay), which is a very promising result for the embedded real-time applications.

6.2. Simulation and Verification

For synthesis, simulation, and verification of the presented converter design, the Matlab–Simulink environment combined with the system generator toolbox [30] were used. The system generator toolbox is an additional part of Vivado IDE connecting synthesis and simulation tools of digital designs targeted for FPGA with Matlab–Simulink. After a successful installation of both environments, it is possible to add system generator blocks to the Simulink simulation scheme. These blocks are not simulated in the Matlab environment but separately in the Vivado simulator. The simulation scheme is shown in Figure 5. Specialized input and output blocks serve as an interface between Matlab and Vivado simulation tools. This approach is more flexible than a standard testbench created in Vivado IDE.

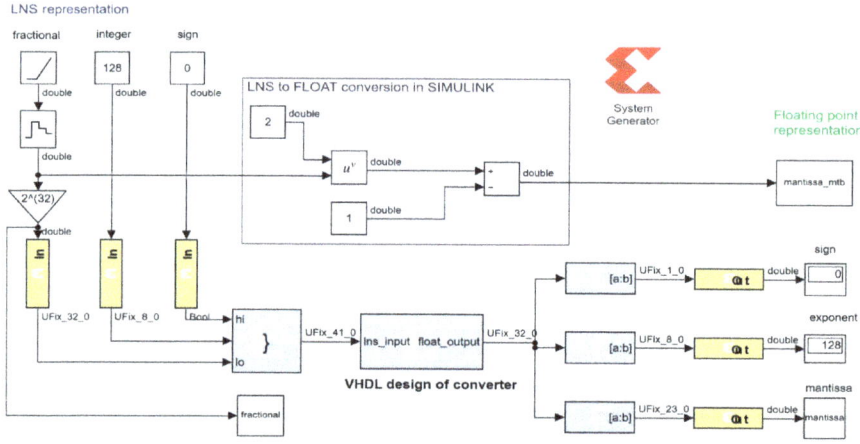

Figure 5. Converter simulation and verification in the Matlab–Simulink environment.

The core of the simulation scheme is the system generator block, which represents the VHDL code of the converter developed in Vivado IDE. The output signal from the converter is split into the sign, exponent, and mantissa parts. For verification, the mantissa signal simulated in Vivado using system-generator-for-DSP blocks and the mantissa_mtb obtained directly in Simulink were compared. The mantissa error was calculated as the difference between the mantissa and the mantissa_mtb.

From the simulated fractional input in the interval $\langle 0, 1 \rangle$, the maximum positive mantissa error is 1.5806×10^{-5} and the maximum negative mantissa error is -1.4047×10^{-5}, which corresponds to the maximum error as mentioned in Section 4. The average error across the input range is 1.9979×10^{-7}.

Overall dependence of the conversion error on the fractional part L_Z is depicted in Figure 6. Dependence of the error focused for one sector on Address 30 in Table 3 is shown in Figure 7.

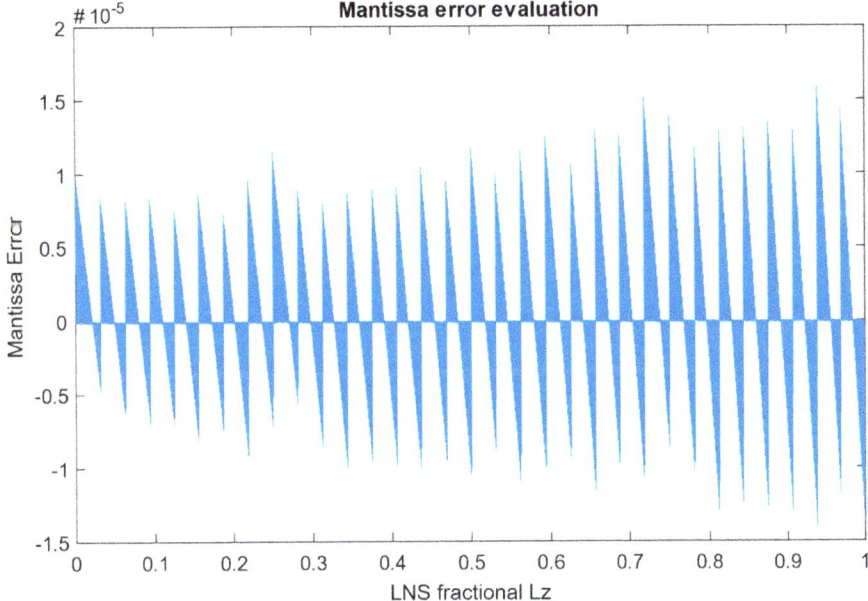

Figure 6. Mantissa error evaluation.

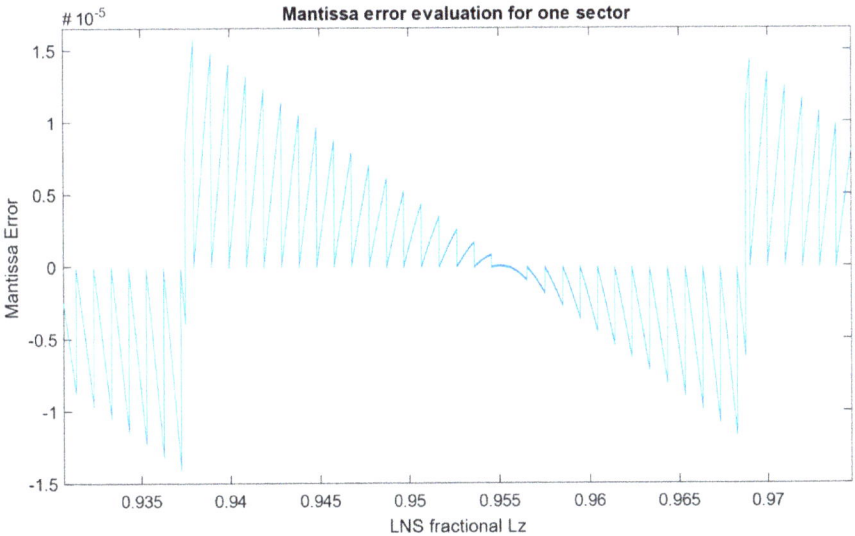

Figure 7. Zoomed mantissa error evaluation for one sector.

In Table 6, accuracy achieved using the proposed LiS method of LNS to FPL converter was compared with selected existing methods.

Table 6. Comparison of accuracies achieved using the proposed LiS method with selected conversion methods.

	Maximum Positive Error	Maximum Negative Error	Error Range	Maximum Positive Relative Error	Maximum Negative Relative Error	Relative Error Range
Mitchell [31]	0.086	0	0.086	5.791%	0	5.791%
Combet [32]	0.0253	−0.0062	0.0315	2.293%	−0.468%	2.761%
SG [33]	0.0292	−0.028	0.0572	2.503%	−1.704%	4.207%
Juang [18]	0.0369	−0.0097	0.0466	2.963%	−0.487%	3.45%
Juang [11]	0.0319	0	0.0319	2.337%	0	2.337%
AS [17]	NA	NA	NA	1.331%	−0,5631%	1,8941%
Kuo [19]	NA	NA	NA	1.2%	−0.1436%	1.3436%
Nandan [20]	NA	NA	NA	0.94%	−0.1436%	1.084%
Chen [23]	NA	NA	NA	NA	NA	0.138%
LiS method	0.00001581	−0.00001407	0.00002988	0.000966%	−0.000783%	0.001749%

It can be observed that the proposed converter has a much better accuracy compared with other presented converters.

The proposed looping-in-sectors (LiS) method can be included in a category in between the approximation methods and the shift-and-add methods, which are very fast and ROM-less, based only on logic circuits and binary operations. The proposed LiS method uses a very simple approximation at the bit-manipulation level, a simple so-called RISC operation, and needs relatively small memory (32 + 40 memory cells).

The advantage of the proposed LiS method over the shift-and-add methods is a much higher relative accuracy (up to 1000 times), the disadvantage is a higher latency. Compared to conventional approximation methods, its advantages are simplicity and a higher speed.

7. Discussion

Conversions of numbers play an important role in the LNS arithmetic, especially in real-time systems. In this paper, equivalence between the FLP (semi-logarithmic) and LNS (fully logarithmic) systems was defined, and conversion between them was reduced to a conversion between the mantissa and fraction performed within the interval ⟨1, 2⟩. Derived correction functions enabled to specify an optimal interval for conversion within the plateau around their maxima where a mathematically simple and accurate interpolation can be performed. According to the developed procedure called looping in sectors (LiS), the converted number is to be moved to the plateau in the vicinity of the correction function peak and back after the interpolation accomplishment, hence reducing memory consumption. Note, that the LiS is performed without loss of accuracy, resulting in effective RISC conversions, which use only elementary computer operations. The developed conversions are then implemented in the designed LNS addition based on RISC operations and can thus be realized without a necessity to use multiplication, division, and other functions.

As a part of the development, conversion algorithms from LNS to FLP and vice versa [25] were implemented on FPGA, and their accuracy was verified by simulation. The presented methodology based on the new correction functions, the looping-in-sectors method, and the optimal choice of a base for an efficient interpolation can further be optimized for different types of applications in embedded systems. For different applications, different attributes are prioritized: in measurement and signal processing from sensors it is accuracy, in complex control algorithms it is speed, in automotive and autonomous systems reliability and credibility of the information obtained are the most essential ones.

The modern very fast "shift-and-add" methods (latencies about 1 ns) prevail only in selected complex algorithms that tolerate low accuracy (only 1%) but are inappropriate in other applications. The proposed LiS method is suitable for control applications in combination with input and output signal processing, as well as one of alternative methods for redundant signal processing and fault-detection due to a deterministically determined high accuracy. The proposed LiS method belongs to faster methods (with a latency of 22 ns) and can be used in real-time control applications even in time-critical applications with a sampling period up to 1 ms.

The future research will be directed on the functional safety of embedded system applications, usually implemented through redundancy and dual-channel technology today. In this sense, we understand LNS not only as an alternative to FLP but also as an SW/HW independent dual method of calculation to eliminate errors and increase the plausibility of results in full compliance with a new paradigm [34]: "It is much more important to know whether information is reliable or not than the accuracy of the information itself."

Author Contributions: Conceptualization, P.D. and O.H.; methodology, P.D.; software, M.K. and E.K.; validation, P.D., M.K. and E.K.; formal analysis, P.D.; investigation, P.D.; resources, M.K.; writing—original draft preparation, P.D.; writing—review and editing, P.D., M.K., and A.K.; supervision, O.H.; project administration, A.K.; funding acquisition, P.D. and A.K. All authors have read and agreed to the published version of the manuscript.

Funding: This research was funded by the SLOVAK RESEARCH AND DEVELOPMENT AGENCY, grant No. APVV-17-0190 and the SLOVAK CULTURAL EDUCATIONAL GRANT AGENCY, grant No. 038STU-4/2018.

Acknowledgments: The paper was partially supported by the Slovak Research and Development Agency, grant No. APVV-17-0190 and the Slovak Cultural Educational Grant Agency, grant No. 038STU-4/2018.

Conflicts of Interest: The authors declare no conflict of interest.

References

1. Lee, B.R. A Comparison of Logarithmic and Floating Point Number Systems Implemented on Xilinx Virtex-II Field Programmable Arrays. Ph.D. Thesis, Cardiff University, Cardiff, UK, 2004.
2. Chugh, M.; Parhami, M. Logarithmic arithmetic as an alternative to floating-point: A review. In Proceedings of the 2013 Asilomar Conference on Signals, Systems and Computers, Pacific Grove, CA, USA, 3–6 November 2013; pp. 1139–1143. [CrossRef]
3. Kingsbury, N.G.; Rayner, P.J.W. Digital filtering using logarithmic arithmetic. *Electron. Lett.* **1971**, *7*, 56–58. [CrossRef]
4. Swartzlander, E.E.; Alexopoulos, A.G. The Sign/Logarithm Number System. *IEEE Trans. Comput.* **1975**, *24*, 1238–1242. [CrossRef]
5. Arnold, M.G. A VLIW architecture for logarithmic arithmetic. In Proceedings of the Euromicro Symposium on Digital System Design, Belek-Antalya, Turkey, 1–6 September 2003; pp. 294–302. [CrossRef]
6. Haselman, H.M.; Beau champ, M.; Wood, A.; Hauck, S.; Underwood, K.; Hemmert, K.S. A comparison of floating point and logarithmic number systems for FPGAs. In Proceedings of the 13th Annual IEEE Symposium on Field-Programmable Custom Computing Machines (FCCM'05), Napa, CA, USA, 18–20 April 2005; pp. 181–190. [CrossRef]
7. Kouretas, I.; Paliouras, V. Logarithmic number system for deep learning. In Proceedings of the 2018 7th International Conference on Modern Circuits and Systems Technologies (MOCAST), Thessaloniki, Greece, 7–9 May 2018; pp. 1–4. [CrossRef]
8. Juang, T.; Lin, C.; Lin, G. Design of High-Speed and Area-Efficient Cartesian to Polar Coordinate Converters Using Logarithmic Number Systems. In Proceedings of the 2019 International SoC Design Conference (ISOCC), Jeju, Korea, 6–9 October 2019; pp. 180–181. [CrossRef]
9. Garcia, J.; Arnold, M.G.; Bleris, L.; Kothare, M.V. LNS architectures for embedded model predictive control processors. In Proceedings of the 2004 International Conference on Compilers, Architecture, and Synthesis for Embedded Systems (CASES '04), New York, NY, USA, 7–16 September 2004; pp. 79–84. [CrossRef]
10. Arnold, M.G.; Collange, S. The Denormal Logarithmic Number System. In Proceedings of the 2013 IEEE 24th International Conference on Application-Specific Systems, Architectures and Processors, Washington, DC, USA, 5–7 June 2013; pp. 117–124. [CrossRef]
11. Juang, T.; Meher, P.K.; Jan, K. High-performance logarithmic converters using novel two-region bit-level manipulation schemes. In Proceedings of the 2011 International Symposium on VLSI Design, Automation and Test, Hsinchu, Taiwan, 25–28 April 2011; pp. 1–4. [CrossRef]

12. Arnold, M.G.; Kouretas, I.; Paliouras, V.; Morgan, A. One-Hot Residue Logarithmic Number Systems. In Proceedings of the 2019 29th International Symposium on Power and Timing Modeling, Optimization and Simulation (PATMOS), Rhodes, Greece, 1–8 March 2019; pp. 97–102. [CrossRef]
13. Zaghar, D.R. Design and Implementation of a High Speed and Low Cost Hybrid FPS/LNS Processor Using FPGA. *J. Eng. Sustain. Dev.* **2010**, *14*, 86–104.
14. Naziri, S.Z.M.; Ismail, R.C.; Shakaff, A.Y.M. The design revolution of logarithmic number system architecture. In Proceedings of the 2nd International Conference on Electrical, Electronics and System Engineering (ICEESE), Kuala Lumpur, Malaysia, 8–9 November 2014; pp. 5–10. [CrossRef]
15. Nam, B.; Kim, H.; Yoo, H. Power and Area-Efficient Unified Computation of Vector and Elementary Functions for Handheld 3D Graphics Systems. *IEEE Trans. Comput.* **2008**, *57*, 490–504. [CrossRef]
16. Arnold, M.G.; Collange, S. A Real/Complex Logarithmic Number System ALU. *IEEE Trans. Comput.* **2011**, *60*, 202–213. [CrossRef]
17. Abed, K.H.; Siferd, R.E. CMOS VLSI implementation of a low-power logarithmic converter. *IEEE Trans. Comput.* **2003**, *52*, 1421–1433. [CrossRef]
18. Juang, T.; Chen, S.; Cheng, H. A Lower Error and ROM-Free Logarithmic Converter for Digital Signal Processing Applications. *IEEE Trans. Circuits Syst. II Express Briefs* **2009**, *56*, 931–935. [CrossRef]
19. Kuo, C.; Juang, T. Area-efficient and highly accurate antilogarithmic converters with multiple regions of constant compensation schemes. *Microsyst. Technol.* **2018**, *24*, 219–225. [CrossRef]
20. Nandan, D. An Efficient Antilogarithmic Converter by Using Correction Scheme for DSP Processor. *Trait. Signal* **2020**, *37*, 77–83. [CrossRef]
21. Chen, C.; Chow, P. Design of a versatile and cost-effective hybrid floating-point/LNS arithmetic processor. In Proceedings of the 17th ACM Great Lakes symposium on VLSI (GLSVLSI '07), Stresa-Lago Maggiore, Italy, 11–13 March 2007; pp. 540–545. [CrossRef]
22. Ismail, R.C.; Zakaria, M.K.; Murad, S.A.Z. Hybrid logarithmic number system arithmetic unit: A review. In Proceedings of the 2013 IEEE International Conference on Circuits and Systems (ICCAS), Kuala Lumpur, Malaysia, 18–19 September 2013; pp. 55–58. [CrossRef]
23. Chen, C.; Tsai, T.C. Application-specific instruction design for LNS addition/subtraction computation on an SOPC system. In Proceedings of the 2009 6th International Conference on Electrical Engineering/Electronics, Computer, Telecommunications and Information Technology, Pattaya, Thailand, 6–9 May 2009; pp. 640–643. [CrossRef]
24. Finnerty, A.; Ratigner, H. *Reduce Power and Cost by Converting from Floating Point to Fixed Point—White Paper: Floating vs Fixed Point*; Xilinx: San Jose, CA, USA, 2017.
25. Drahos, P.; Kocur, M. Logarithmic addition and subtraction for embedded control systems. In Proceedings of the 2020 Cybernetics & Informatics: 30th International Conference, Velke Karlovice, Czech Republic, 29 January–1 February 2020. [CrossRef]
26. Klimo, I.; Kocúr, M.; Drahoš, P. Implementation of logarithmic number system in control application using FPGA. In Proceedings of the 16th IFAC Conference on Programmable Devices and Embedded Systems (PDEeS'19), High Tatras, Slovak Republic, 29–31 October 2019. [CrossRef]
27. Xilinx. 7 Series FPGAs Data Sheet: Overview. Xilinx. 2018. Available online: https://www.xilinx.com/support/documentation/data_sheets/ds180_7Series_Overview.pdf (accessed on 10 April 2020).
28. Xilinx. 7 Series FPGAs Configurable Logic Block: User Guide. Xilinx. 2016. Available online: https://www.xilinx.com/support/documentation/user_guides/ug474_7Series_CLB.pdf (accessed on 10 April 2020).
29. Xilinx. 7 Series DSP48E1 Slice: User Guide. Xilinx. 2018. Available online: https://www.xilinx.com/support/documentation/user_guides/ug479_7Series_DSP48E1.pdf (accessed on 10 April 2020).
30. Xilinx. Vivado Design Suite User Guide: Model-Based DSP Design Using System Generator. Xilinx. 2019. Available online: https://www.xilinx.com/support/documentation/sw_manuals/xilinx2019_2/ug897-vivado-sysgen-user.pdf (accessed on 10 April 2020).
31. Mitchell, J.N. Computer Multiplication and Division Using Binary Logarithms. *IRE Trans. Electron. Comput.* **1962**, *11*, 512–517. [CrossRef]
32. Combet, M.; Zonneveld, H.V.; Verbeek, L. Computation of the base two logarithm of binary numbers. *IEEE Trans. Electron. Comput.* **1965**, *14*, 863–867. [CrossRef]

33. SanGregory, S.L.; Siferd, R.E.; Brother, C.; Gallagher, D. A fast, low-power logarithm approximation with CMOS VLSI implementation. In Proceedings of the 42nd IEEE Midwest Symposium on Circuits and Systems (MWSCAS), Las Cruces, NM, USA, 8–11 August 1999; Volume 1, pp. 388–391. [CrossRef]
34. The European Safety Critical Applications Positioning Engine (ESCAPE). GSA/GRANT/02/2015. Available online: http://www.gnss-escape.eu/ (accessed on 15 June 2020).

© 2020 by the authors. Licensee MDPI, Basel, Switzerland. This article is an open access article distributed under the terms and conditions of the Creative Commons Attribution (CC BY) license (http://creativecommons.org/licenses/by/4.0/).

MDPI
St. Alban-Anlage 66
4052 Basel
Switzerland
Tel. +41 61 683 77 34
Fax +41 61 302 89 18
www.mdpi.com

Mathematics Editorial Office
E-mail: mathematics@mdpi.com
www.mdpi.com/journal/mathematics

www.ingramcontent.com/pod-product-compliance
Lightning Source LLC
LaVergne TN
LVHW070228100526
838202LV00015B/2106